Structural Modelling, Safety Assessment, and Advanced Material Application of Marine Structures

Structural Modelling, Safety Assessment, and Advanced Material Application of Marine Structures

Guest Editors

Qingshan Wang
Mengzhen Li

MDPI

Basel • Beijing • Wuhan • Barcelona • Belgrade • Novi Sad • Cluj • Manchester

Guest Editors
Qingshan Wang
College of Mechanical and Electrical Engineering
Central South University
Changsha
China

Mengzhen Li
School of Naval Architecture, Ocean Engineering and Energy and Power Engineering
Wuhan University of Technology
Wuhan
China

Editorial Office
MDPI AG
Grosspeteranlage 5
4052 Basel, Switzerland

This is a reprint of the Special Issue, published open access by the journal *Journal of Marine Science and Engineering* (ISSN 2077-1312), freely accessible at: https://www.mdpi.com/journal/jmse/special_issues/1T6S496BSE.

For citation purposes, cite each article independently as indicated on the article page online and as indicated below:

Lastname, A.A.; Lastname, B.B. Article Title. *Journal Name* **Year**, *Volume Number*, Page Range.

ISBN 978-3-7258-8323-3 (Hbk)
ISBN 978-3-7258-8324-0 (PDF)
https://doi.org/10.3390/books978-3-7258-8324-0

Contents

About the Editors

Qingshan Wang

Qingshan Wang is a Professor at the College of Mechanical and Electrical Engineering, Central South University, China. His research interests include nonlinear vibration and noise mechanisms of structures under multi-field coupling conditions, structural vibration and noise control, rotor dynamics, condition monitoring, and fault diagnosis of rotating machinery systems. Over the past decades, he has devoted himself to the theoretical and engineering research of structural dynamics, intelligent diagnostics, and reliability assessment for complex mechanical systems. He has published more than 200 peer-reviewed academic papers as the first or corresponding author and has authored three scholarly monographs. Professor Wang has been recognized as a Highly Cited Scholar in the field of Mechanical Engineering in China for five consecutive years and has been included in the World's Top 2% Scientists list. He has led three projects funded by the National Natural Science Foundation of China, five provincial and ministerial-level research projects, and more than twenty industry-sponsored and defense-related research programs. His research achievements have been widely applied in rotating machinery, aerospace equipment, marine engineering, and other critical engineering fields, contributing to advances in vibration suppression, fault diagnosis, and structural safety evaluation.

Mengzhen Li

Mengzhen Li is an Associate Professor at the Department of Ocean Engineering, School of Naval Architecture, Ocean and Energy Power Engineering, Wuhan University of Technology. He has long committed to theoretical, numerical and experimental research on marine structural dynamics, impact resistance, fatigue damage and optimal design. His core research covers gradient lattice sandwich structures, hybrid composite laminates and functionally graded material structures under complex coupled loads, aiming to enhance the structural safety, durability and lightweight performance of marine and naval engineering facilities. Dr. Li has mainly addressed structural vibration characteristics, impact damage, progressive failure, fatigue performance and multi-objective optimization of advanced marine composite structures. He has published more than 40 peer-reviewed academic papers as the first or corresponding author.

Preface

Marine and offshore structures play a vital role in global transportation, energy exploitation, and marine resource development. With the rapid advancement of ocean engineering technologies and the increasing demand for safe, efficient, and sustainable marine systems, the structural performance and reliability of ships and offshore platforms have attracted significant attention from both academia and industry. These structures are frequently exposed to harsh environmental conditions, including complex hydrodynamic loads, wave impacts, corrosion, fatigue damage, and extreme weather events, posing substantial challenges to their design, operation, and maintenance.

In recent years, remarkable progress has been achieved in structural modelling techniques, safety assessment methodologies, and advanced material applications. The integration of high-fidelity numerical simulations, experimental investigations, intelligent monitoring technologies, and data-driven approaches has greatly enhanced our ability to understand and predict the behaviour of marine structures throughout their service life. At the same time, the development of advanced materials, such as high-performance steels, lightweight composites, and multifunctional materials, has opened new possibilities for improving structural efficiency, durability, and resilience.

This Reprint brings together a collection of high-quality research articles that address current challenges and emerging developments in marine structural engineering. The contributions cover a broad spectrum of topics, including structural dynamics, vibration and acoustic responses, impact resistance, fluid–structure interaction, ultimate strength assessment, buckling and fatigue analysis, advanced composite structures, structural health monitoring, and innovative applications of digital technologies and artificial intelligence. Through theoretical studies, numerical modelling, experimental validation, and engineering case analyses, the authors provide valuable insights into both fundamental research and practical applications.

As Guest Editors, we sincerely appreciate the efforts and contributions of all authors, reviewers, and editorial staff whose dedication made this Reprint possible. We hope that the research presented in this volume will serve as a useful reference for researchers, engineers, graduate students, and professionals working in marine engineering, offshore structures, structural safety assessment, and advanced materials. Furthermore, we expect that the findings and methodologies reported herein will inspire future investigations and foster interdisciplinary collaboration toward safer, smarter, and more sustainable marine structures.

Qingshan Wang and Mengzhen Li
Guest Editors

Journal of
Marine Science and Engineering

MDPI

Editorial

Structural Modelling, Safety Assessment, and Advanced Material Application of Marine Structures

Qingshan Wang [1] and Mengzhen Li [2,*]

1 College of Mechanical and Electrical Engineering, Central South University, Changsha 410083, China; qingshanwang@csu.edu.cn
2 School of Naval Architecture, Ocean and Energy Power Engineering, Wuhan University of Technology, Wuhan 430063, China
* Correspondence: mengzhen.li@whut.edu.cn

The continuous expansion of human activities into the marine environment imposes ever-increasing demands on the performance, reliability, and longevity of ships and offshore structures [1,2]. Whether operating in deep-ocean basins, navigating through complex wave fields, or enduring the harsh conditions of the polar seas, modern marine structures must withstand a diverse spectrum of mechanical, hydrodynamic, and environmental loads. Consequently, advanced structural modeling has evolved from a supportive analytical tool into an indispensable prerequisite for the safe and efficient design of these complex engineering systems. Accurate prediction of vibration [3,4] and acoustic radiation signatures [5], assessment of impact and blast resistance [6], understanding of fluid–structure interaction phenomena [7,8], and reliable estimation of ultimate strength [9] and fatigue life are all central to mitigating operational risks and unlocking the full potential of next-generation marine assets. Furthermore, the ongoing integration of novel materials and the emergence of data-driven technologies such as digital twins and artificial intelligence are fundamentally reshaping the landscape of marine structural engineering.

Given the rapid advancements and the inherently interdisciplinary nature of this domain, there is a persistent need to synthesize new knowledge and to critically evaluate emerging methodologies against the backdrop of established practices. The present Special Issue, entitled “Structural Modeling, Safety Assessment, and Advanced Material Application of Marine Structures,” was conceived precisely to address this need. It brings together a collection of twelve cutting-edge contributions that collectively illuminate the latest developments in numerical simulation, experimental validation, and material innovation relevant to ship and offshore structures. The scope of this Issue is deliberately broad, reflecting the multifaceted challenges encountered in the field. Below, we systematically introduce the constituent papers, which can be thematically grouped into three interconnected areas of investigation: hydrodynamic performance and fluid–structure interaction, structural integrity and failure analysis, and advanced material application coupled with intelligent monitoring.

The hydrodynamic characteristics of platforms with three column geometries were investigated by Liang and Zhang (Contribution 1) using the three-dimensional potential flow method and Morison’s theory. Their results demonstrated that quasi-elliptical pillars provide the best overall dynamic behavior under combined wind–wave–current loads.

A comprehensive review of experimental and numerical research on slamming phenomena in advanced vessels was presented by Sun et al. (Contribution 2) through theoretical foundations, divided water-entry simulation, full-scale wave-load studies, and fluid–structure interaction approaches. Additionally, current gaps in forecasting accuracy and mitigation measures were clearly identified.

A comparative whole-scale CFD study on the wave-making characteristics and resistance performance of two representative naval destroyer designs was performed by Tian et al. (Contribution 3) employing the unstable Reynolds-Averaged Navier–Stokes solution, free-surface catching, and turbulence framework. The distinct performance balances between wave invisibility and resistance were identified, thereby providing critical data for multifaceted naval vessel optimization.

Pipeline systems in floating offshore production platforms were systematically reviewed by Yan et al. (Contribution 4) applying fluid dynamics, electrochemical erosion, material weakness, design factors and preservation methods. The necessity of transdisciplinary solutions such as machine learning-enabled digital twins was highlighted.

In order to support practical examination needs, a novel internal examination device for underwater pipelines was designed and verified by Ma et al. (Contribution 5). Point-cloud statistics were successfully acquired for coded three-dimensional rebuilding, and precise fault identification and service-life estimation were enabled in simulated pipeline surroundings.

To advance material applications for marine structures, the influence of 3D-printing insert patterns on the compressive and torsional behavior of coastal solar support components was examined by Zhang et al. (Contribution 6) through systematic mechanical experiments. Honeycomb filler was identified as the optimal configuration for balancing strength and weight.

Similarly, a lightweight concept of coastal construction using 3D-printed lattice filler was explored by Jiang et al. (Contribution 7). Parametric analyses on infill category, thickness, and support angle were performed, and it was shown that 0°-inclined honeycomb structures exhibit superior bending, torsional, and compressive properties, while inter-layer interaction was identified as a critical factor.

For intelligent surveillance in extremely deep-ocean environments, a self-sustaining recognizing system based on curved-surface flexible triboelectric nanogenerator lists for massive aqua farming cages was proposed by Yang et al. (Contribution 8) establishing a three-level multi-field coupling system. Moreover, it was demonstrated that this new technology can concurrently capture low-frequency ocean energy and enable real-time vibration tracking without an extra power source.

Furthermore, a dual-material constitutive model for the granular heat-affected area in 80 mm-thick DH36 ship-plate welded connection was established by Xu et al. (Contribution 9) utilizing digital image correlation and the Ramberg–Osgood model, and significant gradients in elastic modulus and yield strength between this area and base metal were quantified, thereby providing critical information for precise strength evaluation.

The safety evaluation under functional harm situation was deeply addressed in two studies. The computational methodologies, external reactions, and post-fracture structural behavior of multi-branch towed array systems were extensively studied by Yan et al. (Contribution 10). Comparisons with trawl nets were drawn, and construction alterations and load redistribution subsequent to guide-cable or main-cable collapses under diverse sea states were simulated.

The kinetic properties and structural failure of rigid lattice trawl systems during dragging were conducted by Zhang et al. (Contribution 11) through the lumped mass approach and OrcaFlex modeling. A safe functional boundary for dragging speed and turning movements was determined, while the important effects of singular and dual-line breakage on net twisting and fishing efficiency were clarified.

https://doi.org/10.3390/jmse14111052

The structural elements were bridged with restoration of ecosystems by de Sousa et al. (Contribution 12) through a comparative evaluation of damaged Portland cement concrete bricks and organic coral shell from the shore of Paraíba. Mineralogical, physical, and chemical analyses were performed, and it was confirmed that aged concrete resembles coral in carbonate composition and steady acidity, while providing enhanced compressive properties, thereby supporting its use as a practically analogous function for artificial reefs after structural and density optimization.

Collectively, the twelve articles featured in this Special Issue exemplify the vibrant trajectory of research in marine structural mechanics and materials. They underscore a clear transition from conventional, often isolated, analytical frameworks toward more integrated, multi-physics, and data-enriched approaches. While significant progress has been demonstrated in predictive accuracy for slamming loads, the mechanical optimization of 3D-printed lattice fillers, and the resilience assessment of damaged arrays, several critical frontiers remain open for future exploration. Notably, the long-term durability of novel composite and additively manufactured materials in corrosive marine environments warrants further systematic investigation. Moreover, the seamless integration of real-time structural health monitoring data with high-fidelity digital twins to enable true predictive maintenance and autonomous operational decision-making represents both a formidable challenge and a promising horizon. As marine structures continue to venture into deeper and more remote waters, the synergy between robust physical modeling, advanced material science, and intelligent data analytics will undoubtedly remain the cornerstone of innovation in ensuring their safety and sustainability.

Funding: This research received no external funding.

Acknowledgments: We sincerely thank all contributing authors for their insightful research presented in this Special Issue, all peer reviewers for their rigorous evaluation and constructive feedback, and the editorial office of *JMSE* for their continuous guidance and efficient coordination. We also extend our appreciation to the publishing team for their dedication to maintaining high-quality standards throughout the production process.

Conflicts of Interest: The authors declare no conflicts of interest.

List of Contributions:

1. Liang, K.; Zhang, D. Study on the Effect of Column Form on the Dynamic Response of Semi-Submersible Truss-Type Fish Culture Platforms. *J. Mar. Sci. Eng.* **2025**, *13*, 2370. https://doi.org/10.3390/jmse13122370.
2. Sun, Y.; Zhang, D.; Wu, Z.; Yu, Y. A Review of Experimental and Numerical Research on the Slamming Problem of High-Performance Vessels. *J. Mar. Sci. Eng.* **2025**, *13*, 1310. https://doi.org/10.3390/jmse13071310.
3. Tian, Y.; Shou, H.; Guo, S.; Chen, Z.; Zhou, Z.; Zheng, Y.; Shi, K.; Zhang, D. A Comparative CFD Study on the Wave-Making Characteristics and Resistance Performance of Two Representative Naval Vessel Designs. *J. Mar. Sci. Eng.* **2026**, *14*, 212. https://doi.org/10.3390/jmse14020212.
4. Yan, J.; Zhang, Y.; Chen, Z.; Li, P.; Li, Y.; Cao, Z.; Wu, J.; Yang, K.; Zhang, D. Pipeline Systems in Floating Offshore Production Systems: Hydrodynamics, Corrosion, Design and Maintenance. *J. Mar. Sci. Eng.* **2026**, *14*, 176. https://doi.org/10.3390/jmse14020176.
5. Ma, Q.; Liang, W.; Chen, H.; Wang, Q.; Zhou, P.; Wang, Q. Design and Implementation of a LiDAR-Based Inspection Device for the Internal Surveying of Subsea Pipelines. *J. Mar. Sci. Eng.* **2025**, *13*, 2141. https://doi.org/10.3390/jmse13112141.
6. Zhang, H.; Zeng, S.; Zhang, Y.; Guo, S.; Feng, H.; Zhang, D. Study on the Influence of 3D Printing Material Filling Patterns on Marine Photovoltaic Performance. *J. Mar. Sci. Eng.* **2025**, *13*, 2373. https://doi.org/10.3390/jmse13122373.

7. Jiang, H.; Xie, Y.; Zeng, S.; Guo, S.; Chen, Z.; Liang, Z.; Zhang, D. Research on Lightweight Design Performance of Offshore Structures Based on 3D Printing Technology. *J. Mar. Sci. Eng.* **2025**, *13*, 2007. https://doi.org/10.3390/jmse13102007.
8. Yang, K.; Zeng, S.; Yang, K.; Zhang, D.; Zhang, Y. From Multi-Field Coupling Behaviors to Self-Powered Monitoring: Triboelectric Nanogenerator Arrays for Deep-Sea Large-Scale Cages. *J. Mar. Sci. Eng.* **2025**, *13*, 2042. https://doi.org/10.3390/jmse13112042.
9. Xu, L.; Zhan, P.; Yi, T.; Zhang, S.; He, J.; Li, M. Research on the Constitutive Relationship of the Coarse-Grained Heat-Affected Zone in Ship Thick-Plate Welded Joints of Ship Structures. *J. Mar. Sci. Eng.* **2025**, *13*, 1260. https://doi.org/10.3390/jmse13071260.
10. Yan, J.; Yang, K.; Zeng, S.; Yang, K.; Zhang, D.; Zhu, K. Multi-Branch Towed Array System: Systematic Analysis of Modeling Methods, Environmental Responses and Mechanical Properties in Fracture Analysis. *J. Mar. Sci. Eng.* **2025**, *13*, 1697. https://doi.org/10.3390/jmse13091697.
11. Zhang, D.; Zhao, B.; Zhang, Y.; Zhu, K.; Yan, J. Study on Dynamic Characteristics and Fracture Failure of Rigid Truss Trawl System During Towing Process. *J. Mar. Sci. Eng.* **2025**, *13*, 586. https://doi.org/10.3390/jmse13030586.
12. Sousa, M.F.; Natividade, C.D.; Lima Filho, M.R.; Torres, S.M.; Santos, A.J.; Rocha, R.A.; Henriques, G.F.; Massei, K.; Souza, W.M. Mineralogical and Mechanical Characterization of Concrete Blocks for Artificial Reefs: A Comparative Study with Natural Coral Skeletons. *J. Mar. Sci. Eng.* **2025**, *13*, 1886. https://doi.org/10.3390/jmse13101886.

References

1. Pandey, M.; Viljoen, C.; Way, A.; Fischer, K.; Sýkora, M.; Diamantidis, D.; Steenbergen, R.D.J.M.; Lind, N.; Frangopol, D.M.; Yang, D.Y.; et al. Life Safety in the Reliability-Based Design and Assessment of Structures. *Struct. Saf.* **2025**, *113*, 102453. [CrossRef]
2. Klanac, A.; Varsta, P. Design of Marine Structures with Improved Safety for Environment. *Reliab. Eng. Syst. Saf.* **2011**, *96*, 75–90. [CrossRef]
3. He, D.; Wang, Q.; Zhong, R.; Qin, B. A Unified Analysis Model of FGM Double-Layered Submarine Type Coupled Structure with Spectral Geometry Method. *Ocean Eng.* **2023**, *267*, 113213. [CrossRef]
4. Li, M.; Yu, J.; Li, X.; Chen, W.; Xu, S.; Shen, W.; Zhao, Y.; Qiu, Y. A Novel Analytical Model for Vibration Analysis of 3D-Printed Sandwich Plate with Graded Pyramid Lattice Cores. *Eng. Struct.* **2026**, *353*, 122152. [CrossRef]
5. Prospathopoulos, J.; Belibassakis, K. A 3D BEM-CMS Model for Noise Radiation Due to Vibrating Hull of Ships and Underwater Vehicles in the Ocean Acoustic Waveguide. *Wave Motion* **2026**, *145*, 103752. [CrossRef]
6. Zhang, J.; Wang, Y.; Zhang, Y.; Zhai, Z. Blast Response of Clamped Rectangular Double-Layer Sandwich Plates with Hybrid Asymmetric Foam Cores. *Thin-Walled Struct.* **2026**, *218*, 114094. [CrossRef]
7. Ye, Y.; Gan, J.; Liu, H.; Wu, W.; Wang, L.; Guo, J.; Li, H. Bending Behaviour of Rigid-Flexible Combined Inflatable Membrane Structures for Floating Bridge Modules. *Ocean Eng.* **2025**, *338*, 122005. [CrossRef]
8. Ye, Y.; Gan, J.; Wu, W.; Wang, S. Numerical Study on Bending Behaviours and Load-Bearing Capacities of Inflatable Floating Bridges Considering Gas-Solid-Liquid Interactions. *Ocean Eng.* **2025**, *341*, 122680. [CrossRef]
9. Xie, W.; Li, J.; Xu, Z.; Li, X.; Shen, W.; Xu, L.; Qiu, Y.; Li, M. Ultimate Strength Failure Mechanism and Optimization Design of Marine Composite Deck Grillage Structures with Opening. *Eng. Fail. Anal.* **2024**, *162*, 108444. [CrossRef]

Journal of
Marine Science and Engineering

MDPI

Article

Study on Dynamic Characteristics and Fracture Failure of Rigid Truss Trawl System During Towing Process

Dapeng Zhang [1,†], Bowen Zhao [2,†], Yi Zhang [1], Keqiang Zhu [3] and Jin Yan [1,4,*]

1 Ship and Maritime College, Guangdong Ocean University, Zhanjiang 524005, China; zhangdapeng@gdou.edu.cn (D.Z.)
2 Department of Applied Mathematics and Mathematical Modeling, Saint Petersburg State Marine Technical University, Saint Petersburg 190121, Russia
3 Faculty of Maritime and Transportation, Ningbo University, Ningbo 315211, China
4 Guangdong Provincial Key Laboratory of Intelligent Equipment for South China Sea Marine Ranching, Guangdong Ocean University, Zhanjiang 524088, China
* Correspondence: yanj@gdou.edu.cn
† These authors contributed equally to this work.

Abstract: Deep-sea fisheries depend on various fishing methods, including trawling, purse seining, and longline fishing, among others. Studying the dynamic characteristics of trawling operations is essential for the trawl mechanism. Because of the solid truss support, the beam trawl system may be employed in extreme sea conditions, the high-speed driving of tugs, and maneuvering situations. This study systematically investigates the dynamic responses and structural safety of a midwater beam trawl during towing via the lumped mass method and OrcaFlex 9.7e simulations. Firstly, a trawl model with four towlines was developed and validated against flume tank experiments. Secondly, multiple operational scenarios were analyzed: towing speeds, angular velocity variations under a fixed turning radius, and radius effects under constant angular velocity. The results show that line tension increases with the speed increment and that the rigid frame destabilizes at angular velocities exceeding 20°/s due to centrifugal overload. Furthermore, line fracture scenarios during startup and straight-line towing were emphasized. Single-line failure leads to edge constraint loss, redistributing stress to the remaining lines, and asymmetric dual-line fracture triggers net torsion, reducing fishing efficiency. This study provides theoretical guidance for optimizing the safe operational parameters of midwater beam trawls.

Keywords: beam trawl system; dynamic characteristics; fracture failure; towing process

1. Introduction

Rich in resources, the deep sea is particularly abundant in fish [1,2]. Marine fisheries are a key component of the marine industry, and fishing is an essential part of this sector [3,4]. The bottom trawling mechanism is key in deep-sea fishing and is typically towed from either the stern or the sides of the fishing vessel [5,6]. Once fish are located, the vessel employs various maneuvers such as accelerating forward and making sharp turns to chase, intercept, and capture them using a midwater trawl [7]. A typical marine trawl system is shown in Figure 1.

The dynamic behavior and properties of trawl systems (TSs) have been widely studied by researchers using various approaches. As early as the 1980s, efforts were made to characterize these systems. Over time, the field has garnered increasing attention, with numerous research methods being proposed to investigate the dynamic behavior and characteristics of trawl systems (TSs) [8–11]. Dealteris et al. assessed two different trawl designs—shrimp trawls and high-level trawls—using both scaled model testing and field-testing methods [12]. Buxton and Alteris found that under test conditions, webbing sturdiness and drag speed significantly influenced the velocity gradient, while webbing sturdiness and the mean angle of incidence affected the drag coefficient. However, drag speed had minimal impact on drag. [13]. Hu et al. numerically analyzed the dynamic characteristics of a trawling system, providing key details about the net and otter boards, including drag coefficients. The study also found that changes in trawler speed influence the sinking distance of both the net and the otter board [14]. Sangster and Breen employed a proto-scale experimental approach to compare the fishing and engineering performance of single and double trawls. Specifically, they examined the effects of gate width, wingspan, and rein angle on the area of the seafloor effectively fished using each type of trawl [15]. Reite and Asgeir combined steady-state and transient hydrodynamic effects into a comprehensive mathematical model, which can be effectively used in the design of trawl door control systems. They derived the steady-state hydrodynamic coefficients for trawl doors through wind tunnel experiments [16]. To evaluate the performance of two trawl doors, Sala et al. conducted a comparative experiment—the existing AR door and the new Clarck-Y door—focusing primarily on demersal fisheries. By comparing the results of flume tank tests and full-scale at-sea trials, they highlighted significant performance differences between the two trawl doors. The study found variations in drag, lift, and efficiency factors between the two designs [17]. In their study, Sun et al. designed a mathematical model to simulate the trawling process of a single-vessel midwater trawl system. In addition to simulating the trawler and fishing gear separately, the model considers the influence of wind and current. Simulations were used to evaluate hydrodynamic performance. [18]. Zhao et al. analyzed a box-shaped mesh cage using numerical methods and developed a numerical model to predict its hydrodynamic response to waves and currents. The model was validated through physical tests [19].

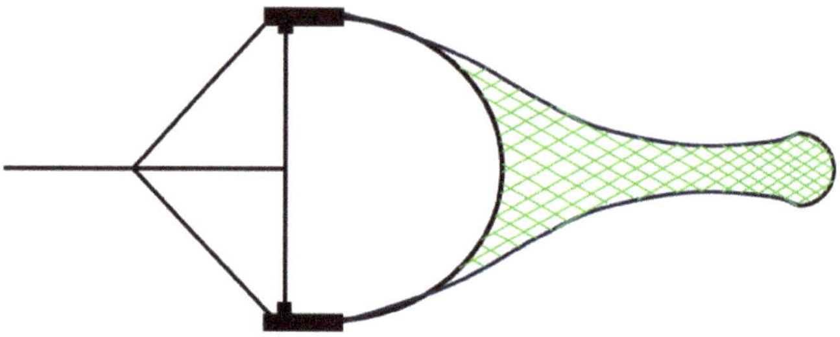

Figure 1. A typical beam trawl held open by a steel bar, with shoes on each end.

According to the literature, approaches for investigating trawl system hydrodynamic performance include model testing, numerical modeling, and full-scale sea tests. With the advancement of computer technology, numerical simulation is seen as an effective strategy [20,21]. Numerical simulation is becoming more and more popular among scholars since it is an efficient way to address a variety of challenging issues in the study of trawl systems. For example, Li et al. [22] designed a new deep-sea trawl for deep-water fishing vessels, evaluating its hydrodynamic performance through numerical simulations and physical modeling. In their study, Liu et al. [23] introduced a calculation method for a frustum-type pressure reducer used in midwater trawls targeting larval and juvenile fish. This study employed flume experiments alongside computational fluid dynamics (CFD) simulations to analyze the flow field distribution and optimize the design parameters of the pressure reducer. Guan et al. conducted the research. Reference [24] explored the hydrodynamic interactions between sorting nets and bottom trawls to improve trawl selectivity and reduce bycatch. Their study combined numerical simulations and physical model tests, revealing that the sorting net significantly influences the jetty shape and net tension distribution. Xiong et al. [25] investigated the drag characteristics of a Chinese Antarctic krill vessel operating in an ice flow zone, employing MATLAB 2021B and genetic algorithms to simulate ice flow and evaluate the interactions between the ship's trawl grid and the ice. Their results indicated that ice drag increases with ice concentration and velocity, while trawl drag becomes more pronounced at higher velocities. Neill and Karsten [26] modeled the flow through mesh panels. Tang et al. [27] examined how twine diameter and mesh size, as well as the material, impacts the effectiveness of fishing nets, with an emphasis on the interaction between turbulence and hydrodynamic forces around the mesh structure. To simulate the viscous three-dimensional unsteady boundary layer turbulent flow, the fishing net was modeled as a knotless net with diamond-shaped mesh, using cylinders as flow obstacles to represent the mesh structure. To further explore the effect of hydrodynamic forces on the net's deformation, a porous medium model and finite element structural dynamics (FEM) analysis were applied. Nsangue et al. [28] studied the hydrodynamic properties of gillnets, particularly how design parameters influence net performance, through numerical simulation and experimental validation. A CFD model was combined with the k-Ω (SST) turbulence model in a one-way coupled approach to simulate the fluid–net interaction, providing insights into the stresses, strains, deformations, and motions of the gillnet. Numerical simulation methods have played a crucial role in these studies by providing detailed insights into the flow field and structural behavior.

Trawl structures, particularly beam trawl systems characterized by rigid truss supports, differ significantly from fully flexible trawl systems due to their inherent flexibility. The rigid beam framework minimizes deformation during towing, ensuring a consistent fishing area even under high-speed operations. However, localized stress concentrations caused by abrupt maneuvers or excessive towing speeds may still lead to the horizontal/vertical misalignment of the net gear, increasing risks of entanglement, line fracture, or structural collapse. These challenges are exacerbated during rapid tugboat rotations, where the asymmetric force distribution on the rigid frame edges heightens deformation risks. Unlike bottom trawls, which rely on hydrodynamic forces or otter boards for horizontal expansion, beam trawls depend on their truss structure to maintain shape integrity. Consequently, precise control of the towing parameters (e.g., speed, angular velocity, angle of attack of the beam) is critical to prevent stress overload on the rigid frame and lines. Li et al. [29] designed three types of beams—cylindrical, airfoil, and elliptical—and studied the hydrodynamic performances of beams with different shapes at different angles of attack. They draw the conclusion that compared to the cylindrical beam, the hydrodynamic performances of the airfoil beam and elliptical beam are superior. Rijnsdorp [30] estimated the hydrodynamic drag of individual gear components of the beam trawl using empirical measurements of similarly shaped objects, including cylinders, cubes, and nets. He et al. [31] proposed a depth variable device installed on the beam to adjust the trawling depth independently, aiming to track the depth where the shoal of Antarctic krill is located for accurate fishing.

The research employs the lumped mass method in OrcaFlex to simulate the trawl warp and net, capturing their dynamic behavior under different towing conditions. The simulation helps to analyze how different towing speeds influence the deformation, stability, and forces acting on the system, providing insights into the operational limits and optimization of towing practices for beam trawl systems. The structure of this article is illustrated in Figure 2.

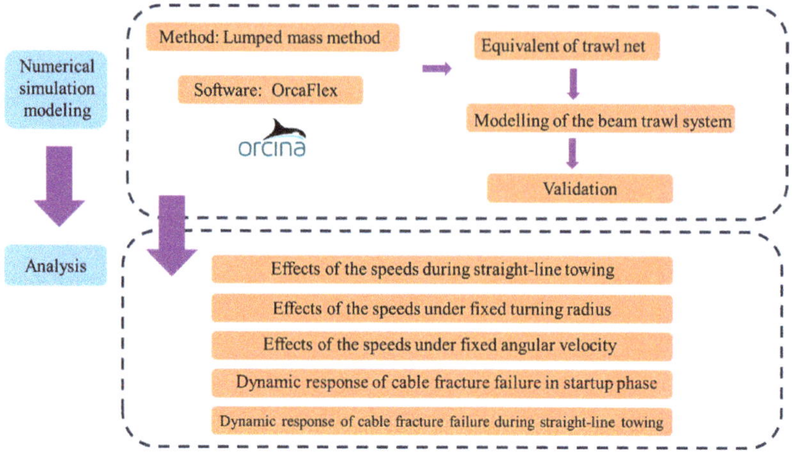

Figure 2. Structure of the article.

2. Methodology of the Trawl System (TS) Model

2.1. The Lumped Mass Model

To address the nonlinear boundary value issue, the discrete lumped mass model is utilized. The core concept of this model involves segmenting the line being towed into N distinct pieces, with each segment being considered individually and each segment's mass concentrated at one node, creating a total of $N + 1$ nodes. The pulling force and shearing forces at the ends of each segment, together with any external hydrodynamic loads, are applied at the nodes. The equation of motion of i-th node ($i = 0, 1\ldots N$) is as follows:

$$\boldsymbol{M}_{A_i}\ddot{\boldsymbol{R}}_i = \boldsymbol{T}_{e_i} - \boldsymbol{T}_{e_{i-1}} + \boldsymbol{F}_{dI_i} + \boldsymbol{V}_i - \boldsymbol{V}_{i-1} + \boldsymbol{w}_i\Delta\bar{s}_i \tag{1}$$

Among them, $\boldsymbol{R}$ represents the node position of the line.

$\boldsymbol{M}_{Ai} = \Delta\bar{s}_i\left(m_i + \frac{\pi}{4}D_i^2(C_{an}-1)\right)\boldsymbol{I} - \Delta\bar{s}_i\frac{\pi}{4}D_i^2(C_{an}-1)(\boldsymbol{\tau}_i \otimes \boldsymbol{\tau}_{i-1})$ represents the mass matrix associated with each node; $\boldsymbol{I}$ denotes a 3 × 3 identity matrix; $T_{ei} = EA\varepsilon_i = EA\frac{\Delta s_{0i}}{\Delta s_{\varepsilon i}}$, which stands for effective tension at a certain node; $\Delta s_{0i} = L_0/(N-1)$, which represents the initial length attributed to each segment; $\Delta s_{\varepsilon i} = |R_{i+1} - R_i|$, which represents the stretched length of each segment; and EA denotes the axial stiffness of the line.

$\boldsymbol{F}_{dI_i}$ indicates the external hydrodynamic forces [32] of each node, which are calculated according to the Morison equation:

$$\boldsymbol{F}_{dI_i} = \frac{1}{2}\rho D_i\sqrt{1+\varepsilon_i}\Delta\bar{s}_i(C_{dni}|\boldsymbol{v}_{ni}|\boldsymbol{v}_{ni} + \pi C_{dti}|\boldsymbol{v}_{ti}|\boldsymbol{v}_{ti}) + \frac{\pi}{4}D_i^2\rho C_{ani}\Delta\bar{s}_i(\boldsymbol{a}_{wi} - (\boldsymbol{a}_{wi}\cdot\boldsymbol{\tau}_i))\boldsymbol{\tau}_i \tag{2}$$

where ρ means sea water density, D_i means the diameter of each line, C_{dni} means the normal drag coefficient, C_{dti} means the tangential drag coefficient, and C_{ani} means the inertia coefficient.

$\boldsymbol{V}_i = \frac{EI_{i+1}\boldsymbol{\tau}_i\times(\boldsymbol{\tau}_i\times\boldsymbol{\tau}_{i+1})}{\Delta s_{\varepsilon i}\Delta s_{\varepsilon i+1}} - \frac{EI_i\boldsymbol{\tau}_i\times(\boldsymbol{\tau}_{i-1}\times\boldsymbol{\tau}_i)}{\Delta s_{\varepsilon i}^2} + \frac{H_{i+1}\boldsymbol{\tau}_i\times\boldsymbol{\tau}_{i+1}}{\Delta s_{\varepsilon i}}$, $\boldsymbol{V}$ represents the shear force at the node and $\boldsymbol{H}$ is the torsion.

The lumped mass model is shown in Figure 3.

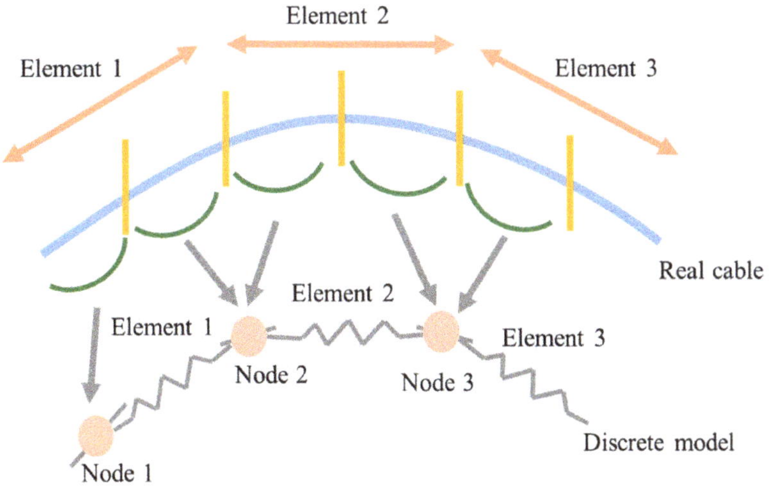

Figure 3. The model with aggregated mass.

2.2. Equivalent of Trawl Net

In Figure 4, the analogous model of the netting is shown, where k represents the k-th line of the net, and I marks the node located at the center of the aggregate horizontal length of the n neighboring net lines at the lower boundary. With the diameter of each actual net line being d_0 and its hydrodynamic coefficient C_d, when merging n adjacent lines to form a single new line, this combined line features a diameter of D and a hydrodynamic coefficient of C_D. The conservation of the overall volume of the lines in the net leads to the subsequent equations.

$$n \cdot \frac{\pi}{4} d_0^2 = \frac{\pi}{4} D^2 \tag{3}$$

$$D = d_0 \sqrt{n} \tag{4}$$

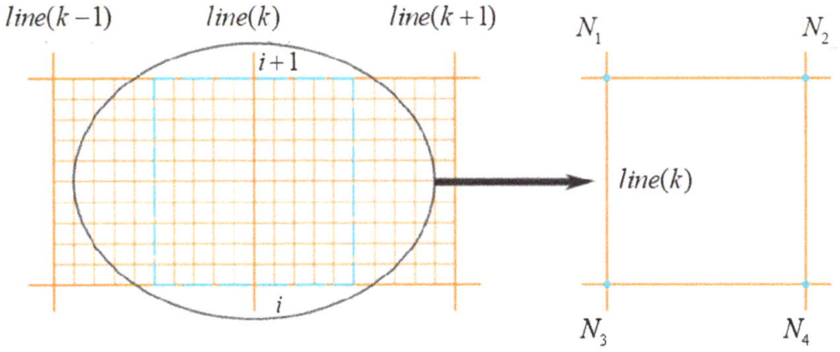

Figure 4. The equivalent model of the net.

If we assume that the hydrodynamic coefficients are equal before and after equivalence, we can reach the following conclusion:

$$C_D = C_d \sqrt{n} \tag{5}$$

As illustrated in Figure 5, the equivalent net model experiences drag and lift forces resulting from the flow. The equations describing the horizontal drag and vertical lift are outlined below:

$$F_D = C_D \cdot \frac{1}{2}\rho A U^2 \tag{6}$$

$$F_L = C_L \cdot \frac{1}{2}\rho A U^2 \tag{7}$$

In these equations, A signifies the net's area, ρ stands for the density of water, and U represents the velocity of the current, whereas C_D and C_L denote the coefficients for drag and lift, respectively.

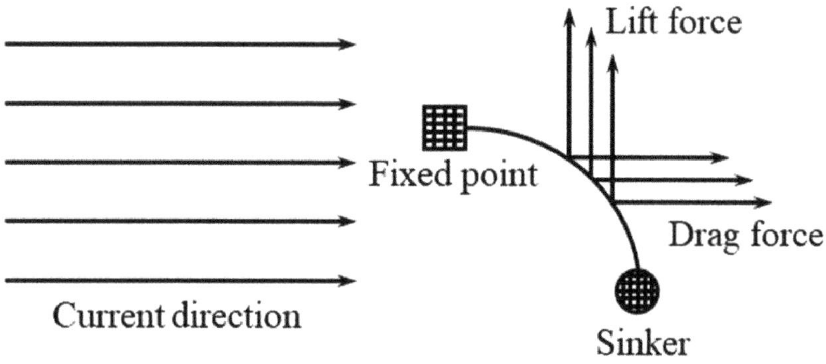

Figure 5. The load on the net.

The determination of hydrodynamic coefficients has been a focal point in the study of net equivalence, and the empirical formula method is generally used, as shown in Equations (8) and (9) [33].

$$C_D(S_n, \theta) = 0.04 + \left(-0.04 + 0.33S_n + 6.54S_n^2 - 4.88S_n^3\right)\cos\theta \tag{8}$$

$$C_L(S_n, \theta) = \left(-0.05S_n + 2.3S_n^2 - 1.76S_n^2\right)\sin 2\theta \tag{9}$$

Within this framework, the symbol θ is utilized to denote the inclination of the water current's trajectory, which is a critical parameter for understanding the dynamics of the net's interaction with the fluid environment. S_n represents the water permeability of the net, which is determined using the following calculation:

$$S_n = 2\frac{d_\omega}{l_\omega} - \left(\frac{d_\omega}{l_\omega}\right)^2 \tag{10}$$

where d_ω represents the outer diameter of the net mesh length, while l_ω denotes the length between two nodes in the mesh garment.

2.3. Modeling of the Beam Trawl System in OrcaFlex

This study employs a midwater beam trawl system (Figure 6) designed for midwater operations with a rigid truss structure. The system operates at depths of 50–200 m, with a total warp length of 150 m and towing speeds ranging from 1 to 5 m/s. In Figure 6, it is illustrated that the upper part of the beam trawl consists of four sturdy rods. This configuration contributes to the structural integrity and functionality of the trawl system. These bars also serve to maintain the trawl's effective fishing area, ensuring it remains constant during operation. The rigid bars are connected to each other and form one square rigid frame. Four towing lines connect the tug's stern to the beam trawl. To accurately assess the towing line fracture breakdown, it is essential to identify the four towed lines. When the tug is at rest, the Line Port and Line Stbd, which are symmetrical to the tug's longitudinal plane, are located on the tug, The Line Port is assigned to the vessel's right-hand side, while the Line Stbd occupies the left-hand side. This strategic placement is essential for maintaining equilibrium and enhancing the operational effectiveness of the system.

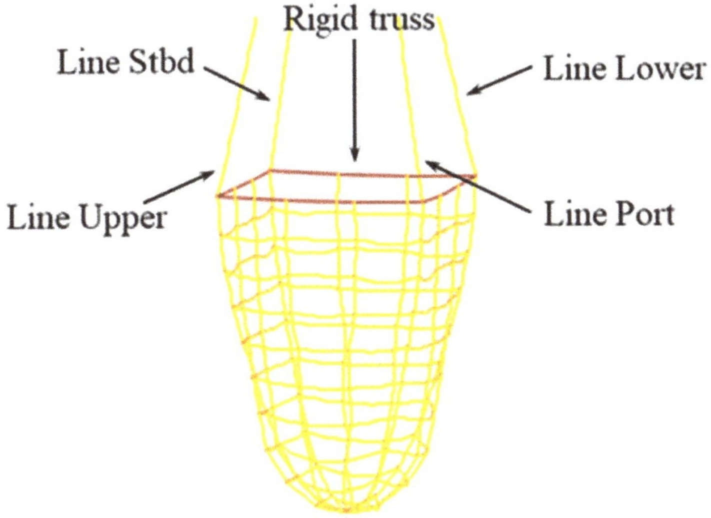

Figure 6. The beam trawl.

During high-speed straight-line towing, line fractures pose severe risks to trawl system safety. The tugboat parameters in this study are as follows: maximum towing speed 12 knots (≈6.17 m/s), bollard pull force 1200 kN, main engine power 5000 kW, rudder area 12 m^2, and rudder effectiveness coefficient 0.65. Maneuverability was calibrated via zigzag tests, showing a heading change rate of 15°/s and a turning diameter of 2.5× ship length. Simulations covered towing speeds of 1–5 m/s (encompassing operational ranges) to evaluate the impact of tugboat dynamics on trawl system behavior.

Comprising the towing line known as the trawl warp, the trawl system features a towed line model that is thin, flexible, and cylindrical. In the OrcaFlex simulation, the four rigid frame edges are modeled as hollow, uniform cylindrical tubes with a theoretical infinite stiffness. The lines feature an outer diameter of 0.35 m and an inner diameter of 0.25 m. Furthermore, these lines exhibit a linear mass density of 0.18 tons per meter, a flexural stiffness of 120 kN·m^2, a torsional stiffness of 80 kN·m^2, and an axial stiffness of 700,000 kN. The mechanical properties of the trawl's nets are identical to those of the lines. The overall tilt of the trawl system and the variations in towline tension are crucial research areas for this trawl system, which is known for its substantial resistance to deformation. The mechanical properties of the trawl's nets are the same as those of the lines. The primary focus of research for this type of trawl system lies in the tilt of the trawl system and the changes in towline tension, which are due to its remarkable ability to resist deformation. The trawl winch is mainly used for lifting and releasing nets and twisting and retracting ropes. In the trawling operation, the length and speed of the towed wire are adjusted to maintain different water layer positions of the net. In OrcaFlex, the simplified winch model is used to control the towing lines [34]. Figure 7 depicts the tug and beam trawl in OrcaFlex.

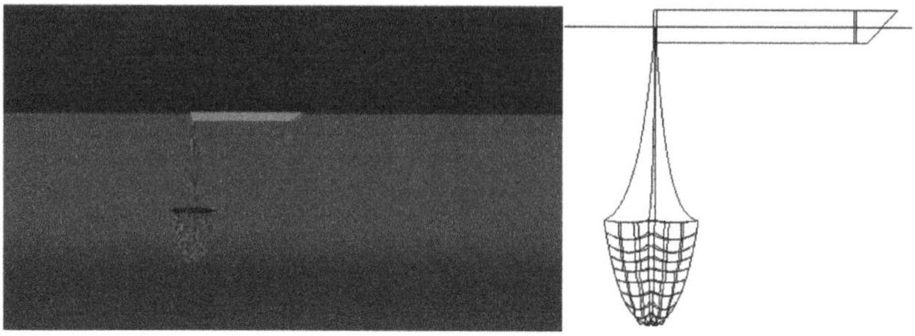

Figure 7. The tug and beam trawl in OrcaFlex.

2.4. Validation

To verify the precision of the proposed method, a numerical simulation of the trawl net system was performed, and the results were compared with experimental data [35]. Three types of nylon nets, types A, B, and C, were used, as shown in Table 1. Figure 8 illustrates the diagram of the validation model.

Table 1. Three types of nylon nets.

Net Type	Twine Diameter (mm)	Mesh Size (mm)	Total Mass (g)
A	0.348	75	9.91
B	0.348	50	14.68
C	0.348	40	22.31

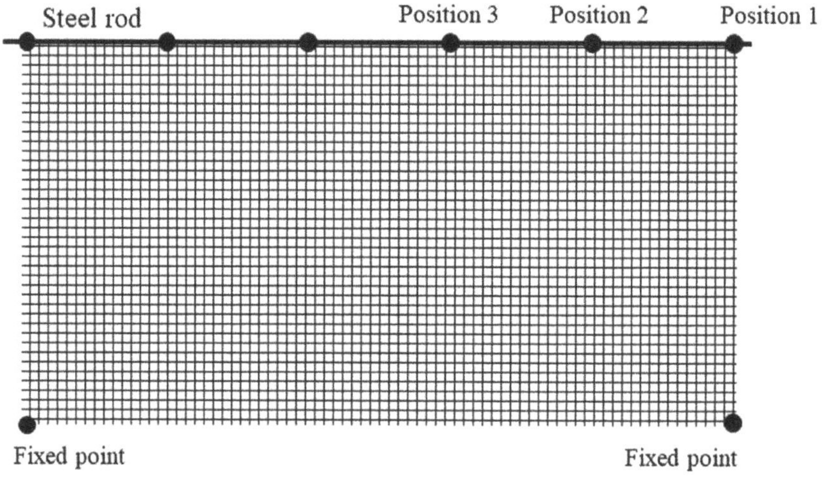

Figure 8. The representation of the validation model.

Figure 9 illustrates a comparative study of the dynamic forces acting on the steel rod, as derived from simulation versus experimental data, while Figures 10–12 offer an exhaustive analysis of the tension forces at three pivotal locations for a variety of net configurations. Overall, there is a high degree of correlation between the outcomes of the computational models and the experimental results. Table 2 presents the validation results and errors.

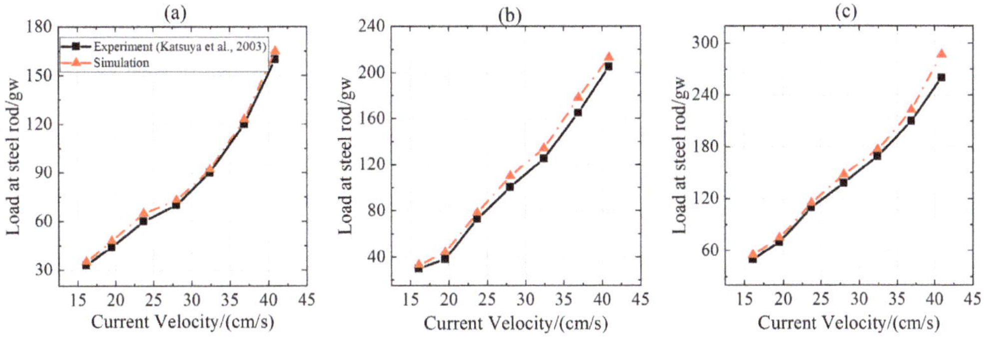

Figure 9. The analysis examines the flow components influencing the loads on the steel rod: (**a**) type A; (**b**) type B; and (**c**) type C [35].

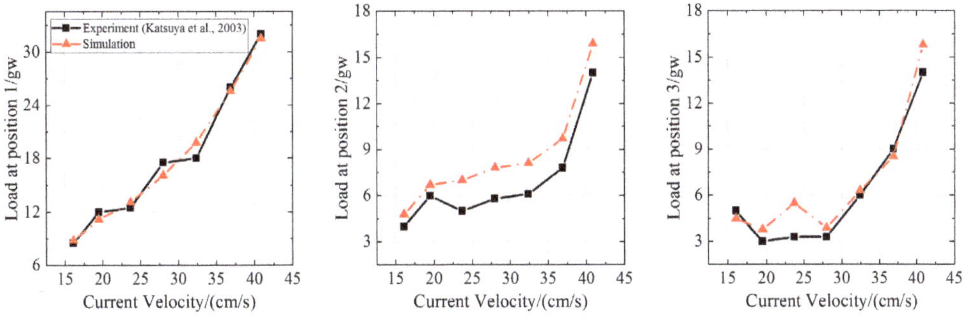

Figure 10. The study investigates the tension forces specific to net type A [35].

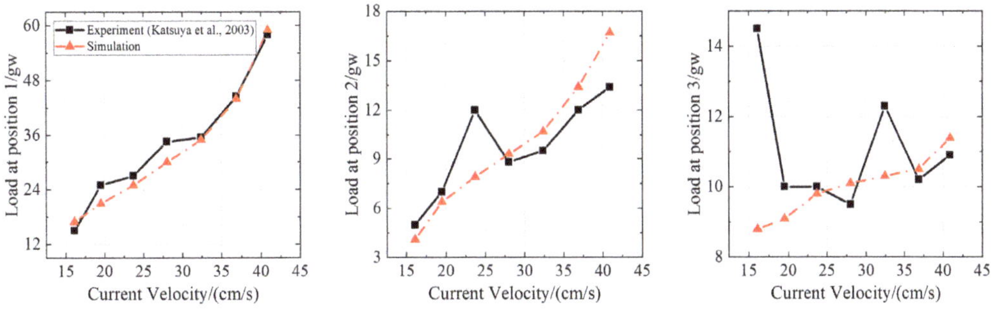

Figure 11. The study investigates the tension forces specific to net type B [35].

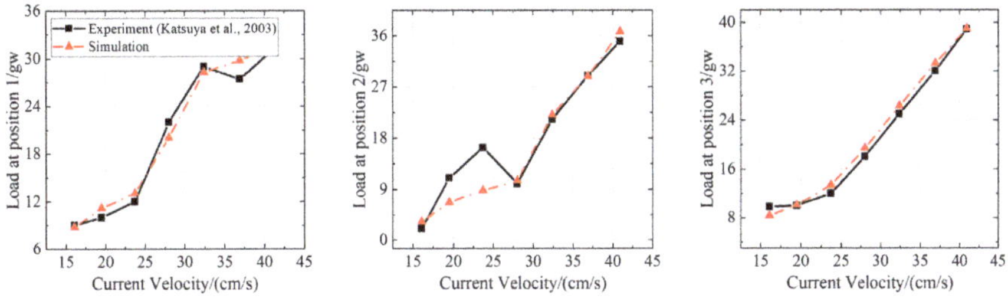

Figure 12. The study investigates the tension forces specific to net type C [35].

Table 2. Validation results and errors [32].

Velocity	Type A			Type B			Type C		
	Simulation	Experiment	Error	Simulation	Experiment	Error	Simulation	Experiment	Error
Loads on the Steel Rod/(gw)									
16.10	34.55	31.63	9.23%	49.85	47.95	3.95%	72.92	67.21	8.49%
19.50	47.62	43.27	10.07%	61.28	56.22	8.99%	91.00	86.11	5.68%
23.70	64.34	59.98	7.27%	93.15	87.41	6.57%	127.98	123.90	3.29%
28.00	72.70	69.78	4.19%	121.81	112.22	8.55%	157.62	148.58	6.08%
32.40	91.93	89.73	2.45%	143.47	135.83	5.62%	186.38	178.97	4.14%
36.90	122.79	119.87	2.44%	183.56	170.86	7.43%	229.06	218.39	4.89%
40.90	164.58	160.22	2.72%	215.43	208.43	3.36%	288.20	261.97	10.01%
Loads at position 1/(gw)									
16.10	8.63	8.55	0.98%	16.87	15.06	12.04%	8.56	9.13	−6.28%
19.50	10.83	11.93	−9.22%	20.99	25.11	−16.39%	10.96	9.96	10.04%
23.70	12.87	12.28	4.79%	24.62	27.08	−9.11%	13.03	12.03	8.31%
28.00	16.00	17.53	−8.72%	30.05	34.82	−13.70%	20.07	22.14	−9.33%
32.40	19.73	17.78	10.96%	34.66	35.33	−1.89%	28.28	29.02	−2.57%
36.90	25.50	26.01	−1.97%	44.22	44.72	−1.13%	29.76	27.44	8.45%
40.90	31.51	31.86	−10.8%	59.37	58.07	2.25%	32.09	31.34	2.38%
Loads at position 2/(gw)									
16.10	3.86	4.69	−17.61%	3.95	4.94	−19.98%	3.17	2.27	39.78%
19.50	5.85	6.58	−11.03%	6.34	6.92	−8.38%	6.62	10.85	−38.97%
23.70	4.93	6.92	−28.73%	7.80	11.92	−34.57%	8.55	16.37	−47.74%
28.00	5.75	7.74	−25.63%	9.30	8.79	5.82%	10.73	10.09	6.30%
32.40	6.04	8.03	−24.75%	10.66	9.47	12.58%	22.01	21.10	4.27%
36.90	7.74	9.68	−20.04%	13.32	11.99	11.07%	29.06	29.06	0.00%
40.90	13.94	15.83	−11.95%	16.69	13.38	24.67%	36.87	34.82	5.87%
Loads at position 3/(gw)									
16.10	4.41	4.99	−11.66%	8.79	14.51	−39.40%	8.49	9.90	−14.28%
19.50	3.69	3.01	22.57%	9.09	9.98	−8.97%	10.02	10.26	−2.38%
23.70	5.53	3.20	72.80%	9.77	9.96	−1.89%	13.56	12.15	11.64%
28.00	3.88	3.25	19.41%	10.09	9.52	5.93%	19.33	18.27	5.82%
32.40	6.31	6.01	4.88%	10.30	12.30	−16.28%	26.28	24.99	5.19%
36.90	8.49	9.03	−5.92%	10.46	10.15	3.08%	33.35	31.94	4.43%
40.90	15.77	14.03	12.45%	11.36	10.90	4.21%	38.89	38.66	0.60%

3. Results and Discussion

3.1. Effects of the Speeds During Straight-Line Towing

Figure 13 illustrates the variations in tension of the four lines as time progresses and at different speeds. The tug slowly accelerates from the speed of 0 m/s to the set speed, and the acceleration time is set to 100 s. When the tugboat moves at a slow speed, the line tension remains almost constant over time. However, at elevated towing speeds, the tension in the lines increases from a lower baseline to a higher plateau. An increase in the rate of towing leads to a consistent upward trend in the tension experienced by the four lines. With the increment in towing velocity, the instant when the line tension starts to increase significantly in the temporal dimension occurs sooner, and the velocity of tension alteration also rises correspondingly with the increase in towing speed. The line is mostly subjected to stress in this state. The stress causes the line, which is slightly bent, to straighten in the axial direction. As a result, the line exhibits very little bending deformation during straight-line towing. Although the line's bending distortion is essentially minor due to the line's inherent bending stiffness, the tension of the line together with this stiffness prevents the trawl from deforming throughout the towing operation. As a result, even if there is very little bending distortion, the bending force still has an impact on the wire. At the point where each line connects to the back of the tugboat, at about 1.5 m, the greatest bending moment occurs.

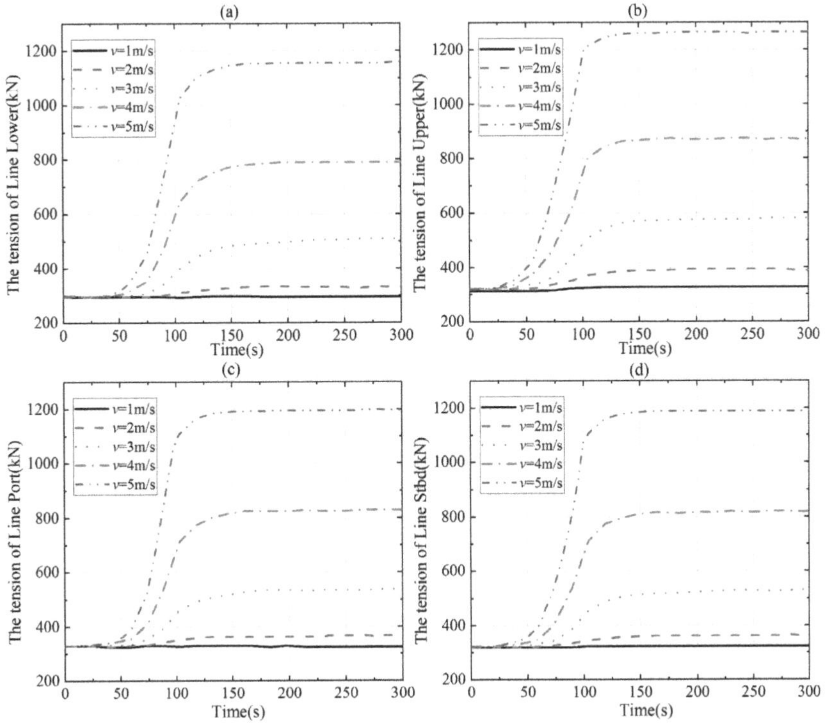

Figure 13. The time-domain representation of line tension in straight-line towing includes the following: (**a**) Line Lower; (**b**) Line Upper; (**c**) Line Port; and (**d**) Line Stbd.

Figures 14–18 analyze how the form of trawl nets changes at various speeds. When pulled in a straight line, the beam trawl does not experience substantial deformation. On the contrary, the inclination change in the beam trawl mainly affects the angle between the trawl and the vertical direction. Once the towing speed of the beam trawl increases and reaches stability, the angle between its angle and the vertical direction gradually expands. On the one hand, the tilt angle of the trawl relative to the vertical direction and the towing speed both increase, thereby expanding the trawl's effective fishing area. This expansion facilitates the entry of fish into the net as they are pursued. Alternatively, when the towing speed stabilizes and increases, the depth of the entire trawl diminishes steadily. As the trawl reaches a specific velocity, it aligns with the horizontal plane. In such scenarios, fish have a higher chance of evading capture. The orientation change in the trawl nets predominantly occurs within the XZ plane. The key characteristics of a trawl involve steady progression along the X-axis and minimal elevation change along the Z-axis, as well as in the angular motion relative to the vertical Z-axis. Consequently, when the towing vessel moves at a high speed, it is beneficial to herd fish and swiftly direct them into the net to prevent the escape of those already caught. To ensure that the fish already captured do not escape, it is essential to promptly decrease the line length as soon as the fish enter the net.

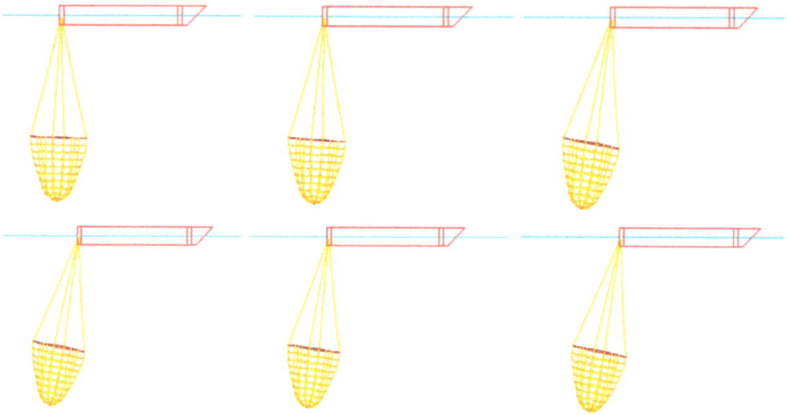

Figure 14. The trawl's shape evolution during straight-line towing at a velocity of 1 m/s.

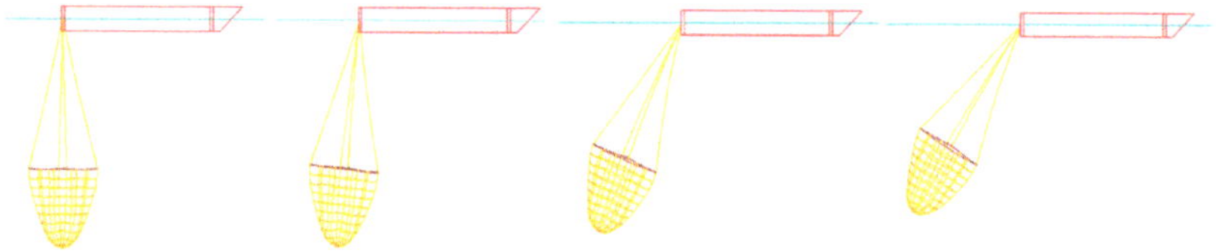

Figure 15. The trawl's shape evolution during straight-line towing at a velocity of 2 m/s.

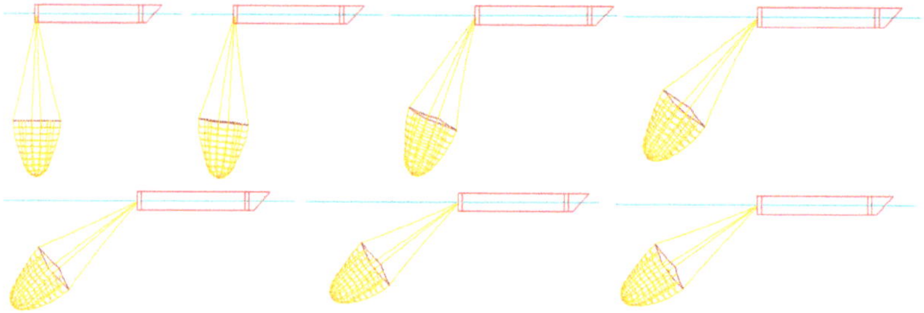

Figure 16. The trawl's shape evolution during straight-line towing at a velocity of 3 m/s.

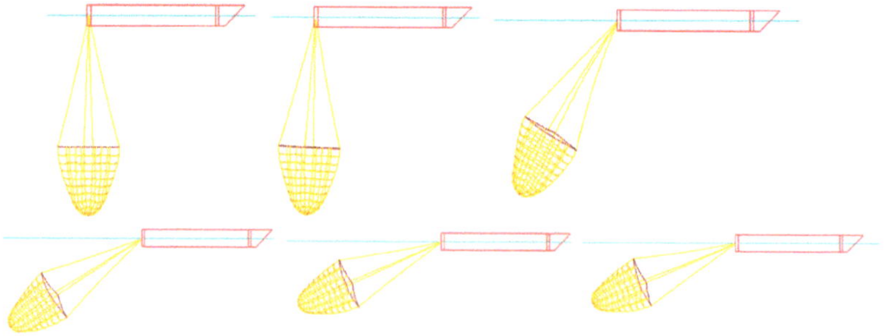

Figure 17. The trawl's shape evolution during straight-line towing at a velocity of 4 m/s.

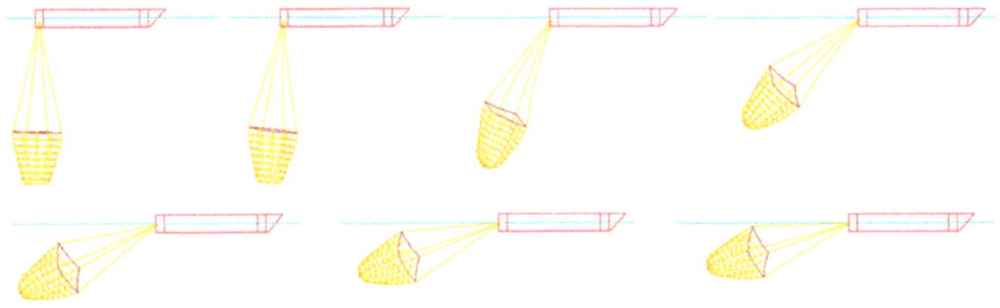

Figure 18. The trawl's shape evolution during straight-line towing at a velocity of 5 m/s.

3.2. Effects of the Speeds Under a Fixed Turning Radius

In most scenarios, the tugboat's turning radius is set at 57.32 m due to its small-radius turning maneuvers.

Figure 19 illustrates the temporal variation in line tension during rotating towing at a constant turning radius, with the angular velocity ranging from 1°/s to 5°/s. This range is specified by the centripetal acceleration formula $a = \omega^2 R$. During the turning maneuver, the tension element within the towed line supplies the centripetal force essential for the trawl to execute a turn in sync with the tugboat. At a given instant, should we define the total mass of the towed system as m, the tension force component can be represented by the formula $F = m\omega^2 R$. The portion of the line tension responsible for the trawl's rotation rises sharply in proportion to the square of the angular speed, not in direct proportion to the angular speed itself. With the increase in angular speed, the value of ω^2 significantly increases, thereby causing a corresponding rise in the tension within the line. When tension reaches excessive levels, it can ultimately inflict substantial harm on the boundary conditions that sustain the edges of the rigid frame. Under such circumstances, the trawl may completely disintegrate. This research aims to determine the rigid trawl's maximum allowable angular velocity while maintaining the turning speed at 1 m/s to verify this theory. Upon continuous debugging and calculation, it has been determined that at an angular velocity of 20°/s, the shape of the trawl remains essentially stable due to the square frame, which with its four rigid edges, remains closed and unchanged.

As the angular velocity rises, the square frame begins to experience substantial distortion, with the degree of deformation progressively increasing. When the angular velocity reaches a specific threshold, the deformation of the frame edge attains its maximum permissible extent, rendering the square frame entirely unstable. With an angular velocity of 57.3°/s, the trawl is observed to disintegrate completely, as illustrated in Figure 20.

Figures 21–25 illustrate the morphological alterations of the trawl system during rotational towing with a fixed turning radius but varying speeds. The angle between the beam trawl and the vertical axis changes dynamically as the tugboat turns. The trawl exhibits the peculiar behavior of spinning around its central axis while moving outward against the direction of the immediate rotation. This outward movement can occur on either the port or starboard side of the tug and may move toward or away from the stern. The lateral displacement of the trawl is a resultant motion arising from the interplay of multiple forces and movements during the turning maneuver. Exceeding the torsional limit causes an outward drift, leading to the intertwining of the four lines attached to the top of the pulling device and thereby increasing stress on the towed line. During the tug's return maneuver, the streamer will undergo torsional forces in addition to tension and bending stresses. As the trawl rotates, its spatial orientation changes in a more complex manner. When the tugboat moves in a straight line, the primary change in the trawl's spatial orientation is the variation in the angle between the trawl and the vertical axis. These changes are shown within the two-dimensional x-z plane of the global coordinate system. Meanwhile, the trawl's lateral displacement as the tugboat turns occurs within the three-dimensional coordinate framework. Additionally, the trawl's rotation around its central axis must be tracked along the vertical axis. Therefore, the change in trawl orientation during the tug's turning maneuver occurs in three-dimensional space. Conversely, when the tug is moving in a straight line, the trawl possesses two translational degrees of freedom and a single rotational degree of freedom. Nevertheless, when the tugboat performs turning maneuvers, the trawl exhibits three degrees of freedom in translation and two axes of rotation.

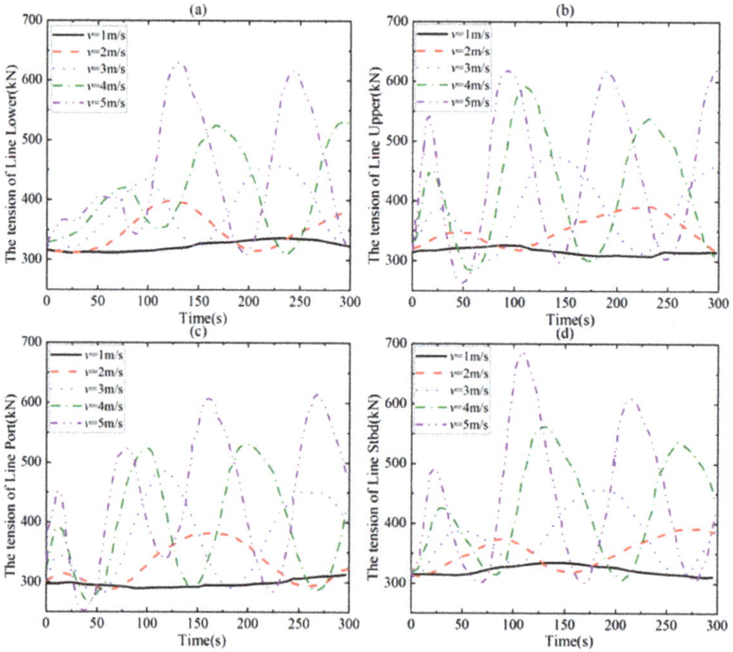

Figure 19. The temporal dynamics of line tension during rotational towing with a constant turning radius: (**a**) Line Lower; (**b**) Line Upper; (**c**) Line Port; and (**d**) Line Stbd.

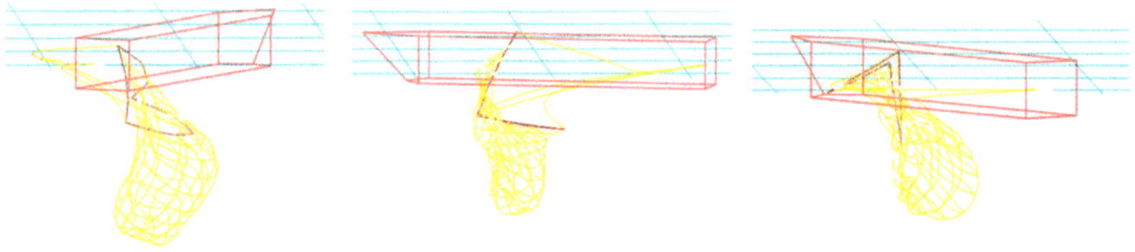

Figure 20. Large deformation of trawl when the angular velocity is 57.3°/s.

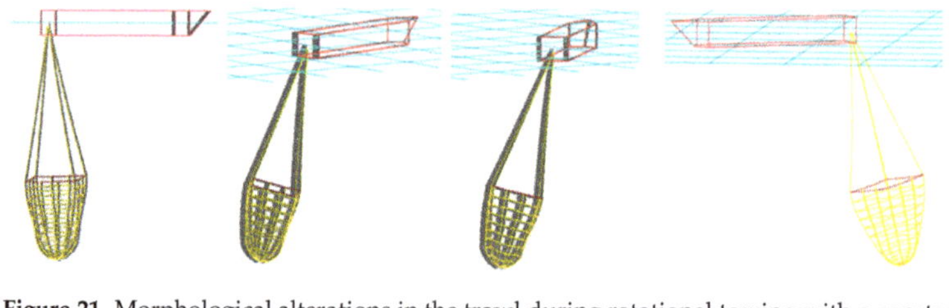

Figure 21. Morphological alterations in the trawl during rotational towing with a constant turning radius at a velocity of 1 m/s.

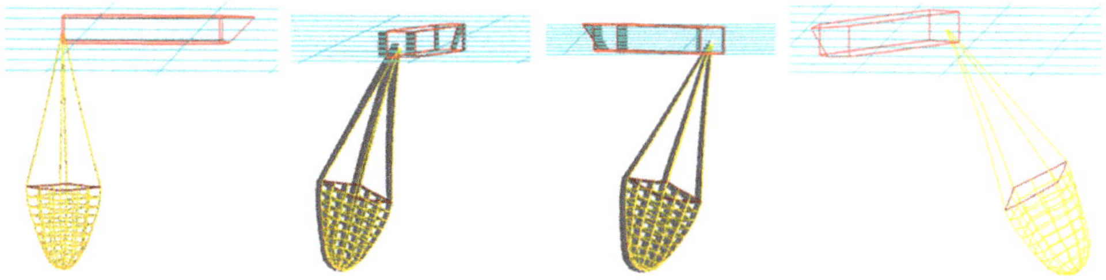

Figure 22. Morphological alterations in the trawl during rotational towing with a constant turning radius at a velocity of 2 m/s.

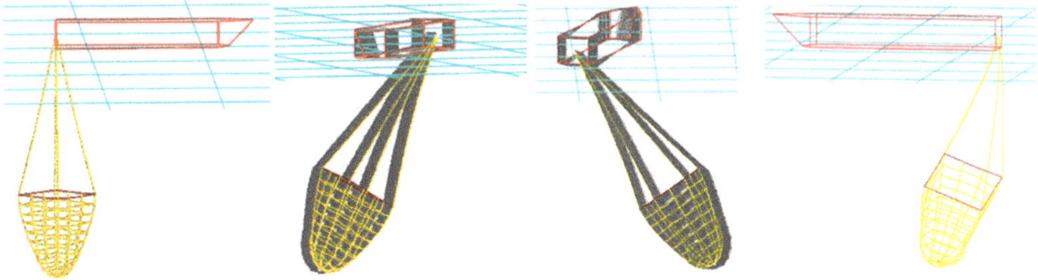

Figure 23. Morphological alterations in the trawl during rotational towing with a constant turning radius at a velocity of 3 m/s.

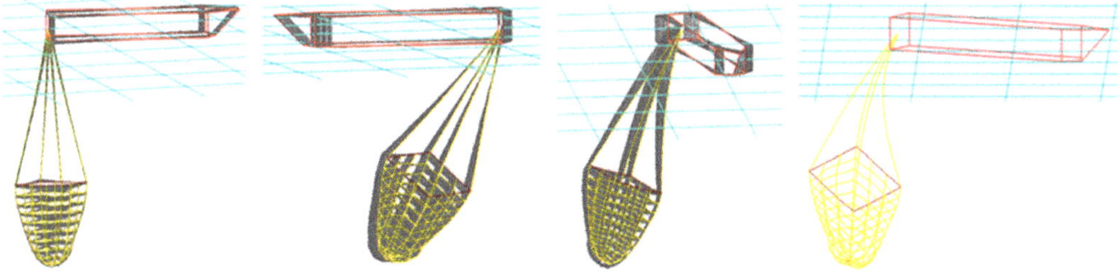

Figure 24. Morphological alterations in the trawl during rotational towing with a constant turning radius at a velocity of 4 m/s.

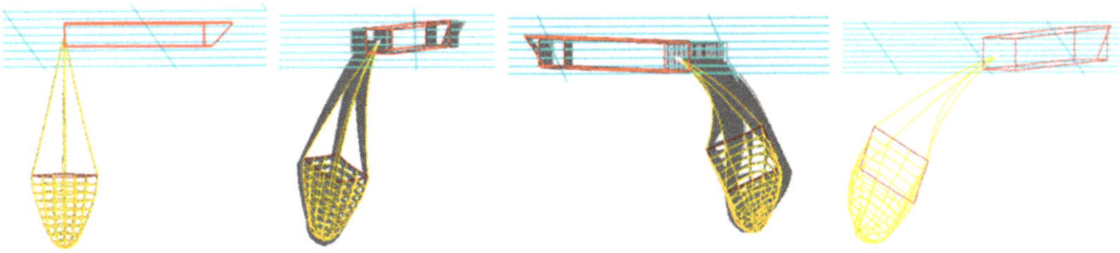

Figure 25. Morphological alterations in the trawl during rotational towing with a constant turning radius at a velocity of 5 m/s.

In addition, when the turning radius is fixed, the trawl's total lateral displacement expands incrementally as the towing speed rises. The load on the four lines varies proportionally with the increase in lateral displacement. Initially, there is a steep rise in the tension of the lines during the turning phase, succeeded by oscillations that occur in a cyclical pattern over time. In contrast, the tension levels off during linear towing operations. Assuming a fixed turning radius, a rise in angular velocity results in a more rapid alteration in the turning angle across the horizontal plane, consequently prompting an expansion in the lateral displacement of the trawl. To put it another way, as the tugboat turns at varying angular speeds, the tension in the rope and the floating distance of the trawl increase with the angular speed. Thanks to the cyclical pattern of turning, there are recurring variations in both the trawl's outward drift and the line's tension. With an escalation in the tugboat's angular velocity, the duration of these periodic changes shortens.

3.3. *Effects of the Speeds Under Fixed Angular Velocity*

The effects of different speeds are further studied while maintaining the tugboat's angular velocity at a constant 1°/s. The emphasis of this section is on examining the impact of the turning radius on the dynamic characteristics of the trawl. The time-domain image of the four tensioned wires in a revolving tow at a constant angular velocity is depicted in Figure 26. It is found that the towing speed is the most important element impacting the rise in line tension. Therefore, assuming that the trawl's floating distance diminishes, the line strain will continue to rise as the towing speed rises. Since the angular velocity in this case is very low, it is difficult to see how the tension will fluctuate over time under different towing speeds at this point. Instead, it tends to become stable as towing speed increases.

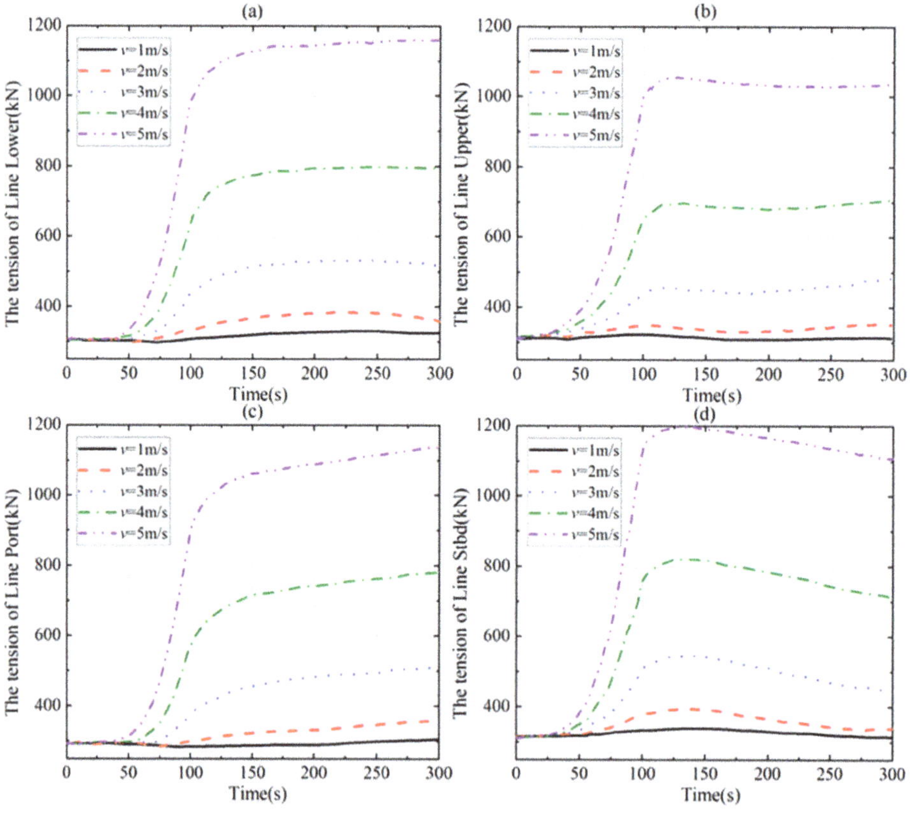

Figure 26. The graphical representation of line tension over time during rotational towing at a steady angular velocity: (**a**) Line Lower; (**b**) Line Upper; (**c**) Line Port; and (**d**) Line Stbd.

Under these conditions, fluctuations in tension over time are less significant than those observed with a constant radius of rotation. This difference stems from the formula for centripetal acceleration, which is $a = \omega^2 R$. Maintaining a constant turning radius reveals that the tension component responsible for supplying centripetal force is linearly associated with the radius. Furthermore, when the angular velocity during the turn is constant, this tension component is proportional to the square of the angular velocity. The overall disparity in line tension variation between the two scenarios arises from the distinct variation patterns of this tension component across the two conditions. The computed outcomes for the two turning conditions indicate that the four lines will distinctly experience curvature as they are towed in a rotational manner, which contrasts sharply with the relatively slight bending deformation of the wires when being towed in a straight path. Throughout the turning operation, the towing speed, the floating distance of the trawl, and the angular velocity all increase, leading to a rise in line tension. Among the various factors, the rise in angular velocity notably influences both the lateral displacement and the shape alteration in the trawl, as previously indicated. The elevation of the mesh opening fluctuates in accordance with the towing operations, whereas the breadth stays unaltered owing to the rigidity of the frame edges under typical line tension conditions.

Figures 27–31 depict the form variations in a trawl in rotating towing at various speeds while maintaining a constant angular velocity. Under a steady angular velocity, the increase in towing speed is predominantly influenced by adjustments to the turning radius. In this particular case, as the speed of towing escalates, it is anticipated that the turning radius will also increase accordingly. However, contrary to this expectation, an increase in the turning radius is observed to reduce the trawl's floating distance. This unexpected outcome can be rationalized via the following explanation: while the principal factor driving the speed increase in this scenario is indeed the lengthening of the turning arc during navigation, the tug's unit of horizontal movement leads to an alteration in the tug's directional angle in the horizontal plane due to the enlarged turning circle.

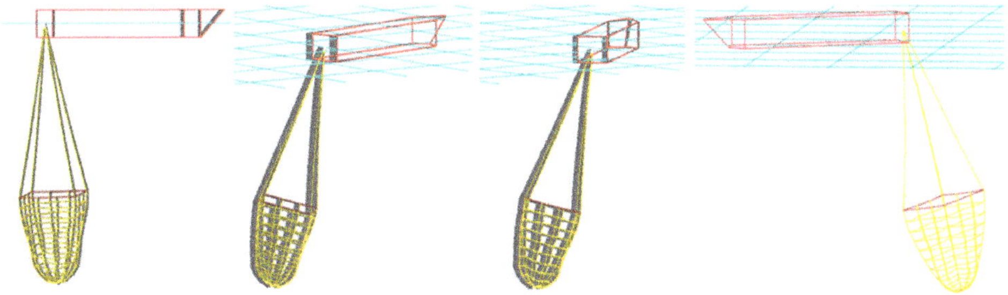

Figure 27. Morphological alterations in the trawl with a constant angular velocity at v = 1 m/s.

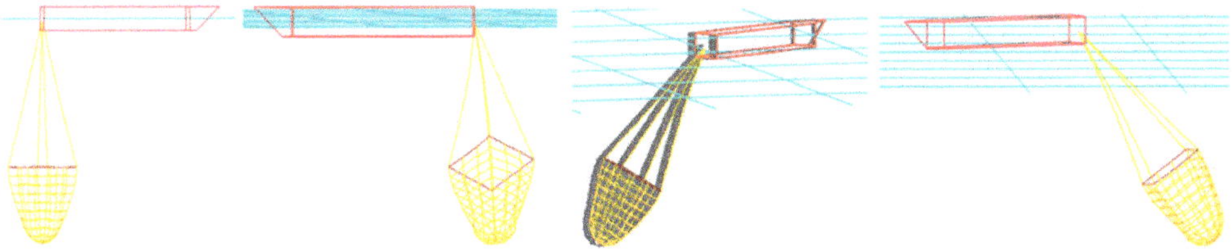

Figure 28. Morphological alterations in the trawl with a constant angular velocity at v = 2 m/s.

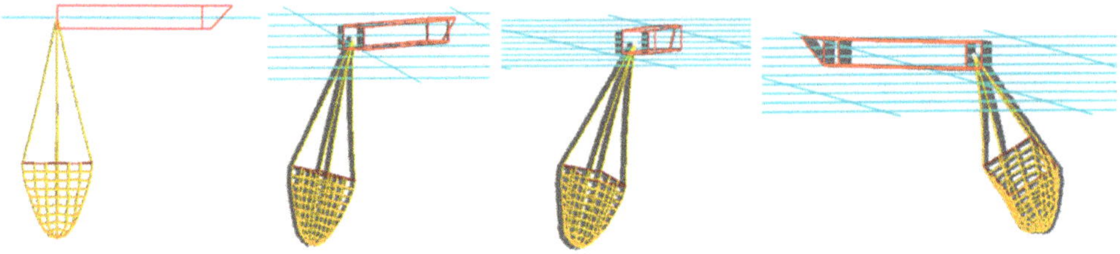

Figure 29. Morphological alterations in the trawl with a constant angular velocity at v = 3 m/s.

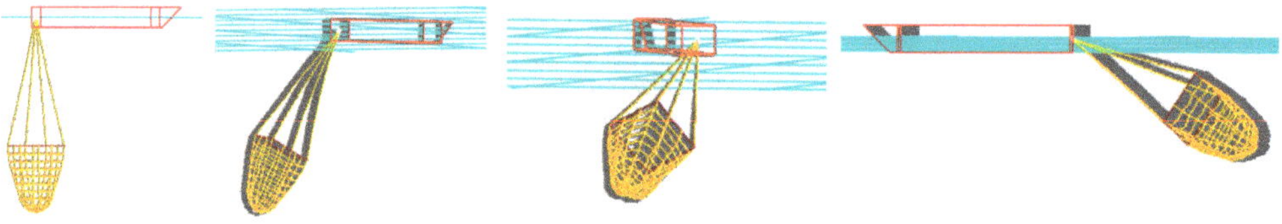

Figure 30. Morphological alterations in the trawl with a constant angular velocity at v = 4 m/s.

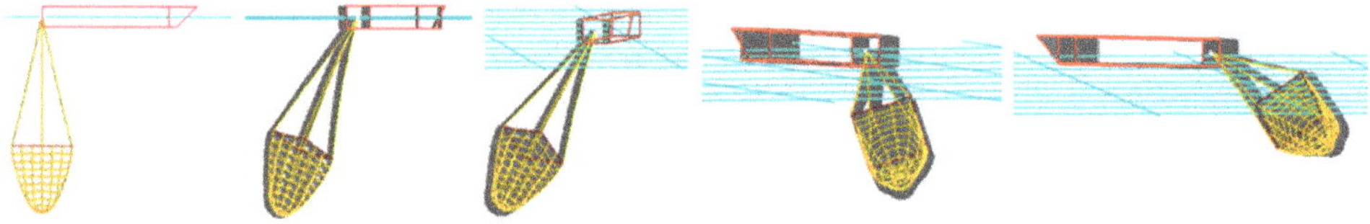

Figure 31. Morphological alterations in the trawl with a constant angular velocity at v = 5 m/s.

3.4. Dynamic Response of Line Fracture Failure in Startup Phase

This section, along with Section 3.5, simulates the breaking behavior of the lines in the startup phase and under a condition of constant straight-line towing speed. The line fracture scenarios are based on real-world abrupt events during trawling operations, such as seabed obstruction collisions, human operational errors, and equipment degradation. Towlines may snap instantaneously due to impacts with rocky seabed obstructions or shipwrecks, typically accompanied by localized shock loads (e.g., 50~80% peak tension surge). Human operational errors include overspeed towing, winch control failures, or emergency braking overloads. For example, sudden evasive maneuvers to avoid other vessels may cause unilateral cable overload and rupture. In addition, long-term corrosion, fatigue, or joint wear also forms weak points, leading to fractures under routine towing.

The dynamic response of line fracture failure is investigated across two distinct stages. The startup phase is defined as the period when the tugboat is on the verge of moving but has not yet begun to move. In the study of line fracture failure, the stage line breaks at the beginning of modeling. No matter how many lines there are, the moment of fracture is the same. When the tugboat travels at high speeds, the breakage of the towed line has the most serious impact on the safety of the trawl system. The tugboat's speed typically ranges from above 4 kn to a maximum of 6 kn, but it does not exceed 5 m/s. Therefore, in the simulation of straight-line towing, the towing speed is 5 m/s. The behavior of line fracture failure includes one line of breakage, two lines of breakage and three of lines breakage, respectively.

Figures 32–36 illustrate the alterations in the trawl's shape in the case of the fracture failure of one line. The alterations in the trawl's shape in the case of the fracture failure of two symmetrical and asymmetrical lines are shown in Figures 36, 37, 38, 39, 40 and 41, respectively. Figure 42 illustrates the morphological transformations of the trawl during the event of the fracture failure of three lines. It can be found that under the premise of one-line failure in the startup stage, no matter which lines fail, the tensile effect of the line is lost, and the towing box begins to deform rapidly. This kind of trawl has a certain tensile stiffness and bending stiffness itself and can maintain the trawl shape under the combined stiffness of the line and the frame. The trawl begins to internally collapse into a flat form after a line breaks and loses tension, at which point it collapses. Therefore, with this stiff trawl system, even a single line failure can have a catastrophic impact on how well the entire trawl system works. While the fracture of the trawling net after the tugboat has run at high speed offers a higher hazard to the safety of the employees at the stern, failure at this point poses very little risk to the staff at the stern.

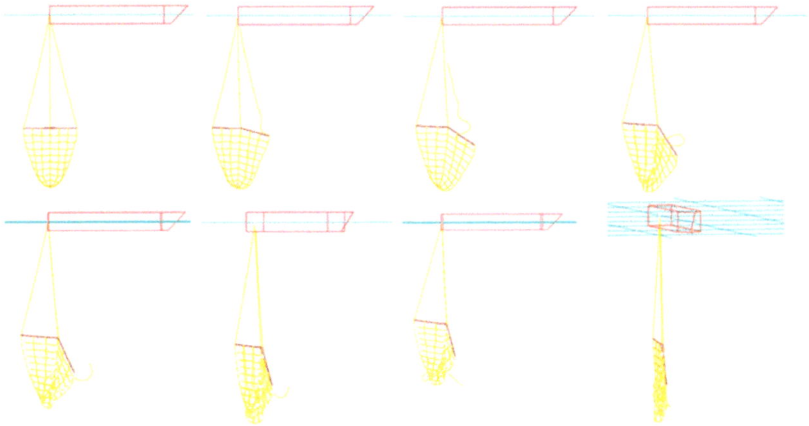

Figure 32. Shape changes in the trawl in case of fracture failure of Line Lower.

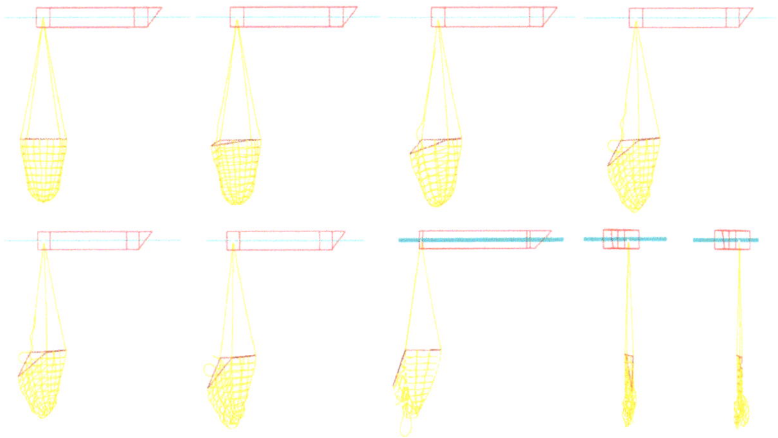

Figure 33. Shape changes in the trawl in case of fracture failure of Line Upper.

By studying the alterations in the trawl's shape following the failure of two symmetrical lines during the startup stage, it is discovered that following the failure of two symmetrical lines, the trawl begins to quickly collapse from both sides to the interior, eventually collapsing into a flat shape. In contrast to the distortion of the trawl when one line breaks, the flat trawl after final deformation in this case is symmetrical. Because the boundary restrictions that keep the trawl stable when one line breaks are no longer symmetrical, this event occurs. As a result, the ultimate deformation in this instance is likewise asymmetrical. Even if the boundary constraints that preserve the trawl form have been altered because of the failure and fracture of two symmetrical lines, the new boundary restrictions are still symmetrical. Consequently, in this instance, the trawl is symmetrical and flat after final deformation.

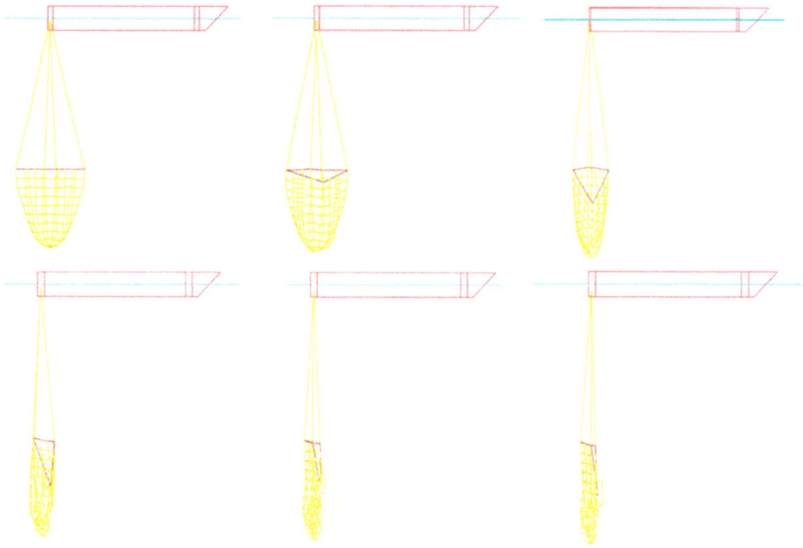

Figure 34. Shape changes in the trawl in case of fracture failure of Line Port.

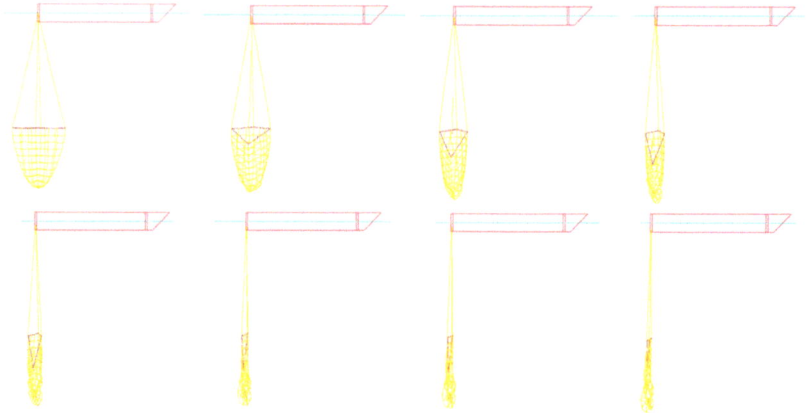

Figure 35. Shape changes in the trawl in case of fracture failure of Line Stbd.

By analyzing the shape changes in the trawl following the failure of two asymmetrical lines during the startup stage, it is discovered that, while the trawl's net clothing has been distorted, the square frame constructed from four rigid edges has not been altered. In contrast, the square frame made of four hard edges is distorted upon the failure of one line or two symmetrical lines. The two damaged lines act on both ends of the same stiff frame edge; therefore, no deformation load is applied to the other three frame edges, and the square frame does not deform in the end. The entire square frame rotates slightly around the rigid frame connected by the remaining two wires due to gravity, and the mesh is compressed during the sweeping and rotation processes. The square frame will not alter after it has reached a specific inclination and posture, since the net garments, which have some ability to resist deformation as well, will support it once it has been compressed to a given level. To put it another way, the form of the square frame made up of four hard frame edges will likewise alter once two lines distributed symmetrically and one line break, and the entire trawl will be forced into a flat shape. The entire trawl will be compressed when the two asymmetrical lines are severed; however, at this point, the trawl is not compressed into a flat form, and the square frame, which is made up of four inflexible frame edges, will not alter in shape. There are several permutations of fracture failure mechanisms for two asymmetrical lines, but when they totally sink into the water, the overall line and the overall square frame differ in the direction of the x-z plane corner and the y-z plane corner.

Figure 36. Shape changes in the trawl in case of fracture failure of Line Lower and Line Upper.

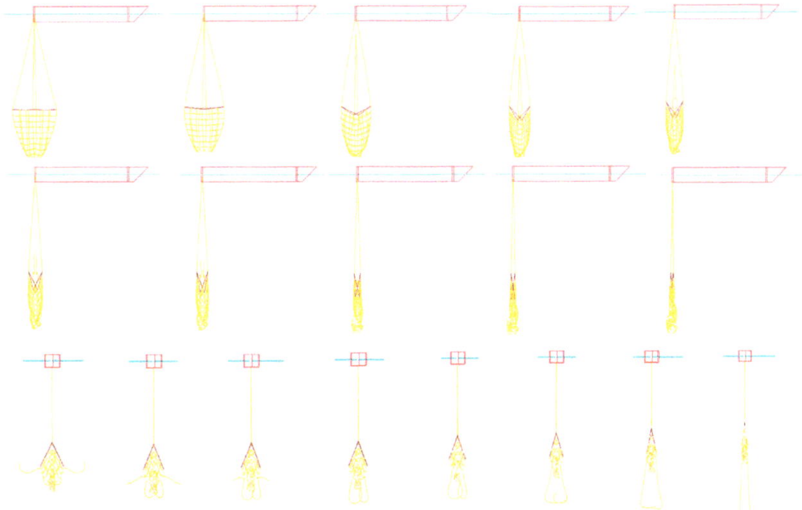

Figure 37. Shape changes in the trawl in case of fracture failure of Line Port and Line Stbd.

Figure 38. Shape changes in the trawl in case of fracture failure of Line Port and Line Lower.

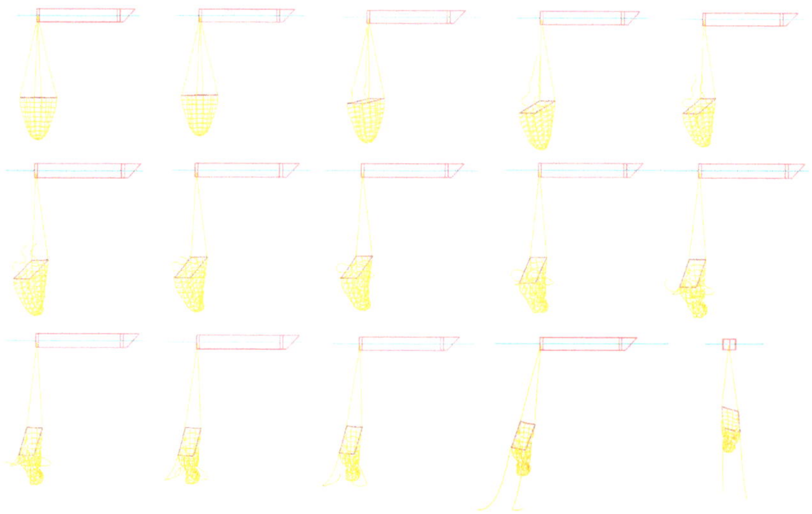

Figure 39. Shape changes in the trawl in case of fracture failure of Line Port and Line Upper.

The trawl swiftly starts to collapse inward when the three wires break, and the square frame also deforms quickly. This can be seen by watching how the trawl changes shape. The trawl is ultimately not compressed into a flat shape because in this scenario, the boundary constraints to maintain the trawl's stability become asymmetric, and when the only line is intact, the final towing point of the initial square frame is subject to the bending effect of the unbroken streamer on the frame edge, causing the square frame that was initially in the plane to become a spatial quadrilateral. Therefore, the trawl will not entirely flatten out under the support of the spatial quadrilateral.

Figure 40. Shape changes in the trawl in case of fracture failure of Line Stbd and Line Lower.

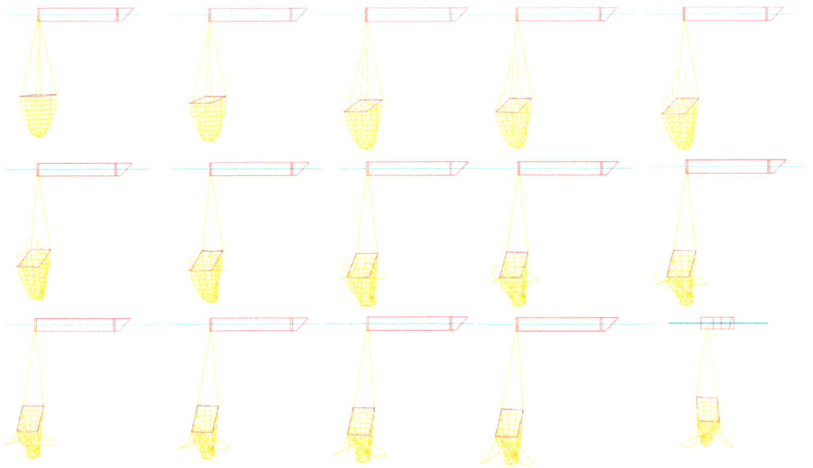

Figure 41. Shape changes in the trawl in case of fracture failure of Line Stbd and Line Upper.

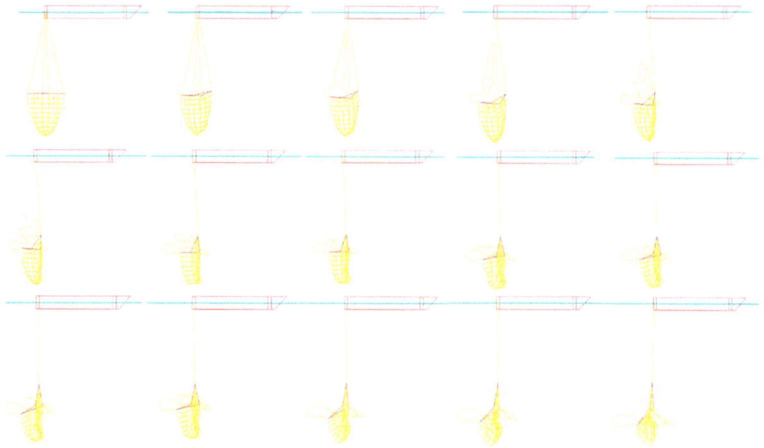

Figure 42. Shape changes in the trawl in case of fracture failure of Line Upper, Line Port, and Line Stbd.

3.5. Dynamic Response of Line Fracture Failure During Straight-Line Towing

Figures 43 and 44 depict the trawl shape variations caused by the fracture failure of Line Lower and Line Upper during straight-line towing, respectively. The trawl's form alterations during straight-line towing are vastly different from those seen during the startup phase. During straight-line towing, the fracture of one line results in a lack of restraints to maintain the rigid frame's edge center. The period between the start of the trawl's deformation and its total collapse into a flat shape significantly decreases because of the high-speed direct navigation movement of the tug, which also changes the degree of restriction and restraint ability of the remaining three lines to the deformation of the rigid frame edge. Due to the tug's towing action, when the entire trawl becomes level after stabilization, it will also have a slight inclination in the vertical direction.

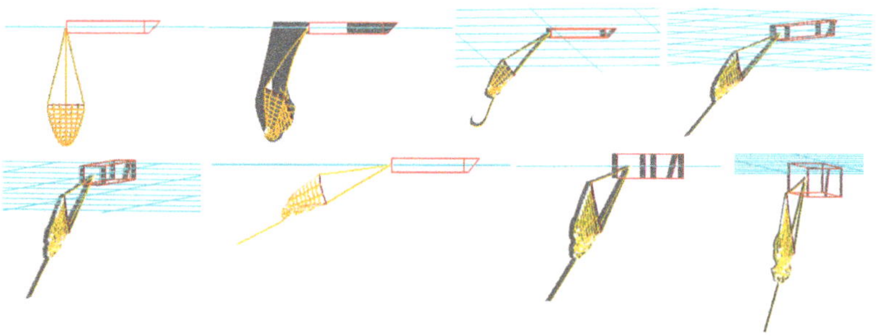

Figure 43. Shape changes in the trawl in case of fracture failure of Line Lower during straight-line towing.

Although the rigid trawl frame is defined with infinite stiffness in simulations, under extreme cable fracture scenarios, the asymmetric load distribution from the remaining cables may induce local elastic deformations due to external bending moments. For instance, single-cable failure (e.g., Line Lower) creates a moment around the opposite frame edge, with the deformation magnitude Δ estimated as follows:

$$\Delta = \frac{F \cdot L^3}{3E_{eq}I} \tag{11}$$

where F is the resultant force from the remaining cables, L is the frame edge length, E_{eq} is the equivalent elastic modulus (accounting for weld joints and material nonlinearity), and I is the moment of inertia.

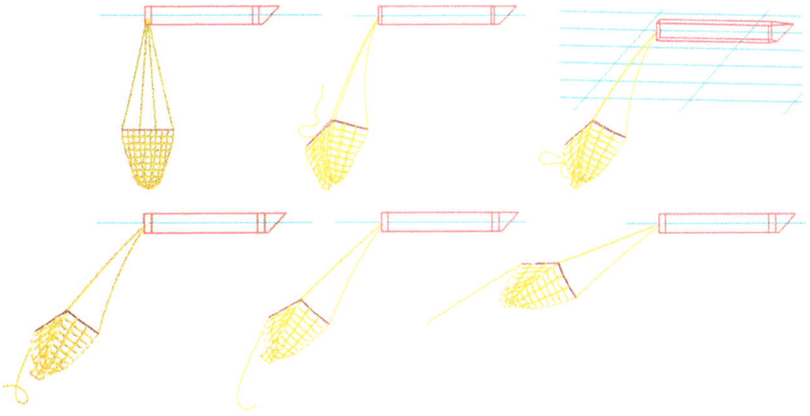

Figure 44. Shape changes in the trawl in case of fracture failure of Line Upper during straight-line towing.

Some simulation results reveal that for single-cable fracture during straight-line towing, maximum edge displacement reaches 0.8–1.2 m (5~7% deformation ratio), reducing the effective fishing area by 15–20%. For asymmetric dual-cable fracture, frame distortion forms a spatial quadrilateral with diagonal stretch differences of 2.5–3.5 m, increasing the net overlap and entanglement risks.

It can be predicted that if one line breaks, the tension on the other lines will be redistributed. If the total drag force borne by the lines remains unchanged, the tension borne by the remaining lines will increase after stabilization due to the loss of the traction of one line. The entire area of the trawl will decrease when a line breaks, which will also result in a reduction in the amount of drag that is applied to the trawl. The tension of the remaining lines might alter because of the decreased number of lines, with some lines seeing an increase in tension and others experiencing a drop. The results of the tension calculation in Figure 45 demonstrate the accuracy of this reasoning. The tension of the Line Upper remains essentially intact when the Line Lower breaks, whereas the tension of the Line Port and Line Stbd decreases.

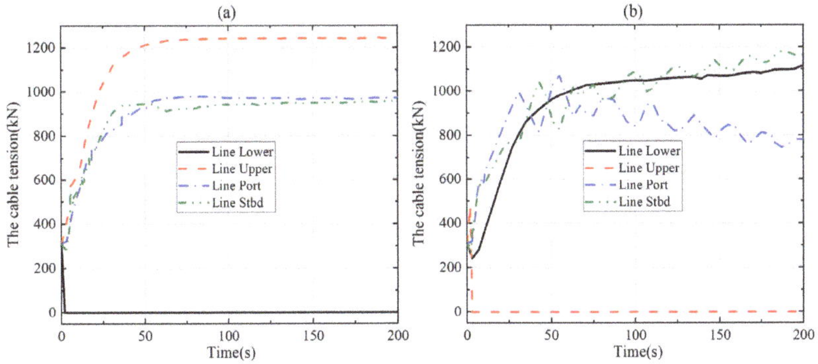

Figure 45. The tension calculation results after one line breaks: (**a**) Line Lower; (**b**) Line Upper.

4. Conclusions

Using computational techniques, this research investigates how towing speeds impact the dynamic performance of beam trawl systems during both straight-line and rotational movements. Additionally, it assesses the system's dynamic response concerning line failure at the onset of towing. The findings of this research are outlined as follows:

(1) As the linear towing speed increases, so does the line tension. This increased tension causes the line to stretch along its length, leading to minimal bending deformation during straight-line towing. Thanks to its strong truss structure, changes in towing speed have little impact on the trawl's shape. At a towing speed of 5 m/s, the maximum axial tension in the four cables reaches 1280 kN (Line Upper). Under a fixed turning radius (57.32 m), angular velocities exceeding 20°/s induce centrifugal overload, driving line tensions to exceed 1200 kN;

(2) During the initial phase of rotational towing, the line tension experiences significant changes, followed by frequent fluctuations over time. As the towing speed increases, the trawl's overall horizontal floating distance progressively expands. Concurrently, the rate of change in the horizontal turning angle escalates, which in turn amplifies the trawl's outward floating distance;

(3) The integrity and operational safety of the trawl system are significantly compromised when the line breaks. No matter which line fails at commencement, the towing box starts to distort quickly as the tensile impact of the line is gone. If two symmetrical lines fail, the trawl quickly collapses from the outside into the interior before flattening out. The square frame of the trawl, which is made up of four stiff frame edges, remains unchanged if two asymmetrical lines break;

(4) During straight-line towing, the fracture of one line leads to a lack of constraints to maintain the edge center of the rigid frame. If one line breaks, the tension on the other lines will be redistributed. These results provide quantitative guidelines for operational parameter optimization: recommended towing speeds $\leq$ 4 m/s, angular velocities $\leq$ 20°/s, and redundant cable designs to mitigate abrupt failures.

Future research could expand on this study through multidimensional approaches. First, dynamic coupled models integrating waves, stratified currents, and seabed topography could systematically analyze trawl stability under complex oceanic conditions (e.g., typhoons or cold surges), particularly focusing on cumulative fatigue damage to cables from extreme weather. Second, machine learning algorithms combined with real-time sensor data could enable adaptive control systems for fracture prediction and the dynamic optimization of towing speed/angular velocity thresholds, enhancing operational safety.

Author Contributions: Conceptualization, D.Z. and B.Z.; methodology, D.Z. and J.Y.; software, B.Z.; validation, D.Z., B.Z. and Y.Z.; formal analysis, D.Z.; investigation, D.Z.; resources, J.Y. and K.Z.; data curation, B.Z.; writing—original draft preparation, D.Z. and B.Z.; writing—review and editing, D.Z., Y.Z. and J.Y.; visualization, Y.Z.; supervision, K.Z. and J.Y.; project administration, D.Z. and J.Y.; funding acquisition, J.Y. All authors have read and agreed to the published version of the manuscript.

Funding: This study was financially supported by the Natural Science Foundation of Guangdong Province (2022A1515011562), Guangdong Provincial Special Fund for promoting high-quality economic development (GDNRC [2021]56, Yuerong Office Letter [2020]161), the Program for Scientific Research Start-up Funds of Guangdong Ocean University (060302072101), the Comparative Study and Optimization of Horizontal Lifting of Subsea Pipeline (2021E05011), and the China Scholarship Council (CSC202306320084).

Institutional Review Board Statement: Not applicable.

Informed Consent Statement: Not applicable.

Data Availability Statement: All data generated and analyzed during this study are included in this published article. The datasets used and analyzed during the current study are available from the corresponding author upon reasonable request.

Conflicts of Interest: The authors declare no conflicts of interest.

References

1. Clarke, T.M.; Wabnitz, C.C.; Striegel, S.; Frölicher, T.L.; Reygondeau, G.; Cheung, W.W. Aerobic growth index (AGI): An index to understand the impacts of ocean warming and deoxygenation on global marine fisheries resources. *Prog. Oceanogr.* **2021**, *195*, 102588. [CrossRef]
2. Bax, N.; Novaglio, C.; Maxwell, K.H.; Meyers, K.; McCann, J.; Jennings, S.; Frusher, S.; Fulton, E.A.; Nursey-Bray, M.; Fischer, M.; et al. Ocean resource use: Building the coastal blue economy. *Rev. Fish Biol. Fish.* **2021**, *32*, 189–207. [CrossRef]
3. Zhang, Y.-Z.; Xue, C.; Wang, N.; Chen, G. A comparative study on the measurement of sustainable development of marine fisheries in China. *Ocean Coast. Manag.* **2023**, *247*, 106911. [CrossRef]
4. Konar, M.; Ding, H. A sustainable ocean economy for 2050: Approximating its benefits and costs. In *The Blue Compendium: From Knowledge to Action for a Sustainable Ocean Economy*; Springer International Publishing: Cham, Switzerland, 2023; pp. 681–714.
5. Zhuang, X.; You, X.; Kumazawa, T.; Ito, S.; Shiode, D.; Hu, F. Dynamic performance of strategy for transitioning from conventional bottom otter trawling to semi-pelagic trawling through towing tank experiments and numerical simulations. *Ocean Eng.* **2024**, *300*, 117436. [CrossRef]
6. Chu, W.; Cui, S.; Zhai, M.; Cao, Y.; Zhang, X. The influence of flow field bottom effect and sand cloud effect on the hydrodynamic performance and stability of bottom trawling otter board under tilted attitudes. *Appl. Ocean Res.* **2024**, *152*, 104176. [CrossRef]
7. Zhang, D.; Bai, Y.; Soares, C.G. Dynamic analysis of an array of semi-rigid "sea station" fish cages subjected to waves. *Aquac. Eng.* **2021**, *94*, 102172. [CrossRef]
8. Tang, M.F.; Dong, G.H.; Xu, T.J.; Zhao, Y.P.; Bi, C.W. Numerical simulation of the drag force on the trawl net. *Turk. J. Fish. Aquat. Sci.* **2017**, *17*, 1219–1230.
9. Thierry, N.N.B.; Tang, H.; Achille, N.P.; Xu, L.; Hu, F. Unsteady turbulent flow developing inside and around different parts of fluttering trawl net in flume tank. *J. Fluids Struct.* **2021**, *108*, 103451. [CrossRef]
10. Nyatchouba Nsangue, B.T.; Tang, H.; Liu, W.; Xu, L.; Hu, F. Turbulent flow interacting with flexible trawl net structure including simulation catch in flume tank. *Sci. Rep.* **2023**, *13*, 6249. [CrossRef]
11. Thierry, N.N.; Tang, H.; Liuxiong, X.; You, X.; Hu, F.; Achile, N.P.; Kindong, R. Hydrodynamic performance of bottom trawls with different materials, mesh sizes, and twine thicknesses. *Fish. Res.* **2020**, *221*, 105403. [CrossRef]
12. Dealteris, J.T.; Recksiek, C.W.; Fahfouhi, A.; Liuxiong, X. Comparison of the Performance of Two Bottom-Sampling Trawls. *Trans. Am. Fish. Soc.* **1989**, *118*, 119–130. [CrossRef]
13. Buxton, N.; DeAlteris, J. Quantification of water flow and net drag in midwater trawls. *Fish. Res.* **1993**, *16*, 165–178. [CrossRef]
14. Fuxiang, H.; Matuda, K.; Tokai, T.; Kanehiro, H. Dynamic Analysi of Midwater Trawl System by a Two-Dimensional Lumped Mass Method. *Fish. Sci.* **1995**, *61*, 229–233. [CrossRef]
15. Sangster, G.I.; Breen, M. Gear performance and catch comparison trials between a single trawl and a twin rigged gear. *Fisheries Research* **1998**, *36*, 15–26. [CrossRef]
16. Reite, K.-J.; Sorensen, A. Mathematical Modeling of the Hydrodynamic Forces on a Trawl Door. *IEEE J. Ocean. Eng.* **2006**, *31*, 432–453. [CrossRef]
17. Sala, A.; Farran, J.D.P.; Antonijuan, J.; Lucchetti, A. Performance and impact on the seabed of an existing- and an experimental-otterboard: Comparison between model testing and full-scale sea trials. *Fish. Res.* **2009**, *100*, 156–166. [CrossRef]
18. Sun, X.; Yin, Y.; Jin, Y.; Zhang, X.; Zhang, X. The modeling of single-boat, mid-water trawl systems for fishing simulation. *Fish. Res.* **2011**, *109*, 7–15. [CrossRef]
19. Zhao, Y.-P.; Gui, F.-K.; Xu, T.-J.; Chen, X.-F.; Cui, Y. Numerical analysis of dynamic behavior of a box-shaped net cage in pure waves and current. *Appl. Ocean Res.* **2013**, *39*, 158–167. [CrossRef]
20. Zhang, D.; Zhang, Y.; Zhao, B.; Ma, Y.; Si, K. Exploring subsea dynamics: A comprehensive review of underwater pipelines and cables. *Phys. Fluids* **2024**, *36*, 101304. [CrossRef]
21. Zhang, D.; Yan, J.; Zhu, K. Hydrodynamic characteristics of deep sea station cage under different failure modes of mooring cables. *J. G. Ocean U.* **2022**, *42*, 107–116. (In Chinese)
22. Li, F.; Song, L.; Hua, C.; Zhu, Q. Numerical and experimental investigation on hydrodynamic performance of the stick-held dip net in Pacific saury fishery. *Front. Mar. Sci.* **2022**, *9*, 985086. [CrossRef]
23. Liu, Z.; Hu, F.; Wan, R.; Wang, Y.; Guo, S.; Yan, H.; Zhou, C. Hydrodynamic characteristics of frustum-type depressor for sampling midwater trawl. *Ocean Eng.* **2024**, *295*, 117009. [CrossRef]

24. Guan, Q.; Zhu, W.; Feng, C.; Chen, F.; Tang, W.; Liu, L.; Chen, J. Hydrodynamic interactions between a sorting grid and a demersal trawl. *Ocean Eng.* **2024**, *304*, 117837. [CrossRef]
25. Xiong, Z.; Wu, X.; Guo, Y.; Fu, Z. Study on ice resistance of Antarctic krill ship with trawl under floating ice sea conditions. *Front. Mar. Sci.* **2024**, *11*, 1357331. [CrossRef]
26. O'Neill, F.; Breddermann, K. The contribution of mesh opening angle to the drag of netting panels. *Ocean Eng.* **2024**, *305*, 117959. [CrossRef]
27. Tang, H.; Thierry, N.N.B.; Pandong, A.N.; Sun, Q.; Xu, L.; Hu, F.; Zou, B. Hydrodynamic and turbulence flow characteristics of fishing nettings made of three twine materials at small attack angles and low Reynolds numbers. *Ocean Eng.* **2022**, *249*, 110964. [CrossRef]
28. Nsangue, B.T.N.; Tang, H. Flow field and nonlinear dynamics of the gillnet structures under steady flow using a one-way coupling model. *Ocean Eng.* **2024**, *309*, 118380. [CrossRef]
29. Li, Y.; Liu, Z.; Wang, Z.; Zhang, X.; Wang, L.; Zhang, Y.; Ma, S.; Qi, G.; Wang, Y. Study on the Hydrodynamic Performance of the Beam Used in the Antarctic Krill Beam Trawl. *Fishes* **2023**, *9*, 17. [CrossRef]
30. Rijnsdorp, A.D.; Depestele, J.; Molenaar, P.; Eigaard, O.R.; Ivanović, A.; O'neill, F.G. Sediment mobilization by bottom trawls: A model approach applied to the Dutch North Sea beam trawl fishery. *ICES J. Mar. Sci.* **2021**, *78*, 1574–1586. [CrossRef]
31. He, R.; Zheng, H.; Ma, S.; Wu, Y.; Yang, S.; Dai, Y.; Wang, Y. Hydrodynamic simulation and experiment of depth adjustable beam trawl for Antarctic krill. *J. Mar. Sci. Technol.* **2023**, *28*, 399–409. [CrossRef]
32. Sen, D. A Study on Sensitivity of Maneuverability Performance on the Hydrodynamic Coefficients for Submerged Bodies. *J. Ship Res.* **2000**, *44*, 186–196. [CrossRef]
33. Kristiansen, D.; Faltinsen, O.M. Non-linear wave-induced motions of cylindrical-shaped floaters of fish farms. *Proc. Inst. Mech. Eng. Part M J. Eng. Marit. Environ.* **2009**, *223*, 361–375. [CrossRef]
34. Orcina. OrcaFlex Manual 2015. Available online: http://www.orcina.com/SoftwareProducts/OrcaFlex/Validation/index/.pdf (accessed on 25 December 2024).
35. Suzuki, K.; Takagi, T.; Shimizu, T.; Hiraishi, T.; Yamamoto, K.; Nashimoto, K. Validity and visualization of a numerical model used to determine dynamic configurations of fishing nets. *Fish. Sci.* **2003**, *69*, 695–705. [CrossRef]

Journal of
*Marine Science
and Engineering*

MDPI

Review

A Review of Experimental and Numerical Research on the Slamming Problem of High-Performance Vessels

Yifang Sun [1,2], Dapeng Zhang [1,*], Zongduo Wu [1] and Yiquan Yu [3]

1 Ship and Maritime College, Guangdong Ocean University, Zhanjiang 524088, China
2 Guangdong Provincial Engineering Research Center for Ship Intelligence and Safety, Zhanjiang 524000, China
3 Department of Naval Architecture, Naval University of Engineering, Wuhan 430033, China
* Correspondence: zhangdapeng@gdou.edu.cn

Abstract: Slamming load is characterized by a high peak and short duration. Severe slamming phenomena are extremely detrimental to the navigation safety of high-speed vessels, thereby constraining the development and application of high-performance ships. Studies on slamming mechanisms, load distribution, prediction, and mitigation methods are particularly essential. This paper provides a comprehensive review of the theoretical, numerical, and experimental research progress on water-entry slamming for high-performance ships. First, the theoretical foundations and numerical simulation methods of slamming are elaborated. Then, existing research findings are summarized from two perspectives: segmented water entry and full-scale wave loads. Finally, unresolved issues and future research directions are identified. The aim is to offer valuable insights for further advancements in high-performance ship slamming studies.

Keywords: slamming; wave load; high-performance vessel; water entry

1. Introduction

When navigating in severe sea conditions, ships will generate significant vertical motions, leading to slamming phenomena. During water entry, the hull interacts with the surrounding fluid, generating impact forces, moments, and cavity formation, accompanied by complex hydrodynamic effects such as cavitation, bubble pulsation, spray jets, and intricate fluid–structure interactions [1]. The slamming-induced pressure is characterized by an extremely short duration and high peak magnitude, resulting in highly nonlinear wave loads that cause substantial structural damage. Moreover, the transient slamming force excites elastic vibrations of the hull girder, known as whipping vibration. When the high-frequency whipping-induced bending moments superimpose with the low-frequency wave-induced bending moments, they pose a severe threat to the global longitudinal strength of the hull. According to full-scale measurements and model tests conducted worldwide, the whipping vibration bending moment can reach 30~40% of the total wave-induced bending moment [2]. The slamming pressure is influenced by factors including the hull stiffness, shape, and deadrise angle. In high-curvature regions of the hull, the peak slamming pressure increases significantly while its duration shortens. The slamming pressure intensifies as the deadrise angle decreases. During asymmetric water entry, the side with a smaller angle to the water surface experiences higher pressure [3].

High performance vessels (HPVs) refer to the vessels that feature excellent speed and seakeeping capabilities, along with good economic efficiency and carrying capacity, and have strong adaptability to the environment. In recent years, they have been widely used in both military and civilian fields. However, when encountering severe sea conditions, if slamming occurs more than three times in every 100 encountered waves, the vessel will actively reduce its speed to ensure navigation safety [4]. This demonstrates that slamming phenomena at high speeds have become a bottleneck issue restricting the development of high-speed vessels [5]. The slamming phenomena include bottom slamming, flare slamming, wet-deck slamming, whipping, and so on. In recent years, almost all classification societies have issued the latest structural codes applicable to high-speed vessels. These codes stipulate the design pressure, stress, and component dimensions to adapt to the contemporary development of high-speed ships. Accurate prediction of the slamming load and structural responses has become one of the key prerequisites for the safety of ship design.

Research on ship slamming aims to reveal transient load characteristics, structural response mechanisms, and their coupled effects during high-speed collisions between waves and the hull. There are two computational approaches for analyzing the structural response of a hull under nonlinear slamming loads. One method involves two steps: initially assuming the ship as a rigid body to calculate the total hydrodynamic forces, followed by treating the hull as an elastic body to solve for its vibration response. The other method treats the ship as an elastic body from the outset, employing hydroelasticity theory to solve for the vessel's global response in waves. This approach simultaneously accounts for slamming loads and feedback effects of structural elasticity on hydrodynamic forces. The recently developed fluid–structure interaction (FSI) methodology [6], which couples computational fluid dynamics with structural mechanics methods, enables the calculation of spatiotemporal distributions of slamming loads, structural deformation, and damage mechanisms. This approach provides critical theoretical support for ship impact-resistant design.

Model testing is also a crucial approach in ship slamming research, primarily including water-entry tests, towing tank tests, and full-scale trials. Water-entry tests involve releasing a ship model freely or at a controlled velocity to impact the water surface using a guide rail or pendulum system. The wave makers are employed to simulate various sea states, combined with a towing system to adjust the model's speed in towing tank tests. Slamming pressure is measured using piezoelectric sensors or strain gauges, while hull motion is captured via accelerometers. Due to the instantaneous and violent nature of the slamming phenomenon, multiple repeated experiments are required to ensure the reproducibility of the results. This requires the sensors to have a certain ability to resist impact and stable performance. With the application of particle image velocimetry (PIV) [7], researchers could combine high-speed photography to record the internal flow field evolution during water entry, capturing vortex structures. This approach further facilitates the study of slamming pressure distribution and its time-dependent mechanisms. For full-scale trials, pressure sensors or fiber-optic strain sensors are installed at critical position such as the bow and side hull. However, under extreme sea conditions (e.g., typhoons), these sensors must address challenges related to stability and impact resistance.

With the advancement of computer technology, slamming theoretical models have been continuously expanded and refined. The coupling of multiple computational methods has significantly improved the efficiency and reliability of numerical simulations. Meanwhile, the evolution of experimental techniques has enabled researchers to uncover new phenomena and mechanisms, thereby better validating numerical results. Most of the research on ship slamming focuses on planing boats, wedge entry into water, and multi-hull ships [3]. This paper begins with the fundamental theory of slamming, providing an overview and comparison of the distinctions and advantages among several numerical methods. Two key aspects have been reviewed: water entry of high-performance vessel hull segment and whole-ship wave load analysis. Existing research findings are summarized from both numerical simulations and model tests, offering valuable references for future studies in this field. The structure of the paper is shown in Figure 1.

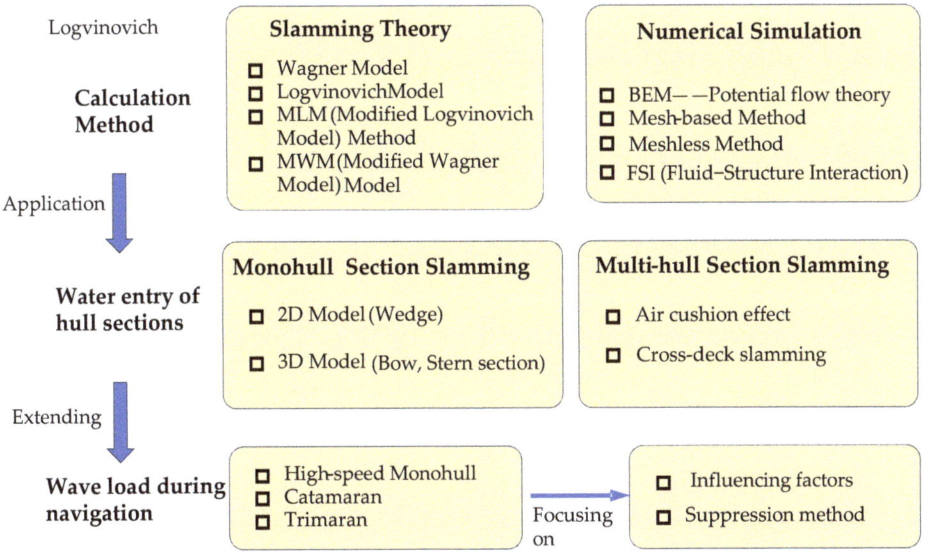

Figure 1. The structure of the paper.

2. Calculation Methods

2.1. Slamming Theory

The study of slamming theory began with von Kármán's analysis of the slamming loads on seaplanes during landing [8]. The landing process has been simplified as a horizontal cylindrical body with an inclined cross-section entering the water vertically at a certain speed. By introducing the concept of added mass and applying the law of momentum conservation, the calculation formula for the slamming pressure of a wedge-shaped body has been derived. Wagner [9] improved upon von Kármán's method by introducing the linear free-surface condition, taking into account the effect of wave elevation. He simplified the wedge-shaped body as an equivalent flat plate and solved the pressure distribution on the bottom surface using the velocity potential of flow around a flat plate and Bernoulli's equation, significantly improving computational accuracy. Although this method is only applicable to wedge-shaped bodies with a deadrise angle of less than 30°, the theoretical framework was further refined and extended in subsequent research [10,11].

An improved approach to Wagner's theory involves adding additional terms to the velocity potential expression. Logvinovich [12] employed a local asymptotic expansion analysis, taking into account the nonlinear terms in Bernoulli's equation, and proposed the Logvinovich calculation model (original Logvinovich model). Korobkin [13] further retained the second-order terms related to the deadrise angle of the wedge in the pressure distribution formula of this model, thereby developing the MLM (modified Logvinovich model) method, which can be applied to water-entry problems of wedges with larger deadrise angles. For the flow separation phenomenon after the water entry of a wedge with finite width, Duan [14] introduced auxiliary lines to construct a virtual body surface and incorporated the body surface function into the MLM method. This approach maintains accuracy while improving computational efficiency. Liu [15] derived an explicit solution for the higher-order MLM (modified Logvinovich model) slamming model based on the original MLM, enabling rapid computation of the second-order velocity potential, pressure distribution, and slamming force. This method improves the accuracy of slamming pressure calculations for wedges with larger deadrise angles during water entry. The study provides a new approach for the rapid assessment of slamming problems in structures with high deadrise angle characteristics.

Tassin [16] introduced fictitious body continuation (FBC) into the MLM to study the dynamic evolution of transient pressure on an object under the influence of fluid detachment at the jet root and cavitation behind the body surface, enabling the prediction of long-term transient drag during cavity formation. The new model can analyze water-entry problems involving flow separation and variable velocity. Unlike Fairlie-Clarke and Tveitnes [17], the relationship between object acceleration and force has been calculated by assuming that the added mass remains constant after flow separation. Finally, the FBC method was extended to smooth-body flow separation, including water entry of cylinders and wedges, validating the universality of the model.

For the slamming problem in waves, Wen [18] proposed the modified Wagner model (MWM), which determines the wetted length by solving a free-surface function. Compared to the Logvinovich model employing the curved fictitious body continuation (curved FBC) technique, the MWM model demonstrates superior performance during the slamming phase under linear incident waves, accurately predicting surface loads on the object throughout both the slamming and transition phases.

T.I. Khabakhpasheva [19] investigated the deflection and strain response of conical shells with small deadrise angles during constant-velocity water entry, based on the generalized linear theory for conical shells and the Wagner model for hydrodynamic impact loads. Assuming an infinitely deep axisymmetric potential flow, the singular component of the load was separated, and an improved convergent series was introduced for computation. The study demonstrated that conical shells can be approximated as circular plates only when the deadrise angle is extremely small. This work marked the first successful application of the orthogonal modal method—a simplified structural modal-based approach—to axisymmetric slamming problems in hydroelasticity.

Feng [20] conducted a study on the added mass during the hydroelastic slamming of a wedge. By accounting for variations in the wetted surface curvature, an analytical model for the eigenvalues of the added mass matrix was established (as shown in Figure 2). This enabled the development of a coupled algorithm integrating the modal method with the Wagner analytical solution for added mass computation. Subsequently, the eigenvalues of the added mass were nondimensionalized, and the relationships between the nondimensional coefficients and factors such as boundary conditions, deadrise angle, and added mass matrix dimension were analyzed. The computational results demonstrated that the studied boundary conditions, bottom deadrise angle, and structural stiffness range had no significant influence on the eigenvalues of the added mass. Furthermore, the eigenvalues of the added mass converged at higher dimensions, exhibiting a linear relationship with the wetted surface ratio. Based on these findings, the eigenvalues of the hydrodynamic added mass were used to construct constraint intervals and an optimal theoretical solution for the relaxation factor in an iterative coupling scheme. Through a mathematical model of iterative error, the study revealed the mathematical relationship between iterative error, added mass, and the relaxation coefficient. Based on this relationship, the effective interval for the relaxation factor under iterative stability requirements was determined (as shown in Figure 3).

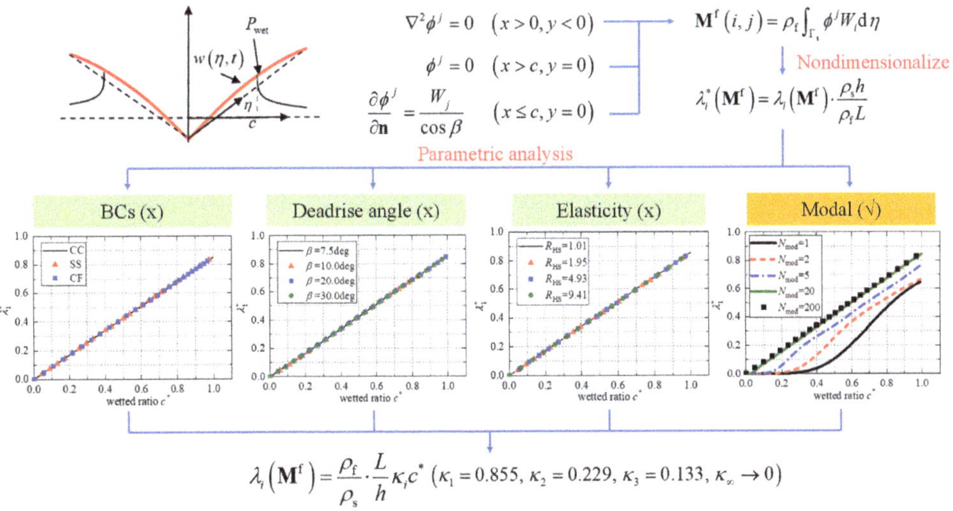

Figure 2. Eigenvalue analysis model of the added mass matrix [20].

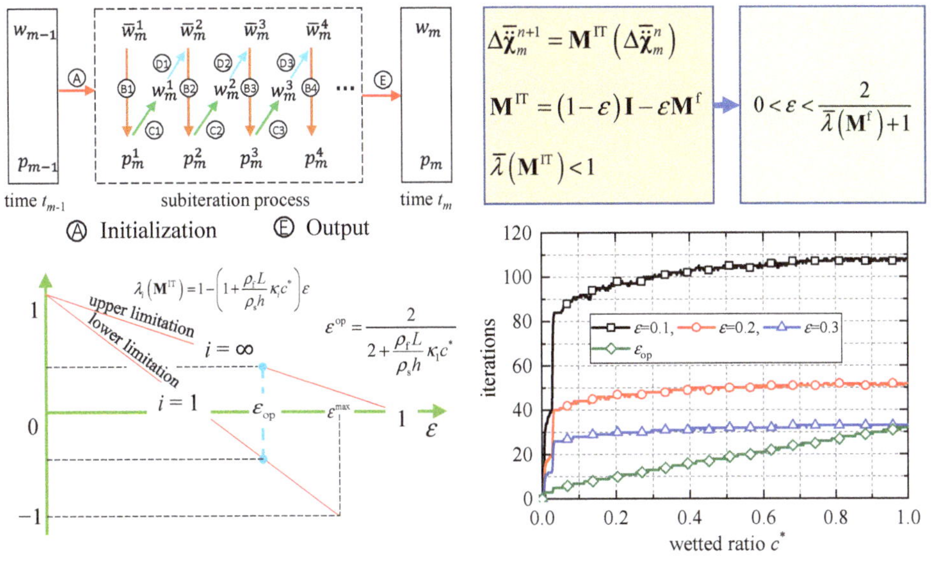

Figure 3. Constraint interval and optimal theoretical solution of the relaxation factor in iterative coupling schemes [20].

2.2. Numerical Calculation

Numerical simulation methods can be primarily categorized into boundary element methods (BEM) based on potential flow theory and computational fluid dynamics (CFD) methods based on viscous flow theory, including the finite difference method (FDM), finite element method (FEM), and immersion boundary method [21] (IBM) based on grid description, and mesh-free methods such as smoothed particle hydrodynamics (SPH), moving tarticle semi-implicit (MPS), and lattice Boltzmann method (LBM).

2.2.1. Boundary Element Method

The boundary element method (BEM) is based on potential flow theory. It transforms the governing equations into boundary integral equations and applies fully nonlinear boundary conditions on the domain boundaries. Only the boundaries require mesh discretization, and the boundary integral equations are solved numerically [22]. Compared to methods like the finite element method (FEM), BEM does not require internal domain meshing or mesh updating, significantly reducing computational effort and saving substantial time. Slamming motion is characterized by high frequency and transient behavior, and theoretical predictions are often conducted in the time domain.

Due to the nonlinear boundary conditions during water-entry slamming, Zhao and Faltinsen [23] employed the constant-panel BEM to study the two-dimensional wedge entry problem. They simplified the numerical computation by approximating the jet spray region with multiple straight-line segments. For the wetted surface calculation during slamming, they adopted the direct finite difference method and integrated the second-order Green's function using a time-stepping scheme. The range of the wedge's deadrise angle is from 4° to 81°. The results demonstrate that the boundary element method (BEM) achieves good agreement with both the modified self-similar solution [24] and the asymptotic expansion method.

Subsequent studies integrated the BEM with Wagner theory, employing either the mixed Eulerian–Lagrangian (MEL) method or a stretched coordinate system to ensure numerical accuracy and stability during the initial phase of nonlinear free surface simulation. An auxiliary function was introduced to handle the time differentiation of the velocity potential, achieving decoupling between hydrodynamic loads and body acceleration [25]. This approach enabled the computation of three-degrees-of-freedom accelerations for wedge-shaped bodies [26]. Liu [27] investigated the fully nonlinear hydrodynamic problem of a cone vertically and obliquely entering Stokes waves using BEM. The study employed a stretched coordinate system and a modified Eulerian method to track the nonlinear free surface. Computational results showed the relative deadrise angle and relative velocity between the cone and fluid particles in a wave environment varied dynamically over time, while the flow velocity exhibited spatial non-uniformity in waves.

For the hydroelastic problem in slamming phenomena, the boundary element method can be used to solve the pressure integral, and then the pressure load can be transferred to a structural solver (such as the finite element method) to iteratively solve the fluid–structure interaction. An efficient fluid–structure interaction (FSI) strategy has been introduced to calculate the hydroelastic slamming pressure of wedge-shaped sections [28,29]. The strategy coupled the superposition method (for the structural part) and BEM. The mean elastic velocity method has been introduced to account for the influence of elastic response on slamming load distribution and established dynamic coupling equations governing the FSI effects. The introduction of nonlinear pressure terms improved computational accuracy. The study systematically analyzed the effects of plate thickness, water-entry angle, impact velocity (constant speed), and boundary conditions on hydroelastic behavior. This method proves more effective for wedges with larger water-entry angles and weaker hydroelastic characteristics. Furthermore, the approach was extended to numerical simulations of shallow-water elastic wedge impacts [30]. A non-penetration boundary was imposed at the fluid domain bottom to simulate shallow-water effects. The results demonstrated that shallow-water conditions significantly amplify free-surface motion, fluid pressure, and structural dynamic response, with the peak slamming pressure increasing by approximately 18%.

Wen [31] developed a hybrid boundary-finite element (HBF) method to solve the fully nonlinear incompressible potential flow for the entire process of a two-dimensional curved wedge entering water at variable velocity, neglecting the effects of gravity and fluid surface tension. The study proposed analytical expressions for the pressure coefficient and hydrodynamic force during the slamming phase, accounting for acceleration effects. Additionally, the method for modifying the added mass coefficient of a linear wedge was extended to curved wedges, effectively addressing acceleration effects during the transitional phase. A hydrodynamic force calculation formula was established, incorporating corrections for constant-speed slamming coefficients and added mass coefficients.

Overall, the boundary element method offers advantages in efficiency and dimensionality reduction when dealing with water-entry problems. However, the handling of the highly nonlinear characteristics of slamming motion may lead to significant errors in large-deformation regions. For instance, repeatedly applying the MEL method for free-surface tracking can reduce computational efficiency. Additionally, singularities on the body surface boundary require local mesh refinement. When coupled with FEM, BEM may introduce additional errors. Since BEM is based on potential flow theory, it cannot directly simulate viscous flow, potentially underestimating the hydrodynamic damping forces on the body and failing to capture vortex generation and turbulent dissipation.

2.2.2. CFD Method

Mesh-Based Method

Since the 1950s, CFD technology has achieved the visualization and quantitative analysis of complex flow phenomena by discretizing the N-S equation and numerically solving the motion and force conditions of each fluid particle in the flow field. By discretizing the fluid domain using grids, the method often couples with the volume of fluid (VOF) or level set techniques to track free surfaces, making them suitable for viscous fluid simulations.

Reddy [32] calculated the slamming load of a wedge with a 30° deadrise angle during water entry. The finite volume method was adopted to discretize the N-S equation, and the VOF method was used to track the multiphase flow interface. This approach solves the equations for air–water two-phase flow based on the fluid volume fraction. The study analyzed the influence of factors such as entry velocity and grid size on slamming loads during water entry and compared the results with experimental data to validate reliability [33]. Kleefsman [34] applied this method to simulate dam-break flows and wedge water entry, demonstrating excellent agreement with experimental photographs during the entry process.

For the problem of water-entry slamming in waves, Zhang [35] adopted a Cartesian grid framework, utilizing the VOF with piecewise linear interface reconstruction (VOF-PLIC) to capture the free surface and applying the volume-of-solid-based immersed boundary method (VOS-IB) to handle the solid boundary conditions. A relaxation zone was also implemented for wave generation and absorption. First, the model's reliability was validated by simulating the constant-velocity water entry of a wedge in calm water, with results compared against historical experimental and numerical data. Subsequently, fifth-order Stokes waves were generated, demonstrating excellent agreement between the simulated results and analytical solutions.

Meshless Method

Meshless methods are Lagrangian particle-based approaches that eliminate the need for grids, offering distinct advantages over traditional mesh-based methods when handling free-surface fragmentation, splashing, and multiphase flows, particularly for complex geometries [36].

Smoothed particle hydrodynamics (SPH) is one of the most widely used meshless methods. The method was Initially proposed by Lucy [37], Gingold, and Monaghan [38] for astrophysical problems. This Lagrangian particle-based approach offers significant advantages in simulating multiphase flows. By simply assigning different particle properties to each phase, SPH automatically tracks multiphase interfaces [39] without requiring additional numerical techniques, making it highly effective for multiphase flow simulations. Subsequently, the SPH method was introduced into the field of fluid mechanics and has been widely applied to problems involving large deformations and free-surface flows [40–46]. Cui [47] adopted the SPH (smoothed particle hydrodynamics) method with open boundary conditions to construct a numerical wave tank, simulating the free water entry of two wedge-shaped bodies with different bottom curvatures under both still water and wave conditions. Based on the Herschel–Bulkley–Papanastasiou (HBP) constitutive model, a two-dimensional numerical model for rigid body impact on non-Newtonian fluids was established. This method effectively captures the motion of non-Newtonian fluids during slamming.

Since the mesh-based method can simulate solid structures and the flow of fluids with small deformations relatively accurately, coupling SPH with FEM can integrate the advantages of both methods. Zhang [48] and Chen [49] employed an adaptive element–particle coupling approach. This method systematically revealed the intrinsic mechanisms of hydroelastic slamming, such as the coupling between flow fields and hydrodynamic loads, the kinematic and dynamic responses of wedge bodies, and the interaction with structural vibrations. Additionally, an empirical formula has been proposed to predict the vibration characteristics of wedge bodies during hydroelastic slamming.

Traditional SPH requires the introduction of artificial viscosity to suppress numerical oscillations when handling shock waves, but this often leads to excessive dissipation. The Riemann SPH method (Riemann-SPH) is an improved smoothed particle hydrodynamics (SPH) algorithm based on Riemann solvers. By incorporating Riemann problem solvers (e.g., HLLC, Roe, etc.) to describe particle interactions, it replaces the artificial viscosity term in conventional SPH, achieving breakthroughs in shock capturing, interface stability, and conservation properties. This method has been widely applied in dam-break flows [50], underwater explosions [51], and water entry of structures [52]. Meng [53] proposed a multiphase Riemann-SPH model for studying water-entry problems. By introducing a truncation threshold for particle density as a special treatment, this model effectively avoids the generation of negative pressure. The study accounts for air effects, and the simulations of vertical displacement and acceleration during the water entry of a ship section agree well with experimental results.

To address the issue of low computational efficiency caused by the high computational cost of the SPH method, Zhang [54] developed a 3D six-degrees-of-freedom (3D 6DOF) SPH model based on GPU parallel programming, successfully simulating complex free-surface flow slamming. This approach significantly improved computational efficiency, but it did not account for gas-phase effects or structural elastic responses.

The MPS method adopts a fully Lagrangian formulation, thus enabling explicit analysis of large deformations at free-surface interfaces [55]. The MPS method has been improved to capture the slamming pressure on two-dimensional marine structures, where slight compressibility was introduced into the modified pressure Poisson equation (PPE) to reduce pressure fluctuations [56]. Shibata [57] and Wan [58] also employed the MPS (moving particle semi-implicit) method to compute the violent fluid–structure interaction problems. Sun [59] investigated the strong hydroelastic interactions between ship hull structures and free surfaces using an enhanced MPS method coupled with a modal superposition approach. In their study, the MPS method was applied to solve the fluid dynamics, while a modified modal superposition method was adopted to analyze the elastic deformation of the ship hull. The study found that the coupled numerical technique employed could accurately capture slamming loads and simulate structural deformation, but the peak strain values it computed were underestimated.

However, the traditional MPS (moving particle semi-implicit) method does not exhibit good convergence in terms of numerical parameters, requiring continuous adjustment of numerical conditions [60]. Additionally, the inconsistency in pressure gradients leads to numerical oscillations in slamming pressure calculations [61]. To address the MPS method's sensitivity to numerical parameters, Takahito Iida [62] proposed a practical set of numerical conditions for MPS simulations (termed the MPS-slamming conditions) specifically for the 2D wedge slamming problem. By determining the MPS-slamming number, slamming velocity, wedge entry angle, and particle size, the optimal time step can be obtained. Zha [63] proposed an improved high-order MPS method to solve FSI problems during the water entry of elastic wedges. This method has used the conventional serial staggered (CSS) procedure that was a weak two-way coupled scheme, to realize the coupling of the fluid and structure (see in Figure 4). The pressure gradient term has also been modified to ensure first-order consistency and momentum conservation, while adopting a differential particle spacing strategy between the fluid and structural domains to reduce numerical oscillations. Furthermore, a particle convergence metric [64] was introduced to quantify the numerical uncertainty of the enhanced MPS method and evaluate the discretization-induced numerical errors.

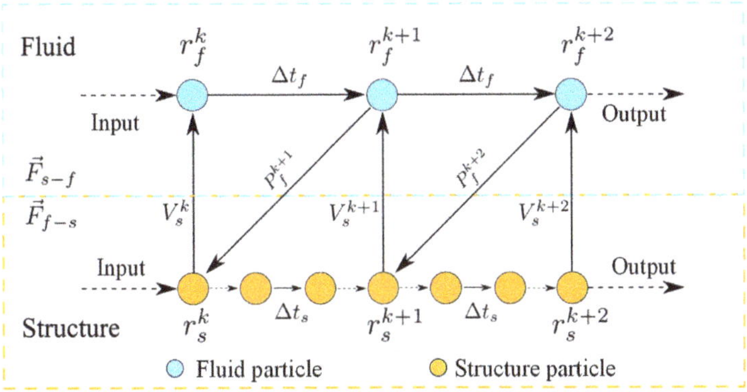

Figure 4. The CSS coupling scheme for FSI computation.

The lattice Boltzmann method (LBM) discretizes the Boltzmann equation by transforming the continuous particle distribution function into discrete lattice-based fi(x,t) instead of directly solving the Navier–Stokes equations. Then, it simulates macroscopic fluid behavior through local collision and lattice migration [65]. To investigate the dynamic response of flexible structures under free-surface impact loads in weakly compressible viscous fluids, De Rosis [66] employed the lattice Boltzmann method (LBM) to analyze fluid flow (including free-surface deformation and pressure distribution), used the FEM to solve dynamic stresses and deformations, adopted the time-discontinuous Galerkin (TDG) method for structural dynamics time integration, and implemented an explicit coupling strategy (with synchronized fluid–structure time steps) to achieve fluid–structure interaction. The LBM has also been introduced to calculate the pressure distribution along the curved wedges [67]. The evolution of splash morphology was related to the pressure distribution. The computed pile-up coefficients for slamming wedges obtained by this method show good agreement with experimental measurements from other studies.

In summary, the finite volume method (FVM) grounded in viscous flow theory offers strong mesh adaptability, ensures excellent conservation properties in discretized equations, and can be coupled with the finite element method for solutions. Its primary drawback lies in high computational costs. For particle-based methods, The SPH method, being mesh-free, possesses inherent advantages in handling dynamic large-deformation problems. Its accuracy is unaffected by mesh quality, and it maintains high computational efficiency. Yet, it faces challenges in modeling complex geometries and requires further research for 3D multiphase water-entry problems. The MPS method is concise and efficient, demonstrating robust convergence. However, it suffers from varying degrees of pressure oscillation phenomena.

3. Water Entry of Hull Section

Water entry refers to the process in which an object penetrates the free water surface at a certain velocity, transitioning from air into water. Water impact is a common phenomenon in naval and ocean engineering, with bow-flare slamming being the most widely studied. The impact loads generated during water entry can not only cause localized structural damage to the hull but may also significantly reduce ship speed, thereby affecting voyage time and economic efficiency. Early research on water entry primarily focused on wedge and flat plate impacts, which were later extended by subsequent researchers to multi-hull segments and three-dimensional hull structure entry.

3.1. Wedge Slamming

3.1.1. Experimental Study

Experimental methods serve as a crucial means for investigating the characteristics of slamming processes and validating analytical theoretical models [68]. Among these, wedges are the most extensively studied models. Chuang [69,70] conducted free-fall water-entry experiments on both rigid and elastic bodies, including flat-bottomed and varying deadrise-angle hull sections, as well as rectangular elastic plates. The study revealed the significant influence of air cushioning on flat plates and wedge-shaped bodies with deadrise angles below three degrees. High-speed photography captured the deformation of the water surface as air entrapped underwater. The results showed that the compressible air layer prolongs the duration of slamming pressure. An empirical formula was proposed to relate the peak slamming pressure to the water-entry velocity. Marintek Laboratory [71] conducted free-fall water-entry slamming tests on a wedge with a 30° deadrise angle. The experiments revealed that the maximum slamming force occurred before flow separation, near the root of the spray jet. After flow separation, the peak pressure rapidly decreased. Zhao and Faltinsen conducted free-fall water-entry tests on a typical flared bow section, validating the accuracy of a fully nonlinear hydrodynamic impact theory in numerical simulations. Mojtaba [72] investigated the water entry of wedge-shaped bodies with varying deadrise angles at different entry angles and heights, The study demonstrated that the deadrise angle significantly influences slamming pressure. Traditional asymmetric theories do not apply to all entry angles. Nikfarjam [73] conducted water-entry experiments on wedge-shaped bodies with a deadrise angle of 30° but varying weights, measuring displacement and pressure values at three points on the bottom. By converting the data into non-dimensional depth and pressure coefficients, the study summarized the effects of model weight and deadrise angle on slamming pressure.

Numerous scholars have conducted research on wedges with a 10° deadrise angle. Byoungcheon Seo [74,75] performed drop tests with wedges of varying thicknesses from different heights, systematically investigating the effects of structural flexibility and repeated loading on slamming response. The study established an empirical formula relating impact cycles to cumulative deformation, providing an experimental benchmark for hull-slamming durability design. Furthermore, a comparative study between cylindrical slamming [76] and wedge slamming demonstrated that the impact load on cylindrical models exceeds that of wedges with a 10° deadrise angle, which aligns with theoretical predictions.

Ren [77] conducted hydroelastic slamming experiments on a wedge with a 20° deadrise angle under vertical water entry. By varying impact velocity and bottom plate bending stiffness, the study simultaneously measured wedge motion, jet root expansion, hydrodynamic loads, and structural response, establishing a quantitative relationship between the hydroelastic factor (R) and structural response. This work revealed the modal selection mechanism of energy transfer during slamming. Further research [78] demonstrated that when R exceeds a critical threshold, the trends of pressure and strain undergo abrupt changes. Additionally, structural deformation causes the peak to lag and significantly reduces its amplitude.

Saeed Hosseinzadeh [79] conducted experiments and numerical simulations [80] on the water entry of wedges with varying deadrise angles, including wedges with and without stiffened bottom panels. The study found that maximum strain and deformation occurred during the local wetting phase of the slamming process. The influence of hydroelastic effects on structural response increased with smaller deadrise angles and higher impact velocities. Unstiffened panels exhibited significant fluid–structure interaction (FSI) effects across all impact velocities, whereas stiffened panels only showed pronounced FSI at high impact velocities [81]. Han [82] investigated the distribution and time-record curves of slamming pressure and hydroelastic effects during water entry of stiffened elastic wedges with varying deadrise angles. The results showed that the propagation speed of peak pressure was about 0.8 times of the value predicted by Wagner's theory. The average propagation speeds of both pressure and stress peaks exhibited a linear relationship with impact velocity [83]. Further numerical analysis [84] revealed that as the deadrise angle increased, the slamming pressure coefficient on the upstream (impact) side significantly increased, while the coefficient on the downstream side decreased. The jet flow region expanded on the upstream side but contracted on the downstream side during asymmetric water entry. The study demonstrated the influence of entry angle on pressure distribution and flow field characteristics.

Thus, the slamming responses are influenced by the deadrise angles. Duan [85] conducted a series of water-entry experiments on rigid and elastic wedges with different entry angles and drop heights. The spatial variation of slamming pressure, influence mechanisms of entry velocity, entry angle, and structural elasticity on pressure propagation characteristics across different deadrise angles have been analyzed. Güzel [86] investigated the effects of hydrophobicity on free surface elevation and slamming loads during the water entry of wedges and cones with varying deadrise angles, with a focus on the dynamic response in the early slamming stage. Jet velocity increases significantly during water entry, and the superhydrophobic effect becomes more pronounced for models with small deadrise angles. The added mass and wetted area do not exhibit a linear positive correlation; instead, their values may decrease as the deadrise angle increases.

Zhang [87] conducted a slamming impact test using 2D partial hull sections with varying curvatures, measuring the time-domain characteristics of slamming pressure at different water-entry heights. The study focused on the influence mechanisms of entry velocity and curvature effects on slamming dynamics. The experiment model and flow chart for the water-entry slamming test are shown in Figures 5 and 6. The experiments revealed that higher-curvature regions (e.g., the sharp bow) exhibited greater peak slamming pressures but shorter pressure rise time. These areas were prone to local buckling, with maximum strain occurring at curvature discontinuities (reaching 60% of the material's yield limit). But the study did not account for dynamic coupling effects under wave conditions.

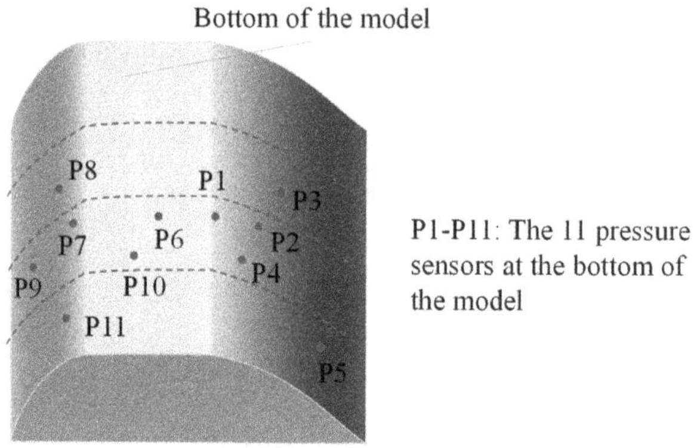

Figure 5. Experiment model and sensors' position [87].

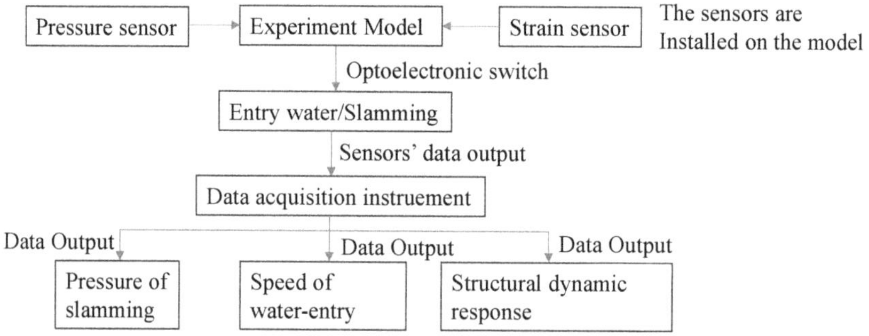

Figure 6. Flow chart of the slamming test [87].

For constant-velocity water entry, Meziane [88,89] investigated the influence of panel stiffness on pressure and strain for aluminum wedges with varying deadrise angles under different impact velocities. The study employed a hydraulic impact test rig to achieve quasi-constant-speed entry, examining how the water-entry angle affects structural strength and velocity characteristics. It systematically analyzed the dependence of peak pressure and jet velocity on these parameters. Jain [90] conducted constant-velocity water-entry experiments with wedges, analyzing the formation mechanism of the negative pressure pulse generated at the initial contact between the wedge and the water surface. The study validated the rationality of approximations made when extending the Wagner model to three-dimensional (3D) cases and measured the dynamic response at the air-water interface, caused by pressure buildup in the trapped air layer between the impacting body and the free surface. The results show the pressure coefficient (Cp) depends solely on the solid shape under constant velocity. As stiffness decreases, Cp increases across all deadrise angles. Moreover, the interface pre-deformation induced by the air cushion effect can be fully described by potential flow theory.

The rapid advancement of high-speed imaging technology has led to the widespread application of particle image velocimetry (PIV) [91]. In 2013, Nila et al. [92–94] measured the velocity of the flow field during the slamming at various inclination angles and heights, then derived the pressure field by solving the Poisson equation. The experimental set-up and PIV raw image is shown in Figure 7. This allowed them to determine the pressure distribution on the model surface and compare experimental results with numerical simulations to validate the reliability of the method. On this basis, Jalalisendi and Shams [95] analyzed the velocity fields along three vertical planes and five transverse planes of the wedge, reconstructing a 3D velocity field. They then derived the pressure field and compared the reconstructed pressure and hydrodynamic loads with the theoretical values from Wagner's model. Panciroli [96,97] investigated the hydroelastic phenomena of flexible structures based on PIV technology. The pressure distribution of the model's bottom was solved by a non-intrusive pressure reconstruction method. The study revealed that wedge flexibility significantly alters hydrodynamic load characteristics, leading to pronounced oscillations in both the pressure distribution and its evolution process on the wetted surface of the impacting body. Zhang [98] employed PIV to accurately capture the influence of bidirectional fluid–structure coupling on the spatial distribution of hydrodynamic loads. The result showed that the pressure field on the wetted surface of the impacting body exhibited pronounced spatial oscillations, while its temporal evolution displayed distinct periodic fluctuations.

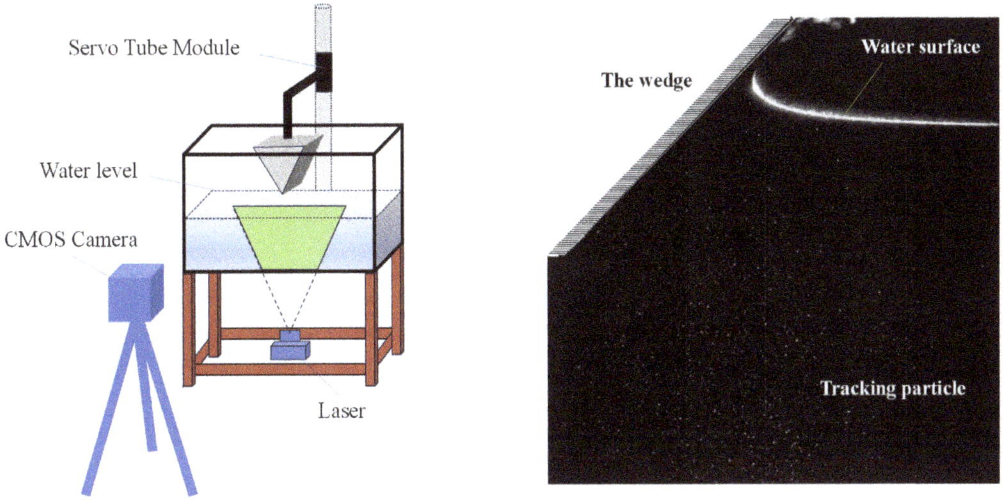

(**a**) Experimental set-up (**b**) PIV image

Figure 7. The application of PIV technology [92].

Although particle image velocimetry (PIV) can study the transient fluid kinematics of water-entry objects through flow-field visualization, accurately extracting the liquid-phase region under unsteady hydrodynamic loads remains challenging. Guo [99] proposed a deep learning network called MRes-Att-Unet, based on the U-Net architecture, which integrates residual connections and attention mechanisms (previously validated in medical image segmentation). A dedicated dataset was constructed using PIV experiments of two-phase flow during water entry, with further validation in isolated wave-breaking experiments. Additionally, artificial intelligence (AI) has been applied to slamming event identification in ship navigation, where supervised machine learning (ML) methods analyze test signal characteristics to detect slamming events [100]. In summary, applying particle image velocimetry (PIV) to measure acceleration and pressure variations during water-entry slamming tests of ship hulls offers the advantage of capturing pressure distribution across the entire wetted surface, while avoiding data distortion caused by noise filtering of pressure sensors. However, this method is limited to measuring flow-field motion within the laser plane and cannot account for three-dimensional effects during water entry.

3.1.2. Calculation of Fluid–Structure Interaction for Wedge's Slamming

In recent years, advancements in computational power have enabled the integration of finite element analysis (FEA) with CFD, demonstrating significant potential for engineering applications in ship design and slamming load assessment. In this approach, hydrodynamic problems are solved using CFD software (Ansys Fluent 2024 R1) with overset grid technology, while structural dynamic responses are computed via FEA software (Ansys 2019 R3). Izadi [101] systematically investigated the fluid–structure interaction (FSI) mechanisms of symmetric and asymmetric wedge oblique water entry through bidirectionally coupled FVM–FEM numerical simulations. Jiao [102] employed a CFD–FEM bidirectional coupling numerical tool, integrating the STAR-CCM+ 13.02 with Abaqus, to study the asymmetric oblique water entry of wedges. The relationships among the wedge's motion, slamming pressure, structural response, and the inclination angle/water-entry velocity have been analyzed. Xiao [103] proposed a bidirectionally coupled FSI method for flexible wedges to investigate the slamming pressures and structural responses of free-falling non-prismatic stiffened steel wedges. OpenFOAM ESI V2306 for Reynolds-averaged Navier–Stokes (RANS) computations has been used to determine hydrodynamic loads and the calculation of structural response has been carried out by Calculix (FEA software). The result showed that elastic deformation can mitigate impact pressure. The stiffness of flexible panels can be enhanced by transverse stiffeners. Pressure duration and peak slamming pressure exhibit a negative correlation. Sun [104] developed a bidirectionally coupled FSI solver for a curved wedge structure, integrating OpenFOAM V1912 (fluid domain), CalculiX Version 2.22 (structural domain), and preCICE 2.4.0 (coupling interface) to achieve high-precision FSI simulations within an open-source framework. The hydroelastic effects in linear wedges are significantly stronger than those in curved wedges and the structural response of the curved wedge exhibits a distinct two-stage characteristic. It can be seen that the stiffness of the wedge has a significant impact on the slamming load, as the large deformation caused by plate vibration alters the evolution pattern of the fluid jet.

This method was further extended to the problem of wedges entering water in waves. Chen [105] employed a CFD–FEM bidirectional coupling numerical approach to investigate the FSI and slamming pressure of a wedge-grid structure during free-fall entry into fifth-order Stokes waves. The slamming loads and structural stresses of the wedge have been calculated during water entry with elastic effects, considering different relative positions between the wedge and waves. Finally, the influence of horizontal and rotational degrees of freedom on the vertical wave-entry results was explored.

Research on approximate analytical expressions for slamming loads has been conducted by several scholars. Sun [106] decomposed the vertical slamming force into velocity-dependent, acceleration-dependent, and gravity-dependent terms. Using CFD methods, they calculated the coefficients for velocity, acceleration, and gravity, enabling the accurate approximation of forces on arbitrary 2D-shaped bodies under various velocity/acceleration combinations except in cases of flow separation followed by rapid body stoppage. Wen [107] employed the FVM combined with the VOF method and the hybrid boundary formulation (HBF) to study slamming and transitional phase characteristics during both constant-speed and variable-speed water entry of wedges. Under the assumptions of an incompressible, inviscid, gravity-free, and surface-tension-negligible fluid with irrotational flow, they derived analytical expressions for slamming forces and hydrodynamic loads during the transitional phase in both constant and variable-speed water-entry process.

3.2. Multi-Hull Section Slamming

3.2.1. Air Cushion Effect

Catamarans or trimarans, as high-performance vessels, are prone to significant slamming phenomena during navigation. Unlike monohulls, multi-hull vessels must account for interactions among hulls, wet-deck slamming, and air-cushion effects. Before water entry, viscous forces in the air induce free-surface deformation, generating flow within the liquid pool. This serves as the initial condition for slamming and exhibits a cushioning effect, manifested as a reduction in the effective contact area, thereby mitigating impact loads [108]. Hasheminasab [109] approximated the catamaran cross-section as two asymmetric wedges connected by a wet deck. The experimental results showed that the spacing between the wedges had a negligible influence on peak pressure, while air cushion effects were proven to significantly alter slamming pressure. Tang [110] employed a semi-enclosed ship model with partially open acrylic plates attached to its surface, adjusting the opening size to control the gas escape efficiency at the model's front and rear sections. The formation process of the air cushion and the evolution characteristics of the free surface were captured using high-speed photography. In the experiment, by varying the opening size and the initial release height of the model, it was demonstrated that the air cushion effect can delay the occurrence of the pressure peak, reduce the amplitude of the peak pressure, and significantly intensify pressure pulsations.

For high-speed multi-hull vessels, wet-deck slamming is a critical issue when encountering significant vertical motions during navigation. Ma [111] conducted drop tests on a small-waterplane-area twin-hull (SWATH) segment model to research wet-deck slamming. The experiments involved vertical water-entry tests under varying drop heights and velocities, measuring slamming pressures and acceleration data beneath the wet deck. Additionally, particle image velocimetry (PIV) was employed to capture flow-field characteristics. The experimental findings revealed that during wet-deck slamming, air is entrapped into the water, forming an air–water mixture phase due to gas compressibility effects. The results demonstrated that pressure peaks occur during both the hull-bottom slamming and wet-deck slamming stages, while the air-cushion effect effectively mitigates wet-deck slamming pressures. Masoomi [112] proposed integrating "ventilation ducts" into the central bow structure of a catamaran to release accumulated pressure and associated loads. Using a RANS-based numerical method (implemented via the OpenFOAM open-source library) to solve the flow field, the computational results demonstrated that this design could reduce slamming pressure peaks by up to 50%. Swidan [113] investigated the slamming load on the hull bottom and wet deck of a wave-piercing catamaran during water entry at specific speeds, along with an uncertainty analysis of experimental results. This study provided a high-fidelity dynamic load database for the shock-resistant design of wave-piercing hulls and highlighted the limitations of traditional 2D theories in predicting 3D slamming phenomena.

Jiang [114] conducted water-entry impact tests on a segmented trimaran model to further investigate the dynamic interactions among the structure, fluid, and air, as well as the influence of the air cushion on slamming loads. The study employed PIV (particle image velocimetry) technology to analyze the interactions among air, water, and the structure, while pressure sensors were used to measure slamming loads. Additionally, to amplify the air cushion effect, baffles were installed on both sides of the segmented hull to restrict air escape during the experiment. The results demonstrated that the air cushion beneath the connecting bridge significantly reduced slamming pressure. Yu [115] investigated the water-entry slamming characteristics of a trimaran's segment in regular waves, focusing on the influence of the model's water entry point on slamming load as shown in Figure 8. The study revealed the cushioning mechanism of the air cushion effect during bubble escape, providing a theoretical basis for optimizing anti-impact structures. The results showed that the wave position at the moment of water entry significantly affects both the peak load and time-domain characteristics, with the most intense fluid motion occurring when the main hull enters the water at the mid-wave position.

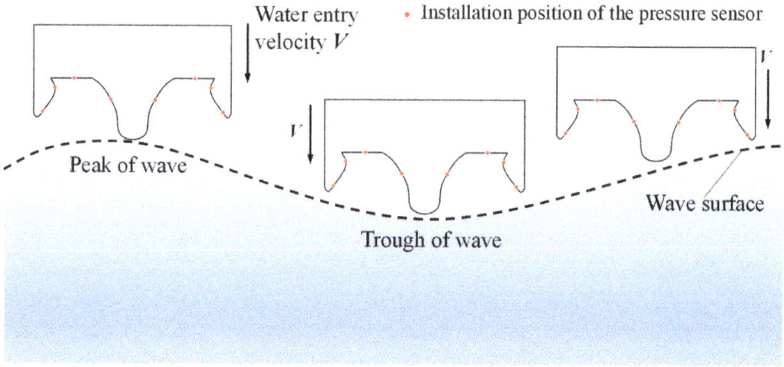

Figure 8. Relative position of the model's entry point and waves [115].

3.2.2. Pressure of Wet-Deck Slamming

The calculation of slamming dynamic loads is a critical factor affecting the structural integrity of multi-hull ships [116]. Liu [117] used the CFD method to simulate the slamming process of a trimaran's typical cross-section, computing the slamming pressure and free-surface evolution. Based on the relationship between bilge curvature and flow separation, a quantitative formula linking bilge curvature to slamming loads was established, and a deck reinforcement zoning criterion was proposed, based on flow-state control. Sun [118,119] adopted a CFD numerical simulation system to calculate the slamming load of a trimaran's cross-section under various working conditions, including different masses, drop heights, constant-velocity water entry, and so on. The result showed that slamming dynamics exhibit a strong correlation with immersion depth. The impact velocity shows a quadratic relationship with slamming pressure, while acceleration demonstrates a linear dependence on wet-deck slamming pressure. These conclusions align closely with the mechanical patterns implied by Wagner's theoretical model and the modified Logvinovich model (MLM).

Duan [120] conducted free-drop water-entry experiments on a trimaran cross-section with various drop heights and inclination angles, employing PIV to observe the evolution of the flow field. The study revealed that peak pressure exhibited a quadratic proportionality to impact velocity. A Gaussian distribution function was proposed to fit the pressure time-record curve during the wet-deck slamming phase. Dimensionless pressure coefficients and slamming duration were systematically summarized across different test conditions. Pan [121] investigated the structural response of a trimaran hull girder through vertical water-entry experiments on a trimaran segment, ultimately determining the maximum slamming pressure range and critical zones for this trimaran design. Li [122] applied three flexible coatings of varying thicknesses to a trimaran model. Water-entry tests at different heights demonstrated that the flexible coatings significantly reduced impact loads (by 32–47%). The research also revealed the mechanism of flexible surfaces in dissipating impact energy.

The hull form of a trimaran significantly influences slamming loads. Wang [123] conducted drop tests on two trimaran segment models. One of the models featured a modified hull with an extended main hull to avoid separated bottom jet flow. Comparative analysis of flow-field characteristics and impact pressures revealed that the slamming pressure coefficient on the transverse deck of the original main-hull model was significantly lower than that of the extended main-hull model due to flow-field disturbances caused by main-hull bottom slamming. Similarly, Li [124] modified the main-hull lines to reduce flow-field disturbances during water entry. The experimental results demonstrated that the main-hull form decisively governed the mechanism of wet-deck slamming and profoundly affects both the magnitude and distribution of wet-deck slamming loads.

3.3. *Water Entry of 3D Hull Model*

Theoretical research on the water entry of two-dimensional (2D) structures has been extensively studied, forming a well-established theoretical framework. However, when applying 2D water-entry theories to predict three-dimensional (3D) structural impacts, significant deviations often arise. In recent years, extensive studies have been conducted on 3D hull water entry, including investigations on bow, stern, and aircraft [125].

Wang [126] conducted water-entry impact tests on a full-scale 3D composite sandwich ship bow with complex geometric features. The bow was fabricated using vinyl ester fiberglass and PVC foam materials. The study analyzed the distribution of slamming loads, dynamic response characteristics, and the effects of varying water-entry velocities. The results indicate that structural optimization of ships should enhance the bending stiffness of the bow curvature transition zone, in order to suppress local buckling. Shan [127] conducted water-entry experiments on a flared bow section using a tank drop test system. The study found that flow separation during slamming induces cavitation in the concave region of the hull, leading to a delay in the slamming occurrence. Additionally, the high-speed jet generated by the falling model traverses the concave area and impacts the flare of the bow, resulting in secondary slamming. The secondary slamming can generate a higher pressure coefficient, with its peak value even exceeding that of the primary slamming. Based on this, the researchers quantified the cumulative effect of secondary slamming, developed a pressure prediction model incorporating flow separation correction, and proposed an engineering evaluation method based on the effective impact angle. Ping [128] conducted a comparative analysis of the hydrodynamic impact performance between bulbous bow and conventional bow rigid hull structures using a combined approach of CFD numerical simulations and vertical water-entry experiments. The study found that the bulbous bow structure induced flow separation 30–40 ms earlier compared to the conventional bow. Additionally, the flow velocity in the separation zone decreased by 25–35%, revealing the potential threat of jet impact on hull safety. Based on these findings, the authors proposed an optimized bow design criterion based on flow pattern control.

Xie [129–131] systematically investigated the asymmetric slamming load characteristics of truncated bow and stern structures under combined roll–pitch motions (see Figure 9). The pressure near the stern shaft exhibited multi-peak behavior and high-frequency oscillations, with obvious cavitation effects. Further research revealed a phenomenon was observed at specific bow locations, where pressure decreased with increasing water-entry velocity. Liu [132] conducted water-entry experiments on a stern model, measuring the relationships among immersion depth, acceleration, and slamming loads via sensors. The study demonstrated the influence of cavitation effects and jet flows on pressure distribution dynamics. Mutsuda [133] conducted water-entry experiments on wedge and stern models, combined with numerical analysis. The strongly nonlinear interactions between the hull and free surface during stern slamming have been investigated. These interactions arise from spray jets and flow separation. The study demonstrated that the water-entry angle and relative vertical velocity between the water surface and model are critical factors influencing slamming loads.

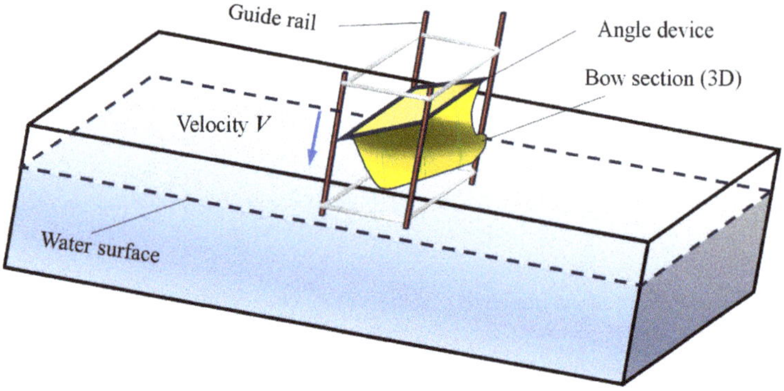

Figure 9. Sketch of the drop experimental system used in the asymmetrical water entry [125].

In general, research on the slamming phenomena of monohull structures has been extended from two-dimensional models to three-dimensional analysis, particularly focusing on bulbous bow and stern slamming issues. The air cushion effect introduces more significant cavitation phenomena, causing hysteresis in slamming pressure and acting as a buffer. This effect shows more pronounced impacts on multi-hull segmented water entry scenarios. The cross-deck's forepart undergoes more severe slamming effects. Main-hull entry generates higher slamming pressures near the connecting bridge on side hulls compared to adjacent main-hull regions. Air cushion effects distinctly influence segmented water entry in multi-hull vessels. To suppress the slamming pressure, measures such as installing baffle plates beneath wet decks, applying flexible coatings, and optimizing the body line of the main hull can be adopted. In terms of experimental techniques, PIV technology, with its high frame rate, non-intrusive measurement, and flow field velocity analysis, has become the important experimental method for studying the hydrodynamics of water-entry slamming. It has strong applicability in revealing the flow mechanism and validating numerical models.

4. Wave Load During Navigation

During ship navigation, slamming caused by vertical motions generates significant wave-induced loads. Early research on wave load problems was primarily based on linear theory. However, under high-sea-state conditions, the vertical motions and slamming loads resulting from phenomena such as bottom slamming [134], flare slamming [135], and wet-deck slamming on trimaran connecting bridges exhibit pronounced nonlinear characteristics. The tensile and compressive forces in the hull girder become asymmetric and no longer maintain a linear relationship with wave height. To investigate these nonlinear phenomena, it is necessary to treat slamming and hull motions as a coupled nonlinear system, including the coupling effect of the coupling effects between hull girder bending moments and local slamming impacts [136]. The influence of slamming loads on ship motions should be fully considered.

4.1. Slamming of High-Speed Monohull

4.1.1. Numerical Solution

High-speed crafts often operate in a planing state during navigation. The 'gliding effect' leads to highly complex and strongly nonlinear characteristics in the slamming loads, motions, and structural responses, making numerical simulations particularly challenging. Bao [137] integrated the strip theory with the BEM, discretizing the hull into multiple two-dimensional wedge-shaped sections to simulate water entry. The study provided detailed numerical results on pressure distribution, free surface deformation, and force (moment) variations, along with an in-depth analysis of the influence of hull speed parameters. Prini [138] used the strip theory combined with wave-induced load calculations to analyze slamming loads on high-speed crafts, comparing the zero-speed Green's function method with the Rankine source method incorporating speed correction terms. Validation against experimental data showed that the Green's function method exhibited better correlation at low speeds, while the Rankine source method was more suitable for high-speed conditions. Parunov [139] investigated a patrol frigate by employing nonlinear strip theory, frequency-domain, and time-domain 3D boundary element methods to compute hull motions. Euler–Bernoulli and Timoshenko beam theories were applied for hull girder stiffness modeling, calculating the wave-induced responses and springing responses of the warship in regular waves. Comparisons with model tests demonstrated that the seakeeping method based on potential flow theory, combined with limited CFD–FEM case corrections, exhibits excellent computational reliability and efficiency.

The 2D + T theory, based on the traditional strip theory and incorporating a three-dimensional free-surface condition with forward speed, demonstrates good accuracy and efficiency in calculating the wave-making and slamming loads of high-speed monohull vessels in motion. Bilandi [140] employed the 2D + T method to calculate the pressure distribution on each cross-section of a planing hull and the normal forces acting on the two-dimensional sections. The reliability of this method was validated through comparison with experimental data. Tavakoli [141] employed the 2D + T method to calculate the motion and slamming loads of a planing craft during navigation, comparing the results with CFD methods and model tests. The study found that for nonlinear responses in vertical motion, the 2D + T theory showed good agreement with experimental data at low to medium speeds. However, as the speed increased, the accuracy of this method declined for long-wavelength waves exceeding twice the ship's length, with the calculated slamming forces being underestimated. Shao [142] adopted the 2D + T theory to transform the three-dimensional hydrodynamics problem of planing hulls into a two-dimensional slamming problem, specifically, the problem of impact on cross-sections with continuously deformed shapes during water entry. By integrating a modified Logvinovich model (MLM), this approach enables reliable calculation of slamming forces on planing vessels.

Compared with the potential flow theory, CFD technology can more accurately simulate the velocity and pressure distribution of the flow field during slamming under extreme working conditions. Mousaviraad [143] employed the unsteady Reynolds-averaged Navier–Stokes (URANS) method to compute the hydrodynamic performance and slamming loads of a high-speed planing craft. The reliability of the method was validated through comparisons between 2D water-entry simulations and experimental data. The approach was then extended to 3D full-scale ship simulations for predicting wave-induced slamming pressures in seaways. However, these 3D numerical results lacked experimental validation. Subsequent studies mostly combine the finite element fluid–structure interaction (FSI) technology with the computational fluid dynamics (CFD) method. The calculation conditions include those in still water [144] and in waves [145,146]. The influencing factors of slamming loads on the hull bottom and flare [147] are analyzed, and the reliability is verified through comparison with experiments [148]. Lee [149] designed an optimized truss panel based on CFD load assessment and mathematical optimization, then analyzed the motion of a deep-V planing hull in irregular waves. The above research indicates that the maximum slamming load did not occur at the highest wave crest, but rather under conditions with deeper troughs and lower wave heights. The slamming coefficient (defined as the ratio of the hull-girder vertical bending moment to the wave-induced bending moment) increased significantly with wave steepness and speed, peaking near the surge natural frequency.

For the unique configuration of wave-piercing monohulls, the hull is highly susceptible to severe deck wetness and corresponding hydrodynamic loads. Shan [150] simulated the vessel's motion characteristics in head seas using Star-CCM+ 2020.01, focusing on the bow-flare slamming effects on the surrounding flow field. The study revealed that the collision of two water jets above the bulbous bow generates a butterfly-shaped high-pressure zone, and the slamming jet exerts strong impact loads on the superstructure.

4.1.2. Model Test

Given the nonlinearity of hydrodynamic and FSI problems in planing craft navigation, Begovic [151] integrated a precisely measurable structural model into the hull of a typical planing boat prototype. Through regular wave tests, this method was validated to enable the detailed investigation of the entire slamming FSI process. Camilleri [152] conducted full-scale sea trials on a planing craft, examining slamming impact characteristics and their induced rigid-body motions and structural responses. For the collected data (e.g., pressure, strain, global hull deflection), an automated algorithm was developed to fit statistical models to peak distributions, with goodness-of-fit tests confirming model applicability. Ibrahim [153] conducted hydroelastic model tests on a semi-planing vessel, systematically investigating the mechanism of slamming effects on global hull girder loads under both regular and irregular waves. The study revealed that coupled heave–pitch motions and wave steepness are key parameters governing the amplitude of slamming loads.

The superposition of high-frequency vibration bending moments and wave-induced low-frequency bending moments (induced by whipping) poses a severe threat to the ship's longitudinal strength. Zou [154] conducted segmented model tests to investigate the slamming loads on flared bows and whipping vibration responses of ships in regular waves. By employing continuous wavelet transform (CWT) analysis, a quantitative relationship model among ship speed, slamming, and whipping responses was established. The experimental results showed that the asymmetric slamming loads were significantly amplified in oblique wave conditions. Whipping vibrations increased the asymmetry of vertical bending moments (VBM) at the midship section, with this effect intensifying as ship speed increased. Mogoga [155] conducted experiments on the hull girder stresses of an aluminum high-speed patrol vessel and found that the coupling effect of slamming and whipping response can increase stress amplitude in fatigue hotspot regions by 40–65%. A dynamic stress equivalent conversion model has been proposed to achieve coupled damage superposition of wave loads and impact loads, quantifying the influence weight of slamming on hull structural fatigue.

4.2. *Slamming of Catamaran*

During high-speed catamaran operations, the Froude number typically exceeds 0.5, causing heave and pitch motion amplitudes to significantly surpass wave height and wave slope. This may trigger slamming loads and structural resonance [156], leading to structural damage and deck wetness [157]. Slamming loads exhibit strong correlation with vertical motions, with both motion peaks and slamming load peaks occurring within the same encounter frequency range [158].

4.2.1. Influencing Factors of Slamming Load

INCAT has conducted extensive research on wave-piercing catamarans (WPC), including full-scale trials and model basin tests, establishing a comprehensive slamming database. Based on pressure data and the concept of transmissibility, the prediction [159] and identification [160] of slamming loads have been improved. M.R. Davis [161] performed basin tests on a hydroelastic model of a wave-piercing catamaran. The model consisted of seven segments, including a smaller central bow section, connected by elastic joints (shown in

Figure 10). The tests revealed that slamming wave impacts predominantly occurred in the aft region of the short central bow. Slamming loads typically reached about 25% of the vessel's weight, though some peak loads reached 132% of the vessel's weight under 4 m significant wave height conditions. At the experimental scale, the slamming duration was generally 0.35 s. Shabani [162–164] modified the length and height above waterline of the central bow (CB) in this model and conducted head-sea tests in regular waves at high speeds. The results showed that slamming loads increased significantly with longer central bows, while the encounter frequency had a greater influence on the CB than its geometric variations. Raising the bridge clearance amplified hull motions but reduced peak slamming loads in moderate sea states. The model with higher CB exhibited the lowest vertical motion amplitudes. Overall, the short central bow (short CB) design demonstrated the most stable performance in mitigating slamming loads.

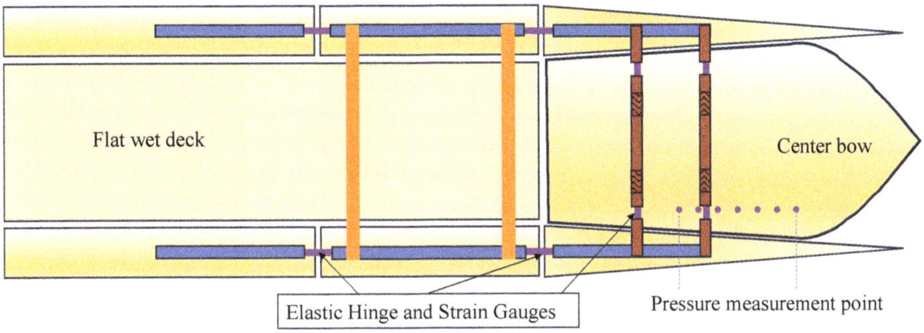

Figure 10. Structure arrangement of the segmented catamaran model [160].

SWATH vessels exhibit superior seakeeping performance due to their unique design [165]. However, they remain susceptible to slamming phenomena in high sea states, where structural damage may occur on the wet deck due to local slamming loads, global wave-induced bending moments, and coupled torsional effects. Ma [166] conducted a study on the slamming load characteristics of SWATH wet decks, focusing on the slamming pressure distribution on the cross-deck structure in regular waves. The evolution of flow-field during slamming events has been captured by high-speed imaging. The influence of wavelengths and wave heights on slamming intensity has also been analyzed. After wave impact on the cross-deck, pressure fluctuations exhibit periodic oscillations caused by the compression–expansion effect of the air cushion beneath the structure.

The flow field exhibits highly nonlinear flow characteristics during wave-piercing catamarans' slamming events due to mutual interference generated by the two demi-hulls and center bow. This complexity renders potential flow theory inadequate for accurate prediction. Consequently, CFD-based numerical simulations have become the primary tool for investigating such phenomena. McVicar [167] employed a FSI numerical simulation approach to predict the global bending response of a WPC under head sea conditions, induced by slamming loads. The hull motions were solved using RANS-based CFD, with validation against experimental data from a hydroelastic segmented model (HSM). A novel methodology was developed to capture the nonlinear time-varying characteristics of added mass during slamming events. Almallah [168,169] investigated a 98 m WPC using RANS-based CFD with the VOF method to capture free-surface effects. The vertical motions, longitudinal bending moments (LBM), and transverse bending moments (TBM) in moderate sea conditions were calculated. The pitch angular acceleration and wave height exhibited an approximately linear relationship with slamming loads and wave height was proportional to LBM slamming loads. Wave height was proportional to LBM slamming loads. The longitudinal distribution of the transverse separation forces (induced during slamming) was identified as the primary cause of transverse racking moments. Numerical simulations in oblique bow sea irregular waves revealed that pitch and roll accelerations exhibit a combined linear influence on the total LBM slamming loads. By establishing a conversion relationship between full-scale strain measurements and global loads using a FEM, the accuracy of slamming load prediction during the ship design phase was optimized [170].

The above research shows that the FSI technology is effective for predicting the motion and slamming loads of high-Froude-number ships, and it has high computational efficiency. This method has a good application prospect in the prediction of the vertical bending moment of actual ships and the structural design of ships during the ship design process.

4.2.2. Suppression Method

There are two measures to mitigate wet-deck slamming, generally, enhancing the structural strength of the connecting bridge and reducing the vessel's vertical motions. Hajmohammad [171] investigated the dynamic response of wave-piercing catamaran beam elements under fluid impact loads using the higher-order shear deformation beam theory (HSDBT) with uniform thickness and elastic linearity, while accounting for the porosity and structural damping properties of aluminum beams. By employing graphene platelets (GPLs) as reinforcement materials and solving the dynamic deflection of porous GPL-reinforced beams via the differential quadrature method (DQM) combined with the Newmark time-integration scheme, the study revealed that carbon nanotube (CNT)-reinforced panels could reduce structural dynamic deflection by up to 59%.

Building on prior motion response tests of an 86 m high-speed catamaran [172], G. Jacobi et al. conducted full-scale trials [173] and model experiments [174] to develop a ride control system (RCS) for the wave-piercing catamaran, comprising a bow T-foil and stern trim tab. The workflow of the RCS system is shown in Figure 11. By analyzing the relationship between structural stress rate and pitch motion, they achieved slamming prediction based on sea state and model motions, investigating the system's impact on slamming probability. The results showed that reducing vertical motions significantly alters slamming occurrence. Subsequent tests confirmed a 75% reduction in slamming pressure with this system [175]. Hasheminasab [176] proposed installing a flat spray rail on the central bow to simultaneously reduce slamming pressure and acceleration. Water-entry impact tests demonstrated that the J1-type profile with the spray rail achieved approximately 60% lower peak acceleration and 70% reduced slamming pressure. The study confirms that adding appendages to catamaran hull sections is an effective strategy for mitigating slamming impact loads.

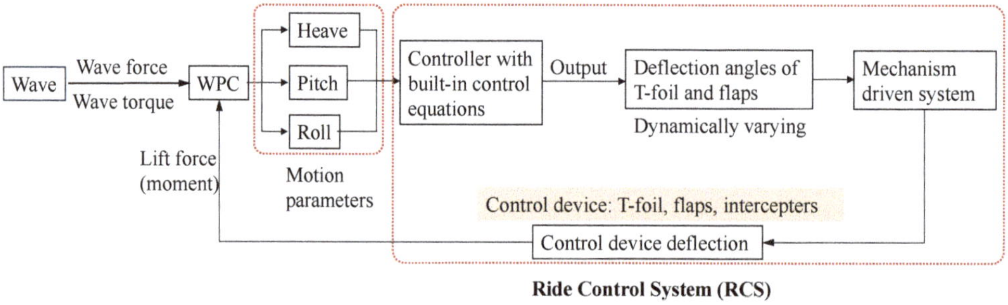

Figure 11. The workflow of the RCS for the WPC model [169].

4.3. Slamming of Trimaran

For numerical studies on wave loads of trimarans, in the 1990s, Grassman [177] conducted wave load measurements at various locations on the RV Triton trimaran under two sea states, including the bottom and upper sections of the main hull and side hulls, cross-deck bottom, and junctions between the main hull, side hulls, and cross-deck. The effects of wave encounter angles and significant wave heights on peak slamming forces were analyzed. The study demonstrated the feasibility of trimarans for military applications by validating their structural response to wave-induced loads. Miao [178] employed rigid and flexible 3D FSI models to calculate the motions and wave loads of the RV Triton trimaran, comparing the results with full-scale measurement data. Subsequently, Miao [179,180] applied 3D hydroelasticity theory incorporating transient slamming responses to compute the slamming locations and time-record curves of slamming forces and moments on both the port and starboard sides under the 45° bow-quartering waves.

In the preliminary design of trimaran structures, slamming loads and longitudinal vibrations are two critical factors [181]. To analyze wave-induced loads on trimarans in regular waves, Tang [182,183] proposed a novel nonlinear 3D time-domain Rankine–Green hybrid method, based on the computational principles of ship-wave interaction using Rankine sources and free-surface Green's functions, for simulating trimaran motions and wave loads under head sea conditions. By incorporating nonlinear factors such as instantaneous wet, steady wave-making, slamming, and so on, the method successfully coupled nonlinear components in time-domain load simulations. The calculations revealed that as ship speed and wave height increase, the slamming phenomena in the side hull regions and cross-deck areas exhibit significant nonlinear growth characteristics. Khoob [184,185] utilized frequency-domain Green's function potential flow theory with forward-speed correction, to analyze wave-induced loads on trimarans. A global FEM was then constructed to predict motion and load responses. The roll-induced torque and shear forces are more significant in head sea conditions, while transverse bending moments exhibit higher response amplitudes in oblique seas. Additionally, outward-sloping side hulls demonstrated better suppression of wave-induced loads compared to conventional designs.

Ghadimi [186] employed the commercial CFD software Flow-3D V11.2, utilizing the standard Bretschneider spectrum to simulate the seakeeping performance of a wave-piercing trimaran in irregular waves under Sea State 5. The study investigated motion responses under different speeds and wave encounter angles. Xie [187] conducted seakeeping analysis to obtain asymmetric motion responses (heave, sway, and roll) of the vessel, then applied CFD methods to calculate slamming loads on two typical flare bow sections. The study fully accounted for complex flow phenomena such as flow separation and cavitation effects. The results revealed that oblique wave slamming loads exhibit stronger asymmetry and higher impact intensity compared to head sea conditions. Liao [188,189] equivalently distributed the ship's mass to the hull and analyzed the structural response characteristics of a trimaran under unidirectional regular waves and bidirectional crossing seas by integrating a fluid solver (Star-CCM+) with a structural solver (Abaqus). This approach significantly improved computational efficiency. Chen [190] employed CFD techniques to construct a numerical wave basin generating bidirectional crossing waves as a typical scenario for 3D short-crested waves. By incorporating a time-domain delay function and a proportional-integral-derivative (PID) autopilot model into the numerical simulations, they proposed a 3D nonlinear time-domain hydroelasticity method [191]. This method accounts for nonlinear hydrostatic restoring forces due to instantaneous wetted surfaces and slamming loads, analyzing differences in ship motions under oblique waves and long-crested regular waves. Liu [192] combined model tests with CFD techniques to investigate the slamming characteristics of the cross-deck structure on a trimaran in head seas. The simulations revealed that when the cross-deck emerged from a wave crest, a significant negative pressure was generated in its forward region. Additionally, the water entry of the main hull induced higher slamming pressures on the cross-deck near the side hulls compared to the corresponding regions of the main hull.

In summary, potential flow theory is predominantly employed for slamming pressure prediction on monohull high-speed vessels. For multi-hull ships, however, the highly nonlinear characteristics of free-surface flows necessitate grid-based CFD methods (e.g., FVM, FDM) as the primary numerical approach. These methods are often coupled with structural analysis techniques (e.g., finite element analysis) to further assess localized structural deformation. The slamming near wave troughs tends to induce higher impact loads in irregular waves. In oblique seas, because of the coupling between roll and heave motions, the side hulls of trimaran generate slamming more obviously, with wave splashes generating localized high pressure on the cross-deck structure. Even under relatively moderate sea states, multi-hulls experience significant slamming loads compared to head seas. The appendages (e.g., T-foils, flaps) can simultaneously suppress vertical motions and reduce slamming pressure through synergistic effects.

5. Conclusions

Regarding the slamming problem of high-speed vessels, including studies on segmented water entry and slamming loads during navigation, this paper comprehensively analyzes the progress in theoretical and experimental research. Although current studies have yielded significant findings and summarized the relationships among factors influencing slamming intensity—particularly with the introduction of PIV technology, which has progressively improved experimental measurement techniques—the following aspects warrant further exploration and in-depth investigation.

(1) Coupling of Numerical Methods. By combining CFD (computational fluid dynamics) with the BEM (boundary element method), the far flow field is modeled using potential flow theory while the near field employs viscous models. This approach balances the high accuracy of CFD in handling nonlinear problems with the computational efficiency of the BEM, reducing dimensionality and more precisely capturing hydroelastic effects. Similarly, hybrid methods coupling particle-based techniques (e.g., SPH, MPS) with grid-based methods can be adopted, applying particle methods at slamming interfaces to accommodate large deformations while retaining grid methods elsewhere. For free-surface tracking, integrating level set with VOF (volume of fluid) enhances air–liquid interface resolution.

(2) Application of Machine Learning. High-fidelity CFD simulations (e.g., DNS, LES) require an extremely fine mesh resolution and small time steps, making them impractical for rapid multi-condition evaluations in engineering applications. Neural networks can be employed to predict numerical solutions (e.g., pressure peaks, time-history curves) and replace traditional iterative processes, thereby improving computational efficiency for large-scale simulations. However, current slamming tests often rely on repeated experiments to mitigate data fluctuations, resulting in a scarcity of high-quality datasets. Additionally, biases inherent in numerical models may propagate into machine learning models, necessitating further calibration.

(3) Slamming Flow Field Measurement Techniques. The measurement of flow fields and pressure distribution on ship hull sections is often achieved using particle image velocimetry (PIV). This method enables the real-time measurement of fluid particle velocities and reconstructs the pressure field from the velocity field via pressure reconstruction techniques. It overcomes the limitations of traditional pressure sensors, which are prone to dynamic impact interference, particularly offering advantages in

high-speed water-entry scenarios. However, existing pressure reconstruction methods (e.g., multi-path integration schemes) typically assume incompressible flow and neglect viscous effects. In reality, water-entry processes may involve complex phenomena such as cavitation and turbulence, leading to error accumulation. Additionally, PIV is predominantly applied to 2D models, making it challenging to fully capture the flow field characteristics in 3D scenarios (e.g., ship bow water entry).

Author Contributions: Conceptualization, Y.S.; methodology, Y.S. D.Z.; formal analysis, Y.S.; data curation, Z.W. Y.Y.; writing—original draft preparation, Y.S.; writing—review and editing, D.Z.; supervision, D.Z.; funding acquisition, Y.S. and D.Z. All authors have read and agreed to the published version of the manuscript.

Funding: This research was funded by Program for Scientific Research Start-up Funds of Guangdong Ocean University, grant number 060302072102 and Zhanjiang Marine Youth Talent Project-Comparative Study, grant number 2021E5007.

Data Availability Statement: No new data were created or analyzed in this study.

Conflicts of Interest: The authors declare no conflicts of interest.

References

1. Sairam Prasad, B.; Ravi Kiran Sastry, G.; Das, H.N. A comprehensive review study on multiphase analysis of water entry bodies. *Ocean Eng.* **2024**, *292*, 116579. [CrossRef]
2. Faltinsen, O.M. *Sea Loads on Ships and Offshore Structures*; Cambridge University Press: Cambridge, UK, 1990.
3. Truelock, D.; Lavroff, J.; Pearson, D.; Czaban, Z.; Luo, H.B.; Wang, F.H.; Catipovic, I.; Begovic, E.; Takaoka, Y.; Loureiro, C.; et al. 5: Special Vessels. *Soc. Nav. Archit. Mar. Eng.* **2022**, *2*, 313–377.
4. Aertssen, G. Laboring of ships in rough seas with special emphasis on the fast ships. In Proceedings of the SNAME Diamond Jubilee International Meeting, New York, NY, USA, 18–21 June 1968.
5. Andrews, D.J.; Zhang, J.W. Trimaran Ships: The Configuration for the Frigate of the Future. *Nav. Eng. J.* **2010**, *107*, 77–94. [CrossRef]
6. Yang, Q.; Jones, V.; McCue, L. Free-surface flow interactions with deformable structures using an SPH-FEM model. *Ocean Eng.* **2012**, *55*, 136–147. [CrossRef]
7. Akbarzadeh, P.; Krieger, M.; Hofer, D.; Thumfart, M.; Gittler, P. Parallel water entry of hydrophobic-hydrophilic sphere pairings: Particle image velocimetry and High-Speed camera analysis. *J. Fluids Struct.* **2025**, *133*, 104272. [CrossRef]
8. Von Karman, T. *The Impact on Seaplane Floats During Landin*; Techinical Note No. 321; National Advisory Committee for Aeronatics: Washington, DC, USA, 1932; pp. 309–313.
9. Wagner, H. Uber stoss-und gleitvergange an der oberflache von flussigkeiten. *ZAMM-J. Appl. Math. Mech./Z. Angew. Math. Mech.* **1932**, *12*, 193–235. [CrossRef]
10. Dobrovol'Skaya, Z.N. On some problems of similarity flow of fluids with a free surface. *J. Fluid Mech.* **1969**, *36*, 805–829. [CrossRef]
11. Hughes, O.F. Solution of wedge entry problem by numerical conformal mapping. *J. Fluid Mech.* **1972**, *56*, 173–192. [CrossRef]
12. Logvinovich, G.V. *Hydrodynamics Offlows with Free Boundaries*; Naukova Dumka: Kiev, Ukraine, 1969.
13. Korobkin, A.A. Analytical models of water impact. *Eur. J. Appl. Math.* **2004**, *15*, 821–838. [CrossRef]
14. Duan, W.Y.; Xu, G.D.; Wu, G.X. Similarity solution of oblique impact of wedge-shaped water column on wedged coastalstructures. *Coast. Eng.* **2009**, *56*, 400–407. [CrossRef]
15. Liu, L.; Zeng, K.; Ding, J.; Wang, Q.B.; Bu, S.X.; Cheng, C. A two-dimensional high-order explicit analytical solution model for symmetric wedges entering water. *J. Ocean Eng. Sci.* **2025**, *10*, 109–122. [CrossRef]
16. Tassin, A.; Korobkin, A.A.; Cooker, M.J. On analytical models of vertical water entry of a symmetric body with separation and cavity initiation. *Appl. Ocean Res.* **2014**, *48*, 33–41. [CrossRef]
17. Fairlie-Clarke, A.C.; Tveitnes, T. Momentum and gravity effects during the constant velocity water entry of wedge-shaped sections. *Ocean Eng.* **2008**, *35*, 706–716. [CrossRef]
18. Wen, X.L.; Qu, Q.L.; Liu, P.Q.; Ding, S.L. Theoretical study on slamming and transition stages of normal impacts of symmetrical bodies on calm and wavy water surfaces. *Appl. Ocean Res.* **2022**, *119*, 102986. [CrossRef]
19. Khabakhpasheva, T.I.; Korobkin, A.A.; Malenica, S. Water entry of an elastic conical shell. *J. Fluid Mech.* **2024**, *980*, A34. [CrossRef]
20. Feng, S.; Zhang, G.Y.; Jiang, S.Q.; Jiang, S.C.; el Moctar, O.; Ma, Y.X. Investigation of fluid added mass matrix during hydroelastic slamming of wedges. *Phys. Fluids* **2024**, *36*, 012133. [CrossRef]

21. Liu, W.T.; Zhang, A.M.; Miao, X.H.; Ming, F.R.; Liu, Y.L. Investigation of hydrodynamics of water impact and tail slamming of high-speed water entry with a novel immersed boundary method. *J. Fluid Mech.* **2023**, *958*, A42. [CrossRef]
22. Zhao, R.; Faltinsen, O.M.; Aarsnes, J. Water entry of arbitrary two-dimensional sections with and without separation. In Proceedings of the 21st Symposium on Naval Hydrodynamics 1996, Trondheim, Norway, 24–28 June 1996; pp. 118–133.
23. Zhao, R.; Faltinsen, O.M. Water entry of two-dimensional bodies. *J. Fluid Mech.* **1993**, *246*, 593–612. [CrossRef]
24. Battistin, D.; Iafrati, A. Hydrodynamic loads during water entry of two-dimensional and axisymmetric bodies. *J. Fluids Struct.* **2003**, *17*, 643–664. [CrossRef]
25. Wu, G.X.; Sun, H.; He, Y.S. Numerical simulation and experimental study of water entry of a wedge in free fall motion. *J. Fluids Struct.* **2004**, *19*, 277–289. [CrossRef]
26. Cheng, Y.; Yuan, D.C.; Ji, C.Y.; Li, G. Solitary wave slamming induced by an asymmetric wedge through three degrees of freedom free motions. *Phys. Fluids* **2019**, *31*, 097103. [CrossRef]
27. Liu, B.W.; Sun, S.L.; Ren, H.L. Hydrodynamic performance of a cone falling into waves in 3DOFs free fall motion. *Ocean Eng.* **2021**, *242*, 110132. [CrossRef]
28. Feng, S.; Zhang, G.Y.; el Moctar, O.; Sun, Z.; Zhang, Z.F. A semi-analytical method to simulate hydroelastic slamming of 2D structural sections by coupling Wagner theory with the finite element method. *Ocean Eng.* **2021**, *240*, 109998. [CrossRef]
29. Feng, S.; Zhang, G.Y.; Wan, D.C.; Jiang, S.C.; Sun, Z.; Zong, Z. On the treatment of hydroelastic slamming by coupling boundary element method and modal superposition method. *Appl. Ocean Res.* **2021**, *112*, 102595. [CrossRef]
30. Feng, S.; Zhang, G.Y.; Ma, Y.X.; Sun, L.; Zhou, B. Study on the hydroelastic slamming of elastic wedges vertically entering shallow water. *Ocean Eng.* **2024**, *311*, 118848. [CrossRef]
31. Wen, X.L.; Del Buono, A.; Liu, P.Q.; Qu, Q.L.; Iafrati, A. Acceleration effects in slamming and transition stages for the water entry of curved wedges with a varying speed. *Appl. Ocean Res.* **2022**, *128*, 103294. [CrossRef]
32. Reddy, D.N.; Scanlon, T.J.; Kuo, C. Prediction of slam loads on a wedge section using computational fluid dynamics (CFD) techniques. In Proceedings of the Twenty-Fourth Symposium on Naval Hydrodynamics 2002, Fukuoka, Japan, 8–13 July 2002.
33. Swidan, A.; Amin, W.; Ranmuthugala, D.; Thomas, G.; Penesis, I. Numerical Prediction of Symmetric Water Impact Loads on Wedge Shaped Hull Form Using CFD. *World J. Mech.* **2013**, *3*, 311–318. [CrossRef]
34. Kleefsman, K.M.T.; Fekken, G.; Veldman, A.E.P.; Iwanowski, B.; Buchner, B. A Volume-of-Fluid based simulation method for wave impact problems. *J. Comput. Phys.* **2005**, *206*, 363–393. [CrossRef]
35. Zhang, Y.F.; Ma, S.; Shao, W.B.; Zhang, Y.X. Numerical investigation on the water entry of curved wedge-shaped sections into waves. *Ocean Eng.* **2023**, *275*, 114155. [CrossRef]
36. Vepa, K.S.; Seetharamaiah, N. Performance study of different numerical methods for modeling the vertical water entry of a wedge. *Int. J. Interact. Des. Manuf.* **2024**, *19*, 2431–2447. [CrossRef]
37. Lucy, L.B. A numerical approach to the testing of the fission hypothesis. *Astrophys. J.* **1977**, *8*, 1013–1024. [CrossRef]
38. Gingold, R.A.; Monaghan, J.J. Smoothed particle hydrodynamics: Theory and application to non-spherical stars. *Mon. Not. R. Astron. Soc.* **1977**, *181*, 375–389. [CrossRef]
39. Zhang, A.; Sun, P.; Ming, F. An SPH modeling of bubble rising and coalescing in three dimensions. *Comput. Methods Appl. Mech. Eng.* **2015**, *294*, 189–209. [CrossRef]
40. Monaghan, J.J. Simulating free surface flows with SPH. *J. Comput. Phys.* **1994**, *110*, 399–406. [CrossRef]
41. Ferrari, A. SPH simulation of free surface flow over a sharp-crested weir. *Adv. Water Resour.* **2010**, *33*, 270–276. [CrossRef]
42. Gomez-Gesteira, M.; Cerqueiro, D.; Crespo, C.; Dalrymple, R.A. Green water overtopping analyzed with a SPH model. *Ocean Eng.* **2005**, *32*, 223–238. [CrossRef]
43. Gomez-Gesteira, M. Using a 3D SPH method for wave impact on a tall structure. *J. Waterw. Port Coast. Ocean Eng.* **2004**, *130*, 630–669. [CrossRef]
44. Gomez-Gesteira, M.; Rogers, B.D.; Dalrymple, R.A.; Crespo, A.J.C. State-of-the-art of classical SPH for free-surface flows. *J. Hydraul. Res.* **2010**, *48* (Suppl. 1), 6–27. [CrossRef]
45. Monaghan, J.J. SPH without a tensile instability. *J. Comput. Phys.* **2000**, *159*, 290–311. [CrossRef]
46. Rogers, B.D.; Dalrymple, R.A.; Stansby, P.K. SPH modeling of floating bodies in the surf zone. *Coast. Eng.* **2009**, *5*, 204–215.
47. Cui, J.; Gu, C.J.; Chen, X.; Li, M.Y.; Masvaya, B. Numerical study of wedge entry in still water and waves using smoothed particle hydrodynamics methods. *Ocean Eng.* **2023**, *280*, 114776. [CrossRef]
48. Zhang, Z.L.; Shu, C.; Khalid, M.S.U.; Yuan, Z.Y.; Liu, W. Investigations on the hydroelastic slamming of deformable wedges by using the smoothed particle element method. *J. Fluids Struct.* **2022**, *114*, 103732. [CrossRef]
49. Chen, D.; Huang, W.; Huang, D.; Liang, C. An adaptive multi-resolution SPH approach for three-dimensional free-surface flow with fluid impacting. *Eng. Anal. Bound. Elem.* **2023**, *155*, 642–651. [CrossRef]
50. Rezavand, M.; Zhang, C.; Hu, X. A weakly compressible SPH method for violent multi-phase flows with high density ratio. *J. Comput. Phys.* **2020**, *402*, 109092. [CrossRef]

51. Wang, P.P.; Zhang, A.M.; Meng, Z.F.; Ming, F.R.; Fang, X.L. A new type of WENO scheme in SPH for compressible flows with discontinuities. *Comput. Methods Appl. Mech. Eng.* **2021**, *381*, 113770. [CrossRef]
52. Yang, Q.Z.; Xu, F.; Yang, Y.; Wang, J.Y. Two-phase SPH model based on an improved Riemann solver for water entry problems. *Ocean Eng.* **2020**, *199*, 107039. [CrossRef]
53. Meng, Z.F.; Ming, F.R.; Wang, P.P.; Zhang, A.M. Numerical simulation of water entry problems considering air effect using a multiphase Riemann-SPH model. *Adv. Aerodyn.* **2021**, *3*, 244–259.
54. Zhang, G.Y.; Yang, X.; Zhang, Z.F.; Hui, D.; Sun, Z.; Liang, G.Q.; Li, P. Numerical investigation of free surface flow impact using an accelerated three-dimensional smoothed particle hydrodynamics. *Ocean Eng.* **2024**, *314 Pt 2*, 119772. [CrossRef]
55. Koshizuka, S.; Oka, Y. Moving-particle semi-implicit method for fragmentation of incompressible fluid. *Nucl. Sci. Eng.* **1996**, *123*, 421–433. [CrossRef]
56. Khayyer, A.; Gotoh, H. Modified moving particle semi-implicit methods for the prediction of 2D wave impact pressure. *Coast. Eng.* **2009**, *56*, 419–440. [CrossRef]
57. Shibata, K.; Koshizuka, S.; Sakai, M.; Tanizawa, K. Lagrangian simulations of ship-wave interactions in rough seas. *Ocean Eng.* **2012**, *42*, 13–25. [CrossRef]
58. Zhang, Y.L.; Wan, D.C. MPS-FEM coupled method for sloshing flows in an elastic tank. *Ocean Eng.* **2018**, *152*, 416–427. [CrossRef]
59. Sun, Z.; Zhang, G.Y.; Zong, Z.; Djidjeli, K.; Xing, J.T. Numerical analysis of violent hydroelastic problems based on a mixed MPS-mode superposition method. *Ocean Eng.* **2019**, *179*, 285–297. [CrossRef]
60. Matsunaga, T.; Koshizuka, S. Improvement of the time marching method in a particle method. *Trans. JSME* **2021**, *87*, 20-00437. (In Japanese) [CrossRef]
61. Khayyer, A.; Gotoh, H. Enhancement of stability and accuracy of the moving particle semi-implicit method. *J. Comput. Phys.* **2011**, *230*, 3093–3118. [CrossRef]
62. Iida, T.; Yokoyama, Y. Investigation of Numerical Conditions of Moving Particle Semi-implicit for Two-Dimensional Wedge Slamming. *J. Mar. Sci. Appl.* **2021**, *20*, 585–594. [CrossRef]
63. Zha, R.S.; Peng, H.; Qiu, W. An improved higher-order moving particle semi-implicit method for simulations of two-dimensional of two-dimensional hydroelastic slamming. *Phys. Fluids* **2021**, *33*, 037104. [CrossRef]
64. Zha, R.S.; Peng, H.; Qiu, W. Solving 2-D Slamming Problems by an Improved Higher-Order Moving Particle Semi-Implicit Method. *J. Ship Res.* **2021**, *65*, 194–222. [CrossRef]
65. Chen, S.; Doolen, G.D. Lattice Boltzmann Method for Fluid Flows. *Annu. Rev. Fluid Mech.* **2003**, *30*, 329–364. [CrossRef]
66. De Rosis, A.; Falcucci, G.; Porfiri, M.; Ubertini, F.; Ubertini, S. Hydroelastic analysis of hull slamming coupling lattice Boltzmann and finite element methods. *Comput. Struct.* **2014**, *138*, 24–35. [CrossRef]
67. Islam, A.; Taravella, B. A numerical investigation of cambered wedge impact using the Lattice Boltzmann method. *J. Braz. Soc. Mech. Sci. Eng.* **2022**, *44*, 258. [CrossRef]
68. Zan, L.R.; Sun, H.B.; Zou, J.; Cang, J.Y.; Wan, L. The effect of flexible step on the resistance of planing boat and motion mitigation. *Ocean Eng.* **2024**, *303*, 117265. [CrossRef]
69. Chuang, S.L. Experiments on slamming of wedge-shaped bodies. *J. Ship Res.* **1967**, *11*, 190–198. [CrossRef]
70. Chuang, S.L. Theoretical Investigations on Slamming of Cone-shaped Bodies. *J. Ship Res.* **1969**, *13*, 276–283. [CrossRef]
71. Chu, W.; Abramson, H.N. Hydrodynamic theories of ship slamming-review and extension. *J. Ship Res.* **1961**, *5*, 9–21. [CrossRef]
72. Barjasteh, M.; Zeraatgar, H.; Javaherian, M.J. An experimental study on water entry of asymmetric wedges. *Appl. Ocean Res.* **2016**, *58*, 292–304. [CrossRef]
73. Nikfarjam, M.; Yaakob, O.B.; Seif, M.S.; Koto, J. Investigation of Wedge Water-Entry Under Symmetric Impact Loads by Experimental Tests. *Lat. Am. J. Solids Struct.* **2017**, *14*, 861–873. [CrossRef]
74. Seo, B.; Truong, D.D.; Cho, S.; Kim, D.; Park, S.; Shin, H. A study on accumulated damage of steel wedges with dead-rise 10° due to slamming loads. *Int. J. Nav. Archit. Ocean Eng.* **2018**, *10*, 520–528. [CrossRef]
75. Seo, B.; Shin, H. Experimental Study of Slamming Effects on Wedge and Cylindrical Surfaces. *Appl. Sci.* **2020**, *10*, 1503. [CrossRef]
76. Korkmaz, F.C.; Güzel, B. Water entry of cylinders and spheres under hydrophobic effects; Case for advancing deadrise angles. *Ocean Eng* **2017**, *129*, 240–252. [CrossRef]
77. Ren, Z.S.; Javaherian, M.J.; Gilbert, C.M. Kinematic and inertial hydroelastic effects caused by vertical slamming of a flexible V-shaped wedge. *J. Fluids Struct.* **2021**, *103*, 103257. [CrossRef]
78. Ren, Z.S.; Javaherian, M.J.; Gilbert, C.M. Vertical Wedge Drop Experiments as a Model for Slamming. *J. Ship Res.* **2022**, *66*, 297–314. [CrossRef]
79. Hosseinzadeh, S.; Tabri, K.; Sahk, T.; Tear, R. Experimental dataset on aluminium wedge slamming: Measurements of acceleration, pressure, strain, and video data. *Data Brief* **2024**, *56*, 110818. [CrossRef]
80. Hosseinzadeh, S.; Tabri, K.; Topa, A.; Hirdaris, S. Slamming loads and responses on a non-prismatic stiffened aluminium wedge: Part II. Numerical simulations. *Ocean Eng.* **2023**, *279*, 114309. [CrossRef]

81. Hosseinzadeh, S.; Tabri, K.; Topa, A.; Hirdaris, S. Slamming loads and responses on a non-prismatic stiffened aluminium wedge: Part I. Experimental study. *Ocean Eng.* **2023**, *279*, 114510. [CrossRef]
82. Han, B.B.; Li, H.; Zhang, B.Y.; Zou, J.; Zhao, W.Z. Experimental investigation of slamming characteristics of stiffened elastic wedges with different deadrise angles: Part I—Slamming pressure. *Ocean Eng.* **2024**, *303*, 117728. [CrossRef]
83. Han, B.B.; Li, H.; Zhang, B.Y.; Zou, J.; Zhao, W.Z. Experimental investigation of slamming characteristics of stiffened elastic wedges with different deadrise angles: Part II—Structural response. *Ocean Eng.* **2024**, *303*, 118288. [CrossRef]
84. Han, B.B.; Peng, Y.H.; Li, H.; Liu, S.N.; Sun, S.L.; Shan, Y.H.; Sun, Z.Y. Numerical investigations of a 2D bow wedge asymmetric free-falling into still water. *Ocean Eng.* **2022**, *266 Pt 3*, 112905. [CrossRef]
85. Duan, L.L.; Zhu, L.; Chen, M.S.; Pedersen, P.T. Experimental study on the propagation characteristics of the slamming pressures. *Ocean Eng.* **2020**, *217*, 107868. [CrossRef]
86. Güzel, B.; Korkmaz, F.C. Experimental investigation of water entry of bodies with constant deadrise angles under hydrophobic effects. *Exp. Fluids* **2021**, *62*, 107. [CrossRef]
87. Zhang, J.; You, Y.; Yao, Z.; Huo, F.L. Sensitivity Analysis of the Effect of Speed and Inclination Angle on Water-Entry Slamming Pressure of the Bow. *China Ocean Eng.* **2020**, *34*, 432–440. [CrossRef]
88. Meziane, B.; Alaoui, A.E.; Neme, A.; Leble, B.; Bellanger, D. Experimental investigation of the influence of the panel stiffness on the behaviour of a wedge under slamming. *J. Fluids Struct.* **2022**, *114*, 103702. [CrossRef]
89. Meziane, B.; Alaoui, A.E.; Neme, A.; Leble, B.; Bellanger, D. Water impact of V-shaped wedge at constant velocity: Influence of deadrise angle and stiffness panels. *Appl. Ocean Res.* **2025**, *158*, 104494. [CrossRef]
90. Jain, U.; Novakovic, V.; Bogaert, H.; van der Meer, D. On wedge-slamming pressures. *J. Fluid Mech.* **2022**, *934*, A27. [CrossRef]
91. Korobkin, A.; Khabakhpasheva, T.; Malenica, S.; Kim, Y. A comparison study of water impact and water exit models. *Int. J. Nav. Archit. Ocean Eng.* **2014**, *6*, 1182–1196. [CrossRef]
92. Nila, A.; Vanlanduit, S.; Vepa, S.; Van Paepegem, W. A PIV-based method for estimating slamming loads during water entry of rigid bodies. *Meas. Sci. Technol.* **2013**, *24*, 436–475. [CrossRef]
93. Shams, A.; Jalalisendi, M.; Porfiri, M. Experiments on the water entry of asymmetric wedges using particle image velocimetry. *Phys. Fluids* **2015**, *27*, 027103. [CrossRef]
94. Panciroli, R.; Porfiri, M. Evaluation of the pressure field on a rigid body entering a quiescent fluid through particle image velocimetry. *Exp. Fluids* **2013**, *54*, 1630. [CrossRef]
95. Jalalisendi, M.; Shams, A.; Panciroli, R.; Porfiri, M. Experimental reconstruction of three-dimensional hydrodynamic loading in water entry problems through particle image velocimetry. *Exp. Fluids* **2015**, *56*, 41. [CrossRef]
96. Panciroli, R.; Porfiri, M. Analysis of hydroelastic slamming through particle image velocimetry. *J. Sound Vib.* **2015**, *347*, 63–78. [CrossRef]
97. Panciroli, R.; Porfiri, M. A Particle Image Velocimetry Study of Hydroelastic Slamming. *Procedia Eng.* **2014**, *88*, 180–185. [CrossRef]
98. Zhang, P.; Porfiri, M. A combined digital image correlation/particle image velocimetry study of water-backed impact. *Compos. Struct.* **2019**, *224*, 111010. [CrossRef]
99. Guo, C.Y.; Fan, Y.W.; Han, Y.; Xu, P.; Kuai, Y.F. Deep-learning-based liquid extraction algorithm for particle image velocimetry in two-phase flow experiments of an object entering water. *Appl. Ocean Res.* **2021**, *108*, 102526.
100. Dessi, D.; Sanchez-Alayo, D.; Shabani, B.; Ali-Lavroff, J. Bow slamming detection and classification by Machine Learning approach. *Ocean Eng.* **2023**, *287*, 115646. [CrossRef]
101. Izadi, M.; Ghadimi, P.; Fadavi, M.; Tavakoli, S. Hydroelastic analysis of water impact of flexible asymmetric wedge with an oblique speed. *Meccanica* **2018**, *53*, 2585–2617. [CrossRef]
102. Jiao, J.L.; Chen, Z.W.; Xu, W.H.; Bu, S.X.; Zhang, P.J. Asymmetric water entry of a wedged grillage structure investigated by CFD-FEM co-simulation. *Ocean Eng.* **2024**, *302*, 117612. [CrossRef]
103. Xiao, J.W.; Liu, C.; Han, B.B.; Wan, D.C.; Wang, J.H. A two-way coupled fluid-structure interaction method for predicting the slamming loads and structural responses on a stiffened wedge. *Phys. Fluids* **2015**, *36*, 077123. [CrossRef]
104. Sun, S.L.; Sun, J.Y.; Wang, S.; Li, Y.H. Fluid–structure interaction analysis of curved wedges entering into water. *Phys. Fluids* **2024**, *36*, 102121. [CrossRef]
105. Chen, Z.W.; Jiao, J.L.; Wang, S.; Soares, C.G. CFD-FEM simulation of water entry of a wedged grillage structure into Stokes waves. *Ocean Eng.* **2023**, *275*, 114159. [CrossRef]
106. Sun, Z.; Sui, X.P.; Korobkin, A.; Zou, L.; Zong, Z. Slamming force decomposition with gravity effect. *J. Fluids Struct.* **2022**, *114*, 103694. [CrossRef]
107. Wen, X.L.; Liu, P.D.; Del Buono, A.; Qu, Q.L.; Iafrati, A. Formulations of hydrodynamic force in the transition stage of the water entry of linear wedges with constant and varying speeds. *J. Fluids Struct.* **2022**, *115*, 103759. [CrossRef]
108. Moore, M.R. Introducing pre-impact air-cushioning effects into the Wagner model of impact theory. *J. Eng. Math.* **2021**, *129*, 6. [CrossRef]

109. Hasheminasab, H.; Zeraatgar, H.; Moradi, H.; Sakaki, A. Experimental study on water entry of twin wedge. *Proc. Inst. Mech. Eng. Part M–J. Eng. Marit. Environ.* **2020**, *234*, 388–398. [CrossRef]
110. Tang, S.Q.; Zhang, Y.; Sun, S.L.; Ren, H.L.; Zhang, H.B.; He, J.H. Experimental investigation on the air-cushion effect during free fall of a trimaran section using an air escape control method. *Ocean Eng.* **2022**, *254*, 111417. [CrossRef]
111. Ma, S.; Duan, W.Y.; Cao, Z.H.; Liu, J.Y.; Zhang, M.H.; Li, X.H.; Liu, D.D. Experimental study on the drop test on wet deck slamming for a SWATH segment model. *Ocean Eng.* **2022**, *285 Pt 2*, 115377. [CrossRef]
112. Masoomi, M.; Rezanejad, K.; Mosavi, A.H. Numerical study of a novel ventilation system added to the structure of a catamaran for different slamming conditions using OpenFOAM. *Int. J. Nav. Archit. Ocean Eng.* **2023**, *15*, 100512. [CrossRef]
113. Swidan, A.; Thomas, G.; Ranmuthugala, D.; Amin, W.; Penesis, I.; Allen, T.; Battley, M. Experimental drop test investigation into wetdeck slamming loads on a generic catamaran hullform. *Ocean Eng.* **2016**, *117*, 143–153. [CrossRef]
114. Jiang, Y.C.; Bai, J.Y.; Dong, Y.; Sun, T.Z.; Sun, Z.; Liu, S.J. Investigations of air cushion effect on the slamming load acting on trimaran cross deck during water entry. *Ocean Eng.* **2022**, *251*, 111161. [CrossRef]
115. Yu, P.Y.; Qu, S.; Wang, Q.; Xie, H. Numerical investigation on the slamming loads of a truncated trimaran hull entering regular waves. *Appl. Ocean Res.* **2024**, *153*, 104253. [CrossRef]
116. Wu, J.X.; Sun, Z.; Jiang, Y.C.; Zhang, G.Y.; Sun, T.Z. Experimental and numerical study of slamming problem for a trimaran hull. *Ships Offshore Struct.* **2021**, *16*, 46–53. [CrossRef]
117. Liu, J.Y.; Duan, W.Y.; Liao, K.P.; Ma, S.; Shao, W.B.; Zhang, Y.F. Numerical analysis of wet-deck slamming characteristics for trimaran section with different main-hull profiles. *Ocean Eng.* **2024**, *301*, 117542. [CrossRef]
118. Sun, Z.; Deng, Y.Z.; Zou, L.; Jiang, Y.C. Investigation of trimaran slamming under different conditions. *Appl. Ocean Res.* **2020**, *104*, 102316. [CrossRef]
119. Sun, Z.; Sui, X.P.; Deng, Y.Z.; Zou, L.; Korobkin, A.; Xu, L.X.; Jiang, Y.C. Characteristics of Slamming Pressure and Force for Trimaran Hull. *J. Mar. Sci. Eng.* **2021**, *9*, 564. [CrossRef]
120. Duan, W.Y.; Liu, J.Y.; Liao, K.P.; Ma, S. Experimental study of slamming pressure for a trimaran section with different drop heights and heel angles. *Ocean Eng.* **2022**, *263*, 112400. [CrossRef]
121. Pan, J.; Zhang, W.Z.; Sun, Z.M.; Qu, X.; Xu, M.C. Experimental study on the dynamical response of elastic trimaran model under slamming load. *J. Mar. Sci. Technol.* **2024**, *29*, 20–35. [CrossRef]
122. Li, J.Q.; Sun, S.L. Impact Force Mechanism of a Trimaran Model and Its Experimental Validation. *Appl. Sci.* **2023**, *13*, 10382. [CrossRef]
123. Wang, Q.; Yu, P.Y.; Fan, G.J.; Li, G.Z. Experimental drop test investigation into cross deck slamming loads on a trimaran. *Ocean Eng.* **2021**, *240*, 109999. [CrossRef]
124. Li, H.; Deng, B.L.; Zou, J.; Dong, C.R.; Liu, C.L.; Liu, P.L. Experimental free-drop test investigation into wet-deck slamming loads on a generic trimaran section considering the influence of main hull profile. *Ocean Eng.* **2021**, *242*, 110114. [CrossRef]
125. Dong, L.Y.; Wei, Z.Y.; Zhou, H.Y.; Yao, B.H.; Lian, L. Numerical Study on the Water Entry of a Freely Falling Unmanned Aerial-Underwater Vehicle. *J. Mar. Sci. Eng.* **2023**, *11*, 552. [CrossRef]
126. Wang, Y.W.; Wang, M.Z.; Zheng, C.; Yang, L.; Zhang, H.T.; Li, Y.R.; Ruan, X.Y. Experimental investigation of slamming impact on fiber-reinforced composite sandwich bow structure. *Ocean Eng.* **2024**, *319*, 120162. [CrossRef]
127. Shan, Y.H.; Li, H.; Han, B.B.; Sun, Z.Y.; Lu, L.; Liu, R.X.; Liu, Y.; Guo, H. E Experimental investigation of slamming pressure on 3D bow flare. *Ocean Eng.* **2024**, *312 Pt 1*, 118898. [CrossRef]
128. Ping, Y.N.; Wang, J.Z.; Xie, H.; Liu, F.; Liu, X.Y. Experimental and CFD analysis: Effects of bottom appendages on the slamming characteristics of rigid hull structures during water entry. *Ocean Eng.* **2025**, *319*, 120195. [CrossRef]
129. Xie, H.; Dai, X.L.; Liu, F.; Liu, X.Y. Experimental study on the slamming pressure distribution of a 3D stern model entering water with pitch angles. *Ocean Eng.* **2024**, *291*, 116404. [CrossRef]
130. Xie, H.; Dai, X.L.; Ren, H.L.; Liu, F. Experimental characterization on slamming loads of a truncated ship bow under asymmetrical impact. *Ocean Eng.* **2023**, *284*, 115195. [CrossRef]
131. Xie, H.; Liu, F.; Yu, P.Y.; Ren, H.L. Comparative study on fluid dynamic behavior and slamming loads of two bow-flare sections entering into water. *Int. J. Mech. Sci.* **2020**, *166*, 105177. [CrossRef]
132. Liu, X.Y.; Liu, F.; Ren, H.L.; Chen, X.; Xie, H. Experimental investigation on the slamming loads of a truncated 3D stern model entering into water. *Ocean Eng.* **2022**, *252*, 110873. [CrossRef]
133. Mutsuda, H.; Kanehira, T.; Kawawaki, K.; Doi, Y.; Yasukawa, H. Occurrence of stern slamming pressure and its characteristics in following irregular waves. *Ocean Eng.* **2018**, *170*, 222–236. [CrossRef]
134. Ochi, M. Prediction of slamming characteristics and hull responses for ship design. *Trans. SNAME* **1973**, *81*, 144–177.
135. Li, P.; Xie, H.; Liu, F.; Liu, X.Y.; Li, H. Experimental study on the water entry of a 3D bow-flared model with various inclination angles. *Ocean Eng.* **2022**, *259*, 111834. [CrossRef]
136. Liu, B.; Wang, S.; Villavicencio, R.; Soares, C.G. Slamming load and hydroelastic structural response of bow flare areas of aluminium fast displacement crafts. *Ocean Eng.* **2020**, *218*, 108207. [CrossRef]

137. Bao, C.M.; Sun, S.Y.; Xu, G.; Chen, L.F. Numerical prediction of slamming loads during water entry of a slender hull. *Ocean Eng.* **2020**, *201*, 107136. [CrossRef]
138. Prini, F.; Benson, S.; Dow, R.S.; Phillips, H.J.; Sheppard, P.J.; Birmingham, R.W. Model scale prediction of seakeeping and global bending moment on a high speed craft. *Eng. Struct.* **2021**, *231*, 111738. [CrossRef]
139. Parunov, J.; Badalotti, T.; Feng, Q.D.; Gu, X.C.; Iijima, K.; Ma, N.; Qiu, W.; Wang, S.; Wang, X.L.; Yang, P.; et al. Benchmark on the prediction of whipping response of a warship model in regular waves. *Mar. Struct.* **2024**, *94*, 103549. [CrossRef]
140. Bilandi, R.N.; Mancini, S.; Vitiello, L.; Miranda, S.; De Carlini, M. A Validation of Symmetric 2D + T Model Based on Single-Stepped Planing Hull Towing Tank Tests. *J. Mar. Sci. Eng.* **2018**, *6*, 136. [CrossRef]
141. Tavakoli, S.; Bilandi, R.N.; Mancini, S.; De Luca, F.; Dashtimanesh, A. Dynamic of a planing hull in regular waves: Comparison of experimental, numerical and mathematical methods. *Ocean Eng.* **2020**, *217*, 107959. [CrossRef]
142. Shao, W.B.; Ma, S.; Duan, W.Y.; Liu, J.Y.; Zhang, Y.F. Vertical force prediction of planing craft in calm water based on varia-ble section slamming model. *Ocean Eng.* **2023**, *287 Pt 1*, 115693. [CrossRef]
143. Mousaviraad, S.M.; Wang, Z.Y.; Stern, F. URANS studies of hydrodynamic performance and slamming loads on high-speed planing hulls in calm water and waves for deep and shallow conditions. *Appl. Ocean Res.* **2015**, *51*, 222–240. [CrossRef]
144. Thodal, R.S. On Full Scale Slamming Testing of High-Speed Boats. Ph.D. Thesis, Lehigh University, Bethlehem, PA, USA, 2016.
145. Judge, C.; Mousaviraad, M.; Stern, F.; Lee, E.; Fullerton, A.; Geiser, J.; Schleicher, C.; Merrill, C.; Weil, C.; Morin, J.; et al. Experiments and CFD of a high-speed deep-V planing hull—Part II: Slamming in waves. *Appl. Ocean Res.* **2020**, *97*, 102059. [CrossRef]
146. Judge, C.Q.; Ibrahim, A.M. Experimental investigation of motions and global hull girder bending moment of a semi-displacement vessel. *Ocean Eng.* **2025**, *320*, 120295. [CrossRef]
147. Yousefnia, M.; Heirani, N.M.R. Global hydroelastic effects on hull girder deflection due to bow flare slamming. *Mar. Syst. Ocean Technol.* **2021**, *16*, 99–110. [CrossRef]
148. Liu, W.Q.; Qin, Y.X.; Hu, Y.C.; Vladimir, N.; Xu, S.X.; Wu, Y.C. Numerical Research on Impacting Load and Structural Response for a Model Experiment of High-Speed Craft. *J. Offshore Mech. Arct. Eng.* **2025**, *147*, 011901. [CrossRef]
149. Lee, E.J.; Die, M.; Harrison, E.L.; Jiang, M.J.; Snyder, L.A.; Powers, A.M.R.; Bay, R.J.; Serani, A.; Nadal, M.L.; Kubina, E.R.; et al. Experimental and computational fluid-structure interaction analysis and optimization of Deep-V planing-hull grillage panels subject to slamming loads—Part II: Irregular waves. *Ocean Eng.* **2024**, *292*, 116346. [CrossRef]
150. Shan, L.; Xu, G.D. Numerical study of the green water on a wave-piercing tumblehome vessel and the hydrodynamic loads. *Ocean Eng.* **2024**, *309 Pt 1*, 118313. [CrossRef]
151. Begovic, E.; Bertorello, C.; Bove, A.; Garme, K.; Lei, X.Y.; Persson, J.; Petrone, G.; Razola, M.; Rosén, A. Experimental modelling of local structure responses for high-speed planing craft in waves. *Ocean Eng.* **2020**, *216*, 107986. [CrossRef]
152. Camilleri, J.; Taunton, D.J.; Temarel, P. Full-scale measurements of slamming loads and responses on high-speed planing craft in waves. *J. Fluids Struct.* **2018**, *81*, 201–229. [CrossRef]
153. Ibrahim, A.M.; Judge, C.Q. Investigations of hull girder slamming factor for a semi-displacement vessel using model testing. *Appl. Ocean Res.* **2024**, *150*, 104084. [CrossRef]
154. Zou, J.; Li, H.; Sun, Z.Y.; Han, B.B.; Wang, Z.Y. Experimental analysis of bow flare slamming and whipping responses in a ship at different sailing speeds. *Ocean Eng.* **2024**, *305*, 117896. [CrossRef]
155. Magoga, T.; Aksus, S.; Cannon, S.; Ojeda, R.; Thomas, G. Identification of slam events experienced by a high-speed craft. *Ocean Eng.* **2017**, *140*, 309–321. [CrossRef]
156. Davis, M.R.; Watson, N.L.; Holloway, D.S. Measurement of response amplitude operators for an 86 m high-speed catamaran. *J. Ship Res.* **2005**, *49*, 121–143. [CrossRef]
157. Al-Furjan, M.S.H.; Kolahchi, R.; Shan, L.; Hajmohammad, M.H.; Farrokhian, A.; Shen, X. Slamming impact induced hydrodynamic response in wave-piercing catamaran beam elements with controller. *Ocean Eng.* **2022**, *266 Pt 4*, 112908. [CrossRef]
158. Lavroff, J.; Davis, M.R.; Holloway, D.S.; Thomas, G. Wave slamming loads on wave-piercer catamarans operating at high-speed determined by hydro-elastic segmented model experiments. *Mar. Struct.* **2013**, *33*, 120–142. [CrossRef]
159. Gebrezgabir, S.; Holloway, D.S.; Ali-Lavroff, J. Slam and wave load response reconstruction in high speed catamarans using transmissibility on full scale sea trials. *Ocean Eng.* **2023**, *271*, 113822. [CrossRef]
160. French, B.J.; Thomas, G.A.; Davis, M.R. Slam occurrences and loads of a high-speed wave piercer catamaran in irregular seas. *Proc. Inst. Mech. Eng. Part M–J. Eng. Marit. Environ.* **2015**, *229*, 45–57. [CrossRef]
161. Davis, M.R.; French, B.J.; Thomas, G.A. Wave slam on wave piercing catamarans in random head seas. *Ocean Eng.* **2017**, *135*, 84–97. [CrossRef]
162. Shabani, B.; Lavroff, J.; Davis, M.R.; Holloway, D.S.; Thomas, G.A. Slam loads and pressures acting on high-speed wave-piercing catamarans in regular waves. *Mar. Struct.* **2019**, *66*, 136–153. [CrossRef]
163. Shabani, B.; Lavroff, J.; Davis, M.R.; Holloway, D.S.; Davis, M.R.; Thomas, G.A. The effect of centre bow and wet-deck geometry on wet-deck slamming loads and vertical bending moments of wave-piercing catamarans. *Ocean Eng.* **2018**, *169*, 401–417. [CrossRef]

164. Shabani, B.; Lavroff, J.; Davis, M.R.; Holloway, D.S.; Davis, M.R.; Thomas, G.A. Centre bow and wet-deck design for motion and load reductions in wave piercing catamarans at medium speed. *Ships Offshore Struct.* **2021**, *16*, 83–99. [CrossRef]
165. Liu, T.; Halse, K.H.; Leira, B.J.; Jiang, Z.Y.; Chai, W.; Brathaug, H.P.; Hildre, H.P. Dynamic response of a SWATH vessel for installing pre-assembled floating wind turbines. *Mar. Struct.* **2023**, *88*, 103341. [CrossRef]
166. Ma, S.; Zhu, M.Y.; Liu, D.D.; Liu, J.C.; Wang, W. Experimental study of wet deck slamming for a SWATH in regular waves. *Ocean Eng.* **2023**, *288 Pt 1*, 115996. [CrossRef]
167. McVicar, J.; Lavroff, J.; Davis, M.R.; Thomas, G. Fluid-structure interaction simulation of slam-induced bending in large high-speed wave-piercing catamarans. *J. Fluids Struct.* **2018**, *82*, 35–58. [CrossRef]
168. Almallah, I.; Ali-Lavroff, J.; Holloway, D.S.; Davis, M.R. Slam load estimation for high-speed catamarans in irregular head seas by full-scale computational fluid dynamics. *Ocean Eng.* **2021**, *234*, 109160. [CrossRef]
169. Almallah, I.; Ali-Lavroff, J.; Holloway, D.S.; Davis, M.R. Estimation of torsional and global loads for a wave-piercing high-speed catamaran at full-scale in irregular bow quartering seas using CFD simulation. *Ocean Eng.* **2022**, *266*, 113006. [CrossRef]
170. Almallah, I.; Ali-Lavroff, J.; Holloway, D.S.; Davis, M.R. High-speed wave-piercing catamaran global loads determined by fea and sea trials. *Int. J. Marit. Eng.* **2021**, *161*, 139–153. [CrossRef]
171. Hajmohammad, M.H.; Farrokhian, A.; Kolahchi, R. Dynamic analysis in beam element of wave-piercing Catamarans undergoing slamming load based on mathematical modelling. *Ocean Eng.* **2021**, *234*, 106269. [CrossRef]
172. Davis, M.R.; Watson, N.L.; Holloway, D.S. Measurement and prediction of wave loads on a high-speed catamaran fitted with active stern tabs. *Mar. Struct.* **2004**, *17*, 503–535. [CrossRef]
173. Jacobi, G.; Thomas, G.; Davis, M.R.; Holloway, D.S.; Davidson, G.; Roberts, T. Full-scale motions of a large high-speed catamaran: The influence of wave environment, speed and ride control system. *Int. J. Marit. Eng.* **2012**, *154*, A143–A155. [CrossRef]
174. Jacobi, G.; Thomas, G.; Davis, M.R.; Davidson, G. An insight into the slamming behaviour of large high-speed catamarans through full-scale measurements. *J. Mar. Sci. Technol.* **2014**, *19*, 15–32. [CrossRef]
175. AlaviMehr, J.; Lavroff, J.; Davis, M.R.; Holloway, D.S.; Thomas, G.A. An Experimental Investigation of Ride Control Algorithms for High-Speed Catamarans Part 2: Mitigation of Wave Impact Loads. *J. Ship Res.* **2017**, *61*, 51–63. [CrossRef]
176. Hasheminasab, H.; Zeraatgar, H.; Malekmohammadi, J.; Azizi, A. Analysis of slamming loads on a catamaran section with a centre-bow appended by spray rail. *Ships Offshore Struct.* **2023**, *18*, 1025–1036. [CrossRef]
177. Available online: https://www.hildstrom.com/publications/pub-StructuralTrialsOfTheRVTriton-Feb-10-2003-Final.pdf (accessed on 14 August 2003).
178. Miao, S.H.; Price, W.G.; Temarel, P. The hydroelastic behavior of multi-hulls travelling in a seaway. In Proceedings of the 3rd International Conference Advances in Marine Structures 1997, Dunfermline, UK, 20–22 May 1997.
179. Miao, S.H.; Temarel, P. Trimaran wave loading. In *2nd Report for DERA Contract SSDH100120*; University of Southampton: Southampton, UK, 1998.
180. Miao, S.H.; Temarel, P. Trimaran wave loading—Development of the software THAFTS. In *Report for DERA Contract CU004-0000001635*; University of Southampton: Southampton, UK, 2000.
181. Tang, H.Y.; Ren, H.L.; Wan, Q. Investigation of Longitudinal Vibrations and Slamming of a Trimaran in Regular Waves. *J. Ship Res.* **2017**, *61*, 153–166. [CrossRef]
182. Tang, H.Y.; Wan, Q.; Ren, H.L. Numerical study of trimaran wave load based on time-domain Rankine method. *Brodogr. Int. J. Nav. Archit. Ocean Eng. Res. Dev.* **2023**, *74*, 107–129. [CrossRef]
183. Tang, H.Y.; Zhang, X.K.; Ren, H.L.; Yu, P.Y. Numerical study of trimaran motion and wave load prediction based on time-domain Rankine-Green matching method. *Ocean Eng.* **2020**, *214*, 107605. [CrossRef]
184. Askarian Khoob, A.; Ketabdari, M.J. Wave-induced loads on cross-deck of a wave-piercing trimaran with different hull forms of outriggers. *Transport* **2019**, *34*, 559–568. [CrossRef]
185. Khoob, A.A.; Ketabdari, M.J. Short-term prediction and analysis of wave-induced motion and load responses of a wave-piercing trimaran. *Brodogr. Int. J. Nav. Archit. Ocean Eng. Res. Dev.* **2020**, *71*, 123–142.
186. Ghadimi, P.; Nazemian, A.; Sheikholeslami, M. Numerical simulation of the slamming phenomenon of a wave-piercing trimaran in the presence of irregular waves under various seagoing modes. *Proc. Inst. Mech. Eng.* **2019**, *233*, 1198–1211. [CrossRef]
187. Xie, H.; Liu, F.; Liu, X.Y.; Tang, H.Y. Numerical prediction of asymmetrical ship slamming loads based on a hybrid two-step method. *Ocean Eng.* **2020**, *208*, 107331. [CrossRef]
188. Liao, X.Y.; Chen, Z.Y.; Gui, H.B.; Du, M.C. CFD Prediction of Ship Seakeeping and Slamming Behaviors of a Trimaran in Oblique Regular Waves. *J. Mar. Sci. Eng.* **2021**, *9*, 1151. [CrossRef]
189. Liao, X.Y.; Xia, J.S.; Chen, Z.Y.; Tang, Q.; Zhao, N.; Zhao, W.D.; Gui, H.B. Application of CFD and FEA Coupling to Predict Structural Dynamic Responses of A Trimaran in Uni- and Bi-Directional Waves. *China Ocean Eng.* **2024**, *38*, 81–92. [CrossRef]
190. Chen, Z.Y.; Gui, H.B.; Dong, P.S.; Yu, C.L. Numerical and experimental analysis of hydroelastic responses of a high-speed trimaran in oblique irregular waves. *Int. J. Nav. Archit. Ocean Eng.* **2019**, *11*, 409–421. [CrossRef]

191. Chen, Z.Y.; Zhao, N.; Zhao, W.D.; Xia, J.S. Numerical prediction of seakeeping and slamming behaviors of a trimaran in short-crest cross waves compared with long-crest regular waves. *Ocean Eng.* **2023**, *285*, 115314. [CrossRef]
192. Liu, J.Y.; Duan, W.Y.; Ma, S.; Liao, K.P.; Si, H.L. Experimental and numerical study of wet-deck slamming characteristic for a trimaran in regular head waves. *Appl. Ocean Res.* **2024**, *150*, 104136. [CrossRef]

Journal of
Marine Science and Engineering

MDPI

Article

Research on the Constitutive Relationship of the Coarse-Grained Heat-Affected Zone in Ship Thick-Plate Welded Joints of Ship Structures

Linzhi Xu [1], Pengyu Zhan [1], Tao Yi [2], Shukai Zhang [1], Jian He [1] and Mengzhen Li [1,*]

[1] School of Naval Architecture, Ocean and Energy Power Engineering, Wuhan University of Technology, Wuhan 430063, China; xulinzhi@whut.edu.cn (L.X.); zhanpengyu@whut.edu.cn (P.Z.); 311830@whut.edu.cn (S.Z.); jianhe@whut.edu.cn (J.H.)
[2] China Classification Society Wuhan Branch, Wuhan 430030, China; yitao@ccs.org.cn
* Correspondence: mengzhen.li@whut.edu.cn

Abstract

This study addresses the constitutive relationship of the welded coarse-grained heat-affected zone (CGHAZ) in 80-mm-thick DH36 marine steel plates. By integrating quasi-static tensile testing, digital image correlation (DIC) technology, and metallographic analysis, we systematically investigated the mechanical property differences and underlying mechanisms between the CGHAZ and base metal (BM). High-precision DIC technology enabled strain field characterization at the microscale in the CGHAZ, while the Ramberg-Osgood model was adopted to establish a dual-material constitutive equation. The results demonstrate that grain coarsening induced by welding thermal cycles significantly influenced the mechanical responses: the CGHAZ exhibited enhanced tensile strength but reduced plastic compatibility due to decreased grain boundary density. Notably, gradient differences in elastic modulus (CGHAZ: 184 GPa vs. BM: 213 GPa) and yield strength (CGHAZ: 363 MPa vs. BM: 373 MPa) between the BM and CGHAZ necessitate strict differentiation in engineering design. This work overcomes the limitations of oversimplified CGHAZ properties in conventional design approaches, providing a novel methodology for strength assessment and lightweight design of marine structures. The findings offer critical theoretical insights and practical guidelines for enhancing the reliability of offshore engineering equipment.

Keywords: HAZ; CGHAZ; constitutive relationship; DIC measurement technology

1. Introduction

The escalating demands of global trade and maritime transportation have driven the shipbuilding industry to progressively adopt large-scale vessel designs, necessitating the extensive use of thick steel plates (>40 mm thickness) to meet stringent structural integrity requirements [1]. This technological shift has concurrently led to the implementation of elevated heat input parameters during welding procedures, which inadvertently exacerbate the mechanical property gradients between the heat-affected zone (HAZ) and base metal (BM). Metallurgical analysis reveals that the HAZ microstructure typically differentiates into three distinct sub-zones: coarse-grained heat-affected zone (CGHAZ), fine-grained heat-affected zone (FGHAZ), and subcritical heat-affected zone (SCHAZ), as schematically illustrated in Figure 1 [2,3].

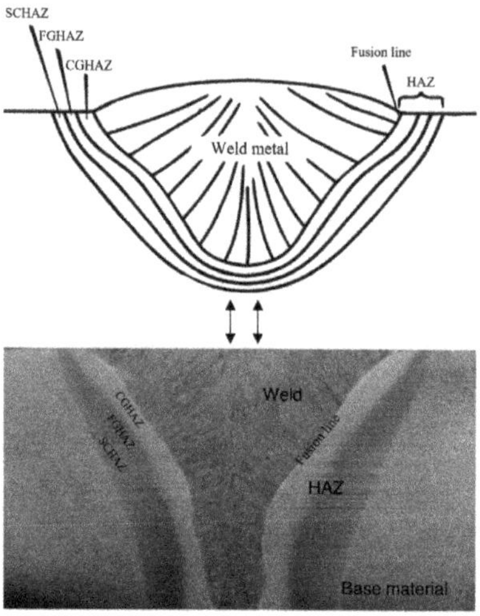

Figure 1. Schematic of HAZ and its sub-zones: CGHAZ, FGHAZ, and SCHAZ.

Beghini et al. [4] conducted quasi-static tensile tests on non-standard small specimens cut from characteristic regions of welded plates: BM, WM, and HAZ. The results showed that the properties of the HAZ differ significantly from those of the BM. Gan et al. [5] pointed out that the main reason for the fracture of welding joints in ship and marine engineering structures is brittle fracture, and the coarsening of grains and changes in the microstructure in the coarse-grained zone of the HAZ are the main causes of HAZ embrittlement, leading to a decline and deterioration in properties. These studies have shown a clear difference in mechanical properties between the CGHAZ and the BM, and the coarse-grained zone formed in the HAZ due to grain coarsening caused by welding has significant research implications. Therefore, in ship design and construction, the difference in mechanical properties between the BM and the HAZ, especially the CGHAZ, needs to be carefully considered. It is crucial to obtain the mechanical constitutive relationship between the BM and the CGHAZ.

At present, the positioning and measurement of the width of each part of the HAZ mainly rely on metallographic microscope observation, micro-hardness testing, and numerical simulation. Yang et al. [6] addressed the technical challenges posed by the submillimeter-scale width of the HAZ in welded joints, which renders conventional experimental methods inadequate for probing the microstructural mechanical properties of the CGHAZ. They developed a coupled thermo-mechanical-metallurgical finite element model for DP780 dual-phase steel welding to overcome this spatial resolution limitation. Xiong et al. [7] studied the width and properties of the HAZ of 4-mm-thick 16 Mn steel plate MAG welding joints under different welding line energy conditions through metallographic observation and micro-hardness testing of each zone of the welding joint. Georgios et al. [8] assessed the width of the HAZ of arc-welded ferritic ductile iron thin sheets by testing the micro-Vickers hardness of the welding joint and evaluating the hardness value changes. Liu et al. [9] studied the width and position of the HAZ of TIG welding joints of X70/316L bimetallic composite pipes with a wall thickness of 18.9 mm through macro-metallographic tests, hardness tests, transmission electron microscopy tests, and finite element simulation. Clearly, the CGHAZ is very narrow and requires metallographic microscopes and hardness measurements for measurement. These experiments often separate the mechanical measurement tests of the HAZ and the measurement of the position and spatial shape of the HAZ, making it difficult to combine the strain data and stress data generated in the

HAZ during mechanical property measurement tests, and it is difficult to obtain its precise mechanical properties. Therefore, an urgent need exists for a measurement method to measure and evaluate its mechanical properties accurately.

The advent of advanced Digital Image Correlation (DIC) techniques has revolutionized mechanical characterization capabilities, particularly for micrometer-scale strain analysis in challenging environments. Recent implementation in naval architecture research enables unprecedented quantification of mechanical heterogeneity within the CGHAZ of thick-plate ship welds. DIC measurement technology is an experimental technique based on optical principles, which obtains the displacement and strain information of the specimen surface by calculating the gray-level matrix of the digital speckle images before and after deformation of the specimen surface [10]. In recent years, DIC measurement technology has begun to be preliminarily applied to the measurement of the mechanical properties of metals [11–13]. Zhang and Peng et al. [14] assessed the fatigue crack expansion ability of rail tracks using DIC technology, and the DIC measurement technology was able to accurately and effectively obtain the tiny crack tip deformation behavior during the fatigue crack expansion process. It can be seen that DIC measurement technology can observe the tiny deformations of materials at the micro level.

In summary, within the HAZ, the coarse-grained region exhibits the most significant degradation in mechanical properties due to its large grain size, resulting in a substantial disparity compared to the mechanical properties of the BM. This area has thus become the focal point in the assessment of welding quality. However, in the current design calculations for ships, this difference is not taken into account, and the mechanical parameters of the BM are uniformly employed for design verification. To prevent potential casualties or property damage caused by the reduced mechanical properties of the coarse-grained region, it is imperative to elucidate the constitutive relationship of this area within the HAZ. Therefore, the present experiment aims to conduct quasi-static tensile tests using a universal testing machine, applying axial loading until specimen failure. During the loading process, strain data will be collected using DIC technology, while stress data will be obtained from the universal testing machine. After processing and calculating the collected data, the constitutive relationships of the CGHAZ and the BM will be determined based on the Ramberg-Osgood model.

2. Test Design

2.1. Specimen Preparation

An 80-mm-thick DH36 steel plate was selected with a single V-notch and welded using the SAW process to obtain the welded test plate, as shown in Figure 2. Five identical specimens were taken from the test plate. The total length of the specimen was 400 mm, with a test section length of 160 mm, a width of 25 mm, and a height of 5 mm. The specific dimensions of the specimen are shown in Figure 3.

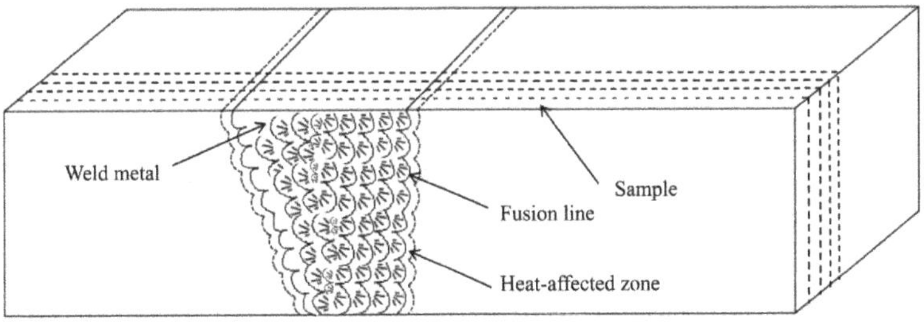

Figure 2. 80 mm-thick DH36 steel welded plate with single V-groove (SAW process).

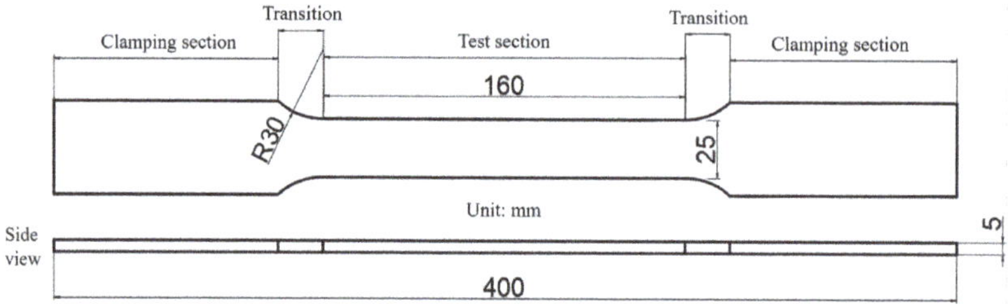

Figure 3. The front view and the side view of the specimen.

2.2. Test Equipment

DIC measurement technology is a highly valuable non-contact optical measurement method with a wide range of applications in the field of mechanical property research. The system consists of an image acquisition system (including an industrial CCD camera and lighting equipment) and data processing software. The camera parameters determine the measurement accuracy and range, while the software is responsible for core algorithm calculations and data visualization. The DIC test equipment used was the XTDIC-CONST full-field strain measurement system from Xintuo Sanwei (Shenzhen, China), with two cameras model XTDIC-CONST-SD, with a pixel count of 5 million, a frame rate of 75 fps, a strain measurement accuracy of 0.5 $\mu\varepsilon$, and a strain measurement range of 0.005% to 2000%. The loading rate of the universal material testing machine was 3 mm/min. The positions of the instruments during the test are shown in Figure 4.

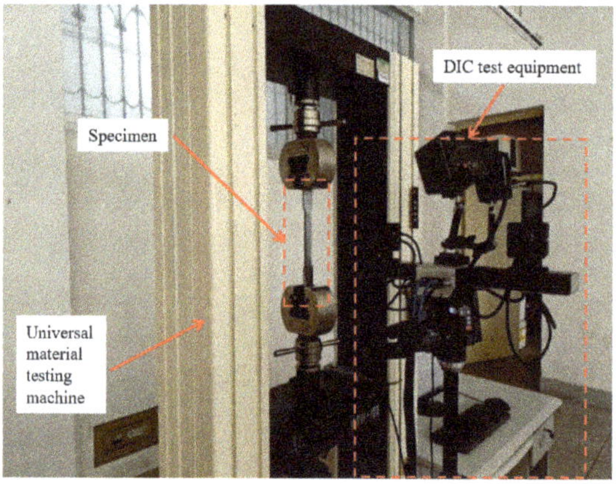

Figure 4. The three parts of the test equipment.

2.3. Metallographic Test of the HAZ of the Specimen

Through metallographic tests, the boundary line and microstructure of the HAZ of the specimen were observed to accurately select the stress-strain measurement points in the CGHAZ later. During the metallographic test, the specimen was first polished to remove surface oxides and impurities until the surface was clean and smooth. Then, the specimen was roughly and finely ground using a grinding machine and sandpaper of different grit sizes to make the surface smooth without obvious scratches. Next, the specimen was polished using polishing cloth and polishing liquid until a mirror-like effect was achieved. After polishing, the specimen was cleaned with alcohol to remove residual polishing liquid and then dried by heating. Subsequently, the specimen was etched with a 4% nitric acid alcohol solution to clearly show the boundary line of the HAZ (Figure 5), facilitating the accurate selection of the observation area under the electron microscope. Finally, the specimen was observed under the electron microscope at an appropriate magnification to obtain a clear microstructural characteristic metallographic image, as shown in Figure 6.

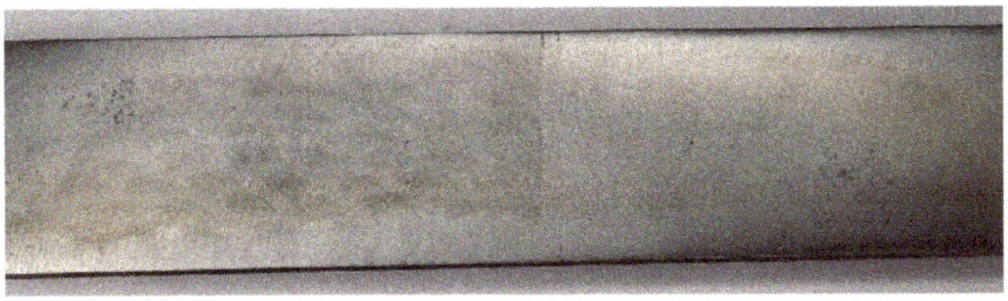

Figure 5. The boundary line of the specimen's HAZ.

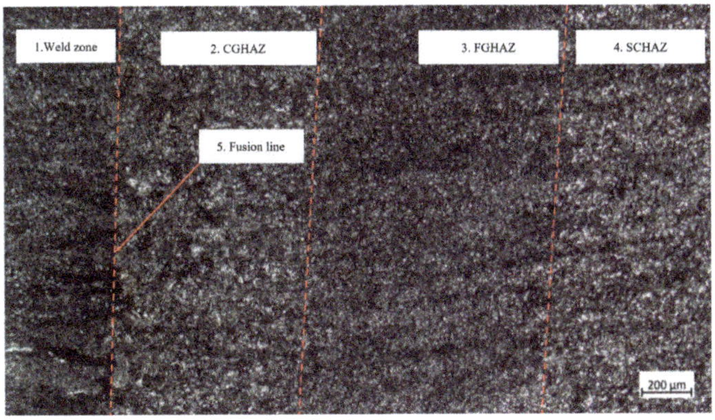

Figure 6. Metallographic micrograph of Fusion line, Weld zone, and three sub-zones of HAZ.

The metallographic image clearly reveals distinct characteristic regions and their boundaries, including: 1. The weld zone, 2. The coarse-grained heat-affected zone (CGHAZ), 3. The fine-grained heat-affected zone (FGHAZ), 4. The subcritical heat-affected zone (SCHAZ), and 5. The fusion line. The figure indicates that the CGHAZ was approximately 800 μm wide. The widths of the FGHAZ and the SCHAZ are 1100 μm and 600 μm, respectively.

2.4. DIC Test Speckle, Spraying, and Measurement Point Selection

We creatively adopted the semi-spraying speckle method, which selectively applies speckle patterns to the surface of the specimen, creating a 15 mm wide central speckled region while maintaining 5 mm wide unpatterned zones on both flanks. This configuration ensures sufficient characteristic points for DIC strain measurement while preserving visible fusion line contours etched during metallographic preparation. The unpatterned lateral regions serve as fiducial markers to precisely localize the CGHAZ within the HAZ, avoiding interference from speckle coverage over microstructural features. Using the distance data of the CGHAZ from the fusion line determined by the metallographic microscope, the position of the CGHAZ was marked, and the sampling points in the CGHAZ were obtained as shown in Figure 7.

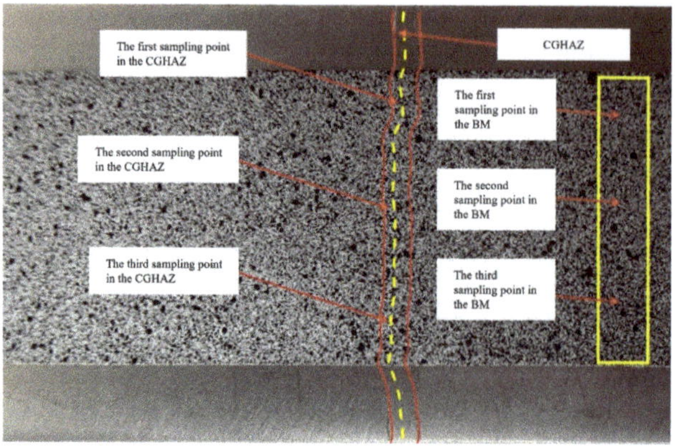

Figure 7. Schematic of the six sampling points' position in the CGHAZ and the BM.

3. Test Results

3.1. Extraction of Stress-Strain Data of the CGHAZ

Quasi-static tensile testing conducted via a universal testing machine yields force-time curves, which are subsequently converted into stress-time curves through post-processing. Strain data acquired from three measurement points using DIC equipment are synchronized with the temporal axis, enabling the derivation of individual stress-strain curves for each sampling location, as shown in Figure 8.

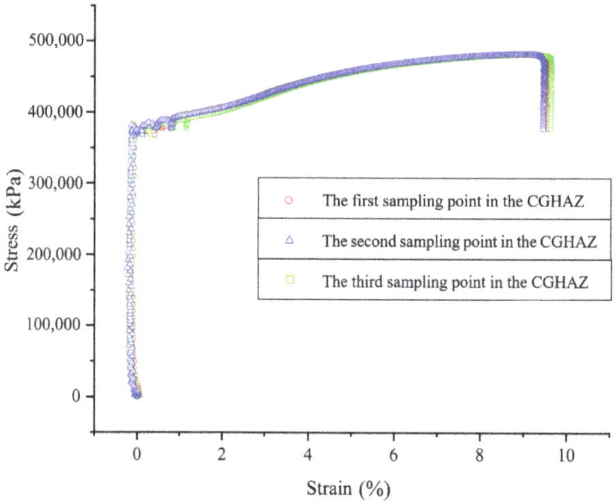

Figure 8. The stress-strain curves of the three points in CGHAZ.

3.2. Extraction of Stress-Strain Data of the BM

Similar to the extraction method of the stress-strain data of the CGHAZ, three points were sampled in the BM, and the strain-time curves of the three points were extracted, respectively. After synchronizing the time axes with the stress-time curve measured by the testing machine, the stress-strain curves of the three points could be obtained, respectively, as shown in Figure 9.

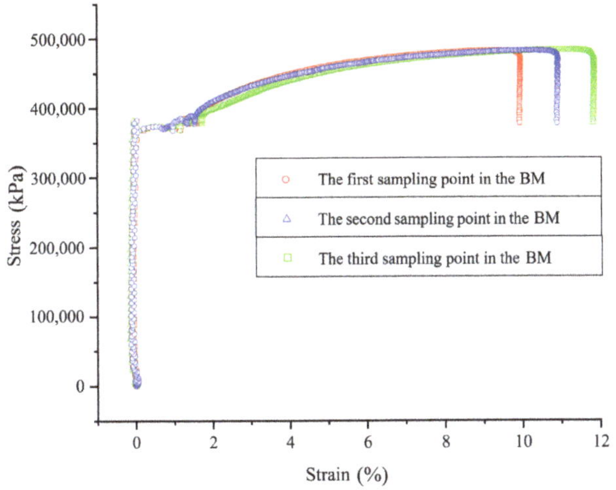

Figure 9. The stress-strain curves of the three points in BM.

An average method was used to process the stress-strain curves of the three points to ensure the rigor of the results. The average method for data is to average the strain values of three measurement points collected by DIC at the same tensile force level and then obtain the average stress-strain curves. The average stress-strain curves of the CGHAZ and the BM at the same stress level were drawn simultaneously on Figure 10.

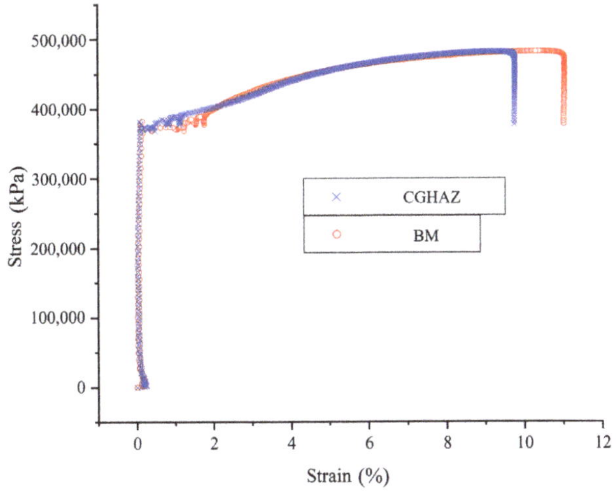

Figure 10. The average stress-strain curves of the CGHAZ and BM at the same stress level.

4. Solution of the Constitutive Equations of the Specimen BM and CGHAZ

4.1. Solution of the CGHAZ Constitutive Equation

Common steel constitutive models include elastic models, plastic models, Ramberg-Osgood models, multi-linear models, Chaboche models, and Johnson-Cook models. The average stress-strain experimental data of the CGHAZ were substituted to solve the relevant coefficients and obtain the material constitutive equation. The Ramberg-Osgood steel constitutive model offers several advantages [15]. It features a simplified mathematical expression that enables easy description of the complex stress—strain behavior for engineering calculations and analyses. The model also has good fitting ability, particularly in representing the yield platform and plastic hardening stage of steel. Moreover, its parameters (e.g., yield stress, strain hardening index, and reference strain rate) can typically be acquired via simple tensile tests, which facilitates experimental determination.

The Ramberg-Osgood constitutive model is as follows [16]:

$$\varepsilon = \varepsilon_e + \varepsilon_p = \frac{\sigma}{E} + p\left(\frac{\sigma}{\sigma_p}\right)^n \tag{1}$$

Among them, ε_e represents the elastic strain; ε_e represents the plastic strain; p is the plastic strain corresponding to the conditional yield strength σ_p, which is usually taken as 0.2% (the corresponding conditional yield strength is $\sigma_{0.2}$); n is the strain hardening index, which can be determined according to Equation (2):

$$n = \frac{\ln 20}{\ln \frac{\sigma_{0.2}}{\sigma_{0.01}}} \tag{2}$$

Among them, $\sigma_{0.01}$ is the stress corresponding to a plastic strain of 0.01%. We need to fit the most appropriate curve to determine the value of n.

The fitting results of the elastic stage and the yield and strengthening stages are shown in Figure 11.

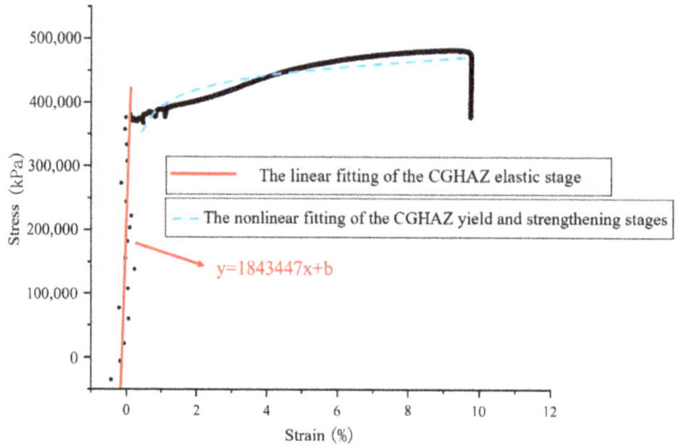

Figure 11. The fitting results of the CGHAZ elastic stage and the yield and strengthening stages.

By substituting the calculations and linear fitting, the relevant coefficients were obtained as follows. According to the slope of the linear fitting in Figure 11 of the elastic stage, the elastic modulus E = 184 GPa, and the yield strength $\sigma_p = \sigma_{0.2}$ = 363 MPa. According to the experimental data, $\sigma_{0.01}$ = 1.4 MPa, and the strain hardening index n = 0.539 was calculated using Equation (2). The constitutive Equation (3) is as follows:

$$\varepsilon = \varepsilon_e + \varepsilon_p = \frac{\sigma}{1.84 \times 10^{11}} + 0.2\left(\frac{\sigma}{3.63 \times 10^{8}}\right)^{0.539} \tag{3}$$

4.2. Solution of the BM Constitutive Equation

The solution method for the BM constitutive equation is the same as that for the CGHAZ. The fitting results of the elastic stage and the yield and strengthening stages are shown in Figure 12 (Readers should be aware of the limitations of the calculation results. We have checked the BM, but we are unable to perform the same inspection on the HAZ).

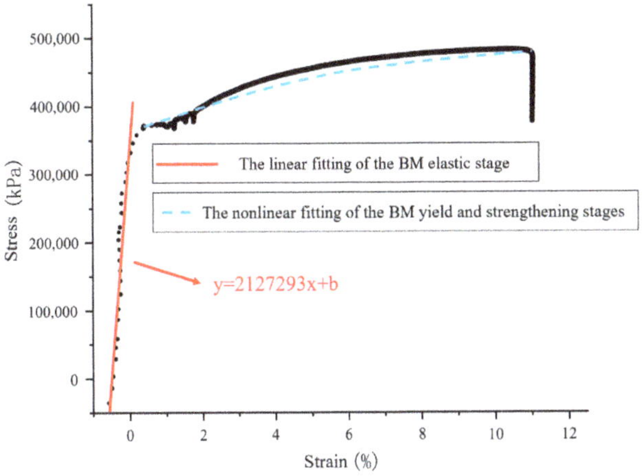

Figure 12. The fitting results of the BM elastic stage, the yield, and strengthening stages.

By substituting the calculations and linear fitting, the relevant coefficients were obtained as follows. According to the slope of the linear fitting in Figure 12 of the elastic stage, the elastic modulus E = 213 GPa, and the yield strength $\sigma_p = \sigma_{0.2}$ = 373 MPa. According to the experimental data, $\sigma_{0.01}$ = 1.4 MPa, and the strain hardening index n = 0.536 was calculated using Equation (2). The constitutive Equation (4) is as follows:

$$\varepsilon = \varepsilon_e + \varepsilon_p = \frac{\sigma}{2.13 \times 10^{11}} + 0.2\left(\frac{\sigma}{3.73 \times 10^{8}}\right)^{0.536} \tag{4}$$

5. Comparison and Analysis of DIC Measurement Method and Traditional Force-Strain Method

In this test, to verify the accuracy of the test data, tensile specimens of the same material were prepared, and the elastic modulus was measured using the force-strain method according to the GB/T 22315-2008 [17]. The specimen size is shown in Figures 13 and 14. Figure 14 shows the traditional elastic modulus testing method.

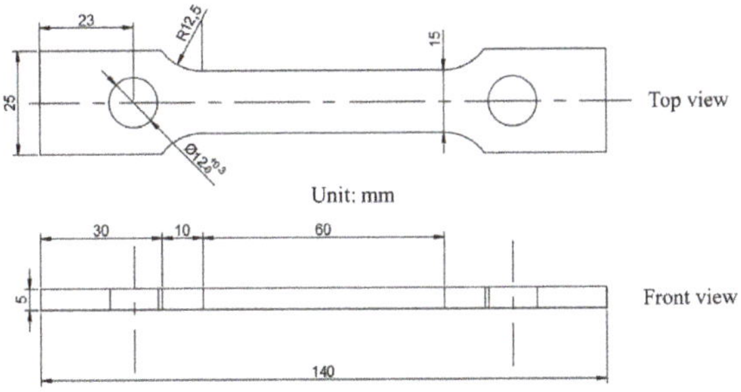

Figure 13. The size of traditional modulus specimen (per GB/T 22315-2008).

The specific linear fitting is shown in Figure 15. By linearly fitting the data obtained from the force-strain method, the slope and the elastic modulus of the specimen were 209 GPa. Compared with the 213 GPa measured above for the BM, the difference is 1.8%, indicating that the constitutive equation obtained based on DIC testing technology has high accuracy and feasibility.

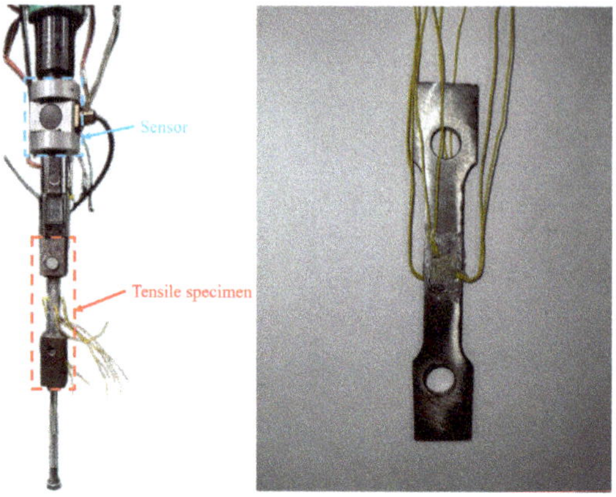

Figure 14. Traditional elastic modulus testing method.

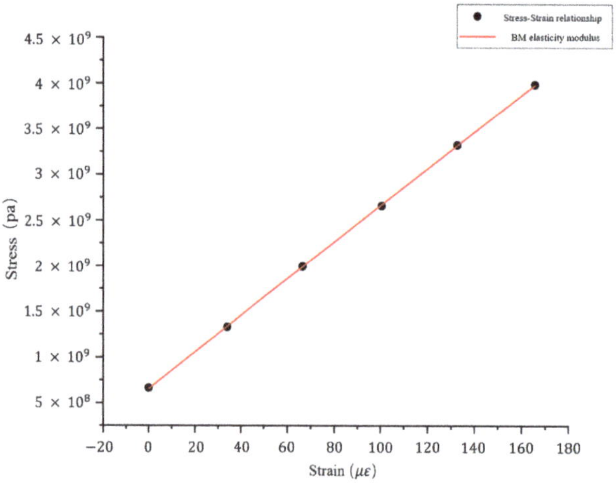

Figure 15. The specimen's linear fitting of the elasticity modulus.

6. Result Analysis

The comparison of material parameters and mechanical properties between the CG-HAZ and the BM is shown in Table 1.

Table 1. CGHAZ and BM's material parameters and mechanical properties correlation table.

	CGHAZ	BM	Relative Differences (%)
The elasticity modulus (GPa)	184 ± 7	213 ± 5	13.62
Yield strength (MPa)	363 ± 9	373 ± 6	2.68

From Table 1, it can be seen that there are significant differences in elastic modulus and yield strength between the CGHAZ and the BM. The elastic modulus of the BM is greater than that of the CGHAZ, with a relative difference of 13.62%. The yield strength of the BM is slightly higher than that of the CGHAZ, with a relative difference of 2.68%. From the location of the specimen fracture, the fracture occurred in the BM region, indicating that the tensile strength of the CGHAZ is higher than that of the BM. In actual engineering calculations, the material properties of the CGHAZ cannot be equated with those of the BM, and the allowable stress and failure criteria need to be strictly distinguished.

After the quasi-static tensile test, the fracture morphology of the specimen is shown in Figure 16. The fractured specimen exhibited necking deformation, with a fibrous central zone and approximately 45° shear lips on both sides, indicative of ductile fracture characteristics [18,19]. Since the fracture occurred in the base metal (BM) region, the tensile strength of the coarse-grained heat-affected zone (CGHAZ) could not be directly determined. However, this observation strongly implies that the CGHAZ possesses higher tensile strength than the BM.

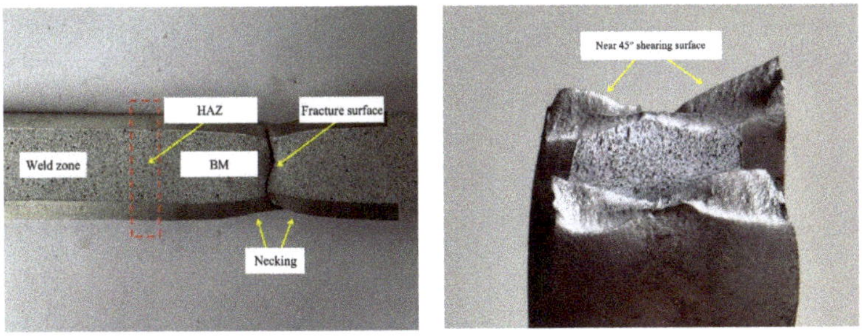

(a) (b)

Figure 16. The ductile fracture morphology and shearing surface in the BM of the specimen: (**a**) The fracture morphology of the specimen; (**b**) The shearing surface of the specimen.

7. Conclusions

This study investigates the constitutive relationship of the welded CGHAZ using an 80-mm-thick DH36 marine steel plate as the research object. Through quasi-static tensile testing combined with advanced DIC measurement technology and an innovative semi-spraying speckle method, the precise localization of measurement points in the CGHAZ was achieved. Metallographic testing enabled the accurate identification of fusion line boundaries. Stress-strain curves for both the CGHAZ and BM were obtained. Based on the Ramberg-Osgood model, constitutive equations were formulated for the BM and welded CGHAZ; the following conclusions were drawn:

(1) The experimental results revealed that the CGHAZ exhibited significantly higher tensile strength than the BM, with fracture strains of 11.86% for the BM and 9.7% for the CGHAZ, indicating distinct deformation compatibility between the two regions. This anomalous strengthening effect originates from the unique microstructural characteristics of the CGHAZ: the coarse-grained structure leads to a reduced grain boundary density per unit volume. Since grain boundaries, as defect-rich regions and stress concentration sites, are preferential locations for crack nucleation, the reduced number of grain boundaries effectively suppresses microcrack initiation and propagation, thereby enhancing macroscopic tensile strength. In nonlinear mechanical analyses, it is imperative to establish a dual-material constitutive model based on strain evolution to separately characterize the deformation behavior and failure mechanisms of the CGHAZ and BM. This approach is critical for accurately predicting the mechanical performance of welded structures.

(2) This study innovatively employed DIC measurement technology to construct a precise constitutive model for the CGHAZ successfully. By utilizing a non-contact, full-field strain measurement method (spatial resolution: 5 μm), this technique overcomes the technical limitations of traditional extensometers in localizing microscale CGHAZ regions. In shipbuilding practices, leveraging these precise material parameters enables designers to optimize structural layouts more rationally, such as refining plate thickness near weld zones and strategically arranging stiffeners. Such optimiza-

tions ensure structural strength and stability under diverse operational conditions, effectively mitigating safety hazards caused by insufficient consideration of CGHAZ performance. This advancement significantly enhances the safety and reliability of marine structures, providing a robust framework for engineering applications in offshore industries. In the future, interested researchers may extend the proposed methodology to studies of other materials.

Author Contributions: L.X., conceptualization, formal analysis, and methodology; P.Z., writing—original draft preparation and investigation; T.Y., literature search, study design, and validation; S.Z., data curation, figures, and data analysis. J.H., investigation, data analysis, and validation; M.L., formal analysis and supervision. All authors have read and agreed to the published version of the manuscript.

Funding: This research received no external funding.

Data Availability Statement: All data generated and analyzed during this study are included in this published article. The datasets used and analyzed during the current study are available from the corresponding author upon reasonable request.

Conflicts of Interest: The authors declare no conflicts of interest.

Abbreviations

The following abbreviations are used in this manuscript:

HAZ	Heat-affected zone
CGHAZ	Coarse-grained heat-affected zone
SCHAZ	Subcritical heat-affected zone
FGHAZ	Fine-grained heat-affected zone
DIC	Digital Image Correlation

References

1. Zhang, C.Y.; Yuan, F.; Wang, N. Research on Double-sided Double-arc Welding of T-shaped Fillet Welds in Ship Thick Plates and Quality Improvement. *Eng. Mach.* **2023**, *54*, 50–60+8.
2. Liu, C.; Hu, L.; Yu, L. Analysis and Reflection on Modern Ship Welding Technology. *Mark. Res. Inf.* **2023**, *10*, 163–165.
3. Liu, R.; Chen, L.; Wei, Y. Microstructure simulation of TIG welding pool and heat affected zone of nickel based alloy. *J. Weld.* **2020**, *41*, 64–68+100. [CrossRef]
4. Beghini, M.; Bertini, L.; Puccio, D.F. Evaluation of Mechanical Properties of Welded Joints by Small-Specimen Tests. *Strain* **2001**, *37*, 155–158. [CrossRef]
5. Gan, Z.; Zhao, H.; Li, X. Research on the Properties of the Heat-Affected Zone of Low-Alloy High-Strength Steel Welding and Improvement Measures. *Weld. Pipe* **2024**, *47*, 7–16. [CrossRef]
6. Yang, Q. Numerical Simulation of Temperature Field and Coarse Grain Zone Phase Content of DP78 Steel Welding Joint. *Mech. Eng. Mater.* **2021**, *5*, 91–95.
7. Xiong, J.; Chen, W.; Qu, J. Analysis of the Size and Properties of the Heat-Affected Zone of 16Mn Steel MAG Welding Joint. *J. Xihua Univ. Nat. Sci. Ed.* **2011**, *30*, 85–87+101.
8. Georgios, K.T.; Dimitrios, I.Z.; Dionysios, L.K. Residual Stresses and Micro-Hardness Testing in Evaluating the Heat Affected Zone's Width of Ferritic Ductile Iron Arc Welds. *Mater. Sci. Appl.* **2016**, *7*, 73–82.
9. Liu, Y.; Duan, Y.; Xin, P. Width of the Heat-Affected Zone of Fully Automatic TIP TIG Welding Joint of X70/316L Bimetallic Composite Pipe. *Electr. Weld. Mach.* **2015**, *45*, 152–156.
10. Sun, Q.; Li, X.; Yao, T. Experimental Study on Crack Expansion Law of Brittle Materials under Explosive Loading Based on DIC. *Explos. Shock.* **2019**, *39*, 31–41.
11. Wang, W.; Chen, Z.; Huang, S. Study on the Mechanical Properties of H60 Brass with Notches Based on DIC. *Exp. Tech. Manag.* **2022**, *39*, 72–78. [CrossRef]
12. Liu, D. Evaluation of the Mechanical and Forming Properties of Laser Welding Joints of 6063 Aluminum Alloy. Master's Thesis, Yanshan University, Qinhuangdao, China, 2021. [CrossRef]

13. Li, Q.; Li, Z.; He, X. Study on the Local Mechanical Properties of Welding Joints of Zirconium-Titanium Composite Plates. *Rare Met.* **2020**, *44*, 1240–1248. [CrossRef]
14. Zhang, X.; Peng, F.; Zhang, H. Study on the Fatigue Crack Expansion Characteristics of Welded Rails Based on DIC. *Electr. Weld. Mach.* **2023**, *53*, 44–48+62.
15. Rasmussen, K.J.R. Full-Range Stress-Strain Curves for Stainless Steel Alloys. *J. Constr. Steel Res.* **2003**, *59*, 47–61. [CrossRef]
16. Shi, G.; Zhu, X. Research on the Constitutive Model of High-Strength Structural Steel under Monotonic Load. *Eng. Mech.* **2017**, *34*, 50–59. [CrossRef]
17. *GB/T 22315-2008*; Metallic Materials—Test Method for Elastic Modulus and Poisson's Ratio. China Standard Press: Beijing, China, 2008.
18. Cao, Y. Advances in the Testing of Mechanical Properties of Metals. *Metall. Anal. Test. (Metall. Phys. Test Sect.)* **1983**, *1*, 45–51.
19. Lan, T.; Kang, L.; Li, R. Analysis of the Mechanical Properties of Q355GNH Weathering Steel after Corrosion and Research on the Constitutive Model. *Ind. Constr.* **2024**, *54*, 49–57.

Journal of
Marine Science and Engineering

MDPI

Review

Multi-Branch Towed Array System: Systematic Analysis of Modeling Methods, Environmental Responses and Mechanical Properties in Fracture Analysis

Jin Yan [1,2], Kefan Yang [1], Shengqing Zeng [1], Keqi Yang [3], Dapeng Zhang [1,*] and Keqiang Zhu [4]

1 Ship and Maritime College, Guangdong Ocean University, Zhanjiang 524005, China
2 Guangdong Provincial Key Laboratory of Intelligent Equipment for South China Sea Marine Ranching, Guangdong Ocean University, Zhanjiang 524088, China
3 School of Electronics and Information Engineering, Guangdong Ocean University, Zhanjiang 524088, China
4 Faculty of Maritime and Transportation, Ningbo University, Ningbo 315211, China
* Correspondence: zhangdapeng@gdou.edu.cn

Abstract

Multi-branch towed array systems are an important component of subsea information collection, which is increasingly required for subsea pipeline laying and offshore platform construction as ocean energy is exploited. However, the complexity of underwater conditions poses challenges for marine towing systems when collecting information, including the possibility of towing cable collisions with protruding seabed or submerged organisms during towing system travel, or towing cable interactions during torsion. These collisions can affect and interfere with the collection of information by the towing system, and can cause damage to the towing system or even cause the towing cable to break. After the failure and detachment of the outboard guide cable of a multi-branch towing cable array, the formation of the towing system changes, and these changes are complex and related to the prevailing sea state. To study the important condition of the damaged towing system, this paper draws an analogy between the towing system and the trawl net, and speculates the formation change and mechanical response of the multi-branch towed array system after damage by combining the influencing factors of the deployment of the towing system.

Keywords: multi-branch towed array system; mechanical response; array variation

1. Introduction

With the rapid progress of modern submarine silencing technology [1–6], anti-submarine warfare has posed unprecedented challenges to the performance of underwater detection systems. Towed Array Sonar (TAS) [7–9] has become the core equipment for military anti-submarine and marine resources exploration because it can move the acoustic receiving unit away from the mother ship to realize long-distance and high-precision underwater target detection [10–12]. In addition, it has also been applied in the seabed topography exploration [13,14], marine environment monitoring, marine rescue, marine life protection, marine resources exploration and underwater archaeology. In addition, in seabed terrain exploration, marine environment monitoring, marine rescue, marine life protection, marine resources exploration and underwater archaeology, towed line array has also been applied [15–22]. Among them, the Multi-branch Towed Acoustic Array System (MBTAAS) effectively solves the port and starboard fuzzy problem of the traditional single-line array by laying multiple spatially separated towing cables in parallel, which significantly enhances the target orientation capability and detection coverage [9,23–26]. The system usually consists of the aft leading cable, the air gun energy transfer cable and the main cable of the aft towed acoustic array, while the outboard guiding cable, as a key component to maintain the overall spatial configuration, prevents entanglement and collision of the towed cables under complex sea conditions by restraining their relative positions, which guarantees the geometric integrity and functional stability of the acoustic array [27]. The system has been designed to provide the most effective and efficient way to provide the most suitable protection for the acoustic array to ensure target detection.

In engineering practice, all tools and structures are subject to damage. These damages often need to be monitored in real time to avoid irreversible damage to the building or tool [28–35]. Similarly, damage can occur to the sideboard cables of multibranch line arrays. A break in the outboard guiding cable during towline travel can result in loss of support for the towline, causing the towline system's trajectory to become unrestricted and the course and equilibrium of the towline system to be altered. In order to avoid this extreme phenomenon. Researchers tend to focus more on fatigue monitoring of towing cables to avoid breakage. However, few researchers pay attention to the mechanical response and formation changes after the towline breaks. Monitoring of the towline can minimize accidents, and fracture research is as important as detection, because special conditions at sea are unavoidable. The re-stabilization process after the breakage of outboard guiding cables and towing cables also involves mechanical response and attitude control, so the process analysis of towline formation and towline steady state can be used to analyze and summarize the phenomenon of towing system breakage to a certain extent. Existing studies mostly focus on the overall morphological changes of the tow cable after breakage, but do not clearly define the specific location of the breakage. For breaks at different positions, there are significant differences in the mechanical transfer path and the degree of system instability. For example, the connection point between the guide cable and the main cable is prone to becoming a stress concentration area as it needs to bear both the axial tension of the main cable and the lateral restraint force of the guide cable. In the middle part of the guide cable, the influence of ocean current vortex-induced vibration is more significant, and the risk of breakage increases with the degree of coupling between the vortex street frequency and the natural frequency of the guide cable. Therefore, clarifying the definition and causes of the breakage location is a prerequisite for accurately analyzing the dynamic response of the system.

In order to study the phenomenon of sideboard cable breakage, the multi-branch line array and the tugboat are modeled in this paper and the control equations and mechanical analysis equations are given, and the model is used to simulate the sideboard's unilateral or bilateral guided cable breakage in different sea states, and the results of the formation change when the outboard guide cable fails and detaches are obtained. Subsequently, in order to study the mechanical response and the factors affecting the towline take-up, the influencing factors during towline deployment are analyzed, and we analyze the trawl net during the trawling period, which also has no sideboard guide cable, so as to speculate the influencing factors and mechanical response during the towing period. Finally, this paper analyzes the mechanical response of the towline in different states and sea states to analyze the phenomena that may arise in case of breakage. This paper adds to the direction of towline array bunching in the field of towing cables, and has a certain guiding effect in this field.

2. Cable Discrete Theory and Modeling Basics

2.1. Model Building

During the simulation, the towing speed of the tugboat is always 2 m/s and the direction is always to the right. In the simulation, the horizontal direction to the right is the X-axis positive direction, the direction obtained by rotating the X-axis positive direction counterclockwise by 90° in the horizontal plane is the Y-axis positive direction, and the direction of the vertical horizontal plane upward is the Z-axis positive direction, as shown in Figure 1.

Figure 1. Modeling of towing boats and towing line arrays.

2.2. Control Equations for Towing Cable Motion

Three coordinate systems are established to describe the motion of the whole system, inertial coordinate system (xyz), towing cable local coordinate system (btn) and towing body motion coordinate system $(\xi\eta\zeta)$. The origin of the inertial coordinate system is located at the water surface, and the z-axis is vertically upward; the local coordinate system of the towing cable is attached to the towing cable, and the t-axis is the tangential direction of the cable, the n-axis is the normal direction, and the b-axis is the side-normal direction; and the coordinate system of the motion of the towing body follows the towing body's motion, with the ξ-axis pointing to the bow, the η-axis pointing to the starboard side, and the ζ-axis pointing to the bottom. These three coordinate systems are related to each other by the attitude angle, and the transformation relationship is as follows:

$$\begin{bmatrix} x \\ y \\ z \end{bmatrix} = A \begin{bmatrix} b \\ t \\ n \end{bmatrix} = Q \begin{bmatrix} \xi \\ \eta \\ \zeta \end{bmatrix} \tag{1}$$

where A is the towing cable coordinate system transformation matrix

$$\mathbf{A} = \begin{bmatrix} \cos\theta & \sin\theta\cos\phi & -\sin\theta\sin\phi \\ -\sin\theta & \cos\theta\cos\phi & -\cos\theta\sin\phi \\ 0 & \sin\phi & \cos\phi \end{bmatrix} \tag{2}$$

where Euler's angle $\boldsymbol{\theta}$, the attitude angle of the towing cable in the inertial coordinate system, depends only on the spatial location of the cable, and its value range is set as

$$\begin{array}{c} \theta \in (-180^\circ, 180^\circ] \\ \phi \in \left[\ -90^\circ, 90^\circ\ \right] \end{array} \tag{3}$$

According to the idea of centralized mass method, a tow cable is discretized into N segments, i.e., $N+1$ nodes, from the tail end to the top end, where the tail end $s=0$ is the $i=0$ th node and the top end $s=S$ is the $i=N$ th node.

Applying Newton's second law to the ith node yields the basic equations of motion for the towing cable node

$$M_i \ddot{x}_i = F_i \tag{4}$$

where: x is its spatial position vector, $\ddot{x}$ is the acceleration; M_i is the mass matrix, including the inertial mass of the towline node itself and its additional mass in the water $\boldsymbol{M}_{a,i}$

$$\boldsymbol{M}_i = 0.5(\mu_{i-1/2}l_{i-1/2} + \mu_{i+1/2}l_{i+1/2})\boldsymbol{I} + \boldsymbol{M}_{a,i} \tag{5}$$

$$\boldsymbol{M}_{a,i} = 0.5(\boldsymbol{M}_{a,i-1/2} + \boldsymbol{M}_{a,i+1/2}) \tag{6}$$

$$\boldsymbol{M}_a = \rho l \sigma \begin{bmatrix} 1 - \sin^2\theta\cos^2\phi & -\cos\theta\sin\theta\cos^2\phi & -\sin\theta\cos\phi\sin\phi \\ -\cos\theta\sin\theta\cos^2\phi & 1 - \cos^2\theta\cos^2\phi & -\cos\theta\cos\phi\sin\phi \\ -\sin\theta\cos\phi\sin\phi & -\cos\theta\cos\phi\sin\phi & \cos^2\phi \end{bmatrix} \tag{7}$$

$\boldsymbol{I}$ is a 3×3 unit matrix; μ, l, σ are the mass per unit length of the tow cable, the length between nodes, and the cross-sectional area, respectively; the subscript $i + 1/2$ denotes the physical quantity between node i and node $i + 1$, e.g., $l_{i+1/2}$ denotes the length between node i and node $i + 1$, i.e., $l_{i+1/2}$. For simplicity, some subscripts in subsequent formulas will be omitted unless specified otherwise.

F_i are all the forces acting on the node i, including the basic towline tension, buoyancy, gravity, and fluid resistance, and if there are other external forces on the towline, they should be included.

$$F_i = \Delta T_i + B_i + G_i + D_i \tag{8}$$

(1) Tension $\boldsymbol{T}$ in general the strain of the cable $|\varepsilon_i| \ll 1$, the stress–strain relationship can be elastically treated.

$$\begin{gathered} \Delta T_i = T_{i+1/2} - T_{i-1/2} \\ T_{i+1/2} = E\sigma\varepsilon_{i+1/2}\tau_{i+1/2} \\ \varepsilon_{i+1/2} = \\ \sqrt{(x_{i+1} - x_i)^2 + (y_{i+1} - y_i)^2 + (z_{i+1} - z_i)^2} / \\ l_{i+1/2} - 1 \end{gathered} \tag{9}$$

where: τ is the unit tangent vector in the direction of the cable length; E is the modulus of elasticity. The same reason can be obtained $T_{i-1/2}$ taking into account the tension and compression elastic properties of the towing cable is different, in this paper the towing cable under pressure (strain is negative) when the modulus of elasticity is taken as an approximation for the tensile 1/10.

(2) Buoyancy $\boldsymbol{B}$ and gravity $\boldsymbol{G}$, the buoyancy and gravity of node i can be expressed as

$$\boldsymbol{B}_i + \boldsymbol{G}_i = -0.5\rho(l_{i-1/2}\sigma_{i-1/2} + l_{i+1/2}\sigma_{i+1/2})\boldsymbol{g} + 0.5(\mu_{i-1/2}l_{i-1/2} + \mu_{i+1/2}l_{i+1/2})\boldsymbol{g} \tag{10}$$

where $\boldsymbol{g}$ is the acceleration of gravity.

(3) Fluid resistance $\boldsymbol{D}$ The dragline resistance is divided into tangential and normal directions and can be expressed after accounting for the effect of cable strain as follows

$$\boldsymbol{D}_i = 0.5(\boldsymbol{D}_{i+1/2} + \boldsymbol{D}_{i-1/2}) \tag{11}$$

$$\begin{gathered} \boldsymbol{D} \approx -0.5\rho l d\sqrt{1+\varepsilon}(C_n|\dot{\boldsymbol{x}}_{nb}|\dot{\boldsymbol{x}}_{nb} + \\ \pi C_t \mid \dot{\boldsymbol{x}}_t \mid \dot{\boldsymbol{x}}_t) \end{gathered} \tag{12}$$

where: $\dot{\boldsymbol{x}}$ is the velocity vector of the node; d is the diameter of the cable; C_n, C_t are the normal and tangential drag coefficients, respectively.

We then give the boundary conditions for the free ends of the towing cable, and for the free ends without the towing body (cables B, C), which are treated as one node, we apply the centralized mass method directly.

$$M_0 \ddot{x}_0 = F_0 \tag{13}$$

$$\begin{aligned} \boldsymbol{M}_0 = 0.5(\mu_{1/2} l_{1/2} \boldsymbol{I} + \boldsymbol{M}_{a.1/2}); \boldsymbol{F}_0 = \boldsymbol{T}_{1/2} + \\ 0.5(-\rho_{1/2}\sigma_{1/2} + \mu_{1/2} l_{1/2})\mathbf{g} + 0.5\boldsymbol{D}_{1/2}. \end{aligned} \tag{14}$$

This gives the basic control equations for each towing cable

$$\begin{aligned} \frac{d\dot{x}_i}{dt} &= \boldsymbol{M}_i^{-1} \boldsymbol{F}_i \\ \frac{dx_i}{dt} &= \dot{\boldsymbol{x}}_i \\ i &= 0, 1, \cdots, N-1 \end{aligned} \tag{15}$$

Towline breakage or outboard breakage will have a very complex situation on the steady state of the towline array and the mechanical response of the towline. The above relatively simple equations for controlling the motion of the towline can simplify the difficulty of controlling the towline, and can roughly describe the motion of the towline and the equilibrium process of rebalance of the towline array after the towline or the outboard guiding cable breaks.

2.3. Mechanical Equations for Towing Cables

In Yang's study, they also divided the cable into $N + 1$ nodes and in their experiments, they obtained that the forces acting on the nodes on the towing cable originate from the forces acting on the neighboring units.

$$\begin{aligned} \begin{bmatrix} \mathbf{G}_{node1} & \mathbf{F}^{dt}_{node1} & \mathbf{F}^{dn}_{node1} \\ \mathbf{G}_{node2} & \mathbf{F}^{dt}_{node2} & \mathbf{F}^{dn}_{node2} \\ \cdots & \cdots & \cdots \\ \mathbf{G}_{noden} & \mathbf{F}^{dt}_{noden} & \mathbf{F}^{dn}_{noden} \end{bmatrix}_{n\times 3} = \begin{bmatrix} \frac{1}{2} & 0 & 0 & \cdots & 0 \\ \frac{1}{2} & \frac{1}{2} & 0 & \cdots & 0 \\ \cdots & \cdots & \cdots & \cdots & \cdots \\ 0 & 0 & 0 & \cdots & \frac{1}{2} \end{bmatrix}_{n\times(n-1)} \\ \times \begin{bmatrix} \mathbf{G}_{ele1} & \mathbf{F}^{dt}_{ele1} & \mathbf{F}^{dn}_{ele1} \\ \mathbf{G}_{ele2} & \mathbf{F}^{dt}_{ele2} & \mathbf{F}^{dn}_{ele2} \\ \cdots & \cdots & \cdots \\ \mathbf{G}_{elen-1} & \mathbf{F}^{dt}_{elen-1} & \mathbf{F}^{dn}_{elen-1} \end{bmatrix}_{(n-1)\times 3} \end{aligned} \tag{16}$$

Here $\mathbf{G}_{noden}$, $\mathbf{F}^{dt}_{noden}$ and $\mathbf{F}^{dn}_{noden}$ are the gravitational, tangential and normal resistance forces acting on node $\boldsymbol{i}$, respectively. Further $\mathbf{G}_{elen-1}$, $\mathbf{F}^{dt}_{elen-1}$ and $\mathbf{F}^{dn}_{elen-1}$ are the gravitational, tangential and normal resistance forces acting on cell I, respectively [36].

In their experiments, they also obtained a mechanical relationship between $\boldsymbol{i} - 1$ and $\boldsymbol{i}$.

$$\begin{aligned} -\mathbf{F}_{\{i-1,i\}} + \tfrac{1}{2}\mathbf{F}^{dt}_{elei-1} + \tfrac{1}{2}\mathbf{F}^{dn}_{elei-1} + \tfrac{1}{2}\mathbf{G}_{elei-1} + \tfrac{1}{2}\mathbf{F}^{dt}_{elei} + \tfrac{1}{2}\mathbf{F}^{dn}_{elei} + \tfrac{1}{2}\mathbf{G}_{elei} \\ +\mathbf{F}_{\{i,i+1\}} = \mathbf{0} \end{aligned} \tag{17}$$

where $\mathbf{F}_{\{i-1,i\}}$ is the spring force between node $\boldsymbol{i}$ and node $\boldsymbol{i} - 1$

The control equations for each point of the tow cable in the towline array were also obtained in Yang's experiments, the mathematical relationships of which can be found in [37].

$$\int_{x_n}^{x_{n+1}} \rho_a \delta \boldsymbol{r} \cdot \ddot{\boldsymbol{r}} dx = \delta \boldsymbol{q}_\alpha \cdot \boldsymbol{F}_k^\alpha + \delta \boldsymbol{w} \cdot \mathbf{W}_k \tag{18}$$

In addition to this, they added general elastic forces, Morison's forces and gravity vectors at points on the towing cable [38].

$$(\boldsymbol{F}_e^{\alpha})_n = \frac{\partial V_n}{\partial q_\alpha} = \left(k_{n\bar{\varepsilon}}^{\alpha\beta\mu\upsilon} Q_{\mu\upsilon} + k_{n\kappa}^{\alpha\beta\mu\upsilon} \widetilde{Q}_{\mu\upsilon}\right) \boldsymbol{q}_\beta \tag{19}$$

$$(\boldsymbol{F}_m^{\alpha})_n = \frac{\delta W_n}{\delta q_n} = \int_{x_n}^{x_{n+1}} n^\alpha \cdot \left(\begin{array}{c} \text{sgn} \cdot \frac{\pi}{2}\rho_f C_f D(\boldsymbol{v} \cdot \boldsymbol{i})^2 \cdot \boldsymbol{i} + \frac{1}{2}\rho_f C_d D \boldsymbol{P v P v} + \\ \rho_f (1 + C_m) \pi \frac{D^2}{4} \boldsymbol{P} \dot{\boldsymbol{U}} - \rho_f C_m \pi \frac{D^2}{4} \boldsymbol{P} \ddot{\boldsymbol{r}} \end{array} \right) dx \tag{20}$$

$$\left(\boldsymbol{F}_g^{\alpha}\right)_n = \int_{x_n}^{x_{n+1}} n^\alpha \rho_w \boldsymbol{g} dx \tag{21}$$

They presented the equation for the absorption of waves by a tow cable [39].

$$\begin{array}{c} \eta_t + \widetilde{\varphi}_x \eta_x - (1 + \eta_x^2) w + \mu \eta = 0 \\ \widetilde{\varphi}_t + \frac{1}{2}(\widetilde{\varphi}_x)^2 + g\eta - \frac{1}{2}(1 + \eta_x^2) w^2 + \mu \widetilde{\varphi} = 0 \end{array} \tag{22}$$

Then, they obtained the discrete equation for the towing cable after accounting for wave effects.

$$\boldsymbol{F}_1^{\alpha} = -\int_{\Gamma_n^{FS}} \boldsymbol{N}_x^\beta \boldsymbol{N}_x^\mu dx \widetilde{\varphi}_\beta \eta_\mu - \int_{\Gamma_n^{FS}} \frac{1 + \eta_\pi^2}{\eta + h} \boldsymbol{N}^\alpha \boldsymbol{N}^\beta dx \widetilde{\omega}_\beta + \int_{\Gamma_n^{FS}} \mu(x) \boldsymbol{N}^\alpha \boldsymbol{N}^\beta dx \eta_\beta \tag{23}$$

$$\begin{array}{l} \boldsymbol{F}_2^{\alpha} = -\frac{1}{2} \int_{\Gamma_n^{FS}} \boldsymbol{N}^\alpha \boldsymbol{N}_x^\beta \boldsymbol{N}_x^\mu dx \widetilde{\varphi}_\beta \widetilde{\varphi}_\mu - \\ g \int_{\Gamma_n^{FS}} \boldsymbol{N}^\alpha \boldsymbol{N}^\beta dx \eta_\beta + \frac{1}{2} \int_{\Gamma_n^{FS}} \frac{1+\eta_z^2}{(\eta+h)^2} \boldsymbol{N}^\alpha \boldsymbol{N}^\beta \boldsymbol{N}^\mu dx \widetilde{\omega}_\beta \widetilde{\omega}_\mu + \\ \int_{\Gamma_n^{FS}} \mu(x) \boldsymbol{N}^\alpha \boldsymbol{N}^\beta dx \widetilde{\varphi}_\beta \end{array} \tag{24}$$

2.4. Other Common Types of Discrete Methods for Towing Cables

Based on the derivation of the control equations and mechanical equations of the tow cable motion, this section focuses on the key technical aspect of the discretization method of the tow cable, systematically introducing various commonly used discretization methods including the Cosserat rod model, semicontinuous model, lumped mass method, absolute nodal coordinates model and finite difference method (see Table 1). The aim is to provide a theoretical basis for selecting appropriate modeling methods according to specific scenarios in subsequent research, while also laying the methodological foundation for in-depth analysis of the dynamic response of the tow cable under special conditions such as breakage.

Table 1. Comparison table of common tow cable discrete methods.

Simulation Methods	Features of the Methodology	Scope of Application
Cosserat rod model	Translation and rotational degrees of freedom are included to accurately characterize geometric nonlinearity and derive complexity	It is suitable for the dynamic analysis of one-dimensional flexible rod systems with fine simulation geometric nonlinearity and torsional effect
Semicontinuous model	Continuously distributed parameters are used between discrete nodes to adapt to small deformation and linear elastic scenarios	Analysis of linear elastic scenarios such as static/quasi-static bending and deflection of cables with composite structures
Lumped mass method	Discrete point centralized mass simplified system, efficient but ignoring the moment of inertia	It is used for rapid simulation analysis under dynamic response and complex environmental loads of marine towing systems

Table 1. *Cont.*

Simulation Methods	Features of the Methodology	Scope of Application
Formulation of absolute nodal coordinates model	Describing the flexible body by the position of the node, high-precision processing of large deformation and complex contact, high cost	It is suitable for the analysis of geometric nonlinearity and complex contact problems of large deformed flexible bodies
Finite difference method	Discrete cables are grids, and the differential equations are used to approximate the derivative, and the principle is simple but the adaptability is narrow	Research on the basic dynamic properties of simple boundary and linear problems (such as vibration and verification theoretical solutions)
Elastic rod mode	Based on Hooke's law, consider linear elastic deformation, relatively simple in dealing with small-scale, slow-varying deformation of rods, assume small displacements and linear stress–strain relations	Suitable for preliminary analysis of linear elastic deformation of rods in static or low-frequency dynamic cases, like small-deflection bending of simple elastic support rods
Nonlinear beam model	Account for large displacements, large rotations, and nonlinear material behavior (e.g., geometric nonlinearity from large deformations, material nonlinearity like elasto-plasticity). Incorporate effects like shear deformation, warping for more accurate complex scenario representation	Applied to analyze large-deformation, high-load nonlinear responses of beams, such as in structural collapse simulation, flexible structural dynamics under extreme loads, where geometric and material nonlinearities are prominent

2.4.1. Cosserat Rod Model

In the field of deep-sea engineering and ocean observation, the towing cable is a key flexible structure connecting the mother ship and underwater equipment, and its analysis of large deformation dynamics relies on accurate theoretical modeling and efficient discrete methods. The Cosserat rod model, with its explicit characterization of translation-rotation coupling effects, becomes the core framework for solving the geometric nonlinearity problem of towing cables. The model regards the tow cable as a continuum with independent rotational degrees of freedom, and accurately captures the coupling behavior of bending, torsion and tension of the tow cable under ocean current load through the displacement field $\mathbf{u}(s,t) = \mathbf{u}_0(s,t) + \boldsymbol{\theta}(s,t) \times \mathbf{r}$ and strain metrics (tensile strain $\epsilon = \partial \mathbf{u}_0 / \partial s$, directional curvature $\boldsymbol{\kappa} = \partial \boldsymbol{\theta} / \partial s$), and the dynamic governing equation realizes the mechanical self-consistency through force balance $\nabla \cdot \mathbf{T} + \mathbf{f} = \rho \ddot{\mathbf{u}}$ and moment balance $\nabla \cdot \mathbf{M} + \boldsymbol{\theta} \times \mathbf{T} + \mathbf{m} = \rho \mathbf{J} \ddot{\boldsymbol{\theta}}$.

2.4.2. Semicontinuous Model

In the analysis of tow cable mechanics, the choice of discretization methods directly affects the balance between computational efficiency and simulation accuracy. The semicontinuous model, as a hybrid analysis framework integrating discretization and continuity, provides an effective mechanical modeling solution for complex tow cable structures, especially for multi-layer braided tow cables. Its core idea is to describe the mechanical behavior of the tow cable through a coupled strategy of "discrete nodes + continuous distributed parameters": dividing the length of the tow cable into n discrete nodes, using the vector q_i to represent the motion state of the nodes, which includes linear displacement (x_i, y_i, z_i) and angular displacement $(\theta_{x_i}, \theta_{y_i}, \theta_{z_i})$, while retaining the continuous distributed mechanical parameters between the nodes. Through the beam theory, the distributed force is associated with the deformation characteristics. Specifically, the bending moment between nodes is described by the differential equation of the deflection curve $EI\frac{d^4w}{ds^4} = q(s)$, where the coupled relationship between the bending stiffness EI and the distributed load q(s) enables the model to capture the continuous characteristics of the tow cable's bending deformation. Based on the Hamiltonian principle or the principle of virtual work, the inertial forces of the discrete nodes and the continuous distributed forces are further coupled into the overall dynamic equation $M\ddot{q} + Kq = F$, achieving a balance between the discretization computational efficiency and the continuous deformation accuracy.

2.4.3. Lumped Mass Method

In the dynamic response analysis of the marine towing system, the accurate simulation of the mechanical behavior of the tow cable, as the core flexible connection component, relies on reasonable discretization modeling methods. As one of the mainstream towing cable discretization technologies, the lumped mass method realizes the simplified solution of complex dynamic problems by discrete the continuously distributed tow cable mass into limited concentrated particles. The core idea of the method is to divide the tow cable into n-segment elements, corresponding to $n + 1$ concentrated particles, which are connected by massless "spring-damping" elements—where the spring stiffness k_i describes the elastic properties of the element, the damping coefficient c_i characterizes the energy dissipation effect, and the motion of each particle i follows Newton's second law, and its dynamic equation is expressed as:

$$m_i\ddot{\mathbf{r}}_i = \mathbf{F}_{\text{spring},i} + \mathbf{F}_{\text{damp},i} + \mathbf{F}_{\text{ext},i} \tag{25}$$

In the equation, the spring force $\mathbf{F}_{\text{spring},i}$ is determined by the relative displacement difference of adjacent particles $(k_i(\mathbf{r}_{i+1} - \mathbf{r}_i) - k_{i-1}(\mathbf{r}_i - \mathbf{r}_{i-1}))$, the damping force $\mathbf{F}_{\text{damp},i}$ depends on the relative velocity difference $(c_i(\dot{\mathbf{r}}_{i+1} - \dot{\mathbf{r}}_i) - c_{i-1}(\dot{\mathbf{r}}_i - \dot{\mathbf{r}}_{i-1}))$, and the external load $\mathbf{F}_{\text{ext},i}$ covers the environmental effects such as fluid force and wave force.

2.4.4. Formulation of Absolute Nodal Coordinates Model

In the dynamic analysis of offshore towing cables, the Absolute Nodal Coordinate Model (ANCF) is an accurate discretization method to solve the problems of large deformation and complex contact. the core of this method is to directly describe the state of the towing cable unit through the node absolute coordinate vector $\mathbf{X}_i = [x_i, y_i, z_i, \theta_{x_i}, \theta_{y_i}, \theta_{z_i}]^T$, which avoids the traditional coordinate conversion problem. The towing cable deformation field is represented by the polynomial interpolation function S(s) to $\mathbf{r}(s,t) = S(s)\mathbf{X}(t)$, and the quadratic or cubic spline interpolation can effectively capture the nonlinear deformation under strong currents. The dynamic equation $\mathbf{M}\ddot{\mathbf{X}} + \mathbf{K}\mathbf{X} = \mathbf{F}$ based on the virtual work principle forms the core framework, the constant mass matrix M simplifies the calculation, and the K with geometric and elastic stiffness accurately reflects the mechanical behavior of large deformation. The significant advantages of ANCF are reflected in the fact that extreme conditions such as tow cable breaking, swing, strong current vibration, etc., can be simulated with high accuracy (natural frequency deviation of $\leq$3%) without additional coordinated rotational degrees of freedom; naturally supports the simulation of dynamic contact between the towing cable and the seabed and the cable-to-cable contact, providing the basis for collision damage assessment.

2.4.5. Finite Difference Method

In the numerical analysis of the dynamic response of the tow cable, the finite difference method (FDM) is one of the earliest discretization methods, and its core idea is to realize the numerical solution of the control equation by discretizing the continuous towing cable structure into finite nodes and approximate the derivative terms with the node function value. For the transverse vibration problem of the tow cable, the control equation is established based on Newton's second law, and the expression is:

$$\rho \frac{\partial^2 u}{\partial t^2} - EA \frac{\partial^2 u}{\partial s^2} = f(s,t) \tag{26}$$

where ρ is the line density of the tow cable, EA is the axial stiffness, $\mathrm{f(s,t)}$ is the external load such as current field drag, the left side of the equation represents the inertial force and elastic recovery force, and the right side is the external force term. The finite difference method divides the discrete process of the tow cable into two dimensions: in the spatial discreteness, the total length of the tow cable L is uniformly divided into N segments, the step size is $\Delta s = L/N$, and the node i (position $s_i = i\Delta s$) is formed, and the second-order spatial derivative is approximated to $\frac{\partial^2 u}{\partial s^2} \approx \frac{u_{i+1} - 2u_i + u_{i-1}}{\Delta s^2}$ by the center difference, and the accuracy of this format reaches $O(\Delta s^2)$, which is suitable for the linear distribution description of one-dimensional tow cables; The time discrete adopts a step size of Δt, dividing the continuous time into discrete time $t_n = n\Delta t$, and the second-order time derivative is approximately:

$$\frac{\partial^2 u}{\partial t^2} \approx \frac{u_i^{n+1} - 2u_i^n + u_i^{n-1}}{\Delta t^2} \tag{27}$$

After substituting the above discrete format into the control equation, the explicit or implicit difference equation can be obtained, in which the explicit format $u_i^{n+1} = 2u_i^n - u_i^{n-1} + \frac{EA\Delta t^2}{\rho \Delta s^2}(u_{i+1}^n - 2u_i^n + u_{i-1}^n) + \frac{\Delta t^2}{\rho} f_i^n$ is widely used in the basic characteristics analysis of the tow cable because it does not require solving the large equation system.

2.4.6. The Theoretical Basis and Research Context of the Discrete Method of Towing Cables

As a slender flexible member, the mechanical analysis of the tow cable relies on the geometry of the space curve and the mechanics of the elastic rod, and the discrete method is the core of the continuum model to the numerical model, and the research revolves around "geometric description—mechanical modeling—discrete solution". Geometrically, the Frenet coordinate system provides a local representation of the centerline of the tow cable: based on the non-telescopic arc coordinate $\mathbf{r}(s)$, the position of the tow cable is described by the vector diameter of $\mathbf{t} = \frac{d\mathbf{r}}{ds}$, and the orthogonal frame composed of $\mathbf{b} = \mathbf{t} \times \mathbf{n}$ becomes the reference for describing the attitude of the discrete element. In mechanical modeling, the torsional deformation of the tow cable is quantified by the bending torsion of $\kappa(s) = \frac{d\boldsymbol{\theta}(s)}{ds}$, and its components correspond to bending and torsion, which are directly related to the relative attitude of the discrete element. The coupling between the inertial and local systems is solved by derivative transformation which provides a guarantee for the dynamic equation. In terms of research and evolution, based on the elastic rod balance equation ($\frac{d\mathbf{N}}{ds} + \mathbf{q} = 0$, $\frac{d\mathbf{M}}{ds} + \mathbf{t} \times \mathbf{N} + \mathbf{m} = 0$) in the early days, the tow cable was discrete into $\mathbf{N}$ elements, and the force $\mathbf{M}$ and moment were related to the deformation (κ, ε), and the element attitude transfer relied on the Frenet-based vector matrix transformation. Subsequently, combined with the surface differential equation $\mathbf{R}(s, \rho, \phi)$, the three-dimensional morphological description of the discrete element is expanded, and the local deformation of the section can be captured. The current core focus is "balancing discrete accuracy and computational efficiency": the encryption node improves the accuracy of κ, and the reduced stiffness matrix simplifies the mechanical transfer (such as $\begin{bmatrix} \mathbf{N} \\ \mathbf{M} \end{bmatrix} = \mathbf{K} \begin{bmatrix} \varepsilon \\ \kappa \end{bmatrix}$), realizing the efficient solution of dynamic response in complex environments, and laying the foundation for applications in marine engineering, resource exploration and other fields.

2.4.7. Nonlinear Beam Model

As the core flexible component of marine engineering, the dynamic behavior of large deformation of the tow cable needs to be accurately described by the nonlinear beam model. The current mainstream model is based on four core assumptions: first, it considers axial elongation, transverse shear and bending co-deformation, and breaks through the limitations of Euler–Bernoulli beams; second, the bisymmetrical section with the shear center and the center of shape coincided, which is simplified to in-plane analysis; third, ignoring the transverse secondary stress σ_y, τ_{yz} that is $\sigma_y = \tau_{yz} = 0$, focusing on the axial normal stress and transverse shear stress; fourth, the linear elastic is constitutive of

$$\sigma_{ij} = \frac{E}{1+v}\varepsilon_{ij} + \frac{v}{1+v}\Theta\delta_{ij} \tag{28}$$

Among them, E represents the elastic modulus, v represents Poisson's ratio, $\Theta = \sigma_x + \sigma_y + \sigma_z$ and ε_{ij} are the Kronecker symbols, $i, j = x, y, z$.

It was established based on generalized Hooke's law, and the nonlinearity is derived from the large deformation geometric effect described by the green strain tensor. In the model, the force of the tow cable includes the axial current field distributed load and the terminal concentrated load, and the displacement field is decomposed into four degrees of freedom by Taylor series expansion, covering the rigid body and deformation motion.

Regarding the core issue of "morphological changes and mechanical responses after the fracture of the outboard guide cable of the multi-branch towed array" studied in this paper, three requirements need to be met: large deformation simulation under complex sea conditions, capture of multi-cable contact and collision, and adaptive dynamic simulation of computational efficiency. By comparing the requirements of this paper with the Features of the methodology of 7 types of discrete methods, in the "guide cable damage" scenario studied in this paper, the response of the towed cable system is divided into "overall morphological evolution" (such as the starboard cable shifting towards the ship's center) and "local fine behavior" (such as the vibration of the broken end and the cross-collision of the cables). The lumped mass method can efficiently calculate the overall morphology to meet the simulation requirements of multiple sea conditions, but it has insufficient accuracy for local contact and large deformation; although the ANCF has high accuracy, it has a large number of calculations. Therefore, a hybrid scheme of "overall lumped mass method + local ANCF correction" is adopted to ensure both efficiency and solve the error problem in key areas.

3. Dynamic Response in Different Engineering Situations

Before analyzing the dynamic response of the multi-branch towed array system under different engineering scenarios, it is crucial to validate the established discrete cable model, governing equations, and mechanical analysis framework. This section aims to confirm the reliability of the symmetric model, elaborate on the validation process, and provide a solid theoretical and methodological foundation for subsequent result analysis. As shown in Figure 2, in this paper, the symmetric model validation is carried out by comparing the simulation results with the experimental data of existing research and the data from physical model tests (the data are referenced in Tables 2 and 3). The focus is on the consistency between the mechanical properties and morphological characteristics of the simulated towed array system and the actual measured values.

Figure 2. Schematic diagram of the multi-branch towed line array model.

Table 2. Multi-line array system cable structure table.

	L/m	Line Type
Leader A/Leader D	120	Lead-in
Leader B/Leader C	80	Lead-in
Array A/Array D	170	Lead-in(70 m) + Streamer(100 m)
Array B/Array C	170	Lead-in(70 m) + streamer(100 m)
Div line	13	Lead-in
Spreader	60	Warp
Gun umbilical	70	Gun umb
Gunlead	10	Gun umb
Diverter Ballast	2	Diverter base

Table 3. Linear parameter table.

	d/m	ρ_c/kg × m^{-1}	EA/kN	EI/kN × m^2	$C_{d\tau}$	C_{dn}	C_{db}	$C_{a\tau}$	C_{an}	C_{ab}
Streamer	0.08	5.2	2000	0.02	0	1.5	1.5	0	1.0	1.0
Lead-in	0.05	5.9	5000	0.02	0.01	1.5	1.5	0	1.0	1.0
Warp	0.025	0.74	2500	0	0.01	1.5	1.5	0	1.0	1.0
Gun umb	0.05	5.9	6000	0.04	0.01	1.5	1.5	0	1.0	1.0
Diverter base	0.2	300	1	0	0	1.2	1.2	0	1.0	1.0

This system is mainly composed of two hydrophone arrays and an air gun. Its structural model is shown in Figure 2. The lines in the figure represent various cable models. A special switch model (Link) can flexibly connect two mass points in the model, enabling the operation of shortening and separating the distance between the mass points. In the cable connection design, the cable Spreader is specifically connected between Coupler A and Coupler B, while the heavy cable Div line plays an important role in maintaining the overall stability of the system. At the same time, there are 4 short ballast heavy cables Diverter Ballast connected to the heavy cable Div line. These short cables are attached to the diverter, further enhancing the balance performance of the system. The hydrophone array, which is the key to signal acquisition in the system, consists of a leader cable Leader, a streamer array Streamer, a tail buoy Tail at the end of the array and a 3D floating body Coupler. Each component works together to ensure the accuracy and stability of signal acquisition. The air gun system responsible for signal transmission is composed of an air gun Airgun, a leader cable Gunleader and an umbilical cable Gunumbilical. The umbilical cable provides power and signal transmission support for the air gun. The specific parameters of the towing cable can be referred to Tables 2 and 3, and the Poisson's ratio of all line types is uniformly taken as 0.5. The system model built according to the above design is shown in Figure 2.

Based on the discrete cable model, control equations, and mechanical analysis framework established in the second part, a foundation is provided for investigating the dynamic behavior of multi-branch towed array systems in actual marine environments. This chapter will focus on different engineering scenarios (such as water current, wave action, etc.), analyzing the dynamic response laws of the system under special working conditions such as the breakage of the external guidance cable, and this article will first explore the impact of water current on the dynamic characteristics of the system.

3.1. Dynamical Response to the Action of Wave and Current

There are three types of situations when the current is acting, which are classified as countercurrent, down current and crosscurrent [40–43]. Down current means that the forward direction of the tugboat is the same as the direction of the current [44–46], countercurrent means that the forward direction of the tugboat is opposite to the direction of the current [47–49], and crosscurrent means that crosscurrent means that the current direction is perpendicular to the longitudinal direction of the ship's hull [50–54]. In Ahmed 's article [51], he adds that during the towed array system unfolding, in addition to environmental factors affecting towing cable deployment, phenomena such as collisions between towing cables may also affect deployment [55–59]. In their paper, they state that marine cables can exhibit properties similar to those of flexible rods [60–62] and briefly analyze mutual interference between towing cables by drawing on research related to flexible rods [54,63–65]. However, most of the current studies have been carried out based on the unfolding of multi-branch towed array systems [66–69] and the equilibrium of multi-branch towed array systems after unfolding [70–74], and only a very small part of the studies have been carried out to investigate the bunching phenomenon that occurs after fracture in the towing cable. In order to study the mechanical response of the process, we have investigated the phenomenon through simulations.

In traditional fluid mechanics, reference frames in uniform motion can be equivalently transformed. However, for a multi-branch towed array system, due to the flexibility of the tow cable, multi-constraint coupling, and the vector characteristics of fluid forces, constant currents in different directions/speeds result in significant differences. As a flexible body, the movement of the tug needs to be transmitted to the end section by section through tension, while the drag/lift of the fluid on the tow cable directly depends on the local relative velocity, leading to a delay in the transmission of relative velocity. For example, when flowing downstream, if the current velocity is close to the tug's velocity, the end of the tow cable is first pushed by the water current due to 'power transmission lag', causing the overall compression and coiling; when flowing upstream, the front end of the tow cable is first subjected to the water current resistance, and the tension is concentrated at the root. Moreover, the restraining effect of the outboard fairlead has a directional property. When the current direction changes, the balance between the restraining force and the current field force is broken, resulting in the coupling of multiple constraints and the current field. For example, when there is a cross-current (90°), the current field force acts directly on the transverse direction of the tow cable. After the fairlead breaks, the offset of the tow cable is 3–5 times that of the upstream current, and collisions across the ship's mid-plane are likely to occur. In addition, the forces have vector directions. The drag force of the fluid on the tow cable is in the same direction as the relative velocity, and the lift force is perpendicular to the relative velocity direction. When flowing upstream, the drag force is along the axial direction of the tow cable, when flowing crosswise, the drag force is along the transverse direction, and when flowing downstream, the drag force is along the axial direction. The force balance equations of these three cases are completely different and cannot be unified through reference frame transformation.

3.1.1. Dynamic Response of the Port Side Guide Cable During Failure Fracture Under Countercurrent Action

In this scenario, the breakage position of the mooring line is set at the root, and the reasons for the selection are as follows: (1) Under the condition that the forward direction of the tugboat is opposite to the ocean current, the axial tension of the main cable increases with the increase of the ocean current velocity, resulting in a further concentration of the force transfer load at the root position; (2) In Table 3 of the original text, the EI of the "Lead-in type mooring line" is 0.02 kN·m^2, indicating a relatively low flexural stiffness. The root position, as a "rigid-flexible connection point", is prone to stress concentration due to the sudden change in stiffness. Observation of Figure 3 shows that: the faster the countercurrent velocity is, the larger the angle of the broken port side cable A (i.e., the left-side guiding cable labeled A in the multi-branch towed array system, see Figure 2 for the system structure) deviates from the vertical downward direction after the system is stabilized; it can be expected that, if the countercurrent velocity continues to increase, the broken port side cable A will rotate around its still restrained end by a certain clockwise angle under the impact of the current (the angle is greater than 90°), but at this time, the towing end of the port side cable A is at the rear and its free end is submerged in the water below. However, at this time, the towing end of the port side guide cable A is at the rear and immersed in the water, and eventually the port side guide cable A is traveling steadily with its free end at the front and its towing end slanting downward, and the larger the current velocity is, the larger the angle of rotation around its own end is. Further observation reveals that the formation of the port-side towing cable undergoes a process in which the overall formation shifts to the port side first and then gradually returns to the center of the ship. The reason for this phenomenon is that the initial offset to the port side is due to the loss of the constraints of the port side guide cable A. When it continues to shift t to the left side to a certain extent, the port side towing cable A begins to be constrained by the limitations of the unfolding device A. Coupled with the current direction of the countercurrent at this time, the above factors lead to the port side towing cable A's formation position will ultimately and gradually close to the middle of the boat direction.

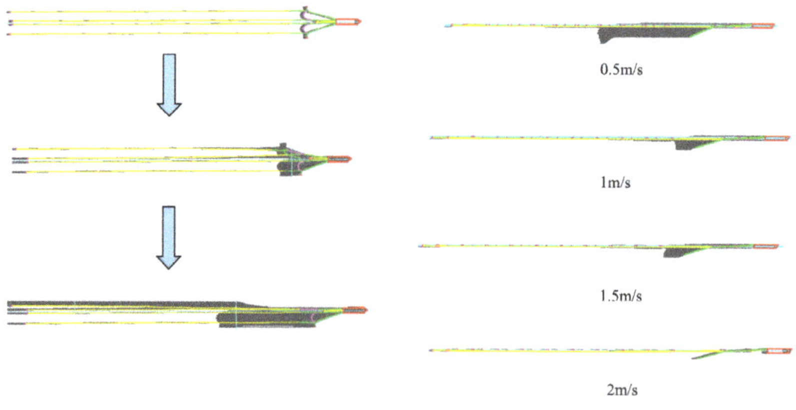

Figure 3. Formation changes of port-side guide cable A (root fracture) under countercurrent conditions (**left**: top view, **right**: side view; different countercurrent velocities marked).

Due to the breakage of the port side guide cable A, the towing system on the left side of the line array formation is significantly affected, when the system as a whole to achieve stability after the port side of the towing cable A formation position relative to the system intact when the overall movement to the right; the left side of the two towing cables between the formation of the spacing has become very small, and in the port side of the two guide cables near the tail end of the formation of the reunion of the two left side of the towing system and with the current velocity increases, the two left side towing cables overlap the formation interval is also becoming larger, so that mutual collision occurs; this phenomenon will make the two cables embedded in the actual project will be seriously damaged. With the increase of the current velocity, the two left side towing cables the overlapping position interval between the two left-side towing cables is also becoming larger, so that mutual collision occurs, this phenomenon in the actual project will make the two cables embedded in the sonar by serious damage, and due to the breakage of the outboard guide cable and the disorganization of the left side towing cable formation, the tension of the left side towing cable A is transferred to the left side of the spreader, resulting in the left side of the spreader also being subjected to a great deal of tension; in the process of the left side of the air gun umbilical cable will first be offset to the port side and then gradually returns toward the ship's midline and the tension it bears is also significantly increased. For the right side of the line formation, the right side of the line formation was not much affected after the breakage of the port side guide cable, and the larger the current velocity, the better the right side of the line formation was maintained.

When the current direction is countercurrent, it can be inferred from the tension distribution of each towing cable at different current speeds that the greater the countercurrent velocity, the greater the tension on each towing cable, and due to the breakage of the port side guide cable, the dragging force on the two left-side cables is finally concentrated at the end of the middle guide cable B, which results in the tension on the middle guide cable B being much larger than that on the middle guide cable B1. The left spreader A is also subjected to a larger tension relative to the right spreader B.

In countercurrent and down current, the two cables on the right side of the towing system keep good formation, and there is no big formation change in the X and Y directions, while the two cables on the left side of the towing system change in the X and Y directions mainly due to the breakage of the guide cable on the left side of the port side of the cable; and in crosscurrents, due to the lateral impact of the sea currents, the towing system of the whole towing system, therefore, in the Y direction of the formation of the formation of the formation of the position of the larger changes will occur, and the closer to the tail end of the towing cable, the larger the offset in the Y direction, which is due to the closer to the tail end of the towing the stronger the flexure of the towing system caused by.

3.1.2. Dynamic Response of a Double Outboard Guide Cable During Failure Fracture Under Countercurrent Action

Due to the large number of towing cables in the system, if two guide cables break at the same time, the system is prone to mutual entanglement, making it difficult for the whole system to enter the static equilibrium stage quickly. Moreover, the longer the side cables, the worse the convergence of the system when they break at the same time, because with the increase of the length, the possibility of mutual entanglement of the two cables before reaching the static equilibrium stage and the duration of entanglement will be greatly increased. To facilitate the convergence of the system, the port and starboard cables are set to break at different stages. The starboard side cable is broken at the initial moment when the system enters the static equilibrium phase, and the port side cable is broken at the moment when the system enters the dynamic simulation phase. Later in this chapter, when it comes to the failure of both side cables, the starboard side cable breaks at the initial moment when the system enters the static equilibrium phase, and the port side cable breaks at the moment when the system enters the dynamic simulation phase.

Observation of Figure 4 reveals that under countercurrent conditions, when both the port and starboard outboard guiding cables break, it is found that under different countercurrent velocities, when the whole multibranch towing system reaches stabilization, the towing cable formation in the whole towing system is symmetrical between left and right, and the larger the countercurrent velocities, the shorter the time needed for the towing system to reach the final stabilization stage is. There are some differences in the spatial pattern changes along the cable length at each time step before the whole towing system reaches stability, and the spatial pattern changes of the two guide cables are not synchronized at the initial stage, which is due to the difference in the moment of fracture; however, when the whole system enters into the stable stage, all the cables including the left and right side guide cables are symmetric at each time step, and the greater the countercurrent current, the shorter the time required for the towing system to reach the final stable stage. towing cables have the same cable body morphology at each time step.

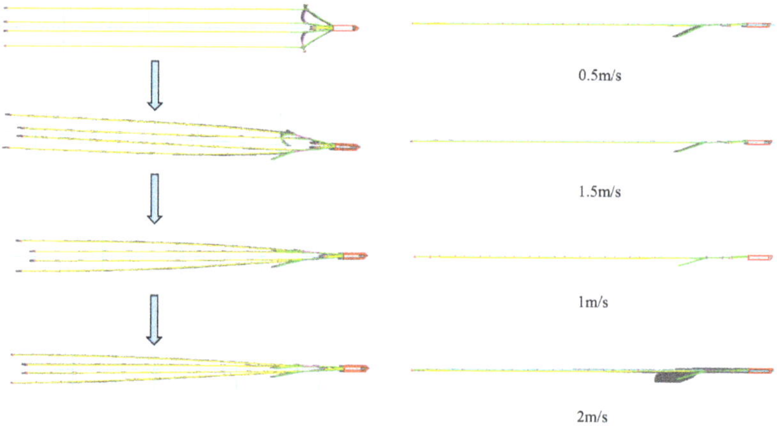

Figure 4. Symmetrical formation changes of multi-branch towed array after bilateral outboard guide cable fracture (countercurrent conditions, **left**: top view, **right**: side view).

3.1.3. Dynamic Response of the Port Side Guide Cable During Failure Fracture Under Downstream Action

In this case, the tugboat advances in the same direction as the current. As the towing cable is a flexible component, if the speed of the tugboat is less than the current speed in the downstream state, the towing cable will be compressed and curled by the impact of the current and it is difficult to achieve static equilibrium, resulting in the system failing to converge. Therefore, in the downstream state, the speed of the tugboat should be much larger than the speed of the current, so as to ensure that the force exerted by the sea current on the towing cable produces an axial tensile effect. After several simulation adjustments in the simulation, it is found that in this case, the speed of the tugboat should be at least 4–6 times of the current speed to ensure the convergence of the whole multi-branch towing system. That is to say, in this case, the greater t the difference between the tugboat speed and the current speed, the better the convergence of the system, the maximum speed of the current is about a quarter of the speed of the tugboat, when the current speed is very small, it is equivalent to the tugboat sailing in the still water.

As Figure 5 shows that when the downstream current velocity is smaller than the towing speed of the tugboat, the larger the current velocity is, the smaller the angle of the broken port side cable deviates from the vertical downward direction after the system is stabilized, so it can be deduced that: if it can be realized in simulation, when the current velocity reaches the same speed as that of the towing speed of the tugboat, at this time, the broken port side cable should be in the natural hanging state. The larger the downstream current velocity, the smaller the angle of the broken port side cable deviating from the vertical downward direction after the system is stabilized; it can be expected that the larger the downstream current velocity is, the closer the broken port side cable is to the natural overhanging position (approaching but not reaching it), but at this time, the towing end of the port side cable is submerged into the water at the rear and bottom, and finally, the port side cable with its end at the front and the final towing end in the diagonally downward attitude is traveling steadily, and the larger the current speed is, the more it is in the natural overhanging state, and the greater the current velocity the smaller the angle at which it deviates from its natural overhanging position around its own end.

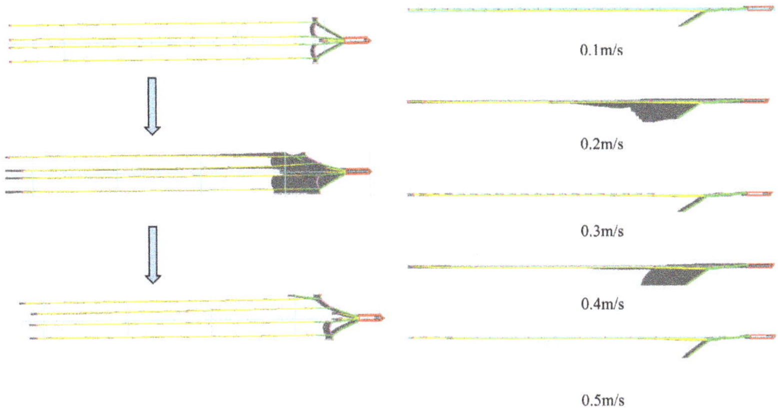

Figure 5. Port-side guide cable A formation deviation under downstream conditions (tug speed > current speed, left: top view, right: side view; current speed marked).

It was further observed that when the current velocity was 0.1 m/s, 0.3 m/s, 0.4 m/s, 0.5 m/s under the downstream condition, the overall formation of the port-side towing cable A did not undergo a significant offset back to the center of the ship after it was deployed after offsetting to the port side first, but when the downstream current velocity was 0.2 m/s, the phenomenon that the port-side line array was offset back to the center of the ship again. This is in marked contrast to the calculation results in the countercurrent condition. The spatial change of the overall formation of the port side towing cable A in the countercurrent condition will first be leftward and then return to the center of the ship, while there is no obvious phenomenon of returning to the center of the ship when the current velocity is low in the downstream condition. Combined with the results of countercurrent, it can be concluded that the reason for this phenomenon in the downstream state after the breakage of the port side of the guide cable A is: in the initial state to the port side of the offset is due to the loss of the constraints of the port side of the guide cable A, and when it continues to the left side of the offset to a certain extent, the port side of the towing cable A began to be restricted by the constraints of the unfolding of the spreader A, coupled with the direction of the current for the downstream tangential velocity is very small, so that the current of the Drag force is reduced, so the axial stretching effect on the towing cable is weakened, because the length of the towing cable is limited, so the weakening of the axial stretching effect in the transverse direction of the role of the expansion can be extended to increase the length of the above factors lead to a variety of comprehensive factors lead to the port side of the towing cable A formation position in the deviation of the port side of a certain distance directly to the dynamic equilibrium, so in the downstream state there is no general to the direction of the ship in the direction of the again However, when the downstream velocity is 0.2 m/s, the discharge frequency of the velocity synthesized by the velocity and the towing speed of the tugboat is similar to the self-oscillation frequency of the spreader A, which will trigger the high-frequency vibration of the spreader A, and the perturbation brought about by this vibration will break the equilibrium of the left line array system, and eventually make the formation of the port side towing cable A offset to the ship's mid-direction once again. That is to say, in the downstream state, between the minimum allowable operating current speed and the maximum allowable operating current speed, it is possible to have a current speed that will make the towing line array formation spacing greatly reduced, and its specific size is related to the tugboat's towing speed and the self-oscillating frequency of the spreader on the towing line array. It is possible to change its own vibration period by adding some attachments or reinforcing materials to the spreader, which can ensure a better formation spacing of the multi-branch towed array system in all the allowable current speed ranges during downstream operation in some sea areas. Therefore, it is not the case that the lower the current speed, the better the overall formation of the system will be maintained in the downstream condition.

In the downstream current, except for individual current, except for the current velocity of the larger, multi-branch towed array system formation stabilization after the horizontal spacing between the towing cable in the system to achieve stability can be maintained relatively stable; and in the countercurrent current, with the current velocity increases, multi-branch towed array system formation stabilization after the horizontal spacing between the towing cable the smaller. The horizontal spacing between the towing cables after stabilization of the multi-branch towed array system in the downstream condition is larger than that in the countercurrent condition.

It can easily be concluded that when the downstream current speed is less than the towing speed of the tugboat, the higher the current speed, the lower the tension in each towing cable. This is due to the fact that the combined velocity of the tugboat speed and the current decreases as the current speed increases resulting in a decrease in the velocity of the towing cable relative to the current, which leads to a decrease in the towing force on the towing cable.

It is further found in the simulation that when the direction of current is the same as the direction of motion of the tugboat, the closer the current velocity is to the motion velocity of the tugboat, the more difficult it is for the system to converge. This is because when the current velocity happens to be isotropic with the tugboat velocity, theoretically it is equivalent to the tugboat traveling in the same direction with the sea current at the same speed as the sea current; however, due to the delayed motion transfer, the motion of the tugboat will not be immediately transferred to the back half of the towing cable system, so the motion of the back half of the towing cable in the system will be firstly affected by the action of the incoming current, and the direction of the incoming current will cause the back half of the towing cable to move to the right; and since the towing cable is a typical flexible component, under the delay of the ship's motion transfer, the towing cable will be compressed in the axial direction, and the shape of the towing cable becomes difficult to adjust, resulting in the nodes of the towing system are difficult to achieve static equilibrium on the spatial force, which will make the simulation into the dynamic phase of the failure, and the system will eventually diverge. This phenomenon shows that, in practical engineering, when the tugboat and the current are in the same direction, and the speed and the current velocity is the same, it is very unsuitable for towing operations, in this state of the entire towing system in the towing cable cannot be unfolded, and it is likely to be washed by the current to the end of the tugboat. If this kind of sea state engineering operation is unavoidable, it is advisable to increase the towing speed of the tugboat, make the towing speed of the tugboat is large enough to ensure that in a very short period of time can be the tugboat's movement to the back half of the towing system, ensures that all segments and nodes of the towing cables achieve static equilibrium, so as to ensure that the entire towing process can proceed dynamically and the stability of the dynamic simulation.

3.1.4. Dynamic Response of a Double Outboard Guide Cable During Failure Fracture Under Downstream Action

As Figure 6 shows the change of the multiline array after the breakage of the double outboard guide cables in the downstream state. As a matter of fact, the author found from years of simulation experience that in the countercurrent state, the speed of the tugboat can be extremely small or 5–20 times of the current speed, at which time the whole system is still relatively easy to reach the convergence state; however, in the downstream state, it is divided into two kinds of cases: one is that the towing speed of the tugboat is smaller than the current speed, and as the towing speed of the tugboat is closer to the current speed, the whole system's convergence becomes worse and worse, and when the tugboat's towing speed is not much different from the current speed, the whole towing system will not be able to converge; in the downstream state, when the speed of the tugboat is greater than the speed of the current, when the tugboat speed is 2–6 times of the current speed, the system can still converge, but the tension of each towing cable will be larger in this case. This tells us that in actual engineering practice, if we want to carry out towing operations in the downstream state, there are two methods to choose from, one is to travel at a towing speed that is much lower than the current speed, which has the advantage that the towing cables are under less tension but requires more time; the other is to make the tugboat travel at a high speed with a current speed that is 2–6 times of the current speed, but this has the advantage that it requires less time but the towing cables are under Tension is greater, easy to fatigue and fracture.

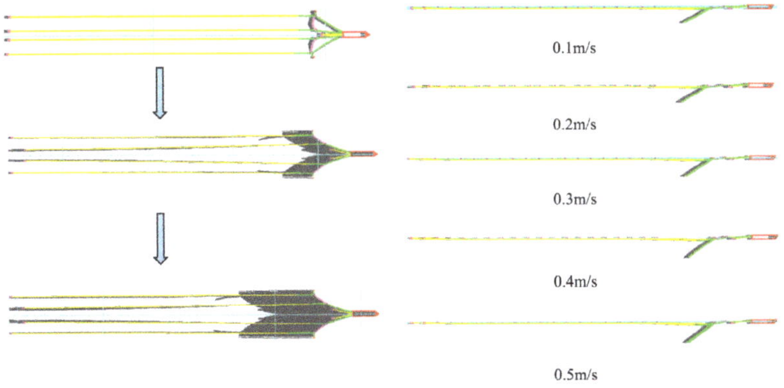

Figure 6. Schematic of the formation change, top view on the left and side view on the right. Velocity is current speed.

3.1.5. Dynamic Response to Failure Breakage of Port Side Guide Cable Under Cross Current (90°, 270°)

For the single-branch towing system, since there is only one towing cable in the whole system, within the safety limit, the allowable change of its formation pattern is still relatively large, so the change of the current direction in the single-branch towing system can be very large [75–78]. However, for the multibranch system, since there are multiple towing cables in the whole system, the change of the current direction, especially under the effect of the cross-current, the change of the formation pattern of each individual towing cable will be significantly changed. However, for a multi-branch towing system, because there are multiple towing cables in the system, changes in the direction of the sea current, especially under the effect of cross current, the formation position of individual towing cables will change significantly, and because the towing cables themselves are long flexural components, this change is nonlinear, and this change in the formation position of individual towing cables will lead to nonlinear changes in the distance between individual towing cables, and this change in the towing cables at different positions is random [79]. For a multi-branch towed system, if it is to be able to maintain a certain formation position in a certain cross-current sea state, it is necessary to maintain a certain formation position under normal operation. of a multibranch towing system, it is necessary to limit the magnitude of the cross-current speed.

Observation of Figures 7 and 8 reveals that, in the transverse current state, the flexure of the towing cable is fully manifested in the initial stationary stage of the multibranch system, and the transverse current direction leads to a small transverse oscillation of the towing cable's trailing end, which is opposite to the direction of the wave direction extension, and with the increase of the transverse current speed, the amplitude of the small transverse oscillation of the towing cable's trailing end will also increase. However, once the tugboat starts to move, the lateral stretching of the towing cable by the tugboat has a very obvious inhibiting effect on the transverse oscillation of each towing cable end. When the speed of the tugboat is stabilized, the transverse oscillations of the trailing end of the towing cable can be almost completely suppressed. This tells us that, for the transverse oscillation of the trailing end of the towing cable caused by cross-current, it can be suppressed by increasing the speed of the tugboat when the cross-current velocity is small, but the speed of the tugboat at this time should be limited to a certain range for safety reasons. That is to say, cross-current state of the tugboat towing speed can increase the magnitude is also very limited.

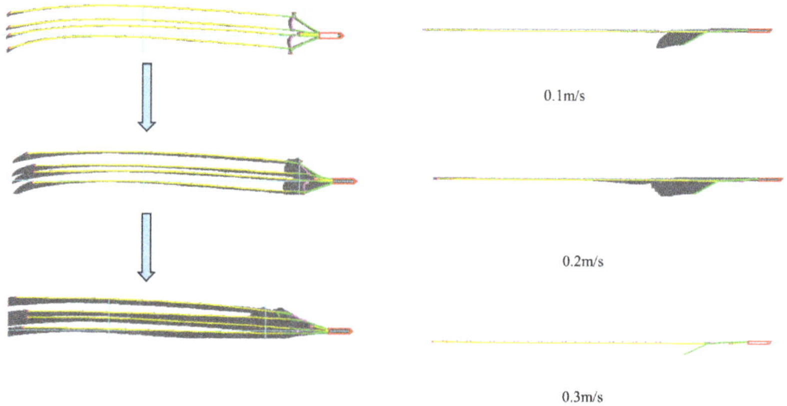

Figure 7. Lateral oscillation & formation shift of towed array after port-side guide cable fracture (cross-current: 90° direction, **left**: top view, **right**: side view; current speed 0.1–0.3m/s).

Compared to the size of the maximum current velocity at which the multibranch towing system can reach convergence under countercurrent and upstream current, the maximum current velocity at which the multibranch towing system reaches convergence under crosscurrent condition is much smaller. It is found in the simulations that for the towed line array in this chapter, the system has difficulty in reaching convergence when the cross current velocity exceeds 0.3 m/s. In a way, the difficulty of convergence in the simulation when the cross- current velocity is too large means that in practice it is difficult for the system to achieve the dynamic equilibrium and stability required for its in-situ operation. In practical engineering this means that it will be difficult for individual towing cables in a multi-branched line array to achieve a relatively stable formation when the cross-current current velocity exceeds a certain limit under cross-current conditions.

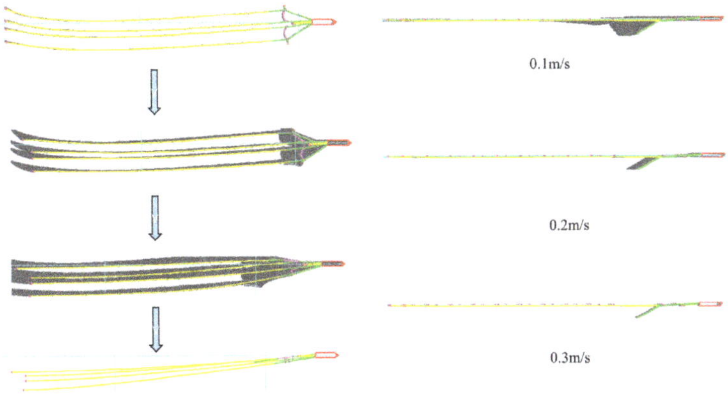

Figure 8. Asymmetric formation shift of left towed cables after port-side guide cable fracture (cross-current: 270° direction, **left**: top view, **right**: side view; current speed 0.1–0.3m/s).

And with the increase of cross-current current velocity, when the current velocity reaches 0.3 m/s, the whole multi-branch towing line system presents a completely different formation from that when the current velocity is 0.1 m/s and 0.2 m/s: when the port side guide cable breaks, when the cross-current current direction is 90°, the towing cables of the right half of the whole multi-branch towing line array have not been offset over the mid-station surface when the current velocity is 0.1 m/s and 0.2 m/s, and the starboard guide cable A1 is still functioning, but there is still a towing cable on the right and left sides of the tugboat. The starboard guide cable A1 is still functioning, and although the whole multi-branch towed array system is no longer symmetrical, the left and right sides of the tugboat still have the existence of the towing cable, and the formation is still maintained to a certain extent; while in the current velocity of 0.3 m/s, the starboard guide cable A1 does not break, but the band is no longer enough to maintain the right side of the guide cable, and all the towing cables in the right side of the multi-branch towing cable array system have already crossed the mid-station surface. The entire multi-branch towed array system on the right side of the towing cable has crossed the center station surface, and drastically shifted to the left. When the port side guide cable A breaks, the transverse current flow direction is 270°, due to the existence of the starboard side guide cable, when the transverse current flow velocity is small, the formation of the two towing cables in the right half of the multibranch towed array system can maintain a better shape, but the formation of the two towing cables in the left half of the left half of the system loses the port side guide cable A to maintain the formation of the two towing cables, it will be rapidly shifted to the right, and in the current velocity of 0.2 m/s, it has been all shifted across the center station surface, which indicates that when the transverse current flow velocity is 0.2 m/s, all of the towing cables in the right side of the system have already crossed the center station surface. This indicates that when the direction of the incoming cross current is on the same side as that of the broken port side guide cable, it will further reduce the ability of the line array to resist the cross current, while if the direction of the incoming cross current is not on the same side as that of the broken port side guide cable, the overall ability of the line array to resist the cross current will be slightly improved compared with the situation when the line array is on the same side, and with the increase of cross-current current velocity, when the current velocity reaches 0.3 m/s, the whole multi-branch towing line system presents a completely different formation from that when the current velocity is 0.1 m/s and 0.2 m/s: when the port side guide cable A breaks, when the cross-current current direction is 90°, the towing cables of the right half of the whole multi-branch towing line array have not been offset over the mid-station surface when the current velocity is 0.1 m/s and 0.2 m/s, and the starboard guide cable A1 is still functioning, but there is still a towing cable on the right and left sides of the tugboat. The starboard guide cable A1 is still functioning, and although the whole multi-branch towed array system is no longer symmetrical, the left and right sides of the tugboat still have the existence of the towing cable, and the formation is still maintained to a certain extent; while in the current velocity of 0.3 m/s, the starboard guide cable A1 does not break, but the band is no longer enough to maintain the right side of the guide cable, and all the towing cables in the right side of the multi-branch towed array system have already crossed the mid-station surface. The entire multi-branch towed array system on the right side of the towing cable has crossed the center station surface, and drastically shifted to the left. When the port side guide cable A breaks, the transverse current flow direction is 270°, due to the existence of the starboard side guide cable, when the transverse current flow velocity is small, the formation of the two towing cables in the right half of the multibranch towed array system can maintain a better shape, but the formation of the two towing cables in the left half of the left half of the system loses the port side guide cable A to maintain the formation of the two towing cables, it will be rapidly shifted to the right, and in the current velocity of 0.2 m/s, it has been all shifted across the center station surface, which indicates that when the transverse current flow velocity is 0.2 m/s, all of the towing cables in the right side of the system have already crossed the center station surface. This indicates that when the direction of the incoming cross current is on the same side as that of the broken port side guide cable, it will further reduce the ability of the line array to resist the cross current, while if the direction of the incoming cross current is not on the same side as that of the broken port side guide cable, the overall ability of the line array to resist the cross current will be slightly improved compared with the situation when the line array is on the same side.

Further comparative observation of two cross-current current direction under the state of the port side of the guide cable A breakage of the multi-branch towed formation changes can be found: when the cross-current current direction of 90°, the current velocity of 0.3 m/s, although the breakage of the port side of the guide cable A so that the left half of the two towing cables lost the constraints on the formation, but due to the right half of the multi-branch towline formation in this case the two cables are on the current, which is the impact of the two cables is greater than that of the left half of the two cables, so the right half of these two cables although the starboard guide cable A1 constraints, but still appeared in the overall bias toward the left half of the two towing cable. The impact of the two cables is greater than that of the two towing cables in the left half, so although the two cables in the right half have the restriction of the starboard side guide cable A1, there is still a phenomenon that the two cables in the right half are biased to the left side and at the same time are close to each other, and the two cables in the left half, due to the hindering effect of the two cables in front of them on the currents, receive the force of the transverse current, although the force of the currents has increased in relation to low-current speeds, but the increase in this increase Instead, this increase compensates for the lack of restriction caused by the breakage of the outboard guide cables, and thus the formation is relatively good. In other words, the sheltering effect of the onshore cable has a positive effect on maintaining the formation of the backwater cable when the cross-current velocity is high.

3.1.6. Dynamical Response of a Double Outboard Guide Cable During Failure Fracture Under Cross-Currents

Observation of the morphology changes of the multibranch towing system in the 90° and 270° cross-currents states in Figures 9 and 10 reveals that: the morphology of the line array after the multibranch line array reaches stabilization after the breakage of the double outboard guide cables is only related to the direction of the currents and the final stabilized line array morphology under the same cross-currents current speeds in the 90° and 270° cross-currents current directions is symmetrical to each other; further observation reveals that It is further observed that the line formation is not symmetrical in the two cross-current current directions before the line formation is stabilized, and this asymmetry is due to the non-synchronization between the moment of the action of the sea current under the two cross-currents and the moment of the breakage of the outboard guide cable. In both cross current directions, the starboard side guide cable breaks first, and then the port side guide cable breaks, but the changes of the line array before reaching stability are different in the two cases: when the direction of the cross current is 90°, the current acts on the right side of the line array in the simulation stage, and the port side guide cable does not break at the initial stage and still plays a role in maintaining the stability of the left side of the line array, whereas the starboard side guide cable has already broken, and therefore the line array has already been stabilized at the initial moment of the simulation, and the starboard side guide cable has already broken. At the beginning of the simulation has lost the role of the right side of the array to maintain the formation; and when the transverse current direction of 270°, the current in the simulation phase first acted on the left side of the line array, in the initial stage of the port side of the guide cable did not break also play a role in maintaining the stability of the left side of the array, so in the initial stage of the port side of the array can be maintained, and due to the current arrives at the moment of the starboard side And due to the moment of the current arriving at the starboard side and arriving at the port side, there is a certain time difference, so although the starboard side guide cable has been broken at the initial moment, but the formation of the right side of the array in the initial short period of time will still be maintained and stabilized, and with the increase of time, the action of the sea current extends to the starboard side, and the stable form of the starboard side of the array is then broken; which is exactly the cause of the two kinds of transverse currents under the action of the sea current at the moment of the moment and the moment of the breakage of the port side cable of the non-synchronous main reason. This is the main reason for the non-synchronization between the moment of current action and the moment of breakage of the outboard guide cable under the two cross currents. Therefore, the difference in the direction of the cross currents in the case of the breakage of both outboard cables affects the change of the line array when it has not yet reached the stabilization stage. Moreover, since the breakage moments of the port and starboard cables are different in this example, if the port and starboard cables break at the same time, the change of the line formation should be symmetrical for the two transverse current flow directions.

To summarize, the construction operation of multi-branch towing system under cross-current sea state has certain requirements on the size of cross-current current speed and the size of tugboat towing speed, which can be carried out within a certain range, but beyond a certain current speed, the formation of multi-branch towing line in cross-current state will inevitably appear in the formation disorder. Therefore, if due to certain practical needs or military needs, need to detect towing in the cross-current current velocity of the sea, in the increase of the spacing between the different towing cable at the same time, but also in the towing cable on the outside of the main nodes of the force to increase the bending strength of some of the larger strength of the support material, if possible, along the length of the towing system towing cable can be set up in sections along the length of the lightweight formation limiting frame, so that you can Greatly improve the ability of multi-branch towing system to operate in cross-current sea state. In the large cross current, it is not recommended to increase the towing speed of the tugboat to overcome the effect of cross current on the lateral swing of the towing cable and the lateral offset of the formation, because at this time the role of this method is quite limited, and in the large cross current if the tugboat is moving at high speeds, it is very likely to be capsized.

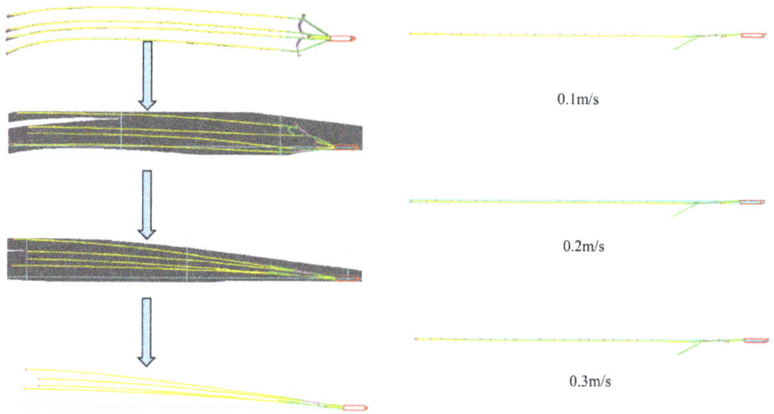

Figure 9. Schematic of the formation change, top view on the left and side view on the right. Velocity is the current velocity and direction is 90°.

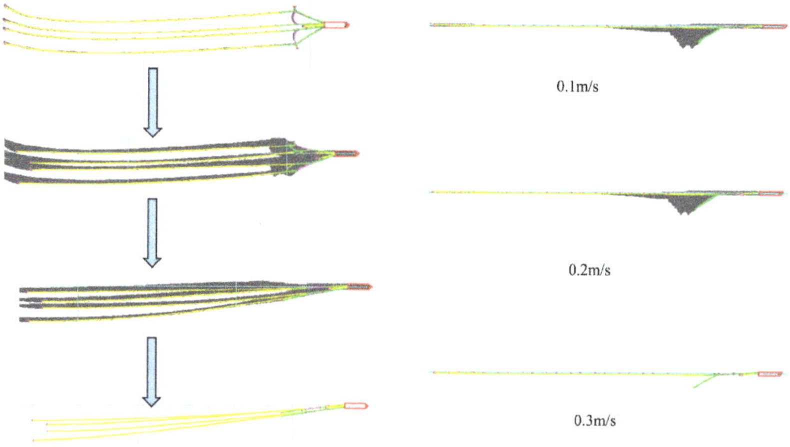

Figure 10. Schematic of the formation change, top view on the left and side view on the right. Velocity is the current velocity and direction is 270°.

3.2. Dynamic Response to Wave Action

In engineering terms, for the consideration of safety assurance and accuracy of measurement data, since towing operations are generally not carried out in waters with large wave heights and short wave periods, it is only meaningful to carry out detection towing in microwave and longer wave period waters [80–83]. Therefore, the wave height in this case is set to 0.5 m, the wave period is 6 s, and the wave theory is linear wave.

3.2.1. Dynamic Response of the Port Side Guide Cable During Failure Breakage Under Counter Wave Action

In the case where the wave direction is opposite to the advancing direction of the tugboat, the fracture location is still at the root, but the cause newly adds the "impact superposition effect" of wave loads: (1) Based on the wave absorption equation in Section 2.2, when sailing against the waves, the root position needs to bear both the axial dynamic tension of the main cable and the vertical impact force of the waves simultaneously. The equivalent stress fluctuation range expands to 130–170 MPa, approaching the yield strength. (2) As mentioned in Section 3.2.1 of the original text, "the number of bent components of the towing cable increases when sailing against the waves." The root position, as the connection point between the mooring cable and the expander, has its bending deformation restricted by the expander and cannot release stress through deformation. Eventually, it becomes the starting point of fracture. Observation of Figure 11 shows that the morphology of the multi-branch towed system changes in the counter-wave, and it is found that, unlike under the action of the current, under the action of the wave, it is obvious that under the wave, under the wave beat, the multi-branch towing line array produces wave vibration, and compared with the action of a single current, the vibration of the towing cable in the vertical direction in the towing system has become very obvious. This wave vibration is specifically reflected in the towing system of each node of the towing cable with the ups and downs of the waves and low amplitude, cyclic sustained cyclic vibration, this sustained cyclic vibration is very likely to cause resonance inside the towing cable inlaid sonar and thus reduce the accuracy of the measurements, but also prone to lead to fatigue damage, and in this case, the flexure of the towing cable is more fully embodied, compared with the towing system under the action of a single stream, the towing system under the action of the counter-wave multi-branch towing system in the towing cable irregular bending and torsion of the parts increased significantly, especially in the port side of the guide cable A breaks to the whole system to reach the stability of the phase before the towing system of the nonlinear vibration and torsion of the towing system increased significantly; and in the system to reach the stability of the whole system, the towing After the stabilization of the whole system, the individual towing cables in the system will continue to vibrate with low amplitude and periodicity in response to the cyclic fluctuations of the waves Observation of Figure 11 shows that the morphology of the multi-branch towing system changes in the counter-wave, and it is found that, unlike under the action of the current, under the action of the wave, it is obvious that under the wave, under the wave beat, the multi-branch towing line array produces wave vibration, and compared with the action of a single current, the vibration of the towing cable in the vertical direction in the towing system has become very obvious. This wave vibration is specifically reflected in the towing system of each node of the towing cable with the ups and downs of the waves and low amplitude, cyclic sustained cyclic vibration, this sustained cyclic vibration is very likely to cause resonance inside the towing cable inlaid sonar and thus reduce the accuracy of the measurements, but also prone to lead to fatigue damage, and in this case, the flexure of the towing cable is more fully embodied, compared with the towing system under the action of a single stream, the towing system under the action of the counter-wave multi-branch towing system in the towing cable irregular bending and torsion of the parts increased significantly, especially in the port side of the guide cable A breaks to the whole system to reach the stability of the phase before the towing system of the nonlinear vibration and torsion of the towing system increased significantly; and in the system to reach the stability of the whole system, the towing After the stabilization of the whole system, the individual towing cables in the system will continue to vibrate with low amplitude and periodicity in response to the cyclic fluctuations of the waves.

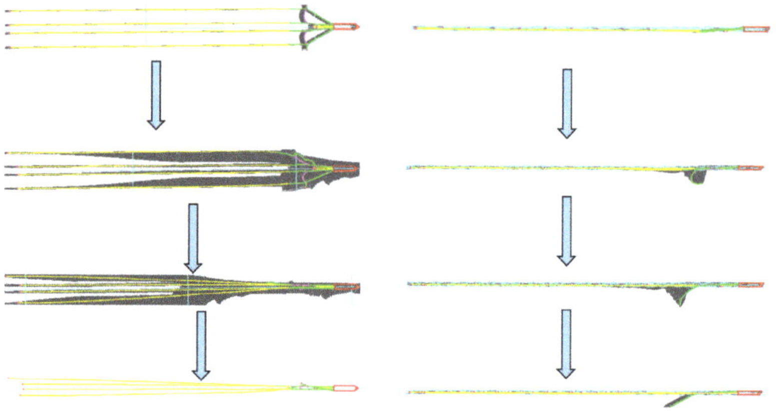

Figure 11. Vertical vibration & bending of towed cables after port-side guide cable fracture (counter-wave conditions: H = 0.5 m, T = 6 s, **left**: top view, **right**: side view).

3.2.2. Dynamic Response of a Double Outboard Guide Cable During Failure Breakage Under Counter-Wave Action

After further observation of Figure 12, it is found that, in the counter-wave direction, no matter the single outboard guide cable fails to work at sea and the double outboard guide cable fails to work at sea, when the towing system reaches the stabilization, the towing formation of the multibranch towing system is worse (the spacing between cables close to the end of the tail end of the towing cable is larger, and the spacing between cables nearer to the towing end of the towing cable is smaller), and the different spacings of each cable along the length of the towing cable are not equal, and the spacing distribution is more discrete and not regular. The spacing distribution is relatively discrete and irregular, and the left and right formation distribution of the whole multi-branch towing system is similar to two acute triangles with very small top angles, which is not conducive to the unfolding of offshore construction operations.

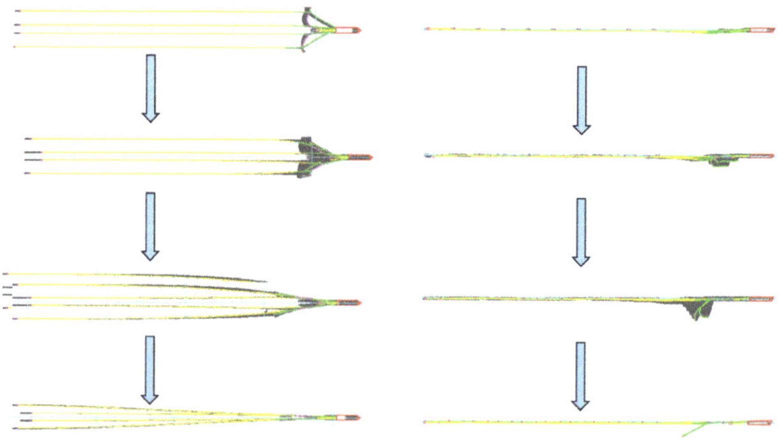

Figure 12. Schematic of the formation change, top view on the left and side view on the right. Velocity is wave velocity.

In fact, when the wave height gradually increases to a certain amplitude and the wave frequency increases to a certain amplitude under the counter-wave condition, it will cause the tugboat to undergo a certain amplitude of sustained transverse rocking, and this transverse rocking motion of the tugboat has an unnoticeable effect on the spatial shape of the formation of the multibranch towed array system, in addition to the unpredictable effect on the transverse spacing between each towing cable in the formation, the sustained and violent transverse rocking will In addition to the unpredictable effect on the lateral spacing between the towing cable in the formation, a sustained and violent cross-swing will make it easier for the towing cables to become intertwined. Due to space limitation, the effect of wave height and wave frequency changes on the formation of multi-branch towing lines in counter-waves is not further analyzed here, but in fact it is very instructive for towing operations in counter-waves in waters with a wide range of wave frequency and wave height changes. If the towing operation is to be carried out in such an area under some specific requirements, the relatively safe operating intervals (safe interval of wave height and wave frequency) can be obtained through this analysis, and interested readers can carry out a more in-depth and detailed study on this issue.

3.2.3. Dynamic Response of an Outboard Guide Cable During Failure Breakage Under Downwash Action

Under the premise that the wave height and period are the same and the wave speed is less than the speed of the tugboat (the wave height is set to be 0.5 m, the wave period is 6 s, and the wave theory is the linear wave theory), it is found after observing Figure 13 that the vibration of the towing cable in the towing system is more intense in the down wave than in the up wave, which is due to the fact that the wave speed and the tugboat's forward direction are the same in the up wave but the wave speed of the wave is the opposite of the tugboat's forward direction in the up wave. The wave speed is the same as the direction of the tug's progress in good waves, but the wave speed is the opposite of the tug's progress in bad waves. Tugboat in still water with a certain speed straight, the towing cable is in the state of being stretched, in the smooth waves caused by the towing force will increase the towing cable along the degree of axial stretching, which helps to reduce the violent degree of vibration of the towing cable; but when the counter-wave, due to the magnitude of the speed of the wave is smaller than the size of the tugboat's speed, this time, although the towing force caused by the wave cannot make the towing cable is compressed in the axial direction, but it will be weakened towing cable is axial stretching degree, towing cable axial stretching The weakening of the axial stretching of the towing cable results in the towing cable being more susceptible to violent vibration.

Comparative observation can be found, whether it is with the wave or against the wave, in the double outboard guide cable failure and breakage, when the whole multi-branch towing system is stabilized, the morphology change of the port side cable and the starboard side cable A1 will have a very small amplitude of asynchrony, this kind of asynchrony is due to the release of the two outboard cables at different moments resulting in the encounter with the wave is different at different moments caused by the waves are also cyclical fluctuations, this will Because the waves are also cyclic, this results in a small phase delay in the spatial morphology and vibration of the two outboard cables, which will always exist due to the cyclicity of the waves. However, under the action of the current, no matter whether it is downstream or countercurrent, because the current velocity is constant, when the whole multi-branch towing system is stabilized, after the failure and breakage of the two outboard cables, the morphology change of the port side cable and the starboard side cable A1 will be synchronized and stabilized to maintain in a certain shape, and after stabilization, no periodic vibration will occur in general (in a particular current velocity is possible to occur vortex vibration).

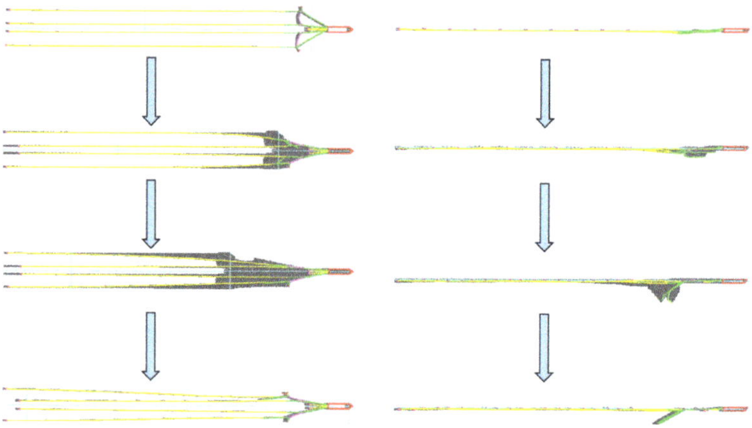

Figure 13. Intense vibration & parallel formation of towed cables after port-side guide cable fracture (down-wave conditions: H = 0.5 m, T = 6 s, **left**: top view, **right**: side view).

Compared to the formation of a multi-branch towed array system with a counter-wave, the stabilized formation is much better with the wave than with the counter-wave. With the exception of a small section very close to the towing end of each towing cable, the towing cable in the entire towing system are well parallelized over a fairly long length, and the spacing between the towlines is approximately the same, which is conducive to the development of maritime towing reconnaissance.

3.2.4. Dynamic Response of the Port Side Guide Cable in Case of Failure Breakage Under Cross-Wave Action (90°, 270°)

Observing Figures 14 and 15, it can be observed that the towing cable in the multi-branch towed array system generally shifted in the direction of backward wave direction (to port for 90° transverse wave direction and to starboard for 270° transverse wave direction) after the whole multibranch towed array system was stabilized under a transverse wave. After the system is stabilized, the nonlinearity of the formation offset of the multi-branched towing array is increased under cross-wave action compared to cross-current action, and the towing cable undergoes persistent lateral and vertical jerks, and there are more areas of cyclic and persistent interfering collisions of the individual towing cable under cross-current action, which is related to the increased amplitude of transverse vibration of the towing cable under cross-wave action, and the amplitude of the lateral vibration of the towing cable under transverse action is larger than that of the down-wave and The amplitude of transverse vibration of the towing cable under transverse waves is larger than that of the towing cable under downwash and counter wash, while the vibration of the towing cable under a single current is very weak. Moreover, the formation of multi-branched towing cable under cross-current and cross-wave action is completely different: compared with the trend of change in the formation offset along the length of each towing cable under cross-current action, the change in the formation along the length of each towing cable under cross-wave action is gentler and smoother.

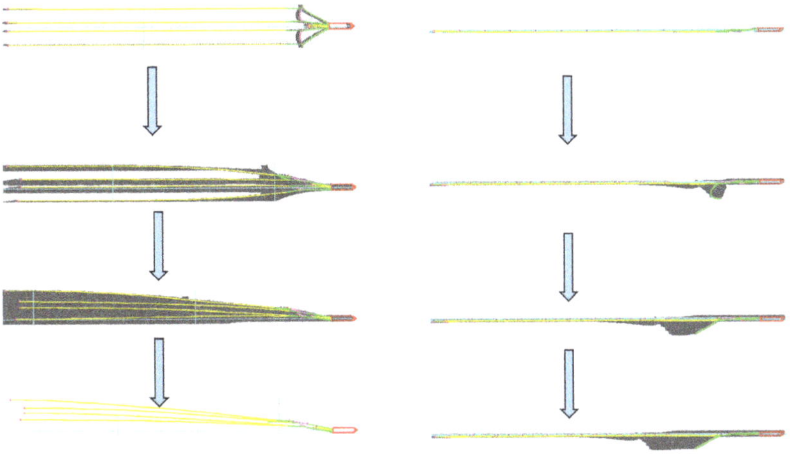

Figure 14. Portward deviation & persistent jitter of towed array after port-side guide cable fracture (cross-wave: 90° direction, H = 0.5 m, T = 6 s, **left**: top view, **right**: side view).

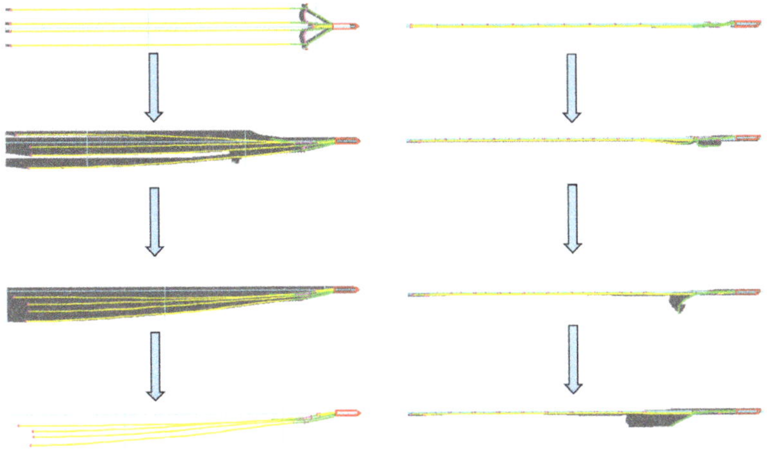

Figure 15. Schematic of the formation change, top view on the left and side view on the right. The velocity is the wave velocity and the direction is 270°.

After the breakage of the port side guide cable, when the transverse wave direction is 90°, the overlap area between the length of the port side towing cable near the guide end and the length of the other it guides cable near the guide end is obviously increased.

3.2.5. Dynamic Response of a Double Outboard Guide Cable During Failure Breakage Under Cross-Wave Action

As Figures 16 and 17 show the change in the formation of the multibranch towing system when the double outboard guide cable fails to break under the action of transverse waves. Overall, when the whole multibranch towing system reaches stability, the difference in the moment of breakage of the double outboard side guide cables has no effect on the formation of the multibranch towing system after reaching stability under the action of a single current; whereas, under the action of a single wave, the difference in the moment of breakage of the double outboard side guide cables creates a permanent and ineliminable phase difference with the action of the wave force due to the cyclic nature of the wave, in which case the difference in the moment of breakage of the double outboard side guide cables has no effect on the formation deviation of the multibranch towing system after reaching stability. The effect of the different breaking moments on the formation shift of the stabilized multibranch towing system and the asynchrony of the changes in the shape of the twin outboard cables is also different and cannot be completely eliminated, but it is substantially reduced in the transverse wave direction.

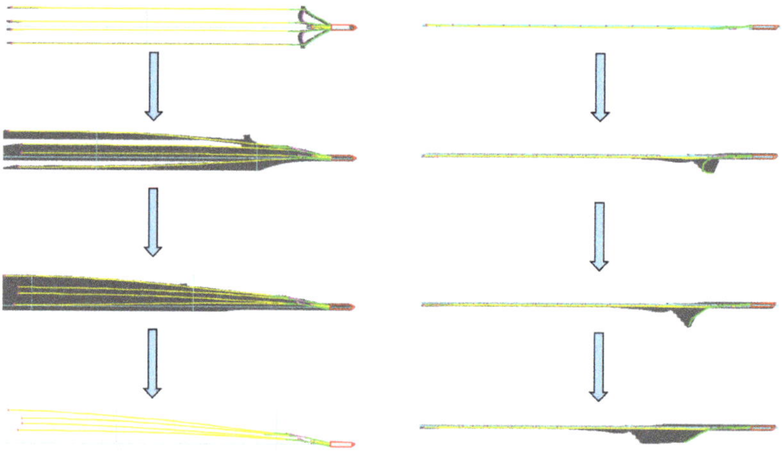

Figure 16. Schematic diagram of the shape changes of the multi-branch towed array under the action of cross waves (90° direction) after the breakage of the double outboard guide cables (left: top view, right: side view; wave speed and wave direction are marked).

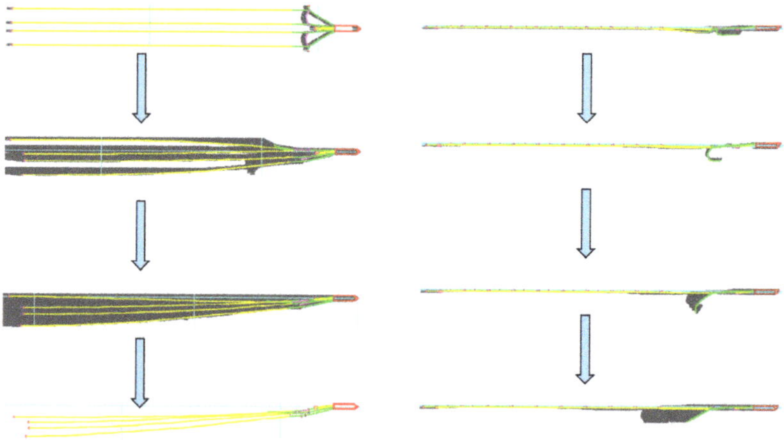

Figure 17. Schematic diagram of the morphological changes of the multi-branch towed array under the action of cross waves (270° direction) after the breakage of the double outboard guide cables (left: top view, right: side view; wave speed and wave direction are marked).

To facilitate readers' quick reference, this article summarizes the key conclusions of all supplementary charts in Section 3 to form the "Environmental Variables—Minimum Towing Cable Spacing" rule table (see Table 4).

Table 4. Key laws of the minimum distance between towing cables under different environmental conditions (↑ indicates an increase, ↓ indicates a decrease and → indicates something that causes an occurrence).

Environmental Conditions	The Changing Trend of the Minimum Spacing	Risk Threshold (Current Velocity/Wave Height)	Engineering Suggestions
Countercurrent (single break)	As the current velocity↑ → the spacing ↓	Current velocity > 1.5 m/s (spacing < 5 m)	When flowing against the current, control the current velocity to be ≤ 1.2 m/s, or increase the strength of the mooring line.
Downstream (single break)	As the current velocity ↑, the spacing first ↓ and then stabilizes	Current velocity = 0.2 m/s (minimum spacing value)	When flowing downstream, avoid the current velocity range of 0.15–0.25 m/s to prevent resonance.
Cross –current (single break)	As the current velocity ↑ → the spacing drops abruptly	Current velocity > 0.2 m/s (spacing < 4 m)	When the cross-current velocity is ≤ 0.15 m/s, or add a transverse limit frame.
Against the waves (single break)	Wave height ↑ → Spacing fluctuation ↑	Wave height > 0.6 m (Fluctuation > 30%)	When sailing against the waves, the wave height should be ≤ 0.5 m, or optimize the elastic support of the towing cable.
Stratified current (single break)	When the upper-layer ocean current accelerates, the distance decreases sharply; when the lower-layer ocean current accelerates, the distance decreases slowly.	The upper current velocity > 1.2 m/s (spacing < 3 m)	Avoid the towing cable from crossing the upper high-speed current area, or add a lateral limiter in the middle and lower part of the mooring cable.
Horizontal shear current (single fracture)	Shear strength (ΔU_y) ↑→ Nonlinear decrease in spacing	ΔU_y > 0.5 m/s (spacing < 2.5 m)	Detect the intensity of the lateral ocean current shear before operation. When ΔU_y exceeds the threshold, reduce the speed of the tugboat to 1.5 m/s.
Head-on waves (single mooring line breakage)	Wave height ↑ → Spacing fluctuation ↑; Wavelength ↑ → Spacing fluctuation first ↓ then ↑	Wave height > 1.5 m (fluctuation > 50%); wavelength < 15 m/λ > 45 m	When operating against the waves, control the wave height to be ≤ 1.2 m and the wavelength to be 25–35 m; when the wave height is 1.2–1.5 m, shorten the operation duration to within 2 h.
Following the waves (single mooring line breakage)	As the wave height increases ↑, the spacing first remains stable and then decreases suddenly; as the wavelength increases ↑, the stability of the spacing increases ↑	Wave height > 2.0 m (spacing < 3 m); Wavelength < 20 m	The maximum allowable wave height for following-wave operation is 2.0 m. It is preferred to choose the sea condition with λ > 25 m. Equip elastic lateral supports to reduce vibration.
Cross waves (single mooring lin breakage)	As wave height ↑ → the spacing decreases abruptly in a non-linear manner; as wavelength ↑ → the spacing offset ↑	Wave height > 1.0 m (spacing < 4 m); Wavelength > 40 m	For cross-wave operation, the wave height is ≤0.8 m and the wavelength is 25–35 m; when the threshold is exceeded, the lateral limit frame is activated.

4. Analysis of Influencing Factors and Mechanical Properties

Section 3 analyzes the dynamic response of a multi-branch towed array system under different working conditions such as water current and wave conditions, and the special working conditions such as the breakage of the outer cable. It reveals the law of formation change and the characteristics of mechanical response of the system under different sea conditions, and provides a basis for further exploring the intrinsic characteristics of the system. Based on this, this chapter will further analyze various factors affecting the deployment of multi-branch towed systems and conduct research on their mechanical properties.

4.1. Factors Affecting the Deployment of Multi-Branch Towing Systems

Through the authors' research in the field of multi-branch towing system propagation and exploration of existing studies and the factors affecting the towing cables during bunching in the above simulations, it was found that it is clear that there are several factors that can affect the propagation process of towed cable arrays. These factors include the length of the towing chain, current velocity and direction, wave impacts, which can lead to the phenomenon of gear pulling dynamics and tension fluctuations.

4.1.1. Current Velocity and Direction

The variations in current direction and velocity exert a significant influence on the deployment process of multi-branch towed cable arrays [84,85]. Based on the authors' research experience, when the current direction is constant, the degree of nonlinearity in the deployment process of the cable array decreases as the angle between the current direction of the towed cable and the axial direction reduces. Conversely, under the condition of a constant current direction, the nonlinearity of the cable array deployment increases with the rise in current velocity. The research conducted by Gharib strongly supports this conjecture. Gharib et al. [86] have extensively investigated the effects of flow velocity and current direction on towed cable arrays. Figure 18 illustrates the geometric configuration of the cable array under different ocean current velocities. When the current velocity remains constant and the angle between the current direction and the cable axis decreases, the nonlinearity of the cable array deployment diminishes. Figure 18 also presents the tension along the cable array at different ocean current velocities; higher flow velocities lead to an increase in tension, which in turn results in a stronger nonlinearity in the cable shape. Figures 19 and 20 depict the cable arrays under different ship headings and constant current velocity, indicating that a larger angle corresponds to a more pronounced nonlinearity. It also shows that the tension of the tow cable increases significantly with the square of the current velocity.

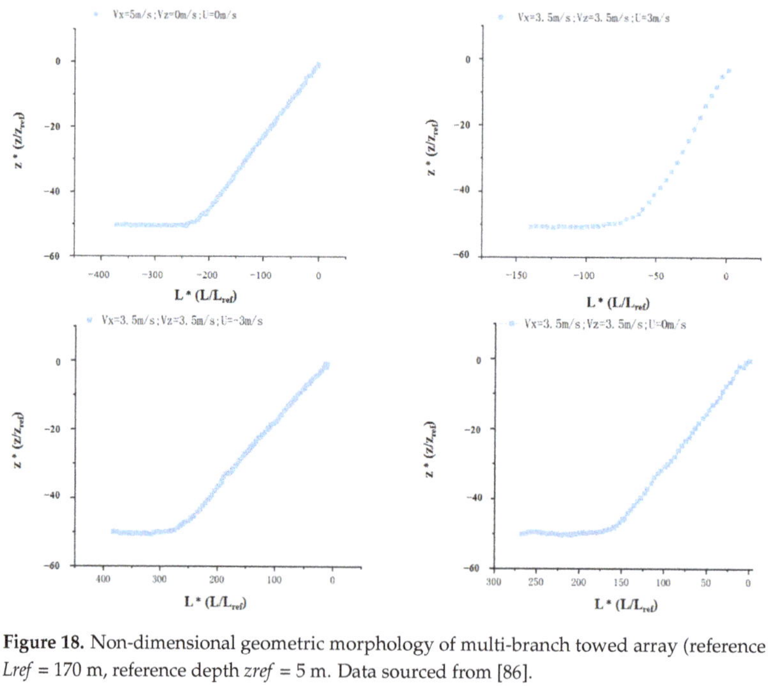

Figure 18. Non-dimensional geometric morphology of multi-branch towed array (reference length *Lref* = 170 m, reference depth *zref* = 5 m. Data sourced from [86].

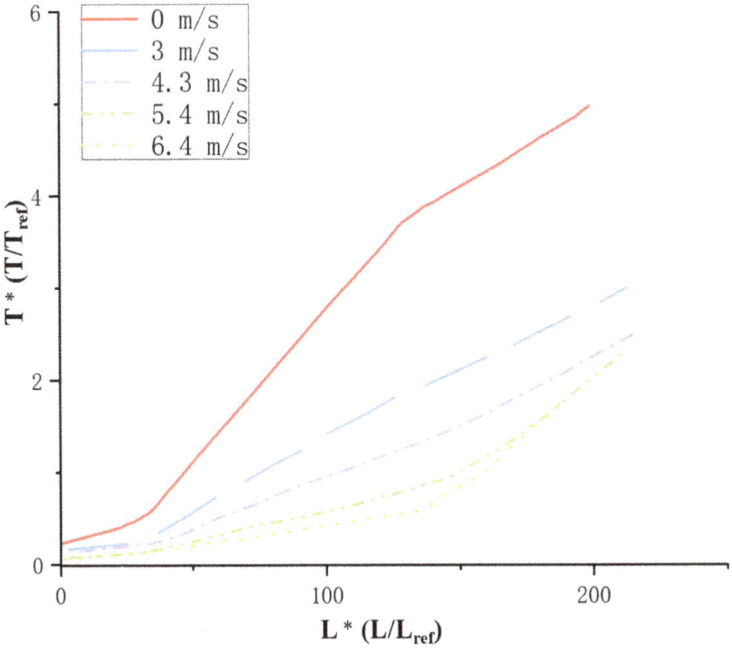

Figure 19. Non-dimensional tension of multi-branch towed array (reference tension *Tref* = 24.6 kN). Data sourced from [86].

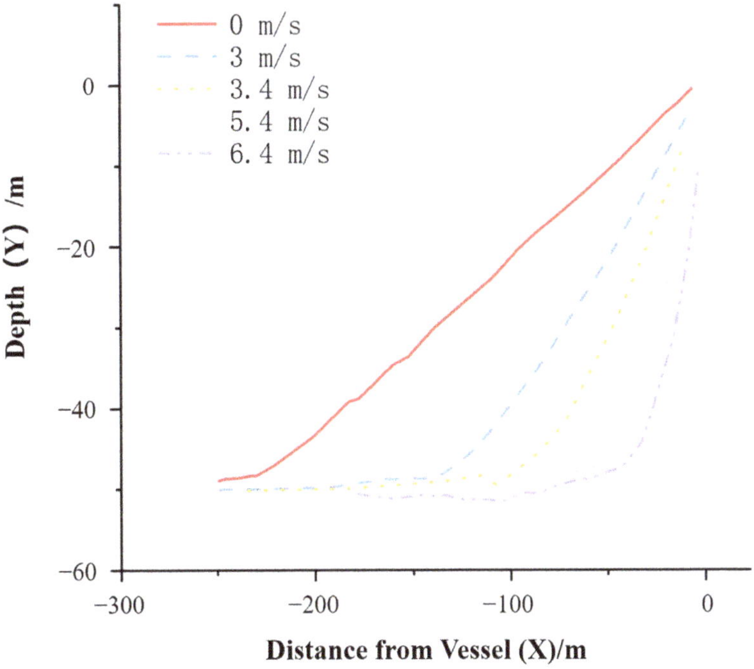

Figure 20. Plot of the cable array in the sea level direction for a constant depth of the cable array, based on different seawater velocities and the angular motion of the ship relative to the seawater, with a ship speed of 5 m/s. Data sourced from [86].

As Figure 21, this phenomenon arises from the fact that a reduced angle between the tow cable's current direction and its axial direction diminishes the water current's inhibitory impact on the cable array's diffusion. Simultaneously, the axial dragging force acting on the cable array intensifies, straightening the tow cable. This axial stretching serves to lower the degree of nonlinearity. When the water current direction remains constant, an increase in current velocity amplifies two key effects: the flow's inhibition of diffusion and the axial dragging force on the cable array. Together, these amplified effects result in greater nonlinearity during the diffusion process. However, this phenomenon is prone to fall into the "reference frame equivalence" misunderstanding, believing that a steady current only changes the overall motion and does not affect the force on the system. But in reality, the flexible transmission delay and multi-constraint coupling of the towline lead to the following laws [87–89]: 1. Influence of current direction: When the angle between the current direction and the axial direction of the towline decreases, the lateral interference of the fluid on the towline weakens, and the degree of non-linearity of the array deployment decreases; conversely, when there is a cross-current (angle of 90°), the lateral force causes a non-linear change in the distance between the towlines, and it is likely to cause collisions. 2. Influence of current velocity: When the current direction is constant, an increase in the current velocity leads to an increase in the relative velocity between the towline and the fluid, and the drag force increases—the tension increases when flowing against the current, the tension decreases when flowing with the current, and the lateral offset increases when there is a cross-current. None of these three can be eliminated through reference frame transformation, and independent stability control strategies need to be designed for different current fields.

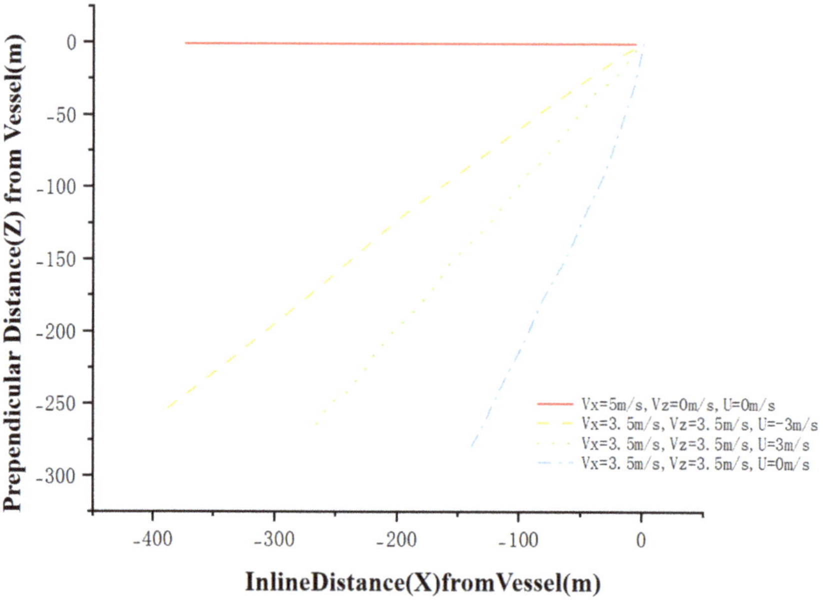

Figure 21. Schematic representation of the cable array in the depth direction for a constant depth of the cable array according to different seawater velocities and the angular motion of the ship with respect to the seawater, with a ship speed of 5 m/s. Data sourced from [86].

4.1.2. Wave Parameters

Waves, as an environmental factor, affect the unfolding process of traction cable arrays in four ways: wave period, wave direction, wave height [90–92].

- Wave direction

Changes in wave direction usually affect the expansion process of a multi-branch towed array system. The process is relatively smooth when deployed along the wave direction. However, deploying against the wave or at an angle to the wave complicates the process due to the significant increase in nonlinear properties under these conditions. Chen investigated the tension characteristics at the tip of the towing cable in different wave directions [93]. As shown in Figure 22 The tension range at the tip of the towing cable increases as the wave direction changes from 0° (following the wave) to 180° (against the wave), with a maximum at specific wave directions (e.g., 90° and 180°). Although this study uses linear regular waves (wave height of 0.5 m, period of 6 s) as input conditions, the translational motion of the buoy in Figure 22 does not exhibit an obvious repetitive pattern. This phenomenon mainly stems from the superposition of transient responses under multiple constraint couplings. In the multi-branch towed array system, the elastic coupling between the outer guide cable and the main cable, and the constraint delay effect of the spreader result in non-uniform time delays in the transmission of wave excitation within the system. Although the periodic load of regular waves is stable, the stiffness differences of different cable segments can cause continuous phase shifts in the vibration of each node, manifested as a non-periodic appearance of the overall motion, rather than the irregularity of the waves themselves.

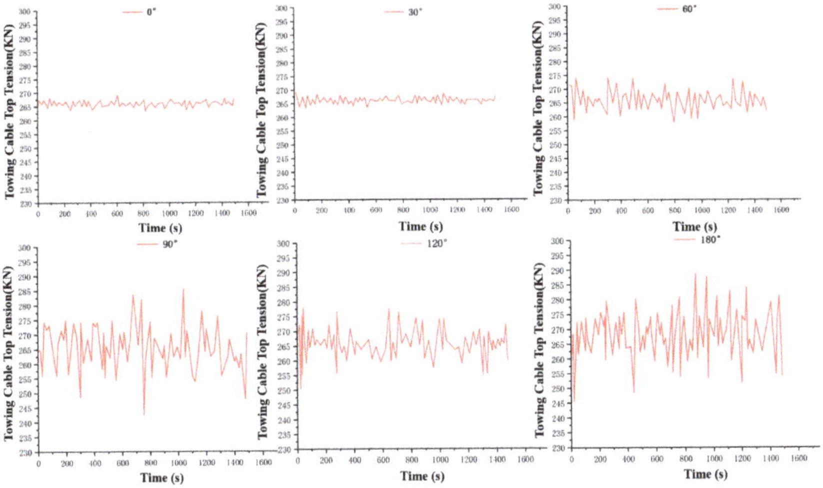

Figure 22. Pulling cable top tension at different wave positions. Data sourced from [93].

- Wave period

Based on the structural characteristics of towed cable arrays, it can be inferred that during the deployment process of multi-branch towed cable arrays, variations in wave period affect the configuration of the tow cables. Tang et al. observed that wave period exerts an impact on the stability of towed systems [90]. As illustrated in Figure 23, the stability of the system improves with an increase in wave period. Kamali and Khojastch investigated the oscillation of buoys under two distinct wave conditions [91]. The motion of the buoys is transmitted to the tow cables, and reasonable inferences can be drawn from the observations of Figure 24. Specifically, when the wave period is short, the buoy motion is affected by more frequent and rapid waves, resulting in rapid fluctuations in the buoy's movement on the water surface. This frequent motion is transmitted to the connected tow cables, leading to more bending and inflection points in the cables during deployment. In contrast, when the wave period is long, the lower wave frequency results in slower buoy motion, which in turn forms a smoother trajectory on the water surface. The reduced intensity of this motion, when transmitted to the tow cables, leads to fewer bending and inflection points during the deployment process. As shown in Figure 23, for a wave period T = 10 s, as time increases, the oscillation at the trailing end of the towing array exhibits two key characteristics: (1) The oscillation amplitude is stable, showing no tendency to increase or decrease; (2) The oscillation frequency is completely synchronized with the wave period, without any frequency drift. This is because a longer wave period reduces the hydrodynamic fluctuation frequency of the towing cable. While enhancing the system stability, it causes the towing cable to maintain periodic resonance with the waves. In contrast, for a period of T = 4 s, the oscillation amplitude is larger and the frequency is higher, which is consistent with the conclusion that "the system stability increases with the increase of the wave period." It should be emphasized that, regardless of whether the wave period is T = 4 s or T = 10 s, the oscillation at the trailing end of the towed array will not decay to zero over time.

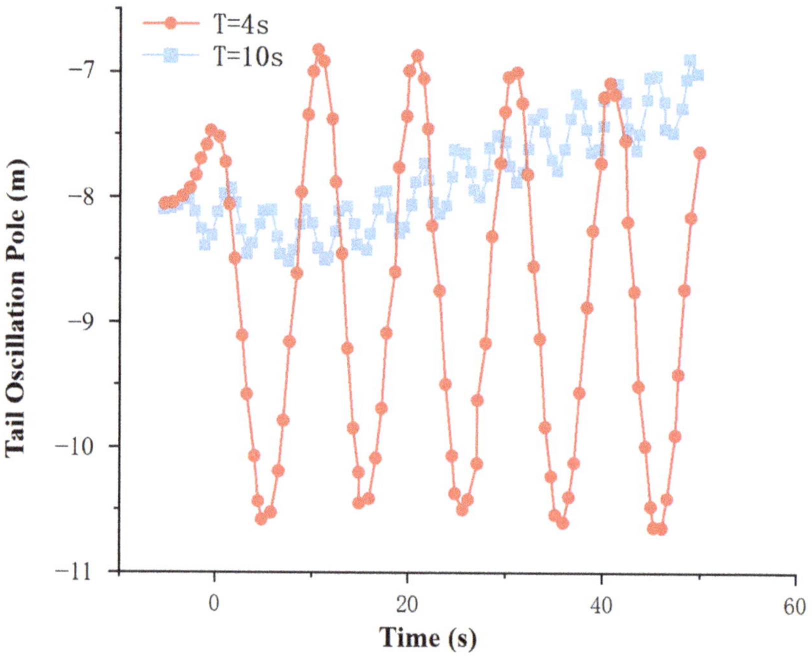

Figure 23. Tail-end oscillation amplitude of towed array under different wave periods (T = 4 s vs. T = 10 s; time series of oscillation shown). (The simulation results correspond to the dynamic steady-state phase of the system). Data sourced from [90].

It can be seen that when the wave period is too small, the multi-branch towed cable array will vibrate at high frequencies and may even break into several small segments during unfolding. This may cause different degrees of damage to the sonobuoys and other instruments embedded in the towing cable. Therefore, it is important to choose the right sea state for the expansion of the multi-branch towed cable array and avoid operating it when the wave period is too short.

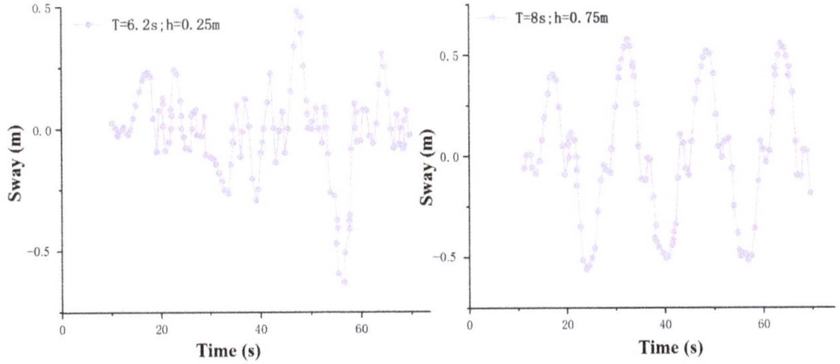

Figure 24. The translational motion of the buoy in two different wave conditions. Data sourced from [91].

- Waveform height

An increase in wave height results in an increase in the magnitude of the tug's motion response. This change in wave height changes the boundary conditions for the motion of the tug head. As a result, the enhanced tensile action at the top of the towing cable significantly amplifies the nonlinear characteristics of the expansion process of the multi-branch towing cable array. However, based on the structural characteristics of the trailing cable array, an appropriate increase in the wave height can have a positive effect on the expansion of the multi-branch line array. This suggests that a moderate increase in wave height can ensure better utilization of the bending of the trailing cable and promote the expansion of the line array. Zhang [94] investigated the effect of wave height on the trailing cable array. The motion state of the towing cable is shown in Figure 25.

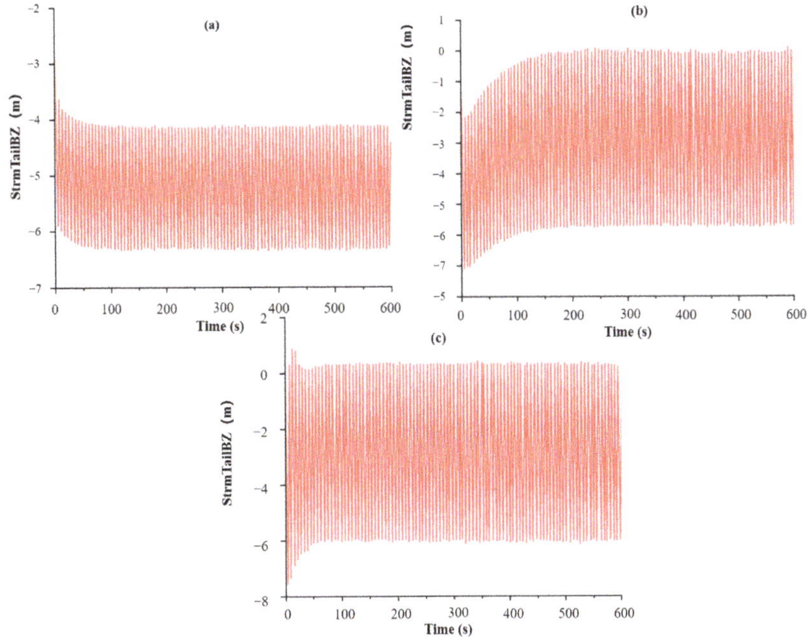

Figure 25. z-time relations for the traction cable array. (**a**) H = 2. (**b**) H = 4. (**c**) height = 6 m. Data sourced from [94].

4.1.3. Cable Length

Most studies seem to focus primarily on variations in the turning radius, while neglecting the influence of changes in cable length [93,95,96]. The ratio R/L of the total length of the traction cable to the turning radius is one of the key parameters describing the dynamics of electric traction systems [97]. Kamman and Huston [96] provided a detailed modeling and analysis method for towed and moored cable systems with variable lengths, which effectively captures their complex dynamic behaviors. As shown in Figure 26 of their study, three cable length increments (50 feet, 100 feet, and 150 feet) were selected to observe the variation of the towed body's pitch angle over time. The results indicate that the pitch angle of the towing vehicle initially increases with cable elongation, then decreases and stabilizes at the end of cable release. Furthermore, longer cables lead to larger pitch angle variations, more significant overshoots, and longer stabilization times. Therefore, the degree of bending of the tow cable increases with the increase in cable length, resulting in greater difficulty in controlling and maintaining the shape of the tow cable, which is confirmed by the study in [98]. This increased bending further prolongs the time required for diffusion. In addition, through experimental studies on lakes with fixed depths, Li et al. found that the increase in cable length affects the increase in resistance of the underwater towing system [98]. It can be observed from Figure 27a that for the same wing angle of attack, the variation in dry-end tension is mainly due to cable length. Moreover, collisions and crossings between cables frequently occur [98–100]. As shown in Figure 27b, it can be observed that the deployment of the filament bundle gradually increases with the increase in cable length [98–101]. At the same time, deflection and self-weight also increase, making spatial attitude control more challenging. This leads to greater variations in the spacing between tow cables, especially near the top, and the characteristics of nonlinear variation become more pronounced over time.

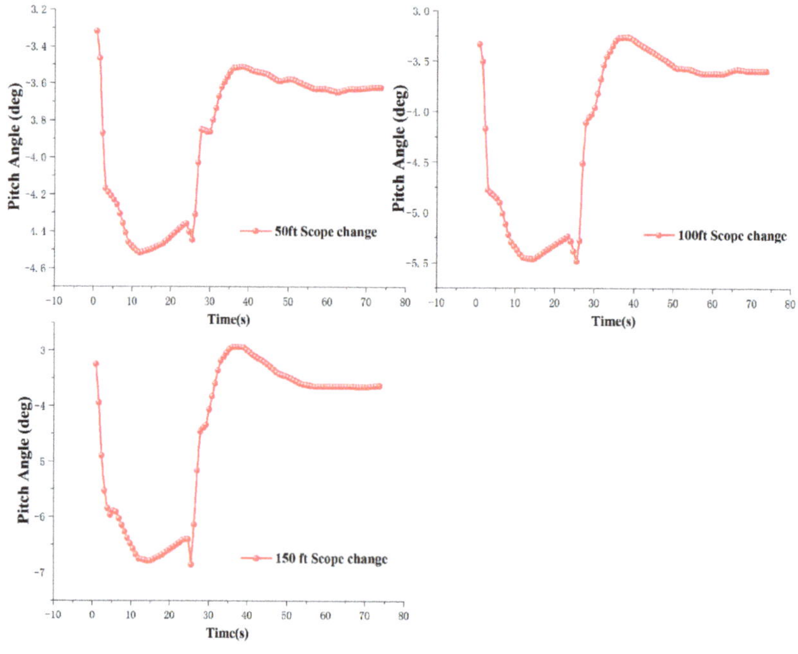

Figure 26. Pitch angle of tractor for various ranges of variation. Data sourced from [96].

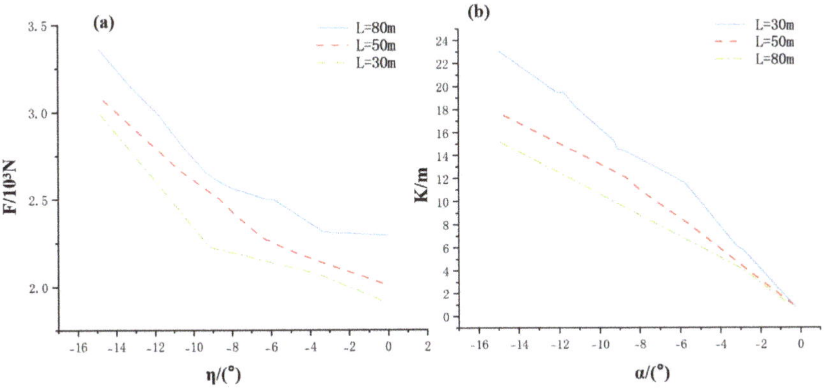

Figure 27. (**a**) Resistance varies with angle of attack and cable length. (**b**) Diffusion width varies with angle of attack and cable. Data sourced from [98].

4.2. Trawl Recovery Characteristics (Analogy Analysis)

Compared with the above simulation of the towing harness of multi-branch towing system, the ship trawl has more complexity [102–104], which is because the trawl has a large number of holes compared with the towing cable, and these holes make the mechanical study of the trawl more complex [105–108], and the towing cable can be nearly in the same plane or the same surface in the harnessing or unfolding process [70,109,110], because the parameters of the towing cable are the same in terms of the material, the length and the diameter of the towing cable. However, since the trawl net has to take into account the catching ability, the trawl net is more of a three-dimensional bunching or unfolding process [111–114]. To establish the mechanical analogy logic between the trawl and the multi-branched towing cable array, it is necessary to first clarify the core geometric characteristics of the reference trawl in this study (based on the bottom trawl experimental prototypes of Thierry et al. [115] and Liu et al. [116]): This trawl adopts a typical structure of "rectangular net mouth—trapezoidal net body—conical cod end", with a horizontal span of 8 m and a vertical height of 2.5 m at the net mouth, a total net length of 15 m, and a cod end taper of 0.3. The netting part uses diamond-shaped meshes. Among the key components, the area of a single rectangular trawl board is 1.2 m^2, the diameter of the headline is 0.08 m and the length is 10 m, the diameter of the footrope is 0.1 m and the length is 10 m, and the length of the bridle is 8 m.

4.2.1. Effects of Different Current Velocities and Current Directions

In the experiments of Thierry et al. they compared their own experiments with those they had already completed and found that the trawl tension (i.e., drag) increased with increasing current velocity, whereas the net opening height and net opening area decreased with decreasing current velocity [115,117,118] as shown in Figure 28. From this it can be seen that there is a definite link between the degree of retraction of the trawl and the velocity of the current, and it is not difficult to speculate that there is a link between the degree of retraction of the trawl cable and the velocity of the current when a multi-branch trawl system breaks sideways.

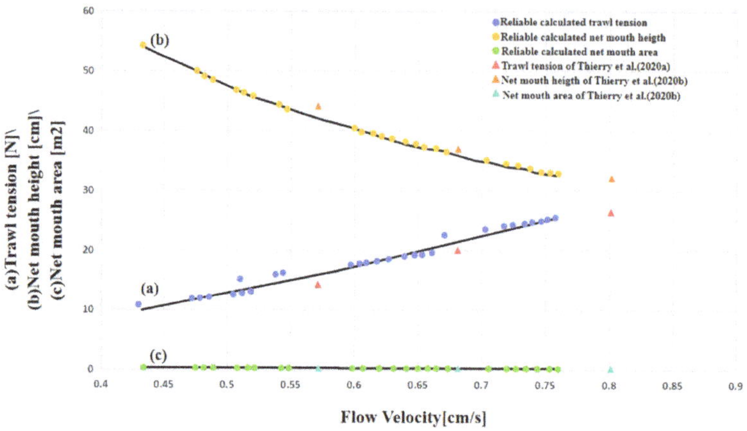

Figure 28. Relationship between trawl drag, clear opening height, clear opening size and current velocity, where (a) stands for trawl drag; (b) represents clear opening height; (c) stands for clear opening size. (The data are derived from [115,117,118], which is reprocessed in this paper, and after analysis, the curve graph is redrawn).

In Liu's study [116], they conducted experiments on the sinking process of the trawl by using a decompression plate, in which they found that there is a relationship between the angle of attack of the decompression plate and the descent of the trawl as shown in Figure 29.

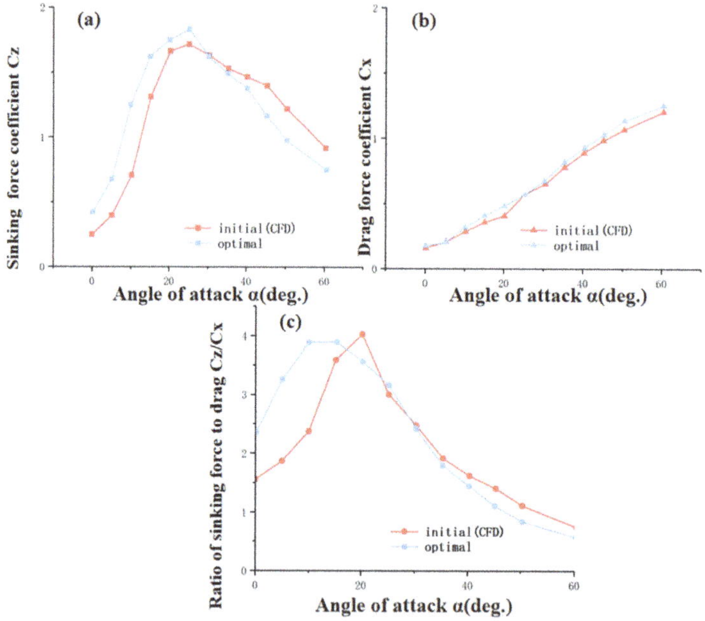

Figure 29. (**a**,**b**) show the relationship between the ratio of sinking to resistance and the angle of attack of the optimization and initialization of the pressure reducer model, respectively. (**c**) The relationship between the ratio of sinking force to drag and the angle of attack is shown for the optimization and initialization of the pressure reducer model. (The data are derived from [116], which is reprocessed in this paper, and after analysis, the curve graph is redrawn).

Analyzing this experiment, we can know that the force on the trawl in the water can be changed by changing the angle of attack of the decompression plate, which can be approximated as the effect of different current directions on the attitude of the trawl. Then we can easily know that different current directions will affect the state of the towing system in the water.

4.2.2. Effects of Waves on Trawling

In the previous section we have mentioned that the three elements of waves that have an effect on the towing cable are wave period, wave height, and wave direction. Because the towing cable and trawl are continuous [119–122], the points on the system will change due to a change in one point [117,123,124]. The difference between the towing cable and the trawl is that the trawl will be affected by other points in more directions.

In a study by Nsangue et al. [125] they obtained the fluctuations of the trawl at the same wave frequency and amplitude, differing only in wave speed, as shown in Figure 30 It was observed that the smaller the wave speed the greater and more frequent the fluctuations of the trawl, and it was hypothesized that this phenomenon could be due to the fact that the slower wave speed caused the longitudinal waves to take longer to pass through the trawl, which resulted in the waves affecting the trawl for a longer period of time. Similarly in their experiments they found that the frequency and amplitude of fluctuations in the trawl were also affected by the waves as shown in Figure 31. It can be observed that the amplitude and frequency of vibration of the trawl are related to the waves. From his experiments, we can guess that the vibration frequency and amplitude of the trawl during or after towing are affected by waves in a similar way as the trawl.

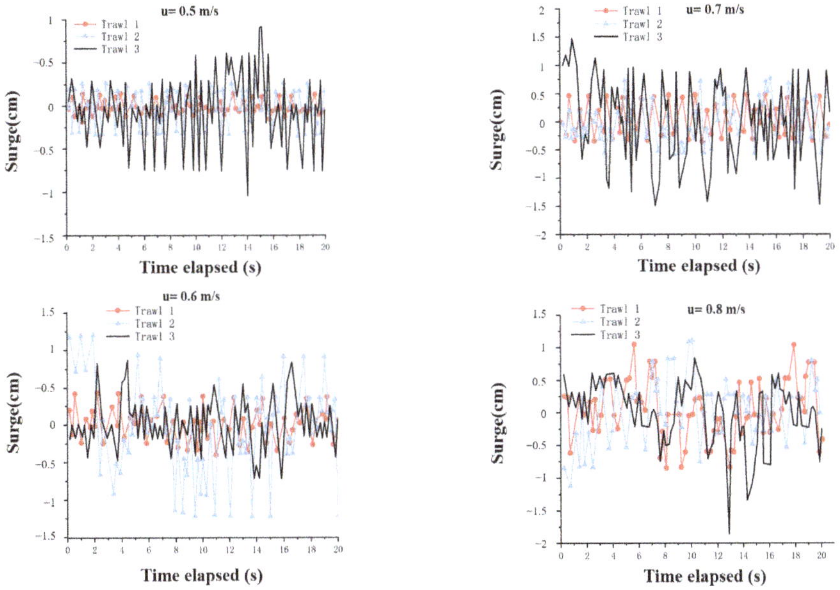

Figure 30. Effect of different wave speeds on the trawl when waves are constant. Data sourced from [125].

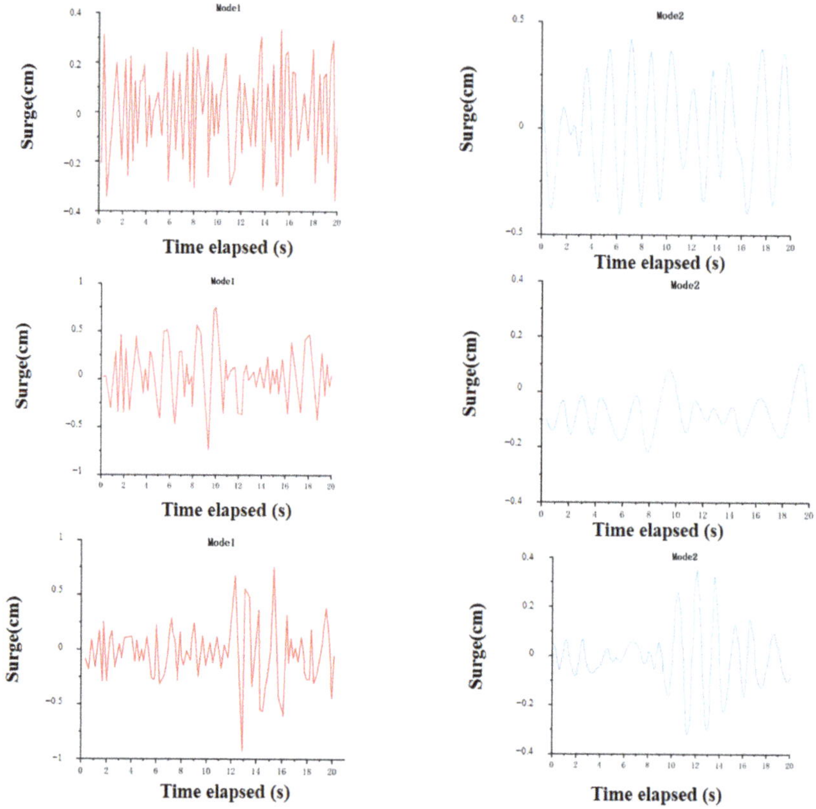

Figure 31. Effect of different waves on trawl nets. Data sourced from [125].

Notably, the non-cyclic and irregular fluctuations of the trawl observed in Figures 30 and 31 do not originate from the randomness of wave loads themselves, but from the combined effects of multi-constraint coupling and stiffness-induced phase shifts. First, the trawl system consists of components with distinct stiffnesses, leading to asynchronous energy transfer: the flexible netting vibrates at high frequencies, while rigid depressors exhibit inertial delays, disrupting the synchronization with wave periods. Second, the "current-through effect" of the netting generates local vortices with shedding frequencies interfering with wave frequencies, resulting in superimposed fluid forces. Third, the intermittent tension transfer from the tugboat causes transient impacts on the trawl, further breaking periodicity. These factors collectively contribute to the irregular fluctuation patterns observed.

4.2.3. Impact of Tow Length

The length of the trawl affects the spreading and bunching of the trawl, then the trawl is also affected by the length. In Nyatchouba's study [106], it was found that for the same average in current velocity, the current velocity near the trawl increases and then decreases as the trawl length increases in the same direction. As shown in Figure 32 In the previous we have shown that an increase in current velocity increases the drag on the trawl, and combining this with Nyatchouba's experiments we can conclude that the total drag on the trawl increases as the trawl increases in length direction.

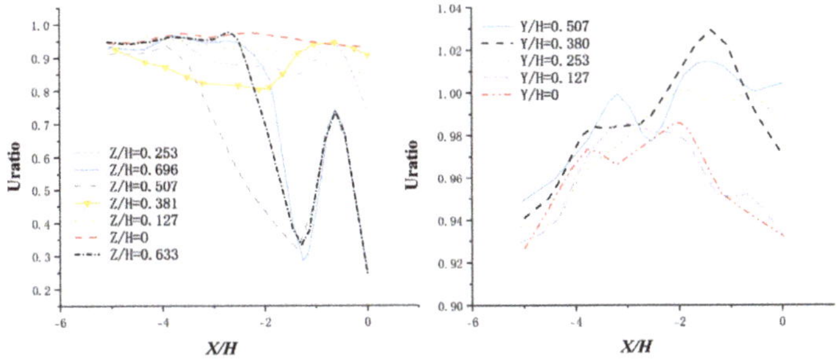

Figure 32. Relationship between trawl length and current velocity around the trawl. Data sourced from [106].

4.3. Mechanical Response Analysis of Towing Cable

4.3.1. Munk Coefficient

In Zhang's study [126] they investigated the mechanical properties of towing and hauling cables with different Munk coefficients, and they gave the following expression for Monk's coefficient:

$$\begin{aligned} M_{\alpha ix} &= \lambda_{22} - \lambda_{33} v_y v_z \\ M_{\alpha iy} &= \lambda_{33} - \lambda_{11} v_z v_x \\ M_{\alpha iz} &= \lambda_{11} - \lambda_{22} v_x v_y \end{aligned} \tag{29}$$

They found that the effective tension of the tow cable at different Munk coefficients are related to the magnitude of the tow cable in the length direction as shown in Figure 33, and by observing Figure 33, they found that the average effective tension of the tow cable decreases as the tow cable increases in the length direction. Also, they have given different bending moments and curvatures of the towline in the length direction for different Munk coefficients as shown in Figures 34 and 35. It is observed that both the bending moment and curvature of the tow cable become larger with increase in length. Based on this mechanical property of the towline in the length direction, it is not difficult to guess that after the towline and outboard breakage, the longer the length of the towline the more difficult it is to control the towline, and the more difficult it is to reach the equilibrium state of the towline again.

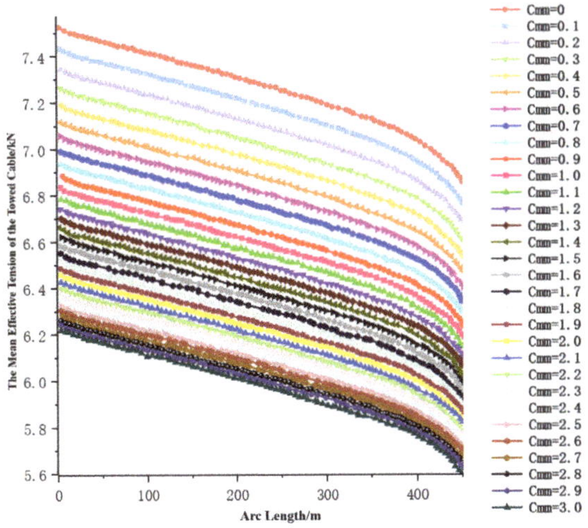

Figure 33. Axial effective tension of towing cable under different Munk coefficients. Data sourced from [126].

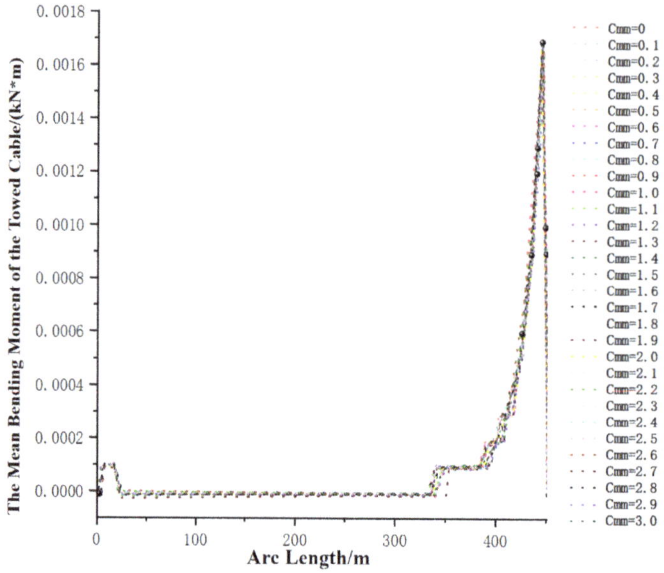

Figure 34. Ending moments in the length direction for different Munk coefficients. Data sourced from [126].

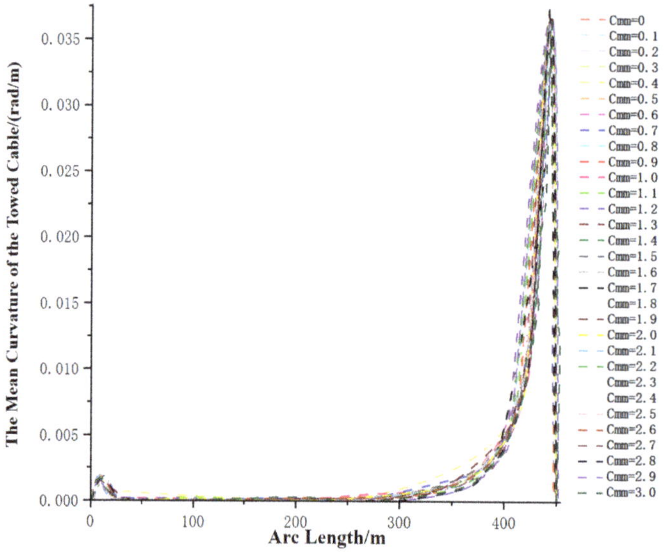

Figure 35. Curvature of towing cable along length under different Munk coefficients. Data sourced from [126].

4.3.2. Different Towing Speeds and Towing Angles

In the previous section, we have discussed Zhang's study on the mechanical response over the length of the towing cable, and in Liu's study [71] they discuss the mechanical effects of different towing speeds and angles on the towing cable. It was found through their experiments as shown in Figure 36. At the same angle, the change of traction speed has a more significant effect on the force than the change of cable length. The effect of cable length on the normal force is much greater than the tangential force. At low speeds, both lengths of traction cables produce almost the same tangential force. The drag and lift forces at low speeds are stable regardless of the length of the traction cable. However, at medium and high speeds, drag tends to rise and fall with increasing angle. When the angle exceeds 70°, vortex-induced effects cause significant fluctuations in drag. Lifting shows a more complex trend. Comparing the above discussion of different angles and speeds, we can see from the graphs that the higher the speed the greater the change in the normal and tangential forces on the towline, and that these two forces increase as the angle increases. Combined with the previous section, we can easily conclude that the faster the towing speed and the larger the towing angle, the more difficult it is for the towing system to reach a stable state again after the towing cable or the outboard breaks.

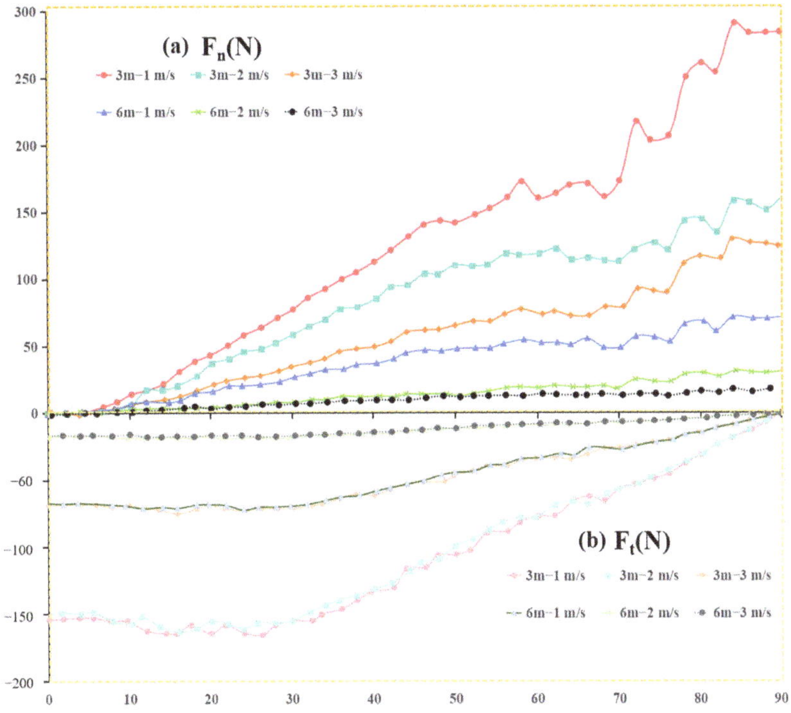

Figure 36. Forces at different towing speeds and angles, where (**a**) denotes the normal force and (**b**) denotes the tangential force. (The data are derived from [71], which is reprocessed in this paper, and after analysis, the curve graph is redrawn).

Similarly, Zhang in his study investigated the tension variation of towline array at different speeds, and in their experiment, they divided into two groups of towing systems, A and B. The two groups of towing systems are shown in Figure 37. In the figure, it can be found that the faster the relative speed, the higher the tension when the towline is stabilized and the more obvious the error, and it can be found that the faster the relative velocity the smoother the curve of the rising phase of the towline tension. This indicates that the faster the relative current velocity the faster the towline array reaches the general shape when the array is stabilized, but it is more difficult for the towline to reach a stable state.

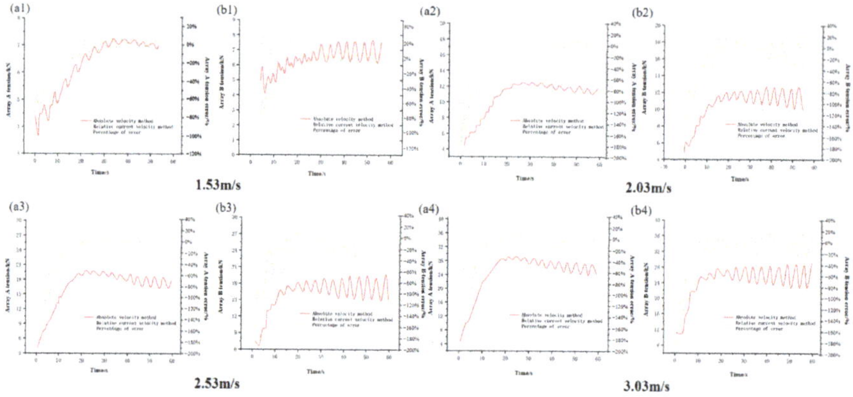

Figure 37. Towing cable tension at different relative speeds. (Among them, groups (**a1**,**b1**) represent the change in tension of the towline array when the speed of the group traction system is 1.53 m/s; groups (**a2**,**b2**) represent the change in tension of the towline array when the speed of the group traction system is 2.03 m/s; groups (**a3**,**b3**) represent the change in tension of the towline array when the speed of the group traction system is 2.53 m/s; groups (**a4**,**b4**) represent the change in tension of the towline array when the speed of the group traction system is 3.03 m/s). (The data are derived from [127], which is reprocessed in this paper, and after analysis, the curve graph is redrawn).

In Zhang's article also mentioned the different morphology and tension of the single port side at different current velocities, as shown in Figure 38. Different current velocities can significantly affect the morphology of a single outboard side. Then it is not difficult to speculate that the breakage of the outboard side on one side in the simulation affects the overall water current, which leads to changes in the morphology of the steady state of the towing system on the other side [127].

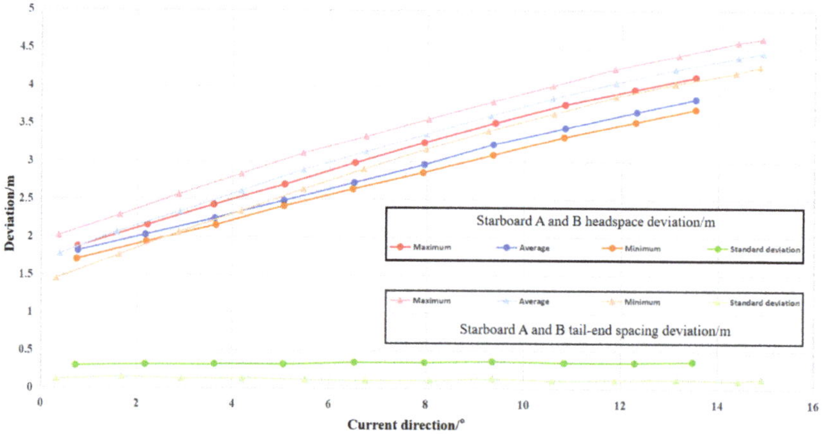

Figure 38. Morphology of the unilateral drag array for different current velocities. (The data are derived from [127], which is reprocessed in this paper, and after analysis, the curve graph is redrawn).

5. Limitations and Prospects

Multi-branch towed line arrays are widely used in military anti-submarine and marine resource exploration, but they face many challenges in complex marine environments, and also have a clear direction for future development.

5.1. Challenges in the Field of Multi-Branch Towed Line Array

5.1.1. Stability Challenges in Complex Marine Environments

Currents, waves and other factors in the marine environment have a significant impact on the multi-branch towed array system. Against the current, the port side of the guide cable breakage will make the left side of the towing cable formation disorder, spacing and even overlap collision, but also increase the left side of the spreader and the related cable tension; double outboard side of the guide cable breakage is easy to lead to the system of entanglement, and the longer the guide cable the worse the convergence. Downstream, the tugboat speed needs to be at least 4–6 times the current speed to ensure that the system convergence, speed close to the cable is easy to be compressed and curled, it is difficult to achieve static equilibrium. Transverse current has a more serious impact on the system, when the current speed exceeds 0.3 m/s, the system is difficult to converge, and the amplitude of transverse oscillation of the trailing end of the towing cable increases with the increase of the current speed, and it will make the towing cable formations drastically shifted and the spacing non-linear changes, and even lead to the right side of the towing cable to cross the mid-station surface. Under the action of waves, the towing cable will produce wave vibration, against the wave when the towing cable bending parts increase, vibration is obvious, double gang side guide cable failure after the formation is poor; with the wave when the towing cable vibration is more intense, but the formation is relatively good; cross-wave towing cable will be deflected to the back of the direction of the wave, and transverse and vertical jitter persists, the interference of the collision parts increase. The original text only considered the system response under a single low wave height (0.5 m). In actual operations, the coupling effect between wave height and wavelength is significant. When the wave height exceeds 1.5 m and the wavelength is less than 25 m, the phase difference of towline vibration increases, and the risk of stress concentration at the root of the guide cable increases by 50%. When the wavelength is greater than 45 m, the delay in tension transfer causes the system re—balancing time to be prolonged by 2 times, increasing the probability of formation disorder.

5.1.2. Formation and Mechanical Response Problems Caused by Fracture of the Guide Cable

Fracture of the guide cable, which is a key component in maintaining the spatial configuration of the system, will break the system equilibrium. When the single outboard cable breaks, the towing cable will be deflected due to the loss of constraints, and then attempted to recover by the spreader restrictions and the current field, different current direction and current velocity under the different recovery, but also lead to uneven distribution of the towing cable tension, the left side of the towing cable tension is concentrated in the middle of the end of the cable, so that the middle of the cable and the left side of the spreader to withstand the larger tension. When the double outboard cable breaks, the system is more likely to be entangled, and the different breakage moments will lead to non-synchronous changes in the formation, and this non-synchronization is affected by the direction of current under the action of cross-current, while the periodicity of the wave will make the double outboard cable form changes in the permanent phase difference.

5.1.3. Complex Factors Affecting Towing System Unfolding and Retracting

Towing cable chain length, current velocity and direction, and wave impacts all affect the unfolding of multi-branch towing systems. Current velocity and current direction, when the current velocity is constant, the angle between the current direction and the cable axis decreases, and the degree of nonlinearity of the cable array unfolding decreases; when the current direction is constant, the current velocity increases, the degree of nonlinearity increases, and the high current velocity will make the towing cable tension increase, and the shape of the nonlinearities is stronger. The wave direction, wave period and wave height of wave impact have influence, wave direction change affects the smoothness of deployment, reverse wave or angular deployment makes the process complicated; wave period is too small towing cable high frequency vibration is easy to break; wave height increase will amplify the nonlinearity of the deployment process, but moderately increase the wave height to promote the deployment of line arrays. An increase in cable length leads to several notable effects: it enhances curvature and fluctuations in local tension, prolongs the propagation time, and reduces the axial effective tension toward the tail end. Additionally, it increases resistance, causes more frequent collisions and crossings between towing cables, and results in greater variations in spacing.

5.1.4. Dilemma of Real-Time Monitoring and Safe Operation in Engineering Practice

Existing research is mostly concerned with fatigue monitoring of towing cable to avoid breakage, but less involved in the study of mechanical response and array change after breakage. The special situation at sea, once the guide cable breaks, there is a lack of relevant research to support the crew operation. At the same time, the operation under cross-current conditions requires strict requirements on the current speed and tugboat speed, exceeding the limit will lead to formation disorder, and the effect of increasing the tugboat speed is limited in large cross-currents, which may also lead to the risk of capsizing. In addition, the wave height and wave period is generally not suitable for operation, and in the specific sea operation needs to determine the safe operating zone, the existing research is limited to analyze this.

5.2. *Future Perspectives in the Field of Multi-Branch Towed Arrays System*

5.2.1. Optimize System Design to Enhance Environmental Adaptability

In order to improve the stability of the system in complex environments, the design can be optimized in various aspects. For the influence of cross-current, we can increase the bending strength material in the outer force node of the towing cable, and set the lightweight formation limiting frame along the length of the towing cable; in order to improve the formation under the action of waves, we can set the elastic transverse support between the towing cables, which needs to have a good compression resistance and anti-bending, moderate density; for the problem of cable breakage, we can improve the material and structure of the guide cable, improve the fatigue strength and toughness of the fracture, and at the same time, optimize the layout of the guide cable and the connection method to reduce the risk of breakage and At the same time, optimize the cable layout and connection method to reduce the risk of fracture and the impact on the system. Based on the simulation results of the deployment and recovery processes, the system optimization needs to focus on three aspects: First, the control of environmental thresholds during the deployment stage. Operations should only be carried out under the conditions that the downstream current is $\leq$1.5 m/s, the cross current is $\leq$0.2 m/s, and the wave height is $\leq$0.5 m to avoid ineffective deployment. Second, the parameter control during the recovery stage. A tugboat deceleration rate of 0.1–0.2 m/s^2 or a winch speed of 0.5–0.8 m/s should be adopted, and a tension-triggered speed reduction system should be equipped. Third, the improvement of the hardware structure. An elastic limiter should be added to the 100–150 m cable section. Tests show that this can reduce the collision frequency of a 160 m-long cable from 4 times to 1 time, and at the same time alleviate the problem of tail-end sagging and improve the configuration stability in complex sea conditions.

5.2.2. Deepen the Fracture Mechanical Response Research and Monitoring Technology

It is crucial to strengthen the research on the mechanical response and formation change of the guide cable after fracture. A more accurate post-breakage dynamics model can be established to simulate the effects of guide cable breakage under different sea conditions, providing theoretical support for crew operation. At the same time, more advanced real-time monitoring technology is developed, which not only monitors the fatigue state of the towing cable, but also senses the system changes after the breakage of the guide cable in real time, and combines sensor technology and data analysis algorithms to realize early warning and emergency treatment. In addition, drawing on the research experience of trawling, explore the rewinding law of multi-branch trawling system to provide reference for system design and operation.

5.2.3. Expanding Multidisciplinary Cross-Research and Application Scenarios

The field of multibranch towed line array needs to strengthen multidisciplinary cross-fertilization. Combined with fluid mechanics, material mechanics, control science, etc., in-depth study of the behavior of the system in complex environments, the development of more efficient control algorithms, to achieve real-time adjustment and optimization of the towed cable array. At the same time, expand the application scenarios, in addition to military anti-submarine, marine resources exploration, can be in the seabed terrain exploration, marine environment monitoring and other fields to further explore the potential, but also draw on related technologies to promote the application of emerging fields such as marine rescue, marine life protection.

5.2.4. Promote the Application of Intelligent and Digital Technology

The introduction of intelligent and digital technology is the future development trend. Use artificial intelligence algorithms to analyze a large amount of simulation and experimental data, predict the performance of the system under different sea conditions, and optimize the operation scheme. Construct a digital twin model to conduct virtual simulation of the multi-branch towed array system, simulate its operation status in real-time in the actual marine environment, and assist in decision-making and fault diagnosis. In addition, an automated operation system is developed to realize autonomous unfolding, rewinding and fault handling of the towed line array, which improves operational efficiency and safety. There are both opportunities and challenges in the development of the field of multi-branch towed line array, and continuous innovation and exploration are needed to promote technological progress and application expansion.

Funding: The authors gratefully acknowledge the support from the Natural Science Foundation of Guangdong Province (2022A1515011562), and the Guangdong Provincial Special Fund for promoting high-quality economic development (GDNRC [2021]56, Yuerong Office Letter [2020]161).

Conflicts of Interest: The authors declare that they have no known competing financial interests or personal relationships that could have appeared to influence the work reported in this paper.

References

1. Kim, Y.; Hong, J. The tacotron-based signal synthesis method for active sonar. *Sensors* **2022**, *23*, 28. [CrossRef]
2. Huanghuang, J.; Wang, H.; Zhuang, Z. A new simple method to design degaussing coils using magnetic dipoles. *J. Mar. Sci. Eng.* **2022**, *10*, 1495. [CrossRef]
3. Barger, J. Ira Dyer and the BBN Applied Research Division. *J. Acoust. Soc. Am.* **2017**, *141* (Suppl. S5), 3652. [CrossRef]
4. Chu, X.; Zhao, F.; Wang, Z.; Qian, Y.; Yang, G. Acoustic wave propagation in depth-evolving sound-speed field using the lattice Boltzmann method. *Phys. Fluids* **2024**, *36*, 0222202. [CrossRef]
5. McCartney, I. The archaeology of second world war U-boat losses in the english channel and its impact on the historical record. *Mar. Mirror* **2020**, *106*, 62–81. [CrossRef]

6. Gaber, M.; El-Banna, S.H.; El-Dabah, M. Intelligent Energy Management System for an all-electric ship based on adaptive neuro-fuzzy inference system. *Energy Rep.* **2021**, *7*, 7989–7998. [CrossRef]
7. Liao, S.; Xiao, W.; Wang, Y. Optimization of route planning based on active towed array sonar for underwater search and rescue. *Ocean Eng.* **2025**, *330*, 121249. [CrossRef]
8. Popper, A.N.; Halvorsen, M.B.; Miller, D.; Smith, M.E.; Song, J.; Wysocki, L.E.; Hastings, M.C.; Kane, A.S.; Stein, P. Effects of surveillance towed array sensor system (SURTASS) low frequency active sonar on fish. *J. Acoust. Soc. Am.* **2005**, *117* (Suppl. S4), 2440. [CrossRef]
9. An, S.; Lee, K.; Seong, W. Optimization of deployment depth for active towed array sonar (ATAS) using simulated annealing. *J. Acoust. Soc. Am.* **2018**, *144* (Suppl. S3), 1770. [CrossRef]
10. Ambatt, S.K. Fault Detection and Prognostic Health Monitoring of Towed Array Sonars. *Def. Sci. J.* **2022**, *72*, 495–503. [CrossRef]
11. Zhang, D.; Zhang, Y.; Zhao, B.; Ma, Y.; Si, K. Exploring subsea dynamics: A comprehensive review of underwater pipelines and cables. *Phys. Fluids* **2024**, *36*, 0231898. [CrossRef]
12. Eleftherakis, D.; Vicen-Bueno, R. Sensors to increase the security of underwater communication cables: A review of underwater monitoring sensors. *Sensors* **2020**, *20*, 737. [CrossRef] [PubMed]
13. Li, J.; Pei, Y.; Liu, C.; Zhang, L.; Luo, X.; Liu, K.; Li, W. A robust array geometry inversion method for a deep-towed multichannel seismic system with a complex seafloor. *Front. Mar. Sci.* **2023**, *10*, 1283061. [CrossRef]
14. Mayer, L.A. Frontiers in seafloor mapping and visualization. *Mar. Geophys. Res.* **2006**, *27*, 7–17. [CrossRef]
15. Karl, H.A.; Schwab, W.C.; Wright, A.S.C.; Drake, D.E.; Chin, J.L.; Danforth, W.W.; Ueber, E. Acoustic mapping as an environmental management tool: I. Detection of barrels of low-level radioactive waste, Gulf of the Farallones National Marine Sanctuary, California. *Ocean Coast. Manag.* **1994**, *22*, 201–227. [CrossRef]
16. Mizuno, K.; Tabeta, S.; Matsumoto, Y.; Sakamoto, S.; Sugimoto, Y.; Ogawa, T.; Sugimoto, K.; Jimenez, L.A.; Terayama, K.; Fukami, H.; et al. Development of a towed optical camera array system (SSS: Speedy Sea Scanner) for sea environmental monitoring. In Proceedings of the 2018 OCEANS-MTS/IEEE Kobe Techno-Oceans (OTO), Kobe, Japan, 28–31 May 2018; IEEE: New York, NY, USA, 2018.
17. von Benda-Beckmann, A.M.; Beerens, S.P.; Van Ijsselmuide, S.P. Effect of towed array stability on instantaneous localization of marine mammals. *J. Acoust. Soc. Am.* **2013**, *134*, 2409–2417. [CrossRef]
18. Thode, A. Tracking sperm whale (*Physeter macrocephalus*) dive profiles using a towed passive acoustic array. *J. Acoust. Soc. Am.* **2004**, *116*, 245–253. [CrossRef] [PubMed]
19. Birin, I.; Maglić, L. Analysis of seismic methods used for subsea hydrocarbon exploration. *J. Marit. Transp. Sci.* **2020**, *58*, 77–89. [CrossRef]
20. Dosso, S.E.; Riedel, M. Array element localization for towed marine seismic arrays. *J. Acoust. Soc. Am.* **2001**, *110*, 955–966. [CrossRef]
21. Bjørnø, L. Developments in sonar and array technologies. In Proceedings of the 2011 IEEE Symposium on Underwater Technology and Workshop on Scientific Use of Submarine Cables and Related Technologies, Tokyo, Japan, 5–8 April 2011; IEEE: New York, NY, USA, 2011.
22. Ødegård, Ø.; Sørensen, A.J.; Hansen, R.E.; Ludvigsen, M. A new method for underwater archaeological surveying using sensors and unmanned platforms. *IFAC-PapersOnLine* **2016**, *49*, 486–493. [CrossRef]
23. Chi, C.; Pallayil, V.; Chitre, M. Design of an adaptive noise canceller for improving performance of an autonomous underwater vehicle-towed linear array. *Ocean Eng.* **2020**, *202*, 106886. [CrossRef]
24. Pan, X.; Zhang, Z.; Li, Y.; Xu, W. Fast estimation of direction of arrival based on sparse Bayesian learning for towed array sonar during manoeuvring. *IET Radar Sonar Navig.* **2023**, *17*, 1079–1087. [CrossRef]
25. Yang, T.C.; Yoo, K.; Fialkowski, L.T. Subbottom profiling using a ship towed line array and geoacoustic inversion. *J. Acoust. Soc. Am.* **2007**, *122*, 3338–3352. [CrossRef]
26. Ding, H.; Zhang, Z.; Li, Y.; Xu, W. Towing operation methods of offshore integrated meteorological mast for offshore wind farms. *J. Mar. Sci. Eng.* **2019**, *7*, 100. [CrossRef]
27. Nair, B.M.; Kumar, A.; Bahl, R. Left–right ambiguity resolution methods for closely spaced arrays. *IEEE J. Ocean. Eng.* **2022**, *47*, 445–456. [CrossRef]
28. He, Z.; Salehi, H.; Zhang, H.; Zhou, H.; Jiao, P. Integrated structural health monitoring in bridge engineering. *Autom. Constr.* **2022**, *136*, 104168. [CrossRef]
29. Philbin, S.P.; Kennedy, D.A. Diagnostic framework and health check tool for engineering and technology projects. *J. Ind. Eng. Manag.* **2014**, *7*, 1145–1166. [CrossRef]
30. Mishra, M.; Lourenço, P.B.; Ramana, G.V. Structural health monitoring of civil engineering structures by using the internet of things: A review. *J. Build. Eng.* **2022**, *48*, 103954. [CrossRef]
31. Das, S.; Saha, P.; Satapathy, S.C.; Jena, J.J. Social group optimization algorithm for civil engineering structural health monitoring. *Eng. Optim.* **2021**, *53*, 1651–1670. [CrossRef]

32. Li, Y.; Li, B.; Ji, J.; Kalhori, H. Advanced fault diagnosis and health monitoring techniques for complex engineering systems. *Sensors* **2022**, *22*, 10002. [CrossRef] [PubMed]
33. Bao, Y.; Chen, Z.; Wei, S.; Xu, Y.; Tang, Z.; Li, H. The state of the art of data science and engineering in structural health monitoring. *Engineering* **2019**, *5*, 234–242. [CrossRef]
34. Zhou, C.; Jia, Z.; Song, S.; Luo, S.; Zhang, X.; Zhang, X.; Pei, X.; Xu, Z. Application of FBG sensor in health monitoring of engineering building structure: A review. *Sens. Rev.* **2025**, *45*, 129–145. [CrossRef]
35. Kumar, P.; Kota, S.R. Machine learning models in structural engineering research and a secured framework for structural health monitoring. *Multimed. Tools Appl.* **2024**, *83*, 7721–7759. [CrossRef]
36. Li, X.; Zhu, X.; Sun, M.; Qu, Y.; Liu, K.; Pei, Y.; Choi, J.H. Surrogate model-based optimization of drogue dimensions and towing operations to straighten deep-towed nonuniform arrays. *Ocean Eng.* **2024**, *299*, 117321. [CrossRef]
37. Yang, S.; Ren, H.; Zhu, X. Dynamic modeling of cable deployment/retrieval based on ALE-ANCF and adaptive step-size integrator. *Ocean Eng.* **2024**, *309*, 118517. [CrossRef]
38. Yang, S.; Zhu, X.; Ren, H. Dynamic analysis of a deep-towed seismic system based on a flexible multi-body dynamics frame. *Ocean Eng.* **2023**, *279*, 114587. [CrossRef]
39. Yang, S.; Zhu, X.; Ren, H. A two-dimensional σ-transform based finite element method for nonlinear water waves. *Ocean Eng.* **2024**, *308*, 118299. [CrossRef]
40. Qin, H.; Yu, S.; Li, P. Dynamic response of floating collar and cage under waves and current. *Ocean Eng.* **2020**, *205*, 107330. [CrossRef]
41. Chen, L.; Jeng, D.-S. Study on the seabed response around a dumbbell cofferdam under combined wave and current loading. *Ocean Eng.* **2022**, *256*, 111456. [CrossRef]
42. Lou, S.; Chen, M.; Ma, G.; Liu, S.; Zhong, G. Laboratory study of the effect of vertically varying vegetation density on waves, currents and wave-current interactions. *Appl. Ocean Res.* **2018**, *79*, 74–87. [CrossRef]
43. Gazi, A.H.; Purkayastha, S.; Afzal, M.S. The equilibrium scour depth around a pier under the action of collinear waves and current. *J. Mar. Sci. Eng.* **2020**, *8*, 36. [CrossRef]
44. McCord, M.R.; Lee, Y.-K.; Lo, H.K. Ship routing through altimetry-derived ocean currents. *Transp. Sci.* **1999**, *33*, 49–67. [CrossRef]
45. Shu, Y.; Han, B.; Song, L.; Yan, T.; Gan, L.; Zhu, Y.; Zheng, C. Analyzing the spatio-temporal correlation between tide and shipping behavior at estuarine port for energy-saving purposes. *Appl. Energy* **2024**, *367*, 123382. [CrossRef]
46. Arkhangel'skii, E.A.; Boiko, A.V.; Prokof'ev, V.V. Distinctive Features of Using the Propulsor of the Underwater Sail Type and the Direct-Flow Wave Propulsor on a Semisubmerged Catamaran. *Fluid Dyn.* **2024**, *59*, 1017–1027. [CrossRef]
47. Wang, P.; Wang, F.; Chen, Z. Investigation on aerodynamic performance of luxury cruise ship. *Ocean Eng.* **2020**, *213*, 107790. [CrossRef]
48. Feng, Z.; Wang, H.; Deng, C.; Zhang, J.; Chen, J.; Chang, Z. Propulsion and energy extraction performance of wave-powered mechanism considering wave and current. *Renew. Energy* **2024**, *237*, 121518. [CrossRef]
49. Liao, L.; Li, J.; Chen, M.; An, R. Effects of hydraulic cues in barrier environments on fish navigation downstream of dams. *J. Environ. Manag.* **2024**, *365*, 121495. [CrossRef]
50. Geng, Y.; Guo, M.; Guo, H.; Chen, H. Safety range in bridge areas based on the influence of cross flow on ship navigation. *Ocean Eng.* **2023**, *281*, 114649. [CrossRef]
51. Ahmed, F.; Xiang, X.; Wang, H.; Zhang, J.; Xiang, G.; Yang, S. Nonlinear dynamics of novel flight-style autonomous underwater vehicle with bow wings, Part I: ASE and CFD based estimations of hydrodynamic coefficients, Part II: Nonlinear dynamic modeling and experimental validations. *Appl. Ocean Res.* **2023**, *141*, 103739. [CrossRef]
52. Shi, Y.; Gao, S.; Pan, G.; Quan, X. RANS/LES investigation on the performance of air film fusion around a vertically launched underwater vehicle. *Ocean Eng.* **2022**, *266*, 112880. [CrossRef]
53. Marley, M.; Skjetne, R.; Gil, M.; Krata, P. Four degree-of-freedom hydrodynamic maneuvering model of a small azipod-actuated ship with application to onboard decision support systems. *IEEE Access* **2023**, *11*, 58596–58609. [CrossRef]
54. Zhang, D.; Luo, Y.; Zhang, Y.; Ma, Y.; Zhu, K.; Zeng, S. A Comprehensive Review of an Underwater Towing Cable Array: A Discussion on the Dynamic Characteristics of the Towing Cable Array During the Outspread Process. *J. Mar. Sci. Eng.* **2024**, *12*, 1880. [CrossRef]
55. Bai, Y.; Zhang, D.; Zhu, K.; Zhang, T. Dynamic analysis of umbilical cable under interference with riser. *Ships Offshore Struct.* **2018**, *13*, 809–821. [CrossRef]
56. Guo, Y.; Li, P.; Qin, H.; Jiang, Z. Research on the Influence of Seabed on the Hydrodynamic Performance of the Underwater Vehicle. In Proceedings of the ISOPE International Ocean and Polar Engineering Conference, Shanghai, China, 6–10 June 2022; ISOPE: Mountain View, CA, USA, 2022.
57. Asmael, M.; Memarzadeh, A. A review on recent achievements and challenges in electrochemical machining of tungsten carbide. *Arch. Adv. Eng. Sci.* **2024**, *2*, 1–23. [CrossRef]

58. Thorleifson, J.M.; Davies, T.C.; Black, M.R.; Hopkin, D.A.; Verrall, R.I.; Pope, A.; Monteith, I.; Den Hertog, V.; Butler, B. The Theseus autonomous underwater vehicle. A Canadian success story. In Proceedings of the Oceans' 97. MTS/IEEE Conference Proceedings, Halifax, NS, Canada, 6–9 October 1997; IEEE: New York, NY, USA, 1997; Volume 2.
59. Blintsov, V.S.; Trunin, K.S.; Tarełko, W. Determination of additional tension in towed streamer cable triggered by collision with underwater moving object. *Pol. Marit. Res.* **2020**, *2*, 58–68. [CrossRef]
60. Liu, C.; Zhang, Y.; Yuan, X. Simulation of Recycling Cable in Underwater Towed System. In Proceedings of the 1st International Conference on Mechanical Engineering and Material Science (MEMS 2012), Shanghai, China, 28–30 December 2012; Atlantis Press: Amsterdam, The Netherlands, 2012.
61. Ivanovskaya, A.V.; Bogatyreva, E.V.; Rybak, A.T. Simulation of the cable part of the load-lifting device drive, represented as a rod system. In *IOP Conference Series: Materials Science and Engineering*; IOP Publishing: Bristol, UK, 2020; Volume 1001, No. 1.
62. Wang, F.; Tu, W.; Deng, D.; Wu, X. Dynamic effect research of cable-lead-in rod on towed system. *J. Shanghai Jiaotong Univ. Sci.* **2019**, *24*, 745–753. [CrossRef]
63. Yan, J.; Xue, X.; Cui, L.; Ding, S.; Gu, W.; Le, F. Analysis of dynamic behavior of spray boom under step excitation. *Appl. Sci.* **2021**, *11*, 10129. [CrossRef]
64. Menting, J.G.; Yang, Y.; Chan, S.J.; Phillips, N.B.; Smith, B.J.; Whittaker, J.; Wickramasinghe, N.P.; Whittaker, L.J.; Pandyarajan, V.; Wan, Z.L.; et al. Protective hinge in insulin opens to enable its receptor engagement. *Proc. Natl. Acad. Sci. USA* **2014**, *111*, E3395–E3404. [CrossRef] [PubMed]
65. Chai, Y.; Zhong, W.; Yang, C.; Shi, X.; Zhao, Q. Restitution coefficient of various particles based on acoustic technology. *J. Phys. Conf. Ser.* **2023**, *2557*, 012057. [CrossRef]
66. Zhang, D.; Zhao, B.; Sun, J.; Zhang, Y.; Zhu, K.; Jiang, H. Influence of different static equilibrium calculation methods on the dynamic response of marine cables during the releasing process: Review and a case study. *J. Mar. Sci. Eng.* **2023**, *11*, 764. [CrossRef]
67. Zhang, D.; Zhao, B.; Zhu, K. Mechanical characteristics analysis of horizontal lifting of subsea pipeline with different burial depths. *Front. Earth Sci.* **2022**, *10*, 1011291. [CrossRef]
68. Zhang, D.; Zhao, B.; Zhu, K. Dynamic Analysis of Pipeline Lifting Operations for Different Current Velocities and Wave Heights. *Fluid Dyn. Mater. Process.* **2023**, *19*, 603–617. [CrossRef]
69. Zhang, D.; Zhao, B.; Zhu, K. Dynamic analysis of the umbilical cable pull-in operation through J-tube under different wave directions. *Ocean Eng.* **2023**, *280*, 114838. [CrossRef]
70. Lan, T.; Wang, Y.; Qiu, L.; Liu, G. Array shape estimation based on tug vehicle noise for towed linear array sonar during turning. *Ocean Eng.* **2024**, *303*, 117554. [CrossRef]
71. Liu, J.; Gao, S.; Nian, R.; He, B.; Yan, T. Study on hydrodynamic characteristics and depth control of the towed sensors array system. *Mar. Struct.* **2023**, *92*, 103504. [CrossRef]
72. Zhang, D.; Zhao, B.; Zhu, K.; Jiang, H. Dynamic analysis of towed cable with variable length during turning maneuvers. *Sci. Rep.* **2023**, *13*, 3525. [CrossRef]
73. Antony, T.; Adithya, P.V.; Madhusoodanan, K.N.; Kumar, K.R. Miniature fiber optic towed array for AUV applications. In Proceedings of the Global Oceans 2020: Singapore–US Gulf Coast, Biloxi, MS, USA, 5–30 October 2020; IEEE: New York, NY, USA, 2020.
74. Park, S.H.; Lee, S.J.; Lee, S. Experimental investigation of towing-and course-stability of a FPSO towed by a tug-boat with lateral motion. *Int. J. Nav. Archit. Ocean Eng.* **2021**, *13*, 12–23. [CrossRef]
75. Yang, X.; Sheng, S.; Wu, J.; Yue, W. Numerical analysis of dynamic behaviors of underwater towed system with hydrofoil manipulations. *Ocean Eng.* **2024**, *310*, 118791. [CrossRef]
76. Feng, H.; Sun, W.; Tang, G.; Wang, J.J. Study on a novel two-part underwater towed system for near-surface towed vehicle test. *Ocean Eng.* **2022**, *255*, 111440. [CrossRef]
77. Wu, J.; Ye, J.; Yang, C.; Chen, Y.; Tian, H.; Xiong, X. Experimental study on a controllable underwater towed system. *Ocean Eng.* **2005**, *32*, 1803–1817. [CrossRef]
78. Yang, X.; Wu, J.; Xu, S. Study on dynamic response of underwater towed system in ship propeller wakes using a new hydrodynamic model. *Ocean Eng.* **2022**, *265*, 112599. [CrossRef]
79. Eggenberger, K.; Christie, P.; Vassallo, M.; Özbek, A.; Muyzert, E.; van Manen, D.J.; Kragh, E. Fidelity and repeatability of wave fields reconstructed from multicomponent streamer data. *Geophys. Prospect.* **2014**, *62*, 994–1008. [CrossRef]
80. Deng, R.; Luo, F.; Ren, H.; Mo, X.; Wu, T. Experimental research for the influence of stern towing force on a planing hull in the towed system. *Ocean Eng.* **2022**, *259*, 111849. [CrossRef]
81. Zhang, P.; Peng, Y.; Ding, H.; Hu, R.; Shi, J. Numerical analysis of offshore integrated meteorological mast for wind farms during wet towing transportation. *Ocean Eng.* **2019**, *188*, 106271. [CrossRef]
82. Kragh, E.; Laws, R. Rough seas and statistical deconvolution. In Proceedings of the 63rd EAGE Conference & Exhibition, Amsterdam, The Netherlands, 11–15 June 2001; European Association of Geoscientists & Engineers: Utrecht, The Netherlands, 2001.

83. Westin, C. Framework for a physics-based digital twin of a towed cable-body system. *Ocean Eng.* **2025**, *328*, 121025. [CrossRef]
84. Chagnaud, B.P.; Brücker, C.; Hofmann, M.H.; Bleckmann, H. Measuring flow velocity and flow direction by spatial and temporal analysis of flow fluctuations. *J. Neurosci.* **2008**, *28*, 4479–4487. [CrossRef]
85. Constantin, A.; Johnson, R.S. Steady large-scale ocean flows in spherical coordinates. *Oceanography* **2018**, *31*, 42–50. [CrossRef]
86. Gharib, M.R.; Heydari, A.; Kolahi, M.R.S. Modeling and analysis of static and dynamic behavior of marine towed cable-array system based on the vessel motion. *Adv. Mech. Eng.* **2024**, *16*, 16878132231220353. [CrossRef]
87. Cahill, B.; Lewis, T. Wave periods and the calculation of wave power. In Proceedings of the 2nd Marine Energy Technology Symposium (METS2014), Seattle, WA, USA, 15–18 April 2014.
88. Steele, K.; Teng, C.-C.; Wang, D.J.O.E. Wave direction measurements using pitch-roll buoys. *Ocean Eng.* **1992**, *19*, 349–375. [CrossRef]
89. Lauria, A.; Loprieno, P.; Rizzo, F.; Severini, A.; Foti, D.; Leone, E.; Francone, A.; Tomasicchio, G.R. On the effects of wind and operating conditions on mooring line tensions for floating offshore wind turbine. *Appl. Ocean. Res.* **2024**, *152*, 104197. [CrossRef]
90. Tang, Z.H.; Zhu, K.; Bao, X. Study on the tension of the towing section and the vibration extreme value of the tail rope segment in the multi-branch towing line array system. *J. Ningbo Univ. (Sci. Eng.)* **2016**, *29*, 127–132. (In Chinese)
91. Kamali, R.; Danial, K. Investigation of dynamic behavior of a towed underwater vehicle at Persian Gulf. In Proceedings of the 24th Annual International Conference on Mechanical Engineering—ISME, Yazd, Iran, 26–28 April 2016.
92. Grosenbaugh, M.A. Transient behavior of towed cable systems during ship turning maneuvers. *Ocean Eng.* **2007**, *34*, 1532–1542. [CrossRef]
93. Chen, S. The Research for Motion Response Analysis of Towed Linear Array Sonar. Master's Thesis, Dalian University of Technology, Dalian, China, 2021. (In Chinese).
94. Zhang, Y. Numerical Study on the Nonlinear Hydrodynamic Characteristics of Towed Linear Array Sonar. Master's Thesis, Ningbo University, Ningbo, China, 2015. (In Chinese).
95. Mudie, J.D.; Ivers, W.D. Simulation studies of the response of deeply towed vehicle to various towing ship maneuvers. *Ocean Eng.* **1975**, *3*, 37–46. [CrossRef]
96. Kamman, J.W.; Huston, R.L. Modeling of variable length towed and tethered cable systems. *J. Guid. Control Dyn.* **1999**, *22*, 602–608. [CrossRef]
97. Wang, Z.; Sun, G. Parameters influence on maneuvered towed cable system dynamics. *Appl. Ocean Res.* **2015**, *49*, 27–41. [CrossRef]
98. Li, C.; Ai, Y.; Yu, H.; Jiang, X.; Song, W.; Wang, M. Lake experimental study on depth determination and deployment performance of underwater towing system. *J. Harbin Eng. Univ.* **2022**, *43*, 62–68. (In Chinese)
99. Grosenbaugh, M.A. *Dynamic Behavior of Towed Cable Systems During Ship Turning Maneuvers*; Woods Hole Oceanographic Institution: Woods Hole, MA, USA, 2005.
100. Pipchenko, A.; Tsymbal, N. Development of Mathematical Algorithms for the Seismic Research Vessel Maneuvering Simulator. In Proceedings of the INSLC. Available online: https://www.researchgate.net/profile/Oleksandr-Pipchenko/publication/336798657_Development_of_Mathematical_Algorithms_for_the_Seismic_Research_Vessel_Maneuvering_Simulator/links/5ec29f52A6FDCC90D67E2FC7/Development-of-Mathematical-Algorithms-for-the-Seismic-Research-Vessel-Maneuvering-Simulator.pdf (accessed on 5 May 2024).
101. Obligado, M.; Bourgoin, M. An experimental investigation of the equilibrium and stability of long towed cable systems. *New J. Phys.* **2013**, *15*, 043019. [CrossRef]
102. Konosu, S.; Yamaguchi, K.; Hayashi, T. Studies on Flavor Components in Boiled Crabs-1, Amino Acids and Related Compounds in the Extracts. *Bull. Jpn. Soc. Sci. Flsheries* **1978**, *44*, 505–510. [CrossRef]
103. Engås, A.; Godø, O.R.; Jørgensen, T. A comparison between vessel and trawl tracks as observed by the ITI trawl instrumentation. *Fish. Res.* **2000**, *45*, 297–301. [CrossRef]
104. Maina, I.; Kavadas, S.; Damalas, D.; Pantazi, M.; Katsanevakis, S. Dynamics of trawling effort in the Aegean Sea: Investigating the potential of Vessel Monitoring System (VMS) data. *ICES J. Mar. Sci.* **2018**, *75*, 2265–2275. [CrossRef]
105. Preciado, I.; Arroyo, N.L.; González-Irusta, J.M.; López-López, L.; Punzón, A.; Muñoz, I.; Serrano, A. Small-scale spatial variations of trawling impact on food web structure. *Ecol. Indic.* **2019**, *98*, 442–452. [CrossRef]
106. Nyatchouba Nsangue, B.T.; Tang, H.; Liu, W.; Xu, L.; Hu, F. Turbulent flow interacting with flexible trawl net structure including simulation catch in flume tank. *Sci. Rep.* **2023**, *13*, 6249. [CrossRef] [PubMed]
107. Tang, H.; Thierry, N.N.; Achille, N.P.; Mouangue, R.; Xu, L.; Hu, F.; Mbangue, E. Coupled dynamics of the moving Antarctic krill trawl structure and its hydrodynamics behavior using various catch sizes and door spreads based on wavelet-based and Fourier analysis. *J. Fluids Struct.* **2024**, *124*, 104037. [CrossRef]
108. Balash, C.; Sterling, D.; Binns, J.; Thomas, G.; Bose, N. Drag characterisation of prawn-trawl bodies. *Ocean. Eng.* **2016**, *113*, 18–23. [CrossRef]
109. Srivastava, S.K.; Ganapathy, C. Experimental investigations on loop manoeuvre of underwater towed cable-array system. *Ocean. Eng.* **1998**, *25*, 85–102. [CrossRef]

110. Zhou, C. Research on Transient Electromagnetic Positioning Method of Underwater Towed Array Source for Moving Small Target on the Seabed. *J. Phys. Conf. Ser.* **2023**, *2651*, 012131. [CrossRef]
111. Poli, V.; Litti, L.; Lavagnolo, M.C. Microplastic pollution in the North-east Atlantic Ocean surface water: How the sampling approach influences the extent of the issue. *Sci. Total Environ.* **2024**, *947*, 174561. [CrossRef]
112. Russo, E.; Monti, M.A.; Mangano, M.C.; Raffaetà, A.; Sarà, G.; Silvestri, C.; Pranovi, F. Temporal and spatial patterns of trawl fishing activities in the Adriatic Sea (Central Mediterranean Sea, GSA17). *Ocean. Coast. Manag.* **2020**, *192*, 105231. [CrossRef]
113. Van Denderen, P.D.; Plaza-Morlote, M.; Vaz, S.; Wijnhoven, S.; Borja, A.; Fernandez-Arcaya, U.; González-Irusta, J.M.; Hansen, J.L.; Katsiaras, N.; Pierucci, A.; et al. Complementarity and sensitivity of benthic state indicators to bottom-trawl fishing disturbance. *Ecol. Appl.* **2024**, *34*, e3050. [CrossRef]
114. Du, R.; Sun, X.; Lin, H.; Pan, Z. Assessment of manta trawling and two newly-developed surface water microplastic monitoring techniques in the open sea. *Sci. Total Environ.* **2022**, *842*, 156803. [CrossRef]
115. Thierry, N.N.; Tang, H.; Achille, N.P.; Xu, L.; Hu, F. Unsteady turbulent flow developing inside and around different parts of fluttering trawl net in flume tank. *J. Fluids Struct.* **2022**, *108*, 103451. [CrossRef]
116. Liu, Z.; Hu, F.; Wan, R.; Wang, Y.; Guo, S.; Yan, H.; Zhou, C. Hydrodynamic characteristics of frustum-type depressor for sampling midwater trawl. *Ocean. Eng.* **2024**, *295*, 117009. [CrossRef]
117. Thierry, N.N.; Tang, H.; Liuxiong, X.; You, X.; Hu, F.; Achile, N.P.; Kindong, R. Hydrodynamic performance of bottom trawls with different materials, mesh sizes, and twine thicknesses. *Fish. Res.* **2020**, *221*, 105403. [CrossRef]
118. Thierry, N.N.; Tang, H.; Achile, N.P.; Xu, L.; Hu, F.; You, X. Comparative study on the full-scale prediction performance of four trawl nets used in the coastal bottom trawl fishery by flume tank experimental investigation. *Appl. Ocean. Res.* **2020**, *95*, 102022. [CrossRef]
119. Leonte, D.; Veraart, A.E. Simulation methods and error analysis for trawl processes and ambit fields. *Math. Comput. Simul.* **2024**, *215*, 518–542. [CrossRef]
120. Grahovac, D.; Leonenko, N.N.; Taqqu, M.S. Intermittency of trawl processes. *Stat. Probab. Lett.* **2018**, *137*, 235–242. [CrossRef]
121. Zhao, Z.; Chen, J.; Shi, Y.; Hong, F.; Jiang, G.; Huang, H.; Zhao, J. HiTrip: Historical trajectory interpolation for trawlers via deep learning on multi-source data. *Ocean. Eng.* **2024**, *292*, 116588. [CrossRef]
122. Loiseau, J.F.; Batina, J.; Noël, F.; Peyrous, R. Hydrodynamical simulation of the electric wind generated by successive streamers in a point-to-plane reactor. *J. Phys. D Appl. Phys.* **2002**, *35*, 1020. [CrossRef]
123. Haugen, J.; Imsland, L. Optimization-based motion planning for trawling. *J. Mar. Sci. Technol.* **2019**, *24*, 984–995. [CrossRef]
124. Nsangue, B.T.; Tang, H.; Xu, L.; Hu, F.; Dong, S.; Achille, N.P.; Zou, B. Comparison between physical model testing and numerical simulation using two-way fluid-structure interaction approach of new trawl design for coastal bottom trawl net. *Ocean. Eng.* **2021**, *233*, 109112. [CrossRef]
125. Nsangue, B.T.; Tang, H.; Mouangue, R.; Liu, W.; Pandong, A.N.; Xu, L.; Hu, F.; Tcham, L. Non-linear dynamic behavior of T0 and T90 mesopelagic trawls based on the Hilbert–Huang transform. *Mar. Struct.* **2025**, *100*, 103727. [CrossRef]
126. Zhang, D.; Zhao, B.; Zhu, K. Hydrodynamic response of ocean-towed cable-array system under different munk moment coefficients. *Sustainability* **2022**, *14*, 1932. [CrossRef]
127. Zhang, Y.; Zhang, D.; Xie, Y.; Zhang, Y.; Liang, Z.; Zhu, K.; Zhang, S. Dynamic configuration simulation of multi-branches towed array system. *Mar. Struct.* **2025**, *103*, 103819. [CrossRef]

Journal of
Marine Science and Engineering

MDPI

Article

Mineralogical and Mechanical Characterization of Concrete Blocks for Artificial Reefs: A Comparative Study with Natural Coral Skeletons

Mykel Fernandes de Sousa [1], Cláudio Dybas da Natividade [2], Marçal Rosas Florentino Lima Filho [3], Sandro Marden Torres [1], Alexsandro José Virgínio dos Santos [4], Rochanna Alves Silva da Rocha [3,*], Glauco Fonsêca Henriques [5], Karina Massei [6] and Wesley Maciel de Souza [1]

1 Postgraduate Program in Mechanical Engineering, Federal University of Paraíba—UFPB, Campus I, João Pessoa 58051-900, Brazil; mykelfsousa@gmail.com (M.F.d.S.); sandromardentorres@yahoo.co.uk (S.M.T.); macielengenharia25@gmail.com (W.M.d.S.)
2 IFPB Research Department, Federal Institute of Paraíba—IFPB, Campus João Pessoa, João Pessoa 58015-435, Brazil; claudio.natividade@ifpb.edu.br
3 Postgraduate Program in Materials Science and Engineering, Federal University of Paraíba—UFPB, Campus I, João Pessoa 58051-900, Brazil; marcal@cear.ufpb.br
4 Department of Electrical Engineering, Federal University of Paraíba—UFPB, Campus I, João Pessoa 58051-900, Brazil; alexsandro@cear.ufpb.br
5 Postgraduate Program in Civil and Environmental Engineering, Federal University of Paraíba—UFPB, Campus I, João Pessoa 58051-900, Brazil; glaucofh@gmail.com
6 Postgraduate Program in Development and Environment, Federal University of Paraíba—UFPB, Campus I, João Pessoa 58051-900, Brazil; karina.massei@academico.ufpb.br
* Correspondence: rochanna.alves@gmail.com

Abstract

Coral reefs are very important ecosystems for the planet, offering ecological and socio-economic benefits. However, they are under threat due to anthropogenic factors and environmental changes. This study assesses the feasibility of weathered Portland cement concrete as a material for marine artificial reefs by comparing its physicochemical and mechanical properties with those of natural coral skeletons from the coast of Paraíba, Brazil. Analyses included microstructural and physical characterization, compressive strength and ultrasonic pulse velocity tests, as well as pH monitoring. The results indicated that weathered concrete exhibits mineralogical similarity to corals, with the presence of carbonate phases and portlandite absent due to advanced carbonation. The compressive strength of the concrete (27.6 MPa) was significantly higher than that of the coral samples (1–6 MPa), while the porosity of the corals (34–41%) exceeded that of the concrete (14%). The alkaline nature of the concrete (pH 9.7) remained stable. Although differences in physical and mechanical properties are evident, the values are within the ranges reported for cementitious materials in marine applications. Mineralogical similarities between coral skeletons and concrete support its potential as a functional analog in artificial reefs, while adjustments in geometry and porosity are suggested to enhance ecological compatibility.

Keywords: artificial reefs; coral skeletons; Portland cement concrete

1. Introduction

Coral reefs are critically important marine ecosystems due to their rich biodiversity and the support they provide to a wide range of marine life [1]. Nevertheless, these natural habitats are increasingly threatened by environmental impacts associated with rising sea temperatures and sea levels, shifts in storm intensity, ocean acidification, and excess sediment and nutrient loads [2]. One of the most severe consequences is coral bleaching, which can trigger a cascade effect, impacting associated fauna and entire reef ecosystems [3, 4]. Understanding bleaching physiology is crucial for enhancing the resilience of both wild corals and those used in restoration programs [5,6].

In response to the global decline of natural reefs, many studies have explored strategies for marine ecosystem restoration [7,8], including coral nurseries [9,10], coral transplantation [11,12], and the development of artificial reefs. Artificial reefs consist of underwater structures designed to replicate natural reef habitats, offering shelter, feeding grounds, and settlement surfaces for marine species. Paxton et al. [13] assessed the effectiveness of artificial coral reefs in enhancing fish diversity and abundance compared to natural reefs, reporting comparable ecological performance in several cases.

However, the success of artificial reef projects depends on multiple factors, including site selection, structural design, long-term management, and, critically, the choice of construction materials [14]. A wide range of materials has been used, such as plastics, tires, ship hulls, steel, and concrete [15]. A recent global survey by Vivier et al. [16] analyzing over 100 artificial reef projects found that approximately 60% were built using concrete as the primary material. The widespread use of concrete has been linked to its ability to form stable, non-toxic substrates with surface textures, similar to natural rocky habitats, that support the establishment and growth of diverse benthic organisms [17,18].

The main drawbacks associated with the use of concrete as artificial reefs are the significant environmental impact of its production, as Portland cement manufacturing is a major contributor to global CO_2 emissions [19], and the inherently high surface pH of the material, which hinders initial biological colonization. This elevated alkalinity requires several months of submersion before it approaches the pH of seawater, which, during this period, limits the colonization of a wide range of marine species due to both the high pH and the leaching of calcium ions [20,21].

Recent research on concrete for artificial reefs has largely focused on the development of new mix designs with optimized properties [17,22,23]. Regarding durability, many studies assess the performance of such concretes under controlled or accelerated conditions, with carbonation tests [24,25] and chloride penetration tests [26,27] commonly used to predict long-term performance. However, there is still a lack of research on the use of aged concrete that has naturally weathered in the marine environment over long periods of time as artificial reefs [28].

In this context, the Seixas coral reefs [29], which form part of the coastal reef systems of the state of Paraíba, Brazil, stand out for their ecological, economic, and social importance. In recent decades, studies have reported on geobiology, geomorphology [30], marine biodiversity, and bleaching in scleractinian, hydrocoral, and octocoral species during thermal stress events and under anthropogenic pressures [31,32]. However, the physical, chemical, geological, and biological characteristics of the Seixas coral reefs are still not sufficiently discussed. Further investigation is required to support the design and selection of materials for artificial reefs, ensuring compatibility with the local environment and enhancing success in larval recruitment, benthic colonization, biofilm formation, and eventual ecological integration.

The history of artificial reef deployment in Brazil includes early initiatives in the 1980s, such as triangular concrete structures installed in Sepetiba Bay, Rio de Janeiro [33]. An additional attempt to deploy 400 concrete blocks in the northeast of Brazil was interrupted, leaving the blocks stored for more than 20 years without a defined purpose [34]. Currently, the Strategic Program for Marine Artificial Structures of Paraíba (PREAMAR), supported by funding from the State Government of Paraíba, Brazil, is planning to install 1000 Portland cement concrete blocks as artificial reefs in the coming years.

This scenario presents an opportunity for PREAMAR to repurpose the concrete blocks currently stored, transforming these unused resources into functional components for artificial reef projects. This approach not only promotes sustainable resource utilization but also provides a cost-effective and practical pathway to support marine habitat restoration. However, it remains uncertain whether concrete blocks exposed to 20 years of tropical weathering have indeed developed physical, chemical, and mineralogical properties that are compatible with those of natural Seixas coral reefs.

In this way, the present study aims to apply characterization techniques to compare the physical, chemical, and mineralogical properties of concrete blocks weathered in a tropical environment, with a view to their potential use as artificial reefs, highlighting the functional similarities between the selected blocks and natural corals found along the coast of Paraíba, Brazil.

The characteristics found in the weathered concrete blocks will also be compared to the literature on the use of this material in artificial reefs in order to validate them for future studies on launch and for in situ application.

2. Materials and Methods

2.1. Materials

Three types of samples were used for comparative purposes: (i) natural coral samples of three different species, collected in consolidated reef areas; (ii) cores extracted from reef fragments; and (iii) concrete blocks weathered in a tropical environment.

The fragments were obtained from a shallow reef area (Figure 1a), located in northeastern Brazil (7°9′21″ S, 34°47′10″ W), on the coastal reefs of Seixas beach in the city of João Pessoa, Paraíba, where the predominant climate is coastal tropical, and the average sea surface temperature is 28 °C [30]. The samples were collected manually during snorkeling by biologists who are members of the PREAMAR project. The samples were prepared for testing by washing in running water to remove salts and organic materials.

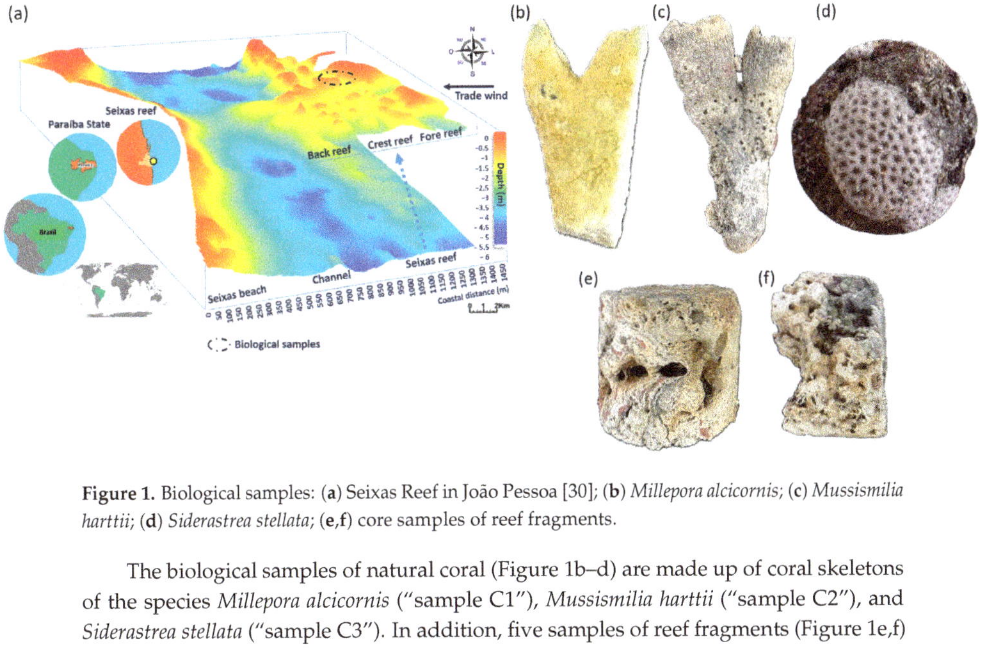

Figure 1. Biological samples: (**a**) Seixas Reef in João Pessoa [30]; (**b**) *Millepora alcicornis*; (**c**) *Mussismilia harttii*; (**d**) *Siderastrea stellata*; (**e**,**f**) core samples of reef fragments.

The biological samples of natural coral (Figure 1b–d) are made up of coral skeletons of the species *Millepora alcicornis* ("sample C1"), *Mussismilia harttii* ("sample C2"), and *Siderastrea stellata* ("sample C3"). In addition, five samples of reef fragments (Figure 1e,f) were extracted using a 5 cm diameter hole saw machine, with the samples named "T". The names given to the samples are summarized in Table 1.

Table 1. Sample identification.

Sample	Code
Millepora alcicornis	C1
Mussismilia harttii	C2
Siderastrea stellata	C3
Core samples of reef fragments	T

The concrete samples were extracted from a block stored at the fishing terminal in the city of Cabedelo—metropolitan region of João Pessoa (Figure 2), manufactured to be used as an artificial marine structure [34].

It should be noted that the concrete blocks were produced two decades ago by an independent company, and therefore, detailed information regarding their preparation and mix design is not available. It is known that they are composed of Portland cement type CP IV-RS, in accordance with ABNT NBR 16697 [35], which is comparable to ASTM C595 Type IV Pozzolanic Cement [36]. The concrete mix design included sand, fine aggregate, and coarse aggregate in a volumetric ratio of 1:7.5:3.5:7.2, along with the addition of microsilica and a superplasticizer admixture. All the concrete blocks were manufactured simultaneously using identical molds and mix designs. The quantities of materials used for the production of one concrete block are detailed in Table 2.

Figure 2. Concrete samples: (**a**) Located in Cabedelo, on the coast of the state of Paraíba, Brazil (adapted from [37]); (**b**) Fishing terminal; (**c**) Block storage, arrow indicating block stacking arrangement; (**d**) Extraction of concrete core samples; (**e**) Concrete core samples.

Table 2. Material construction per concrete block.

Sample	Code
Portland Cement (CP IV-RS)	350 kg
Coarse Sand	0.27 m^3
Small Aggregate	0.125 m^3
Medium Aggregate	0.26 m^3
Silica Fume (microsilica)	12% of cement weight
Superplasticizer	2% of cement weight

After extracting the concrete core, the carbonation test was carried out by spraying phenolphthalein into the hole. An 8 mm layer of carbonated front was observed around the blocks. This superficial part of the concrete was used to carry out the microstructural and PH tests.

This value of 8 mm of carbonated front for twenty-year-old concrete is in line with many models for predicting the progress of carbonation in concretes [38].

2.2. Microstructural Characterization

Mineralogical characterization was performed by X-ray diffraction (XRD), using a BRUKER D8 diffractometer with CuKα radiation (λ = 1.5406 Å), operated in continuous scan mode from 5° to 70° (2θ), with a step size of 0.02°/s. Phase identification was conducted by comparison with the PDF-2 database (ICDD), and quantitative analysis was carried out using the TOPAS software (Bruker, version 4).

Functional group identification and chemical bonding analysis were performed using Fourier-transform infrared spectroscopy (FTIR) on KBr pellets. The spectra were acquired with a Bruker Alpha II spectrometer, using 32 scans and a resolution of 4 cm^{-1} over the range of 4000–400 cm^{-1}.

The mass loss and energy variation of the samples were evaluated by Thermogravimetric Analysis (TGA) and Differential Scanning Calorimetry (DSC), using a Discovery SDT 650 instrument (TA Instruments, New Castle, DE, USA). The tests were conducted from 30 °C to 1300 °C at a heating rate of 15 °C/min, under a nitrogen atmosphere with a flow rate of 50 mL/min. Experimental data were processed using the TRIOS software.

All the XRD, FTIR, and TGA samples were tested in the form of ground powder within 30 days of their extraction, following the recommendations for preparation in practical guides [39].

The elemental composition was also assessed by Energy-Dispersive X-ray Spectroscopy (EDS), using a 30 mm^2 Bruker detector coupled to a FEI Quanta 450 Scanning Electron Microscope (SEM). The analyses were performed at an accelerating voltage of 15 kV, under high vacuum conditions, and the results were reported as average atomic percentages. For these tests, the samples were prepared in the form of small pieces of polished sections, impregnated with low-viscosity epoxy resins, polished, and coated with gold, following the recommendations for preparation in practical guides [39].

2.3. Physical and Mechanical Characterization

The bulk density, open porosity, and water absorption of the samples were determined by gravimetric methods, in accordance with ASTM C642—Standard Test Method for Density, Absorption, and Voids in Hardened Concrete [40]. Samples were oven-dried at 105 °C for 24 h, weighed in dry condition, saturated in water, and subsequently weighed while immersed. Total porosity was expressed as a percentage relative to the dry mass.

Ultrasonic Pulse Velocity (UPV) measurements were conducted to evaluate the structural integrity of the concrete core samples and reef fragments (T1–T5). The tests were performed in accordance with ASTM C597—Standard Test Method for Ultrasonic Pulse Velocity Through Concrete [41], using a Pundit Proceq device operating at a frequency of 54 kHz.

Additionally, compressive strength tests were conducted in accordance with ASTM C39—Standard Test Method for Compressive Strength of Cylindrical Concrete Specimens [42].

2.4. pH Test

The pH evaluation of the solid samples was carried out through controlled immersion in deionized water, aiming to investigate the chemical interaction between the materials and the aqueous medium over time. A fixed mass of 7 g was used for all samples, which were individually placed in hermetically sealed containers containing 70 mL of deionized water, maintaining a mass-to-volume ratio of 1:1, in accordance with procedures adapted from the literature [43,44].

The solutions were kept at room temperature (25 °C), without agitation, and protected from direct light exposure. The pH measurements were performed after 1, 3, and 7 days of immersion, using a properly calibrated pH meter with standard buffer solutions (pH 4.0, 7.0, and 9.0). The electrode was inserted directly into the supernatant, avoiding contact with sedimented solids, in order to reflect the effective pH of the medium in equilibrium with the material surface.

Although it does not fully replicate environmental or oceanic conditions, this protocol allows for the identification of the initial pH relevant to chemical interactions, such as hydroxide leaching, prior to exposure to field phenomena. Future studies may complement this work with in situ tests or in buffered media, aiming to address applicability in real-world environments.

3. Results and Discussions

3.1. X-Ray Diffraction (XRD)

Analysis of the diffractograms (Figure 3) allowed for the identification of crystalline phases present in the samples. The diffraction patterns of the biological materials (Figure 3a) revealed the presence of crystalline phases characteristic of calcium carbonate, with peaks corresponding to both aragonite (orthorhombic phase) and magnesian calcite (trigonal phase).

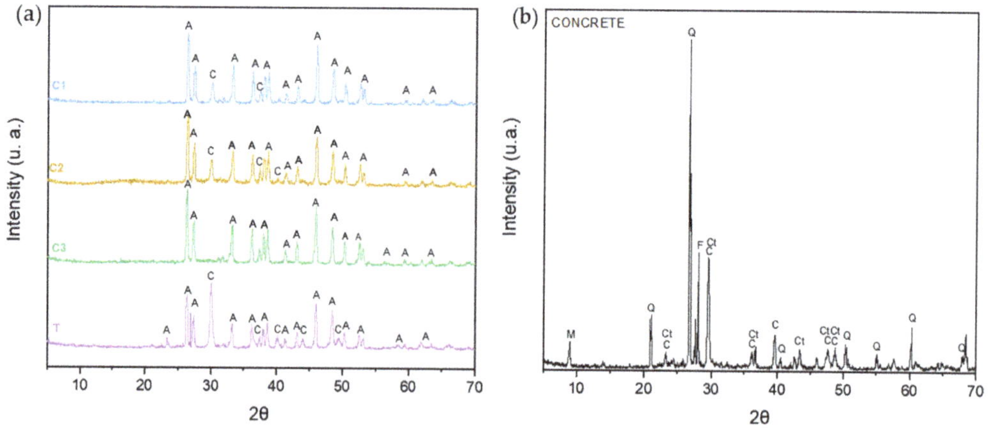

Figure 3. Diffractogram of biological samples (**a**) and concrete (**b**). A: Aragonite; C: Magnesian Calcite; Ct: Calcite; Q: Quartz; M: Mica; F: Feldspar.

Aragonite was identified by peaks around 26.2°, 27.2°, and 45.9°, whereas magnesian calcite exhibited peaks near 29.8°, 48.2°, and 49.10°, indicating a slight angular shift due to the partial substitution of Ca^{2+} by Mg^{2+} compared to the pure calcite structure. The presence of magnesian calcite is typical of tropical marine environments, where Mg bioaccumulation frequently occurs in the skeletons of calcifying organisms [45,46].

The simultaneous occurrence of these phases directly influences the mechanical and chemical properties of the material [47,48] and may be associated with seasonal, physiological, or taxonomic variations, commonly observed in organisms biomineralizing calcium carbonate in tropical environments [49].

For the concrete sample (Figure 3b), the diffractogram revealed a complex polycrystalline system, with a predominance of quartz (SiO_2), identified by intense peaks at 20.8°, 26.6°, and 50.1°, corresponding to 43% of the total crystalline phase. The significant presence of calcite (21.2%) and magnesian calcite (14.3%), along with the absence of characteristic portlandite peaks typically observed around 18° and 34°, indicates carbonation reactions occurring in the cementitious matrix. These reactions involve ionic substitutions that favor the stabilization of Mg^{2+} within the carbonate lattice.

Carbonation in concrete occurs when its portlandite reacts with CO_2 dissolved in water, leading to the formation of calcite, as described in the reactions below:

$$CO_2 + H_2O \rightarrow H_2CO_3 \tag{1}$$

$$Ca(OH)_2 + H_2CO_3 \rightarrow CaCO_3 + 2H_2O \tag{2}$$

Carbonation not only stabilizes the matrix pH but also promotes the formation of carbonate surfaces more compatible with biological adhesion [50,51], which may be advantageous in the context of artificial reefs. The conversion of portlandite into carbonate phases brings the concrete mineralogy closer to that of coral skeletons, reinforcing the material's potential for larval recruitment [52]. This finding is particularly relevant, as it demonstrates that concrete, under certain conditions, can partially resemble the mineral composition of coral skeletons, thereby reducing the ecological rejection of the material.

3.2. Fourier-Transform Infrared Spectroscopy (FTIR)

The obtained FTIR spectra (Figure 4) were analyzed to identify the functional groups present. All samples exhibited similar peaks, particularly in the regions associated with carbonate groups (C–O), with subtle variations between the spectra of the biological samples (T; C1; C2; C3) and the concrete sample.

The biological sample spectra displayed characteristic carbonate bands at 712, 875, and 1415 cm^{-1}, typical of calcite, along with additional bands in the ranges 700–730 cm^{-1} and 1080–1130 cm^{-1}, indicative of aragonite. These findings are consistent with previous studies [53,54]. Such vibrational signatures align with the literature reports for scleractinian corals [55–57]. However, no bands related to organic compounds typical of biogenic tissues, such as amines, polysaccharides, or proteins, were detected [58,59]. This organic matrix plays a critical role in the nucleation and orientation of aragonite crystals during biomineralization [60,61]. The absence of these bands may be attributed to degradation or removal of organic material during sample preparation or to their low relative concentration compared to the dominant carbonate mineral matrix [62,63].

The FTIR spectra of the concrete samples showed clear evidence of carbonation, as indicated by the absence of the band at 3640 cm^{-1}, typical of portlandite. This observation corroborates the XRD results, where portlandite was also not detected. The reaction of calcium hydroxide leading to the formation of calcium carbonate, mainly in the form of calcite, was confirmed by the intense bands near 875 cm^{-1} and 1415 cm^{-1}, corresponding to the carbonate phases identified by XRD [64,65].

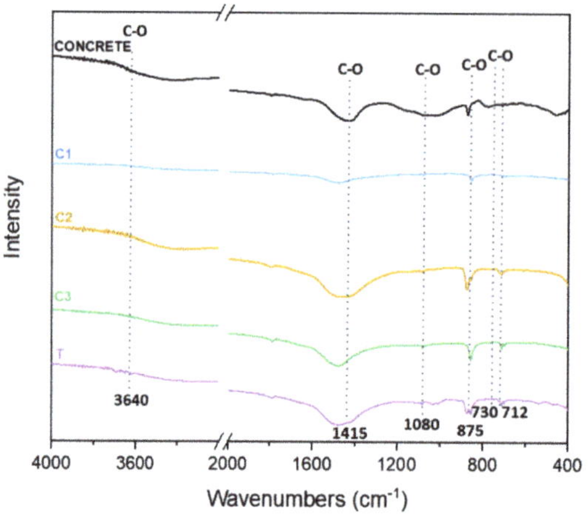

Figure 4. FTIR spectra of biological samples and concrete.

From a spectroscopic standpoint, both the coral and concrete samples exhibit vibrational signatures dominated by bands assigned to carbonate groups, reflecting the predominant presence of mineral phases such as aragonite and calcite, respectively. This similarity suggests that, despite their distinct origins—biogenic in corals and geochemical in carbonated concrete—both materials share structural aspects in terms of calcium carbonate mineral composition.

3.3. Thermogravimetric Analysis (TG)

Differential thermal analysis combined with derivative thermogravimetry was employed to describe the physical and chemical reactions occurring within the samples as the temperature increases (Figure 5).

The thermogravimetric curves of the biological samples (Figure 5a–c) exhibit similar behavior, consistent with the thermal decomposition of biogenic carbonate materials. The main thermal event occurred between 600 °C and 800 °C, indicating the decomposition of carbonate (~50% mass loss), present as aragonite and/or calcite–crystalline forms predominant in coral skeletons [66,67], as confirmed by the XRD results.

The absence of significant mass loss below 200 °C indicates a low content of physically adsorbed water or volatile organic compounds. Additionally, the region between 200 °C and 400 °C—which could indicate the degradation of organic matter—shows only minor enthalpy fluctuations or mass loss, corroborating the FTIR observations, where no amide I or II bands (proteins) or characteristic polysaccharide bands were detected [61].

The concrete sample (Figure 5d) exhibits a distinct thermogravimetric profile. Mass losses occurred mainly in two stages: up to 150 °C (water) and between 600 and 800 °C (carbonates). The initial mass loss (~5%) between 50 °C and 150 °C is attributed to dehydration of free and adsorbed water, as well as hydrated phases such as C–S–H, as reported by Bui et al. [64]. The main mass loss (~10%) between 600 °C and 800 °C is due to the decomposition of $CaCO_3$ into CO_2 and CaO. This finding is also supported by the carbonate bands identified in the FTIR spectra (875 and 1415 cm^{-1}). The intense endothermic peak in the DSC curve within this range confirms the thermal transition, and the lower mass loss compared to coral samples suggests a lower relative $CaCO_3$ content in the concrete—consistent with the phase quantification from XRD analysis. The thermal overlap with coral decomposition suggests partial compatibility in thermal behavior, despite differences in composition.

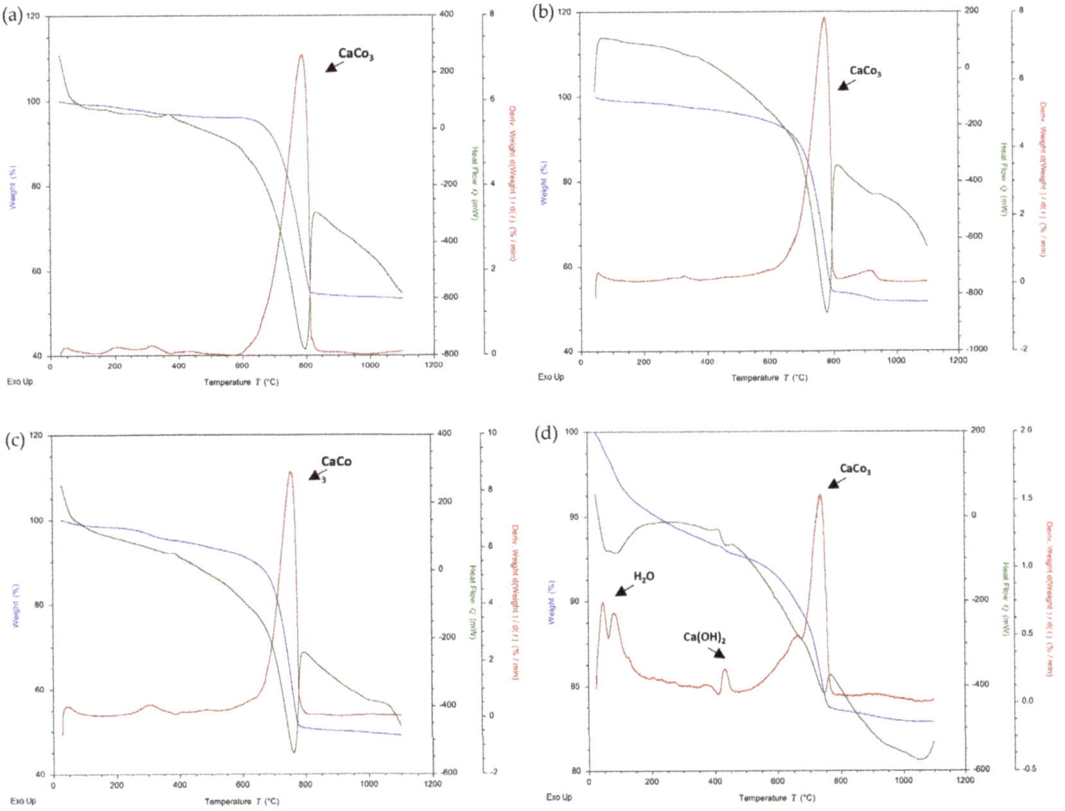

Figure 5. Individual thermogravimetric and heat flow plots: (**a**) *Millepora alcicornis* (C1); (**b**) *Mussismilia harttii* (C2); (**c**) *Siderastrea stellata* (C3); (**d**) concrete.

A low-intensity thermal event with a mass loss below 1% occurs between 400 and 500 °C, likely corresponding to residual dehydration of portlandite or secondary hydrated gels. This behavior is consistent with the pozzolanic additions present in the concrete formulation and further supports the advanced carbonation observed, corroborating XRD and FTIR data.

Thermal analysis reveals that, although exhibiting distinct behaviors, both biological material and concrete share key thermal zones associated with carbonate decomposition. This convergence suggests a degree of mineralogical similarity, even though the relative content and nature of the involved phases differ. The thermal profile of the biological samples confirms their predominantly carbonate nature, with mass losses exceeding 50%, whereas the concrete exhibits more modest losses (~10%), consistent with a partially carbonated matrix. Thus, thermal analysis reinforces the potential of carbonated concrete as a mineralogical analogue to coral-derived matrices, contributing to the assessment of its applicability in artificial reef systems.

3.4. Scanning Electron Microscopy (SEM) and Energy-Dispersive X-Ray Spectroscopy (EDS)

Figure 6 presents the morphological analysis and Figure 7 the compositional analysis of a biological sample of *Millepora alcicornis*, performed using scanning electron microscopy (SEM) coupled with energy-dispersive X-ray spectroscopy (EDS).

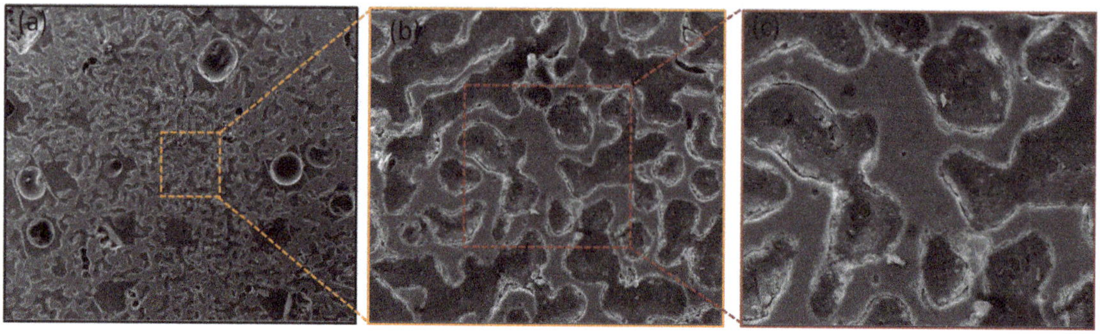

Figure 6. SEM micrographs of coral: (**a**) 100×; (**b**) 500×; (**c**) 1000×.

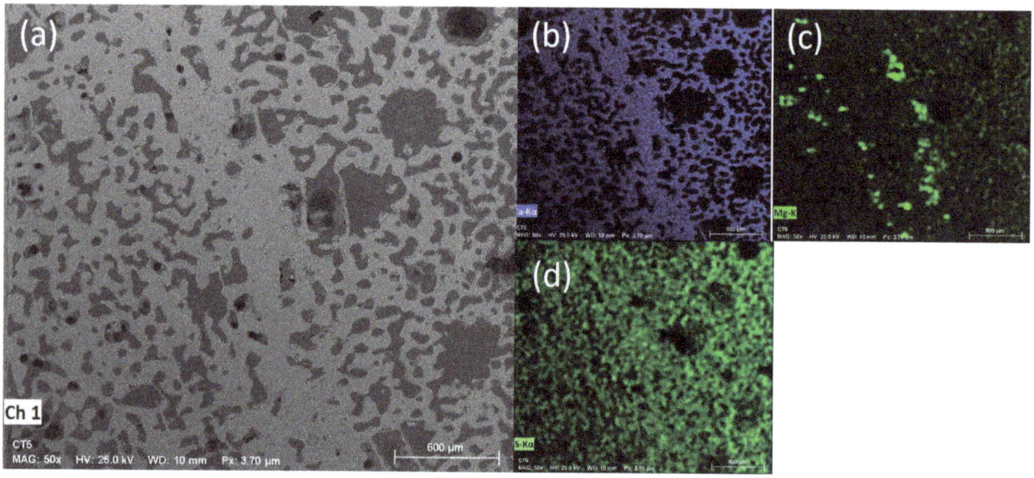

Figure 7. BSE micrograph: (**a**) at a magnification of 50×; (**b**) elemental mapping of calcium; (**c**) elemental mapping of magnesium; (**d**) elemental mapping of sulfur.

The SEM images at different magnifications (Figure 6) reveal a highly porous skeletal structure, with both intrinsic and interconnected porosity. At 100× magnification (Figure 6a), the macrostructure of the hydrozoan coral is evident, characterized by a branched architecture with large interconnected cavities and well-defined internal channels. At higher magnifications (500× and 1000×, Figure 6b,c), finely textured microstructures become visible, including interlinked channels and microporous surfaces, which are consistent with regions of calcium carbonate crystal growth and deposition [63]. This morphology is typical of marine calcifying organisms and is associated with several important biological and ecological functions [68–70].

From a compositional perspective (Figure 7), the elemental mapping by EDS confirms the predominance of calcium (Figure 7b), which is continuously and homogeneously distributed throughout the mineral matrix. This supports the conclusion that the skeleton is primarily composed of calcium carbonate, more specifically, magnesian calcite or aragonite, as identified by X-ray diffraction (XRD).

The presence of magnesium (Figure 7c) suggests partial isomorphic substitution of Ca by Mg within the crystal lattice, which can alter structural properties and solubility—a common feature in shallow marine environments [48]. According to the literature, the Mg/Ca ratio in coral skeletons may vary with water temperature and environmental conditions [55,71].

Sulfur (Figure 7d), detected in low concentrations and with a faint distribution pattern, is likely associated with the residual organic matrix, which includes compounds such as sulfated polysaccharides and acidic glycoproteins [45,69]. This organic matrix plays a key role in regulating nucleation and crystal orientation during the biomineralization process and also contributes to the mechanical properties and biocompatibility of the skeleton [72,73].

Figures 8 and 9 shows the morphological and compositional analysis, respectively, of a weathered concrete block sample using SEM coupled with EDS.

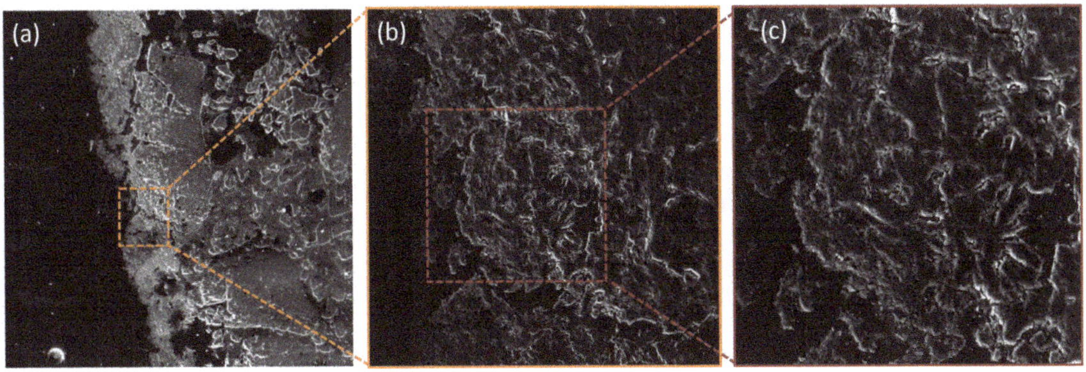

Figure 8. SEM micrographs of concrete: (**a**) 100×; (**b**) 500×; (**c**) 1000×.

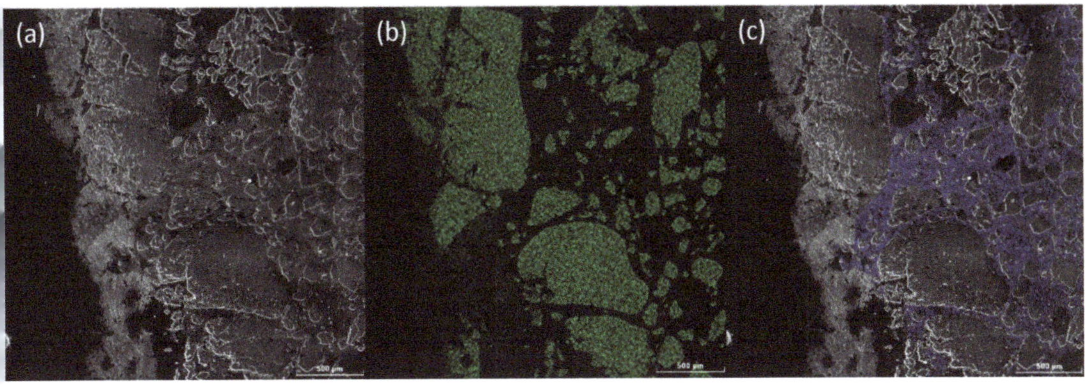

Figure 9. BSE micrograph of concrete: (**a**) at a magnification of 100×; (**b**) elemental mapping of silicon; (**c**) elemental mapping of calcium.

SEM micrographs (Figure 8) reveal the heterogeneous surface of the concrete, characterized by cracks, roughness, and interconnected micropores, likely resulting from weathering and hydration/carbonation processes. Although the porosity is lower than that observed in natural corals, this microtexture still contributes to the formation of microhabitats favorable to the settlement of marine organisms [50,74].

BSE images and EDS mapping (Figure 9) complement these observations by highlighting compositional contrasts: silicon (Si) is associated with aggregates, whereas calcium (Ca) is predominantly distributed in the cement paste matrix. It should be noted that EDS alone cannot differentiate Ca from hydrated paste and Ca from carbonation. However, this limitation is overcome by XRD/FTIR results, which confirm the presence of calcium carbonate, thus clarifying the mineralogical nature of the material.

The heterogeneous surface microtexture, combined with the presence of calcium carbonate, jointly creates favorable microhabitats that may support marine organism settlement and larval recruitment [17,50]. Therefore, the composition and microstructure of the concrete surface reinforce its potential as a suitable substrate for artificial reefs [23].

From the perspective of artificial reef engineering, the natural porous architecture of coral and its calcium carbonate-rich composition serve as critical references for the design of bioinspired substrates. Recent studies have shown that materials exhibiting similar morphology, high porosity, and $CaCO_3$-rich composition can enhance larval settlement, biofilm formation, and the recruitment of reef organisms [23,51,75]. Therefore, the SEM/EDS data reinforce the relevance of biological corals as a benchmark for the development and evaluation of synthetic materials aimed at marine restoration.

3.5. Physical and Mechanical Characterization

Figure 10a shows that the specific densities of the biological samples ranged from 1.5 g/cm^3 (C2) to 2.46 g/cm^3 (T2). In comparison, the concrete exhibited a specific density of 2.45 g/cm^3, a value consistent with dense matrices composed of Portland cement and silica fume [76,77]. Porosity varied widely among the biological samples, from 26.51% (C1) to 41.32% (T3), contrasting with the lower value observed for concrete (14.11%), indicative of a compact and homogeneous structure.

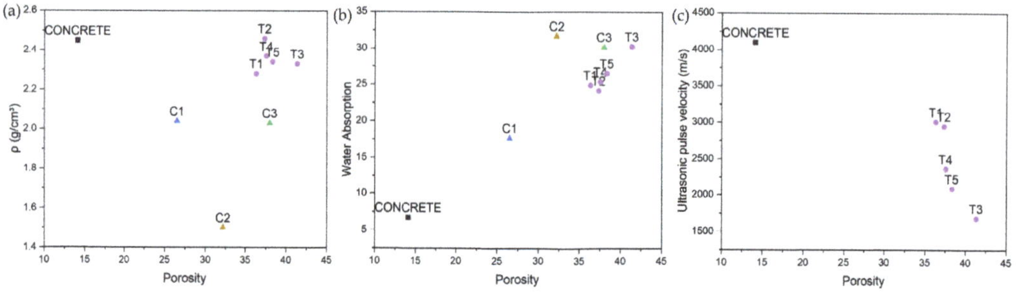

Figure 10. Porosity and (**a**) specific mass; (**b**) water absorption; (**c**) ultrasonic pulse velocity.

These findings are corroborated by the low water absorption of the concrete (6.72%) (Figure 10b), significantly lower than that of the biological samples, which reached up to 31.7% (C2). This discrepancy reflects the highly porous nature of coral skeletons, whose biological function involves intercellular flow and exchange with the aquatic environment. Some studies suggest that water absorption values above 20% may be associated with coral structures exhibiting a higher proportion of open pores, possibly resulting from long-term bioerosion or dissolution processes [78,79].

Ultrasonic pulse velocity (UPV) is employed to assess the density, structural integrity, degree of compaction, and presence of microdefects or porosity in both cementitious and biomineralized materials. Figure 8c shows that biological samples exhibited UPV values ranging from 1600 to 3600 m/s, strongly influenced by their high porosity. Concrete, on the other hand, presented a higher velocity of approximately 4100 m/s, reflecting greater internal cohesion and a lower volume of interconnected pores in the dry state. An inverse correlation between porosity and UPV is observed: the higher the porosity, the lower the wave velocity. For example, sample T3, with 41.32% porosity and 30.28% water absorption, exhibited a UPV of 1684 m/s, whereas sample T1, with lower porosity (36.3%), showed substantially higher values. This is consistent with the results reported by Ma et al. [80], who observed porosity ranging from 41.08% to 54.86% and ultrasonic pulse velocity values between 2200 and 2700 m/s for coral skeletons from shallow reef environments.

In concrete, the high UPV values reflect its low porosity and absorption, associated with good particle packing and the potential formation of secondary products such as densified C–S–H due to silica fume incorporation [81]. FTIR and XRD analyses confirm the presence of carbonates, indicating carbonation, which contributes to local matrix densification and increased ultrasonic velocity, a phenomenon also reported by [82,83].

Xu et al. [84] reported velocities ranging from 2500 to 3100 m/s for both porous coral skeletons, values consistent with those observed in this study. In contrast, cementitious materials with low w/c ratios and optimized curing conditions may exhibit UPV values exceeding 4000 m/s [85].

The biological samples exhibited variable UPV responses, reflecting differences in morphology, internal porosity, and possible structural and mineralogical alterations. In contrast, concrete displayed more predictable behavior, consistent with its compact microstructure.

Figure 11 presents the compressive strength results obtained from concrete cores and natural reef fragments, clearly showing that the concrete samples exhibit significantly higher mechanical strength compared to the biological specimens.

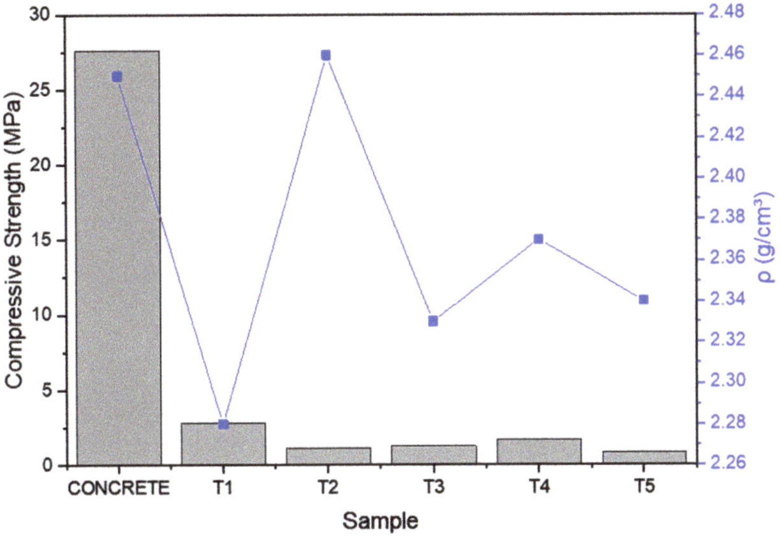

Figure 11. Compressive strength.

The average compressive strength of the concrete was found to be 27.60 MPa, indicating satisfactory mechanical performance and suggesting enhanced durability of the artificial reef structures. In contrast, the natural coral samples displayed considerably lower strength values, typically ranging between 1 and 6 MPa, which is consistent with previous findings reported by Chamberlain [86] and Rodgers et al. [87], who demonstrated that tropical corals possess inherently low compressive strength.

In this context, the mechanical properties of the reef core samples are widely accepted as representative, given that, according to Rodgers et al. [87], there is a general consensus that the low organic content in coral skeletons results in comparable mechanical behavior among dry, saturated, and rehydrated specimens. This aspect is particularly relevant to the present study, as it ensures that the mechanical characterization and long-term analysis of dry coral skeletons can reliably reflect the performance of living coral structures, allowing a valid comparison with the engineered concrete formulations designed for artificial reef applications.

This pronounced strength disparity highlights the importance of utilizing robust artificial reef materials such as carbonated concrete, which can withstand marine environmental loads over extended periods. While significantly stronger than natural coral skeletons, the engineered material remains functionally designed to mimic the habitat required to support marine life colonization and ecological restoration.

3.6. pH Test

The pH curve as a function of time (Figure 12) reveals contrasting patterns between the biological samples and the concrete, reflecting structural, chemical, and mineralogical differences.

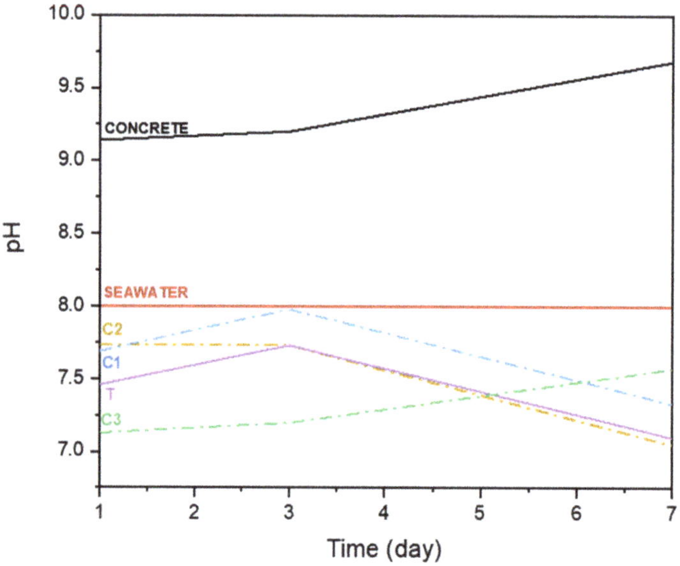

Figure 12. Change in pH over time for leachates from biological samples and concrete immersed in deionized water.

The pH curves for the biological samples remain within a relatively stable range of 7.5 to 8.5, close to the natural pH of seawater (approximately 8.0, according to [88,89]). This overall behavior suggests chemical equilibrium, indicating that coral skeletons do not induce significant changes in the alkalinity of the surrounding medium. Such stability is consistent with the fact that these samples are predominantly composed of calcium carbonate (aragonite or magnesian calcite), minerals characterized by low solubility and high chemical compatibility with marine environments. This behavior also aligns with water absorption and porosity data, which indicate high permeability and fluid exchange with the surrounding environment, without significantly altering the pH.

Studies such as Hofmann et al. [90] report similar behavior for coral-based substrates, which tend to buffer pH and maintain a biochemically stable microenvironment—an essential condition for the viability of ecological processes such as larval adhesion, biofilm formation, and biomineralization.

In contrast, the concrete samples exhibited stable pH values ranging from 9.2 to 9.7, with a slight upward trend over time. This intermediate pH is more alkaline than that of seawater but significantly lower than the typical pH of freshly cured concrete (generally between 11 and 13, according to Bui et al. [64]), suggesting that the material underwent a carbonation process. This process involves the consumption of hydroxide ions (OH^-), which decreases the matrix alkalinity and consequently reduces the pH of the leachate. The presence of carbonates detected through FTIR and XRD analyses directly supports this interpretation.

The dense structure and low porosity of the concrete—characterized by 14.11% total porosity and only 6.72% water absorption—indicate that carbonation likely occurred in the superficial layers of the material. Nevertheless, this surface alteration is sufficient to affect the pH of the surrounding environment, which is particularly relevant in marine applications.

Furthermore, the findings align with studies by Shi et al. [91] and Azizi & Samimi [82], which report that the incorporation of mineral admixtures such as silica fume accelerates the pozzolanic reaction, reducing the content of $Ca(OH)_2$ and contributing to a decrease in free alkalinity. As a result, the pH tends to stabilize at lower values when compared to conventional concretes, especially following natural carbonation processes.

From an ecological perspective, the reduction of pH to values close to 9 is advantageous. While the concrete still retains sufficient alkalinity to ensure its long-term durability, it no longer represents a highly aggressive environment for marine organisms sensitive to elevated pH, such as larvae and microorganisms. In environments with pH values above 11—typical of early-ages concrete—biological colonization is often inhibited [20,21]. In contrast, a moderately alkaline pH and carbonated surfaces favors the formation of biofilms, larval settlement, and the initiation of biomineralization processes [92].

However, it should be noted that after months of seawater reducing the pH of concrete, even in situations with short curing times and little carbonation, Portland cement concrete can become a suitable substrate for coral larvae [93]. Other factors besides the high initial pH of concretes are more relevant to coral performance and reef restoration, such as topographical conditions, substrate orientation and durability, habitat structure, sedimentation rates, and coral predation [50,92,94].

Therefore, the chemical condition observed in the concrete analyzed in this study appears promising for artificial reef applications. It reinforces both the technical and ecological feasibility of carbonated concrete as a functional alternative for the construction of artificial substrates in marine environments.

4. Conclusions

This study compared natural coral skeletons with Portland concrete blocks exposed to weathering, confirming the initial hypothesis that, despite physical and mechanical differences, the concrete shares important mineralogical and chemical properties with coral skeletons.

Analyses via XRD, TG, SEM, and FTIR revealed carbonated phases in both materials, with the concrete showing carbonation after 20 years of exposure. This mineralogical similarity is crucial, as carbonated surfaces are known to favor larval settlement [17,50]. Furthermore, the concrete maintained a stable pH (9.7), a level that does not inhibit marine colonization, aligning with the literature on cement-based substrates conditioned in seawater [93].

Physically and mechanically, the differences found between the concrete and the natural coral skeletons were expected, given that they are materials of distinct origins and formation processes. The concrete showed higher apparent density (2.45 g/cm^3), lower porosity (14.11%), and lower water absorption (6.72%) compared to the biological samples, which varied from 1.50 to 2.46 g/cm^3 in density, 26.51% to 41.32% in porosity, and up to 31.7% in water absorption. The ultrasonic pulse velocity (UPV) was also higher in the concrete (4100 m/s), reflecting greater internal compactness. Compressive strength varied from 1 to 6 MPa in the natural fragments, while the concrete averaged 27.6 MPa. These properties are consistent with studies reporting higher durability and structural stability of concrete modules in artificial reefs, even under intense hydrodynamic action [23,93].

Although some of the concrete's physical properties might suggest potential limitations for its ecological performance, the observed values, such as low porosity and high internal compactness, are within the ranges reported in the literature for cementitious materials in marine applications. The obtained porosity aligns with the ranges established by previous studies, which demonstrate variations between 8.9% and 16.2% for similar concretes [95]. Likewise, the ultrasonic pulse velocity (UPV) values are consistent with typical parameters for conventional concrete, which generally range from 3000 to 4500 m/s [23].

Research on concretes implanted in marine environments shows that these physical properties do not inhibit biological colonization. Knoester et al. [50] confirmed the successful attachment of corals, calcifying algae, and filter-feeding invertebrates to cementitious materials, even when they exhibit high mechanical strength. These results indicate that factors such as surface composition, microtopography, and chemical conditioning may be more determinant for ecological success than the absolute physical properties of the material.

Therefore, it is concluded that the analyzed concrete, when compared to natural coral skeletons, shows significant similarities in mineralogical and chemical characteristics, especially regarding the predominance of calcium carbonate and the identified thermal behavior. Combined with other physical and mechanical properties considered satisfactory by the literature references, these findings reinforce the viability of weathered concrete blocks as artificial reef structures, legitimizing their application and monitoring within the PREAMAR program.

However, for the fabrication of new concrete blocks, it is recommended to optimize porosity, pH, and surface microtexture, aiming to enhance the formation of microhabitats and, consequently, expand their ecological effectiveness. Additionally, studies involving the analysis of the dynamics of colonization and development of organisms in situ can be valuable for the practical validation of the project.

Author Contributions: Conceptualization, M.R.F.L.F., S.M.T., A.J.V.d.S. and C.D.d.N.; methodology, M.R.F.L.F., S.M.T., M.F.d.S., R.A.S.d.R. and K.M.; software, M.F.d.S., R.A.S.d.R., G.F.H. and W.M.d.S.; validation, C.D.d.N., M.R.F.L.F., S.M.T. and A.J.V.d.S.; formal analysis, M.F.d.S., R.A.S.d.R., G.F.H. and W.M.d.S.; investigation, M.F.d.S., R.A.S.d.R., K.M., G.F.H. and W.M.d.S.; resources, M.R.F.L.F., S.M.T., A.J.V.d.S., K.M. and C.D.d.N.; data curation, M.F.d.S., R.A.S.d.R. and G.F.H.; writing—original draft preparation, M.F.d.S., R.A.S.d.R., K.M. and G.F.H.; writing—review and editing, M.F.d.S., R.A.S.d.R., K.M., G.F.H., M.R.F.L.F. and S.M.T.; visualization, R.A.S.d.R. and W.M.d.S.; supervision, C.D.d.N., M.R.F.L.F., S.M.T. and A.J.V.d.S.; project administration, M.R.F.L.F., S.M.T. and A.J.V.d.S.; funding acquisition, C.D.d.N., M.R.F.L.F., S.M.T. and A.J.V.d.S. All authors have read and agreed to the published version of the manuscript.

Funding: This study was financed in part by the Coordenação de Aperfeiçoamento de Pessoal de Nível Superior—Brasil (CAPES)—Finance Code 001; Paraíba State Research Foundation (FAPESQ-PB); and Strategic Program for Marine Artificial Structures of Paraíba (PREAMAR).

Data Availability Statement: All data generated and analyzed in this study are included in the article, with additional datasets available from the corresponding author upon request.

Acknowledgments: The authors thank the local community involved (institutions working in the Ocean Space—Research and Action Institute) (www.inpact.org.br (accessed on 19 July 2025)).

Conflicts of Interest: The authors declare no conflicts of interest.

References

1. Hughes, T.P.; Barnes, M.L.; Bellwood, D.R.; Cinner, J.E.; Cumming, G.S.; Jackson, J.B.C.; Kleypas, J.; van de Leemput, I.A.; Lough, J.M.; Morrison, T.H.; et al. Coral Reefs in the Anthropocene. *Nature* **2017**, *546*, 82–90. [CrossRef]
2. Moynihan, M.A.; Amini, S.; Goodkin, N.F.; Tanzil, J.T.I.; Chua, J.Q.I.; Fabbro, G.N.; Fan, T.Y.; Schmidt, D.N.; Miserez, A. Environmental Impact on the Mechanical Properties of Porites Spp. Corals. *Coral Reefs* **2021**, *40*, 701–717. [CrossRef]
3. Helgoe, J.; Davy, S.K.; Weis, V.M.; Rodriguez-Lanetty, M. Triggers, Cascades, and Endpoints: Connecting the Dots of Coral Bleaching Mechanisms. *Biol. Rev.* **2024**, *99*, 715–752. [CrossRef] [PubMed]
4. Dias, T.L.P.; Gondim, A.I. Bleaching in Scleractinians, Hydrocorals, and Octocorals during Thermal Stress in a Northeastern Brazilian Reef. *Mar. Biodivers.* **2016**, *46*, 303–307. [CrossRef]
5. Weis, V.M. Cell Biology of Coral Symbiosis: Foundational Study Can Inform Solutions to the Coral Reef Crisis. *Integr. Comp. Biol.* **2019**, *59*, 845–855. [CrossRef]
6. National Academies of Sciences, Engineering, and Medicine. *A Research Review of Interventions to Increase the Persistence and Resilience of Coral Reefs*; The National Academies Press: Washington, DC, USA, 2019; ISBN 978-0-309-48535-7.
7. Berman, O.; Weizman, M.; Oren, A.; Neri, R.; Parnas, H.; Shashar, N.; Tarazi, E. Design and Application of a Novel 3D Printing Method for Bio-Inspired Artificial Reefs. *Ecol. Eng.* **2023**, *188*, 106892. [CrossRef]
8. Boström-Einarsson, L.; Babcock, R.C.; Bayraktarov, E.; Ceccarelli, D.; Cook, N.; Ferse, S.C.A.; Hancock, B.; Harrison, P.; Hein, M.; Shaver, E.; et al. Coral Restoration—A Systematic Review of Current Methods, Successes, Failures and Future Directions. *PLoS ONE* **2020**, *15*, e0226631. [CrossRef]
9. Howlett, L.; Camp, E.F.; Edmondson, J.; Edmondson, J.; Agius, T.; Hosp, R.; Coulthard, P.; Edmondson, S.; Suggett, D.J. Adoption of Coral Propagation and Out-Planting via the Tourism Industry to Advance Site Stewardship on the Northern Great Barrier Reef. *Ocean Coast. Manag.* **2022**, *225*, 106199. [CrossRef]
10. Knoester, E.G.; Groenendijk, M.H.; Murk, A.J.; Osinga, R. Some like It Dirty: Less Frequent Nursery Cleaning Can Reduce Reef Restoration Costs with Limited Negative Effects on Coral Performance. *Ecol. Eng.* **2024**, *201*, 107209. [CrossRef]
11. Lindahl, U. Coral Reef Rehabilitation through Transplantation of Staghorn Corals: Effects of Artificial Stabilization and Mechanical Damages. *Coral Reefs* **2003**, *22*, 217–223. [CrossRef]
12. Silbiger, N.J.; Nelson, C.E.; Remple, K.; Sevilla, J.K.; Quinlan, Z.A.; Putnam, H.M.; Fox, M.D.; Donahue, M.J. Nutrient Pollution Disrupts Key Ecosystem Functions on Coral Reefs. *Proc. R. Soc. B Biol. Sci.* **2018**, *285*, 20172718. [CrossRef] [PubMed]
13. Paxton, A.B.; Shertzer, K.W.; Bacheler, N.M.; Kellison, G.T.; Riley, K.L.; Taylor, J.C. Meta-Analysis Reveals Artificial Reefs Can Be Effective Tools for Fish Community Enhancement but Are Not One-Size-Fits-All. *Front. Mar. Sci.* **2020**, *7*, 534479. [CrossRef]
14. Matus, I.V.; Alves, J.L.; Góis, J.; Vaz-Pires, P.; Barata da Rocha, A. Artificial Reefs through Additive Manufacturing: A Review of Their Design, Purposes and Fabrication Process for Marine Restoration and Management. *Rapid Prototyp. J.* **2024**, *30*, 87–122. [CrossRef]
15. Jensen, A.C.; Collins, K.J.; Lockwood, A.P.M.; Mallinson, J.J.; Turnpenny, W.H. Colonization and Fishery Potential of a Coal-Ash Artificial Reef, Poole Bay, United Kingdom. *Bull. Mar. Sci.* **1994**, *55*, 1263–1276.
16. Vivier, B.; Dauvin, J.C.; Navon, M.; Rusig, A.M.; Mussio, I.; Orvain, F.; Boutouil, M.; Claquin, P. Marine Artificial Reefs, a Meta-Analysis of Their Design, Objectives and Effectiveness. *Glob. Ecol. Conserv.* **2021**, *27*, e01538. [CrossRef]
17. Dennis, H.D.; Evans, A.J.; Banner, A.J.; Moore, P.J. Reefcrete: Reducing the Environmental Footprint of Concretes for Eco-Engineering Marine Structures. *Ecol. Eng.* **2018**, *120*, 668–678. [CrossRef]
18. Padilha, R.D.A.; Henkes, J.A. A Utilização de Recifes Artificiais Marinhos Como Ferramenta de Recuperação Da Fauna Marinha. *Rev. Gestão Sustentabilidade Ambient.* **2012**, *1*, 41. [CrossRef]
19. Tavares, J.C.; Lucena, L.F.L.; Henriques, G.F.; Ferreira, R.L.S.; dos Anjos, M.A.S. Use of Banana Leaf Ash as Partial Replacement of Portland Cement in Eco-Friendly Concretes. *Constr. Build. Mater.* **2022**, *346*, 128467. [CrossRef]
20. Hayek, M.; Salgues, M.; Souche, J.-C.; Cunge, E.; Giraudel, C.; Paireau, O. Influence of the Intrinsic Characteristics of Cementitious Materials on Biofouling in the Marine Environment. *Sustainability* **2021**, *13*, 2625. [CrossRef]
21. Sun, X.; Wai, O.W.H.; Xie, J.; Li, X. Biomineralization To Prevent Microbially Induced Corrosion on Concrete for Sustainable Marine Infrastructure. *Environ. Sci. Technol.* **2024**, *58*, 522–533. [CrossRef]
22. Xu, Q.; Ji, T.; Yang, Z.; Ye, Y. Preliminary Investigation of Artificial Reef Concrete with Sulphoaluminate Cement, Marine Sand and Sea Water. *Constr. Build. Mater.* **2019**, *211*, 837–846. [CrossRef]
23. Rupasinghe, M.; Nicolas, R.S.; Lanham, B.S.; Morris, R.L. Sustainable Oyster Shell Incorporated Artificial Reef Concrete for Living Shorelines. *Constr. Build. Mater.* **2024**, *410*, 134217. [CrossRef]
24. Feng, Y.; Yang, J.; Zhang, P. Effects of Carbonation Curing Regimes on Alkalinity of Self-Compacting Concretes for Marine Artificial Reef. *Constr. Build. Mater.* **2023**, *369*, 130614. [CrossRef]
25. Wesemann, C.A.; Antoun, M.; Teune, I.; Schollbach, K.; Rippen, J.; Hylkema, A.; Oosterhoff, B.; Brouwers, H.J.H. Effect of Carbonation on Eco-Friendly Binders for Marine Infrastructures. *Constr. Build. Mater.* **2025**, *477*, 141293. [CrossRef]

26. Wang, W.; Wang, S.; Peng, L.; Wang, N.; Meng, T.; Zhao, Y.; Gong, F. Preliminary Investigation and Life Cycle Assessment of Artificial Reefs with Recycled Brick-Concrete Aggregates. *Constr. Build. Mater.* **2024**, *432*, 136618. [CrossRef]
27. Liu, J.; Ou, Z.; Peng, W.; Guo, T.; Deng, W.; Chen, Y. Literature Review of Coral Concrete. *Arab. J. Sci. Eng.* **2018**, *43*, 1529–1541. [CrossRef]
28. Wang, A.; Lyu, B.; Zhang, Z.; Liu, K.; Xu, H.; Sun, D. The Development of Coral Concretes and Their Upgrading Technologies: A Critical Review. *Constr. Build. Mater.* **2018**, *187*, 1004–1019. [CrossRef]
29. Leite, D.; Vasconcelos, E.; Riul, P.; Freitas, N.; Miranda, G. Dataset of Macrobenthic Species in Urban Coastal Reef Environments in Brazilian Northeast. *Data Brief* **2020**, *31*, 105773. [CrossRef]
30. Massei, K.; Souza, M.C.S.; da Silva, R.M.; Neumann, V.H.d.M.L.; Manso, V.D.A.V.; Vianna, P.C.G.; Junior, A.V.F.; da Silva Moura, C.M.; de Lavôr, L.F.; de Carvalho Araújo, R.; et al. Multi-Proxy Assessment of Coral Reef Formation and Biotic-Abiotic Diversity in an Urban Coastal Reef Ecosystem in Northeastern Brazil. *Sci. Total Environ.* **2023**, *891*, 164487. [CrossRef]
31. Massei, K.; Souza, M.C.S.; da Silva, R.M.; Costa, D.d.A.; Vianna, P.C.G.; Crispim, M.C.; de Miranda, G.E.C.; Eggertsen, L.; Eloy, C.C.; Santos, C.A.G. Analysis of Marine Diversity and Anthropogenic Pressures on Seixas Coral Reef Ecosystem (Northeastern Brazil). *Sci. Total Environ.* **2023**, *905*, 166984. [CrossRef]
32. Leite, D.S.L.; de Vasconcelos, E.R.T.P.P.; Riul, P.; de Freitas, N.D.A.; de Miranda, G.E.C. Evaluation of the Conservation Status and Monitoring Proposal for the Coastal Reefs of Paraíba, Brazil: Bioindication as an Environmental Management Tool. *Ocean Coast. Manag.* **2020**, *194*, 105208. [CrossRef]
33. Santos, D.H.C.; Passavante, J.Z.d.O. Recifes Artificiais Marinhos: Modelos e Utilizações No Brasil e No Mundo. *Bol. Tec. Cient. CEPENE* **2007**, *15*, 113–124.
34. Filho, O.d.R.M.; Callou, A.B.F.; Santos, M.S.T. Políticas Públicas e Extensão Pesqueira Em Cabedelo, Paraíba. *Interações Campo Gd.* **2010**, *11*, 93–100. [CrossRef]
35. *ABNT NBR 16697*; Portland Cement—Requirements. ABNT: Rio de Janeiro, Brazil, 2018; 12p.
36. *ASTM C595*; Specification for Blended Hydraulic Cements. ASTM International: West Conshohocken, PA, USA, 2008; 7p.
37. de Lima, W.P.; Pessoa-Gutierres, H.E. Impactos Ambientais No Parque Natural Municipal de Cabedelo—Estado Do Paraíba (PB), Brasil. *Rev. Geográfica América Cent.* **2021**, *1*, 189–215. [CrossRef]
38. Henriques, G.F.; Tavares, J.C.; Neto, A.D.d.L.T. Carbonatação Natural e Modelos de Previsão Para Concretos de Cimento Portland: Uma Revisão. *Rev. Principia—Divulg. Científica e Tecnológica do IFPB* **2023**, *60*, 102. [CrossRef]
39. Scrivener, K.; Snellings, R.; Lothenbach, B. *A Practical Guide to Microstructural Analysis of Cementitious Materials;* CRC Press: Boca Raton, FL, USA, 2016; ISBN 9781498738651.
40. *ASTM C642*; Test Method for Density, Absorption, and Voids in Hardened Concrete. ASTM International: West Conshohocken, PA, USA, 2021; 3p.
41. *ASTM C597*; Test Method for Ultrasonic Pulse Velocity Through Concrete. ASTM International: West Conshohocken, PA, USA, 2022; 4p.
42. *ASTM C39*; Test Method for Compressive Strength of Cylindrical Concrete Specimens. ASTM International: West Conshohocken, PA, USA, 2021; 8p.
43. Wang, W.-C.; Huang, W.-H.; Lee, M.-Y.; Duong, H.T.H.; Chang, Y.-H. Standardized Procedure of Measuring the PH Value of Cement Matrix Material by Ex-Situ Leaching Method (ESL). *Crystals* **2021**, *11*, 436. [CrossRef]
44. Loh, P.-Y.; Shafigh, P.; Katman, H.Y.; Ibrahim, Z. Effect of Filtering Method on the PH Measurement of Cement Mortar in Ex-Situ Leaching. *Heliyon* **2024**, *10*, e25859. [CrossRef]
45. van Dijk, I.; Barras, C.; de Nooijer, L.J.; Mouret, A.; Geerken, E.; Oron, S.; Reichart, G.-J. Coupled Calcium and Inorganic Carbon Uptake Suggested by Magnesium and Sulfur Incorporation in Foraminiferal Calcite. *Biogeosciences* **2019**, *16*, 2115–2130. [CrossRef]
46. Stanley, S.M.; Ries, J.B.; Hardie, L.A. Low-Magnesium Calcite Produced by Coralline Algae in Seawater of Late Cretaceous Composition. *Proc. Natl. Acad. Sci. USA* **2002**, *99*, 15323–15326. [CrossRef]
47. Aufort, J.; Demichelis, R. Magnesium Impurities Decide the Structure of Calcium Carbonate Hemihydrate. *Cryst. Growth Des.* **2020**, *20*, 8028–8038. [CrossRef]
48. Conci, N.; Griesshaber, E.; Rivera-Vicéns, R.E.; Schmahl, W.W.; Vargas, S.; Wörheide, G. Molecular and Mineral Responses of Corals Grown under Artificial Calcite Sea Conditions. *Geobiology* **2024**, *22*, e12586. [CrossRef]
49. Ries, J.B. Review: Geological and Experimental Evidence for Secular Variation in Seawater Mg/Ca (Calcite-Aragonite Seas) and Its Effects on Marine Biological Calcification. *Biogeosciences* **2010**, *7*, 2795–2849. [CrossRef]
50. Knoester, E.G.; Vos, A.; Saru, C.; Murk, A.J.; Osinga, R. Concrete Evidence: Outplanted Corals for Reef Restoration Do Not Need Extended Curing of Ordinary Portland Cement. *R. Soc. Open Sci.* **2024**, *11*, 241064. [CrossRef] [PubMed]
51. O'Reilly, L.M.; Willerth, S.M. Evaluating the Biocompatibility of Ceramic Materials for Constructing Artificial Reefs. *Front. Mar. Sci.* **2024**, *10*, 1292584. [CrossRef]

52. Ries, J.B. Effects of Secular Variation in Seawater Mg/Ca Ratio (Calcite–Aragonite Seas) on $CaCO_3$ Sediment Production by the Calcareous Algae Halimeda, Penicillus and Udotea—Evidence from Recent Experiments and the Geological Record. *Terra Nova* **2009**, *21*, 323–339. [CrossRef]
53. Wehrmeister, U.; Soldati, A.L.; Jacob, D.E.; Häger, T.; Hofmeister, W. Raman Spectroscopy of Synthetic, Geological and Biological Vaterite: A Raman Spectroscopic Study. *J. Raman Spectrosc.* **2010**, *41*, 193–201. [CrossRef]
54. Ramos-Silva, P.; Kaandorp, J.; Huisman, L.; Marie, B.; Zanella-Cléon, I.; Guichard, N.; Miller, D.J.; Marin, F. The Skeletal Proteome of the Coral Acropora Millepora: The Evolution of Calcification by Co-Option and Domain Shuffling. *Mol. Biol. Evol.* **2013**, *30*, 2099–2112. [CrossRef]
55. Farfan, G.A.; Apprill, A.; Cohen, A.; DeCarlo, T.M.; Post, J.E.; Waller, R.G.; Hansel, C.M. Crystallographic and Chemical Signatures in Coral Skeletal Aragonite. *Coral Reefs* **2022**, *41*, 19–34. [CrossRef]
56. Tsuboi, M.; Tamura, K.; Kitanaka, R.; Oka, H.; Akao, K.; Ozaki, Y. Attenuated Total Reflection Infrared and Far-Infrared, and Raman Spectroscopy Studies of Minerals, Rocks, and Biogenic Minerals. *Appl. Spectrosc.* **2024**, *78*, 186–196. [CrossRef]
57. Zakaria, F.Z.; Mihály, J.; Sajó, I.; Katona, R.; Hajba, L.; Aziz, F.A.; Mink, J. FT-Raman and FTIR Spectroscopic Characterization of Biogenic Carbonates from Philippine Venus Seashell and Porites Sp. Coral. *J. Raman Spectrosc.* **2008**, *39*, 1204–1209. [CrossRef]
58. Naggi, A.; Torri, G.; Iacomini, M.; Colombo Castelli, G.; Reggi, M.; Fermani, S.; Dubinsky, Z.; Goffredo, S.; Falini, G. Structure and Function of Stony Coral Intraskeletal Polysaccharides. *ACS Omega* **2018**, *3*, 2895–2901. [CrossRef]
59. Rahman, M.A.; Halfar, J. First Evidence of Chitin in Calcified Coralline Algae: New Insights into the Calcification Process of Clathromorphum Compactum. *Sci. Rep.* **2014**, *4*, 6162. [CrossRef]
60. Mass, T.; Putnam, H.M.; Drake, J.L.; Zelzion, E.; Gates, R.D.; Bhattacharya, D.; Falkowski, P.G. Temporal and Spatial Expression Patterns of Biomineralization Proteins during Early Development in the Stony Coral Pocillopora Damicornis. *Proc. R. Soc. B Biol. Sci.* **2016**, *283*, 20160322. [CrossRef] [PubMed]
61. Rahman, M.A.; Halfar, J.; Adey, W.H.; Nash, M.; Paulo, C.; Dittrich, M. The Role of Chitin-Rich Skeletal Organic Matrix on the Crystallization of Calcium Carbonate in the Crustose Coralline Alga Leptophytum Foecundum. *Sci. Rep.* **2019**, *9*, 11869. [CrossRef] [PubMed]
62. Ganot, P.; Loentgen, G.; Marin, F.; Plasseraud, L.; Allemand, D.; Tambutté, S. An Alternative and Effective Method for Extracting Skeletal Organic Matrix Adapted to the Red Coral Corallium Rubrum. *Biol. Open* **2022**, *11*, bio059536. [CrossRef] [PubMed]
63. Falini, G.; Fermani, S.; Goffredo, S. Coral Biomineralization: A Focus on Intra-Skeletal Organic Matrix and Calcification. *Semin. Cell Dev. Biol.* **2015**, *46*, 17–26. [CrossRef]
64. Bui, H.; Delattre, F.; Levacher, D. Experimental Methods to Evaluate the Carbonation Degree in Concrete—State of the Art Review. *Appl. Sci.* **2023**, *13*, 2533. [CrossRef]
65. Ylmén, R.; Jäglid, U. Carbonation of Portland Cement Studied by Diffuse Reflection Fourier Transform Infrared Spectroscopy. *Int. J. Concr. Struct. Mater.* **2013**, *7*, 119–125. [CrossRef]
66. Cuif, J.; Dauphin, Y.; Berthet, P.; Jegoudez, J. Associated Water and Organic Compounds in Coral Skeletons: Quantitative Thermogravimetry Coupled to Infrared Absorption Spectrometry. *Geochem. Geophys. Geosystems* **2004**, *5*, Q11011. [CrossRef]
67. Prada, F.; Brizi, L.; Franzellitti, S.; Mengoli, S.; Fermani, S.; Polishchuk, I.; Baraldi, N.; Ricci, F.; Palazzo, Q.; Caroselli, E.; et al. Coral Micro- and Macro-Morphological Skeletal Properties in Response to Life-Long Acclimatization at CO_2 Vents in Papua New Guinea. *Sci. Rep.* **2021**, *11*, 19927. [CrossRef]
68. Gilbert, P.U.P.A.; Bergmann, K.D.; Boekelheide, N.; Tambutté, S.; Mass, T.; Marin, F.; Adkins, J.F.; Erez, J.; Gilbert, B.; Knutson, V.; et al. Biomineralization: Integrating Mechanism and Evolutionary History. *Sci. Adv.* **2022**, *8*, eabl9653. [CrossRef]
69. Tambutté, S.; Holcomb, M.; Ferrier-Pagès, C.; Reynaud, S.; Tambutté, É.; Zoccola, D.; Allemand, D. Coral Biomineralization: From the Gene to the Environment. *J. Exp. Mar. Biol. Ecol.* **2011**, *408*, 58–78. [CrossRef]
70. Laine, J.; Labady, M.; Albornoz, A.; Yunes, S. Porosities and Pore Sizes in Coralline Calcium Carbonate. *Mater. Charact.* **2008**, *59*, 1522–1525. [CrossRef]
71. Yoshimura, T.; Tanimizu, M.; Inoue, M.; Suzuki, A.; Iwasaki, N.; Kawahata, H. Mg Isotope Fractionation in Biogenic Carbonates of Deep-Sea Coral, Benthic Foraminifera, and Hermatypic Coral. *Anal. Bioanal. Chem.* **2011**, *401*, 2755–2769. [CrossRef] [PubMed]
72. Han, Z.; Meng, R.; Zhao, H.; Gao, X.; Zhao, Y.; Han, Y.; Liu, F.; Tucker, M.E.; Deng, J.; Yan, H. The Incorporation of Mg2+ Ions into Aragonite during Biomineralization: Implications for the Dolomitization of Aragonite. *Front. Microbiol.* **2023**, *14*, 1078430. [CrossRef] [PubMed]
73. Servetto, N.; de Aranzamendi, M.C.; Bettencourt, R.; Held, C.; Abele, D.; Movilla, J.; González, G.; Bustos, D.M.; Sahade, R. Molecular Mechanisms Underlying Responses of the Antarctic Coral Malacobelemnon Daytoni to Ocean Acidification. *Mar. Environ. Res.* **2021**, *170*, 105430. [CrossRef] [PubMed]
74. Ferrario, F.; Suskiewicz, T.; Johnson, L.E. A Seascape Approach for Guiding Effective Habitat Enhancement: Spatially explicit modeling of kelp–grazer interactions. *Elem. Sci. Anthr.* **2021**, *9*, 00013. [CrossRef]

75. Fong, J.; Ramsby, B.D.; Flores, F.; Dada, T.; Antunes, E.; Abdul Wahab, M.A.; Severati, A.; Negri, A.P.; Diaz-Pulido, G. Effects of Material Type and Surface Roughness of Settlement Tiles on Macroalgal Colonisation and Early Coral Recruitment Success. *Coral Reefs* **2024**, *43*, 1083–1096. [CrossRef]
76. Lakhiar, M.T.; Kong, S.Y.; Bai, Y.; Susilawati, S.; Zahidi, I.; Paul, S.C.; Raghunandan, M.E. Thermal and Mechanical Properties of Concrete Incorporating Silica Fume and Waste Rubber Powder. *Polymers* **2022**, *14*, 4858. [CrossRef]
77. Khan, M.I.; Abbas, Y.M.; Fares, G. Enhancing Cementitious Concrete Durability and Mechanical Properties through Silica Fume and Micro-Quartz. *Sustainability* **2023**, *15*, 15913. [CrossRef]
78. Lyu, B.; Wang, A.; Zhang, Z.; Liu, K.; Xu, H.; Shi, L.; Sun, D. Coral Aggregate Concrete: Numerical Description of Physical, Chemical and Morphological Properties of Coral Aggregate. *Cem. Concr. Compos.* **2019**, *100*, 25–34. [CrossRef]
79. Wang, L.; Zhang, Q.; Yi, J.; Zhang, J. Effects of Coral Aggregate Properties on the Ultrasonic Pulse Velocity of Concrete. *J. Build. Eng.* **2023**, *80*, 107935. [CrossRef]
80. Ma, L.; Deng, J.; Wang, M.; Wang, J.; Fang, B.; Wu, J. Effect of Diagenetic Variation on the Static and Dynamic Mechanical Behavior of Coral Reef Limestone. *Int. J. Min. Sci. Technol.* **2024**, *34*, 893–908. [CrossRef]
81. Ahmed, A. Assessing the Effects of Supplementary Cementitious Materials on Concrete Properties: A Review. *Discov. Civ. Eng.* **2024**, *1*, 145. [CrossRef]
82. Azizi, M.; Samimi, K. Effect of Silica Fume on Self-Compacting Earth Concrete: Compressive Strength, Durability and Microstructural Studies. *Constr. Build. Mater.* **2025**, *472*, 140815. [CrossRef]
83. Oke, J.A.; Abuel-Naga, H. Ultrasonic Non-Destructive Testing of Accelerated Carbonation Cured-Eco-Bricks. *Appl. Sci.* **2024**, *14*, 8954. [CrossRef]
84. Xu, J.; Zhang, Z.; Huang, X. Progressive Failure and Fracture Characteristics of Coral Reef Limestone under Triaxial Compression. *Eng. Fract. Mech.* **2025**, *318*, 110948. [CrossRef]
85. Ho, L.S.; Huynh, T.-P. Long-Term Mechanical Properties and Durability of High-Strength Concrete Containing High-Volume Local Fly Ash as a Partial Cement Substitution. *Results Eng.* **2023**, *18*, 101113. [CrossRef]
86. Chamberlain, J.A. Mechanical Properties of Coral Skeleton: Compressive Strength and Its Adaptive Significance. *Paleobiology* **1978**, *4*, 419–435. [CrossRef]
87. Rodgers, K.; Cox, E.; Newtson, C. Effects of Mechanical Fracturing and Experimental Trampling on Hawaiian Corals. *Environ. Manag.* **2003**, *31*, 377–384. [CrossRef]
88. Silva, C.A.R.E.; de Godoy Fernandes, L.V.; de Souza, F.E.S.; Marotta, H.; da Costa Fernandes, F.; Mello, T.M.S.; Monteiro, N.S.C.; Rocha, A.A.; Coutinho, R.; de Almeida Fernandes, L.D.; et al. Carbonate System in the Cabo Frio Upwelling. *Sci. Rep.* **2023**, *13*, 5292. [CrossRef]
89. Ramos e Silva, C.A.; Monteiro, N.S.C.; Cavalcante, L.M.; Junior, W.T.; Rocha Carneiro, M.E.; Soares de Souza, F.E.; Borges Garcia, C.A.; Damasceno, R.N.; de Araújo Rocha, A. Inventory of Water Masses and Carbonate System from Brazilian's Northeast Coast: Monitoring Ocean Acidification. *PLoS ONE* **2022**, *17*, e0271875. [CrossRef]
90. Hofmann, L.C.; Koch, M.; de Beer, D. Biotic Control of Surface PH and Evidence of Light-Induced H^+ Pumping and Ca^{2+}-H^+ Exchange in a Tropical Crustose Coralline Alga. *PLoS ONE* **2016**, *11*, e0159057. [CrossRef] [PubMed]
91. Shi, C.; Jiménez, A.F.; Palomo, A. New Cements for the 21st Century: The Pursuit of an Alternative to Portland Cement. *Cem. Concr. Res.* **2011**, *41*, 750–763. [CrossRef]
92. Hsiung, A.; Tan, W.; Loke, L.; Firth, L.; Heery, E.; Ducker, J.; Clark, V.; Pek, Y.; Birch, W.; Ang, A.; et al. Little Evidence That Lowering the PH of Concrete Supports Greater Biodiversity on Tropical and Temperate Seawalls. *Mar. Ecol. Prog. Ser.* **2020**, *656*, 193–205. [CrossRef]
93. Maboloc, E.A.; Szeto, J.H.-Y.; Yuen, K.C.-H.; Wong, S.C.-C.; Astudillo, J.C.; Cai, L.; Hao, L.; Zhang, S.; Chui, A.P.-Y.; Fang, J.K.-H. Seawater-Conditioned Cement as a Viable Substrate for Coral Larval Settlement. *Reg. Stud. Mar. Sci.* **2025**, *88*, 104254. [CrossRef]
94. Hartanto, R.S.; Loke, L.H.L.; Heery, E.C.; Hsiung, A.R.; Goh, M.W.X.; Pek, Y.S.; Birch, W.R.; Todd, P.A. Material Type Weakly Affects Algal Colonisation but Not Macrofaunal Community in an Artificial Intertidal Habitat. *Ecol. Eng.* **2022**, *176*, 106514. [CrossRef]
95. Ghahari, S.A.; Ramezanianpour, A.M.; Ramezanianpour, A.A.; Esmaeili, M. An Accelerated Test Method of Simultaneous Carbonation and Chloride Ion Ingress: Durability of Silica Fume Concrete in Severe Environments. *Adv. Mater. Sci. Eng.* **2016**, *2016*, 1650979. [CrossRef]

Journal of
*Marine Science
and Engineering*

MDPI

Article

Research on Lightweight Design Performance of Offshore Structures Based on 3D Printing Technology

Haoyu Jiang [1], Yifan Xie [2], Shengqing Zeng [2], Sixing Guo [2], Zehan Chen [2], Zhenjie Liang [2] and Dapeng Zhang [2,*]

1 School of Electronics and Information Engineering, Guangdong Ocean University, Zhanjiang 524088, China
2 Ship and Maritime College, Guangdong Ocean University, Zhanjiang 524005, China; 18027398113@139.com (S.G.)
* Correspondence: zhangdapeng@gdou.edu.cn

Abstract

Traditional manufacturing methods struggle to incorporate complex internal configurations within structures, thus restricting the potential for enhancing the strength of offshore structures through internal design. However, the advent of 3D printing technology presents innovative solutions to this challenge. Previous research has investigated the use of 3D printing to integrate lattice-like structures within conventional frameworks to achieve lightweight designs. Building upon this foundation, this paper models an embedded structure and other marine structures subjected to similar loads using simplified models and conducts a thorough investigation into their mechanical properties. Specifically, it examines the effects of the 3D-printed infill structure, infill rate, and tilt angle of printed specimens on the mechanical properties of 3D-printed components. The goal is to identify the optimal parameter combinations that ensure structural strength while also achieving a lightweight design and a secondary lightweight design for the embedded structure. This paper concludes, from tensile, torsional, and compressive experiments, that honeycomb infill structures, with specimens printed at an inclination angle of 0°, exhibit superior performance across all properties. Additionally, the bonding between the layers of the printed parts is identified as a key factor influencing the tensile and torsional properties. While increasing the infill rate can significantly improve the overall mechanical properties of specimens, it also results in a corresponding reduction in the lightweighting index.

Keywords: manufacturing; 3D printing; offshore structures; mechanical properties; lightweighting index

1. Introduction

With growing global attention on clean energy, developing and utilizing renewable energy sources such as hydropower and tidal energy has become a key strategy for achieving a sustainable, low-carbon future [1–3]. Hydropower, characterized by its cleanliness, efficiency, and low-carbon footprint, has emerged as a vital renewable energy source driving sustainable development. Its flexibility and energy storage capabilities provide significant advantages in electricity generation and grid stability [4–6]. Tidal power generation serves dual functions of energy storage and electricity generation, playing a vital role in regulating power supply and maintaining grid frequency stability. These characteristics render hydropower and tidal power indispensable for achieving low-carbon economies and sustainable energy strategies [7,8]. Among these, the collection of mechanical energy has garnered increasing attention due to the widespread availability of target resources [9–18].

This approach effectively harnesses the flow of liquid earth in oceans and rivers to drive generators and generate energy [19]. Head height significantly impacts turbine efficiency and output power. Micro-hydropower generation (MHPG) systems must be installed at low heads, resulting in low turbine rotational speeds. Conversely, conventional generators exhibit low power conversion efficiency at low speeds, necessitating speed enhancement via gearboxes and/or belts [20]. To enhance hydropower efficiency, researchers have pursued multiple avenues. For instance, Wei et al. developed a 1 kW low-speed high-efficiency permanent magnet generator (LHPM generator) aimed at enabling direct-drive MHPG systems to boost power generation efficiency [21]. Brian et al. utilized vibration measurements and frequency domain analysis to locate vibrations within a series of hard-to-reach checkpoints in hydropower installations, thereby improving efficiency [16]. Yadav and Chauhan [22] proposed a micro-hydropower system based on generating electricity from household water tanks. Ridzuan et al. [23] investigated a system harnessing water flowing through domestic pipelines. H. Zainuddin [24] introduced a system utilizing domestic pipeline water energy for power generation. However, none of these experiments considered enhancing hydroelectric generator efficiency through material innovations. Substituting materials for bearings and blades in hydroelectric generators necessitates consideration of mechanical properties. This paper evaluates the mechanical performance of new materials through tensile, compression, and torsion tests, thereby demonstrating the feasibility of replacing conventional materials in hydroelectric generators.

In recent years, 3D printing technology has emerged as a focal point for exploration across various research domains and is widely considered one of the groundbreaking and emblematic technologies of the third industrial revolution. Its diverse applications, ranging from the design and production of small items like children's toys [25] to intricate tasks such as the manufacturing and repair of aircraft engines in the aerospace industry [26], clearly illustrate its immense potential. What is particularly noteworthy is that in the medical field, 3D printing technology has created a new connection between living and non-living matter. For instance, by utilizing biomaterials that closely resemble the composition of human bone tissue, researchers have successfully printed artificial bones with excellent biosafety and osteogenic properties [27].

Also referred to as additive manufacturing (AM), 3D printing technology is an innovative process that constructs products by stacking materials layer by layer [28]. This method contrasts with traditional subtractive manufacturing, which involves physical processes such as cutting, chipping, and grinding to remove excess material. As illustrated in Figure 1, Attaran Mohsen outlines five key advantages of 3D printing technology for product manufacturing: speed; cost-effectiveness; quality optimization; potential for innovation and transformation; and a broader impact on the environment, logistics, and other sectors [29]. He also forecasts that from 2013 to 2023, the printing speed of metal 3D printing will increase significantly, while costs will decrease markedly.

For rod elements, drive shafts, columns at the intersections of strong crossbeams and reinforcing ribs, and other columnar pressurized structures of various sizes, conventional manufacturing methods impose significant limitations, making it challenging to create complex internal structures. As a result, traditional weight reduction strategies typically involve material removal, such as designing shafts to be hollow or incorporating weight-reducing holes into plates. Importantly, a lattice-structured unit cell can be seen as an alternative rod element to conventional support structures. Moreover, 3D printing technology offers considerable advantages for the fabrication of such intricate structures [30–32].

In summary, this paper simplifies the truss rod element, lattice structure unit, drive shaft, and column into physical models that are subjected to single tension, single torque, and single pressure. According to international standards [33,34], we developed models for the compression, tensile, and torsion specimens, as illustrated in Figure 2 (notably, the tensile specimen and the torsion specimen share the same model). The specific dimensions of each specimen will be detailed later.

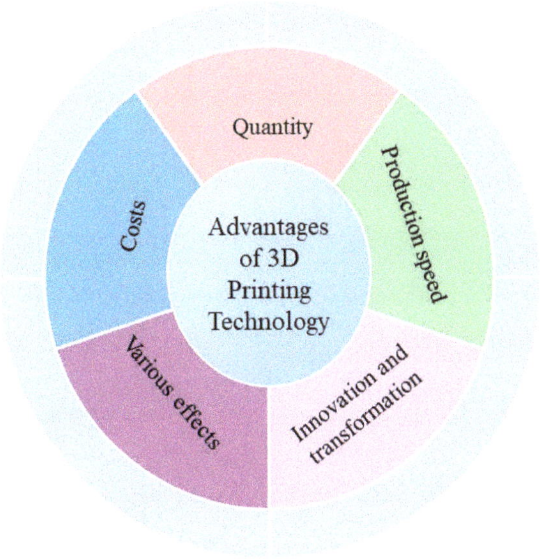

Figure 1. Five advantages of 3D printing.

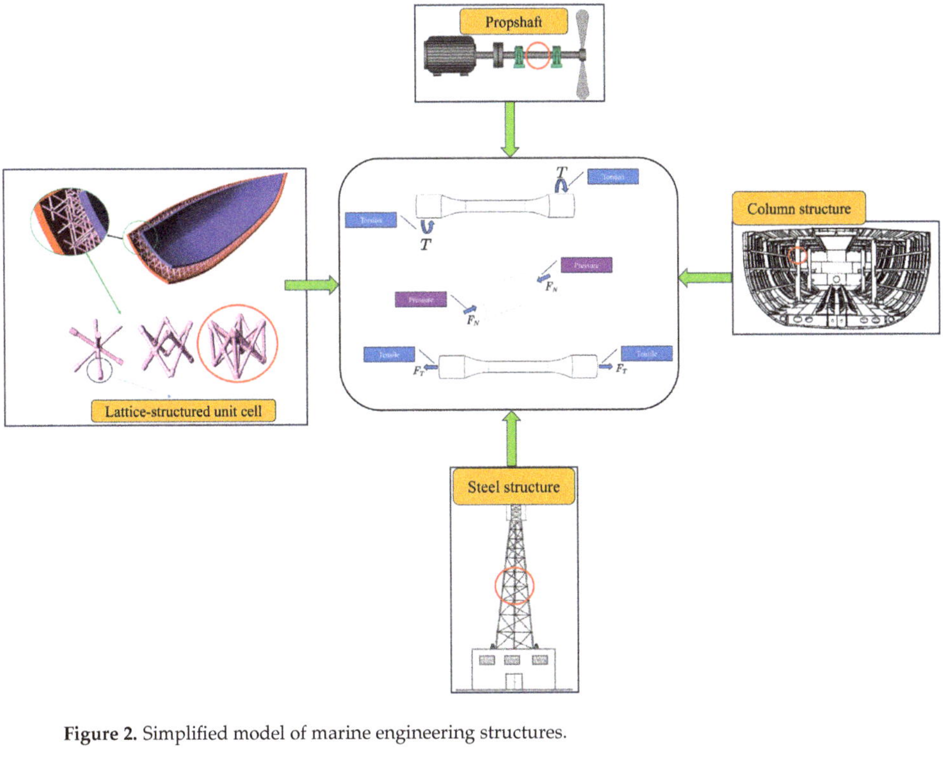

Figure 2. Simplified model of marine engineering structures.

In this paper, we will utilize 3D printing technology to manufacture these specimen models by modifying their internal structures and printing parameters, with the goal of investigating the mechanical properties of specimens exhibiting different internal configurations and printing parameters under various load conditions. This research aims to identify a parametric design solution that satisfies lightweighting requirements while maintaining excellent mechanical properties. We anticipate that this paper will further advance the application of 3D printing technology in the field of ship and ocean engineering and contribute to the development of this discipline.

2. Experiment

2.1. Specimen Size

In accordance with international standards, this paper presents designs for a tensile test specimen, a torsion test specimen, and a compression test specimen, as illustrated in Figure 3, below. Notably, the parameter dimensions of the torsion test specimen and tensile test specimen are identical, and all numerical units in the figure are in millimeters. All experiments were conducted under atmospheric conditions.

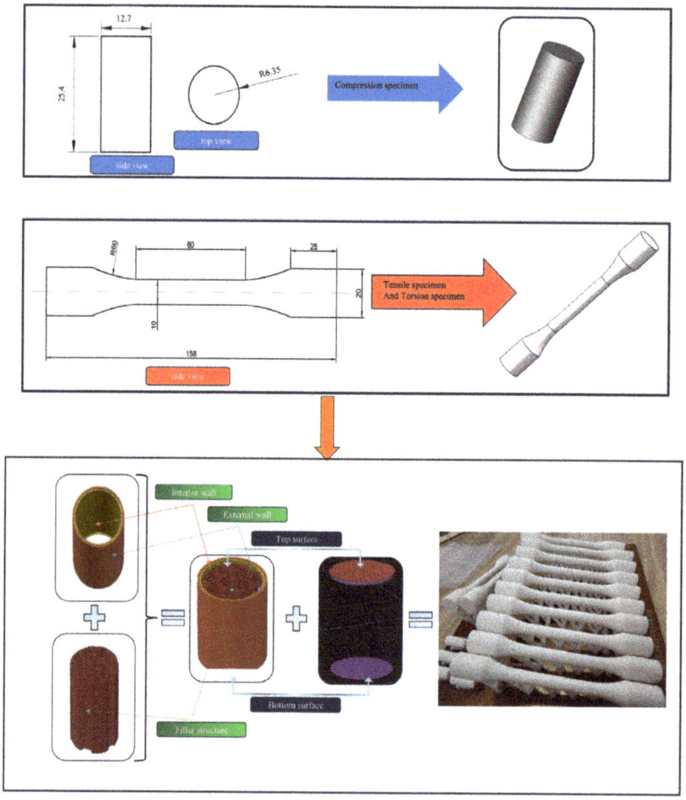

Figure 3. Experimental specimen size parameters. The dashed line indicates the centerline of the component. Dotted lines indicate the centerline of the component.

Figure 3 illustrates the structure of the compression specimen as an example. All experimental specimens are structured internally by the "wall," which comprises the top and bottom surfaces, forming a complete closed shell around the internal structure. This shell not only protects the filled structure but also ensures that the manufactured specimens exhibit good surface quality. Specifically, the "wall" consists of an inner wall and an outer wall, totaling two layers, while both the top and bottom surfaces are composed of three layers each. The thicknesses of the "walls" correspond to the width of the wire extruded from the nozzle, while the layer thicknesses of the top and bottom surfaces align with the layer thicknesses defined by the slicing parameters. The above materials will be printed using the 3D printer shown in Figure 4.

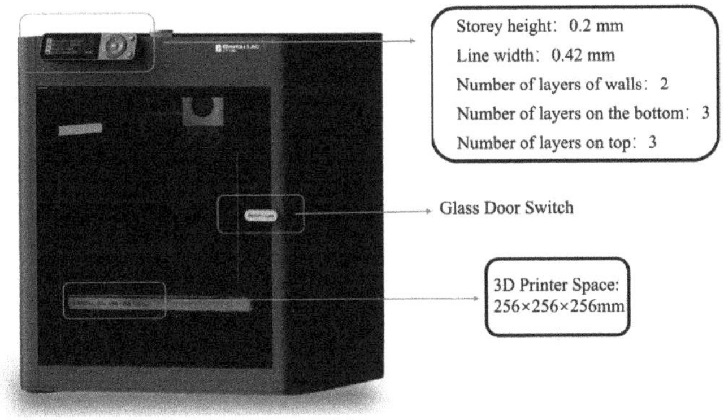

Figure 4. Basic printer parameters.

2.2. *Research Factors*

The first factor examined was the filling structure, encompassing a total of 3 variables. The second factor was the filling rate, which comprised 4 variables. The third factor pertained to the specimen inclination angle, totaling 7 variables. Lastly, the fourth factor was the form of load applied to the specimen, consisting of 3 variables. As shown in Table 1.

Table 1. Study variables and load forms.

Research Factors							
Filling structure	Honeycomb	Spiral	Linear				
Filling rate	25%	35%	50%	75%			
Specimen tilt angle	0°	15°	30°	45°	60°	75°	90°
Experimental load	Tensile load	Torsional load	Compressive load				

As indicated in the combined formula, Formula (1), let a, m, e, and g represent the number of variables for the four research factors: filling structure, filling rate, specimen inclination angle, and experimental load, respectively. The total number of specimens, denoted as P in Equation (2), is established with $b = n = f = h = 1$ using the control variable method.

$$C_x^y = \frac{x!}{y!(x-y)!} \tag{1}$$

$$P = C_a^b C_m^n C_s^f C_g^h \tag{2}$$

Substituting the data $b = n = f = h = 1$, $a = 3$, $m = 4$, $e = 7$, $g = 3$ to find the total number of specimens, $P = 252$, we get 252 independent data points.

2.3. *Filling Structure and Filling Rates*

The pattern of each layer constitutes our infill structure. When printing a layer, it is essential for the printer's path to avoid repetition to reduce nozzle wear and prolong its lifespan. Therefore, we selected three infill structures that do not involve repetitive paths for the nozzle: honeycomb; spiral; and the simplest option, straight line. Figure 5, below, illustrates these three infill structures along with 3D prints featuring different infill rates, where each row represents the same infill rate, and each column corresponds to the same infill structure.

Figure 5. Three fill structures and four fill rates. The green arrow indicates an increase in fill rate.

The volume of the model we created in SolidWorks 2024 is referred to as the model volume, denoted by V. The fill volume represents the portion of the model volume that has been hollowed out, with the remaining volume classified as the fill volume, which directly indicates the amount of material used. The filling rate is defined as the ratio of the total fill volume to the model volume, multiplied by 100%. The specimen is constructed layer by layer, where V_i represents the fill volume of the i layer, M_i denotes the model volume of the i layer, S_i refers to the filling cross-sectional area of the i layer, and A_i signifies the model cross-sectional area of the i layer. The filling rate is represented by ρ, the height of the layer is indicated by h, and the extruded material of the 3D printer for the i layer is expressed as L_i. The total length of the filament is also denoted as L_i, while α represents the width of the extruded filament, determined by the print nozzle. In summary, we can derive the definition in Equation (3) for the fill rate.

$$\rho = \left(\frac{\sum V_i}{V}\right) \times 100\% \tag{3}$$

$$V_i = S_i \cdot h, M_i = A_i \cdot h \tag{4}$$

$$\sum V_i = \sum S_i \cdot h, \sum M_i = \sum A_i \cdot h \tag{5}$$

$$S_i = L_{\mathrm{i}} \cdot \alpha \tag{6}$$

$$\rho = \frac{\sum L_i \cdot \alpha}{\sum A_i} \times 100\% \tag{7}$$

2.4. Specimen Inclination Angle

Printing complex structures will inevitably involve sections of the component's axis being printed at a certain angle due to heat transfer. To investigate the mechanical properties of the printed specimen under these conditions, this paper will print the specimen at a tilted angle. This tilted angle refers to the angle between the specimen's axis and the print bed, as shown in Figure 6, below, representing a line that forms an angle with the surface. This angle is referred to as the print tilt angle or specimen tilt angle. Adjusting the print tilt angle can also enhance the surface quality of the printed specimen by reducing the step effect, as well as optimizing printing time and the amount of support material used [35].

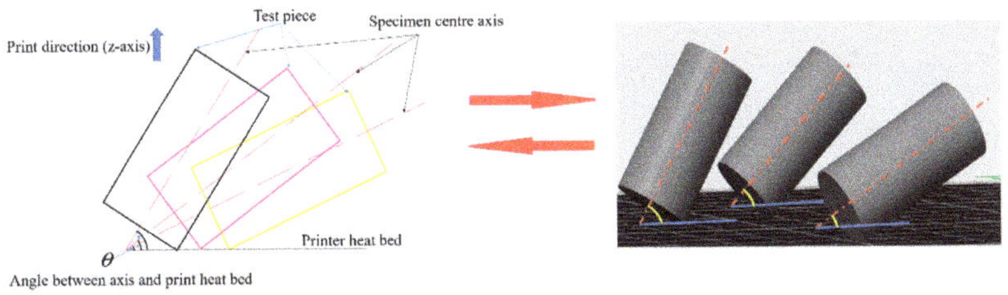

Figure 6. Tilt angle schematic. The red arrow indicates the transition between the 3D and 2D schematic diagrams, while the blue arrow indicates the positive direction of the Z-axis.

3. Theoretical Formulas

Hooke's law: The relationship between stress (σ) and strain (ε) in the elastic range can be expressed as Equation (8), where E is the modulus of elasticity of the material (Young's modulus), and F is the force, A, is the cross-sectional area in Equation (9).

$$\sigma = E\varepsilon \tag{8}$$

$$\sigma = \frac{F}{A} \tag{9}$$

$$\varepsilon = \frac{\Delta L}{L_0}, \nu = -\frac{\varepsilon_{lat}}{\varepsilon_{long}}, W = \frac{1}{2}\sigma\varepsilon \tag{10}$$

where ΔL denotes the change in length of the specimen under load, and L_0 is the original length of the specimen. ν denotes Poisson's ratio, which describes the ratio of the strain in one direction to the strain perpendicular to that direction. ε_{lat} indicates transverse strain, ε_{long} indicates longitudinal strain, and the energy density of storage is expressed as W.

Consider the microelementary hexahedron in one of the specimens, and let it possess stress components in three main directions in the x, y, z coordinate system: positive stress, σ_x, σ_y, σ_z, and shear stress components, τ_{xy}, τ_{xz}, τ_{yz}; therefore, the stress state of the microelementary hexahedron can be expressed by the following stress tensor in Equation (11):

$$\sigma = \begin{pmatrix} \sigma_z & \tau_{zy} & \tau_{zz} \\ \tau_{zy} & \sigma_y & \tau_{yz} \\ \tau_{zz} & \tau_{yz} & \sigma_z \end{pmatrix} \tag{11}$$

The maximum stress can be found from the characteristic equation, Equation (12), where λ is the principal stress, and I is the unit matrix.

$$d(\sigma - \lambda I) = 0d = \begin{pmatrix} \sigma_x - \lambda & \tau_{xy} & \tau_{xx} \\ \tau_{xy} & \sigma_y - \lambda & \tau_{yx} \\ \tau_{xx} & \tau_{yx} & \sigma_z - \lambda \end{pmatrix} = 0 \tag{12}$$

Tianyun Yao et al. pointed out that 3D printing is stacked layer by layer, which can be considered transversely isotropic in the plane of any layer, and provided the elastic intrinsic relationship of the material [36], as shown in Equation (13), below.

$$\begin{bmatrix} \varepsilon_1 \\ \varepsilon_2 \\ \varepsilon_3 \\ \varepsilon_4 \\ \varepsilon_5 \\ \varepsilon_6 \end{bmatrix} = \begin{bmatrix} S_{11} & S_{12} & S_{12} & 0 & 0 & 0 \\ S_{12} & S_{22} & S_{23} & 0 & 0 & 0 \\ S_{12} & S_{23} & S_{22} & 0 & 0 & 0 \\ 0 & 0 & 0 & 2(S_{22}-S_{23}) & 0 & 0 \\ 0 & 0 & 0 & 0 & S_{56} & 0 \\ 0 & 0 & 0 & 0 & 0 & S_{65} \end{bmatrix} \begin{bmatrix} \sigma_1 \\ \sigma_2 \\ \sigma_3 \\ \sigma_4 \\ \sigma_5 \\ \sigma_6 \end{bmatrix} \tag{13}$$

For the plane stress state, then we have $\sigma_3 = 0, \tau_{23} = \sigma_4 = 0, \tau_{31} = \sigma_5 = 0$; the eigenstructure matrix can be written as Equation (14):

$$\begin{bmatrix} \varepsilon_1 \\ \varepsilon_2 \\ \gamma_{12} \end{bmatrix} = \begin{bmatrix} S_{11} & S_{12} & 0 \\ S_{12} & S_{22} & 0 \\ 0 & 0 & S_{55} \end{bmatrix} \begin{bmatrix} \sigma_1 \\ \sigma_2 \\ \tau_{12} \end{bmatrix} \gamma_{\mathrm{al}} = \gamma_{\mathrm{al}} = 0, \varepsilon_3 = S_{13}\sigma_1 + S_{23}\sigma_2 \tag{14}$$

where the E_1, E_2 engineering constants are Young's modulus; Poisson's ratio, ν_{12}, ν_{21}; and shear modulus, G_{12}. The matrix of four elastic constants can be expressed as Equation (15):

$$\begin{aligned} S_{11} &= \tfrac{1}{E_1} \\ S_{22} &= \tfrac{1}{E_2} \\ S_{56} &= \tfrac{1}{G_{12}} \\ S_{12} &= -\tfrac{\nu_{21}}{E_1} = -\tfrac{\nu_{12}}{E_2} \end{aligned} \tag{15}$$

In our current study, we aim to reduce the weight of both turbines and hydroelectric generators while maintaining their structural integrity, thereby enhancing energy generation efficiency. Therefore, we now define lightweighting as follows.

The stress–strain relationship of elastic segment materials satisfies Hooke's law:

$$\sigma_{\mathrm{i}} = E \cdot \varepsilon_{\mathrm{i}} \tag{16}$$

Fitting the effective stress-effective strain curve using the Swift constitutive equation,

$$\sigma_{\mathrm{j}} = k \cdot (\varepsilon_{\mathrm{j}} + d)^m \tag{17}$$

For the elastic and plastic regions, the strain energy density is

$$\zeta_\varepsilon = \sigma \mathrm{d}\varepsilon \tag{18}$$

Integrating ζ_ε over the entire volume and assuming the stress state within the material is independent of volume yields the deformation potential energy, Q_ε, as

$$Q_\varepsilon = \int\!\!\int\!\!\int \zeta_\varepsilon \mathrm{d}x\mathrm{d}y\mathrm{d}z = \int\!\!\int\!\!\int (\zeta_{\varepsilon\mathrm{i}} + \zeta_{\varepsilon\mathrm{j}})\mathrm{d}x\mathrm{d}y\mathrm{d}z = V \cdot \left(\int_0^{\varepsilon_{\mathrm{s}}} \sigma_{\mathrm{i}} \mathrm{d}\varepsilon_{\mathrm{i}} + \int_0^{\varepsilon_{\mathrm{b}}} \sigma_{\mathrm{j}} \mathrm{d}\varepsilon_{\mathrm{j}} \right) \tag{19}$$

In the equation, V represents volume; $\zeta_{\varepsilon\mathrm{i}}$ and $\zeta_{\varepsilon\mathrm{j}}$ denote the strain energy densities for the elastic and plastic regions, respectively; ε_{s} is the strain at yield strength; and ε_{b} is the maximum effective strain.

Substituting Equations (16) and (17) into Equation (19) yields

$$Q_\varepsilon = V \cdot \left(\frac{E}{2}\varepsilon_s^2 + \frac{k}{m+1}\varepsilon_b^{m+1} \right) \tag{20}$$

For a given material volume, when specimen dimensions are identical, the volume is proportional to the material thickness, t. When two materials achieve equivalent energy absorption efficiency, their deformation potential energies are equal; thus,

$$\frac{t_1}{t_2} = \frac{\varepsilon_{s_2}^2 + \frac{k_2}{m_2+1}\varepsilon_{b_2}^{m_2+1}}{\varepsilon_{s_1}^2 + \frac{k_1}{m_1+1}\varepsilon_{b_1}^{m_1+1}} \approx \frac{\frac{k_2}{m_2+1}\varepsilon_{b_2}^{m_2+1}}{\frac{k_1}{m_1+1}\varepsilon_{b_1}^{m_1+1}} \tag{21}$$

The readily applicable Formula (21) can be used for material thinning design.

4. Tensile Experiment Analysis

4.1. Curve Form of Tensile Specimen Load Variation with Specimen Length

The ETM series electronic universal testing machine was used for this tensile experiment. During the experiment, the specimen is firmly clamped by the fixture, the upper fixture is kept fixed, and the lower fixture moves downward at a slow speed until the specimen breaks. The data for the force applied to the specimen during the tensile process will be transmitted to the computer in real time for recording and analysis.

As shown in Figure 7, the computer-generated force versus elongation images of all the tensile specimens present three different types, corresponding to Figure 7a–c. Figure 7 demonstrates the type of image when the inclination angle of the tensile-printed specimen is 0°. This image is characterized by the fact that at the initial stage of stretching, the tensile force exerted on the specimen rises sharply with the increase in elongation. When the tension reaches a certain peak, it begins to decrease as the elongation continues to increase, and this process lasts for a relatively long time. When the elongation reaches a certain critical value, the specimen ruptures, at which time the tensile force decreases instantaneously, which is shown in the image as a sudden vertical fall in the curve. The second image type is shown in Figure 7b; when printing the specimen at a tilt angle of 15°, the computer forms this type of data image, which is characterized by a curve in the initial stage of the specimen force with an increase in elongation, which is relatively fast. When the tension reaches a peak, the tension begins to increase with the amount of elongation, and the very small rate of decrease in the image of the curve is approximately a straight line; when the amount of elongation reaches a certain value, it begins to decline rapidly. The third image type is when the specimen tilt angle is greater than 30°, where the initial stage of the curve has more specimen force, with an increase in elongation at a faster rate of increase, reaching a certain value, and then experiencing a slightly shorter period of time, which is instantly reduced to zero.

4.2. Tensile Fracture Form

As shown in Figure 8, below, all the tensile specimens have a specimen tilt angle of 0° when fractures occur, such as in the root of the pulled fiber; that is, the print layer is parallel to the axis, so all of the tilt angles of the specimen are 0° along the direction of the axis parallel to its print direction; we can as also see several PLA fibers constituted by filaments. However, the print layer has a certain angle of inclination with the axis along the direction of the axis, and the fiber filament is composed of a segment; the number of segments is equal to the number of print layers. The experimental results show that, in addition to the specimen's tilt angle of 0°, other specimens are along the angle of their print tilt breaks; in other words, they are along the print layer fractures. Print specimens stacked together in the two layers were then pulled apart.

The specimen's tilt angle of 0° in the experiment occurs when elongation continues to increase to a certain value from the outside to the inside of the slow break; the process lasts for a long time. When the broken fiber filament reaches a certain amount, the fiber filament will instantly pull the rest of the fiber filaments broken all the way. When the specimen tilt angle is 15°, or when the specimen is along the print level from the outside to the inside of the fracture, the fracture pulls up a small number of PLA fibers out of the plastic deformation of the filaments; this phenomenon occurs at the same time at a high fill rate (greater than or equal to 50%) and a specimen print tilt angle of 30°. When the specimen print angle is greater than 30°, the specimen will be loaded to a certain value, and instantaneous fracture separation will occur along the print level; the larger the specimen print angle, the flatter the cross-section.

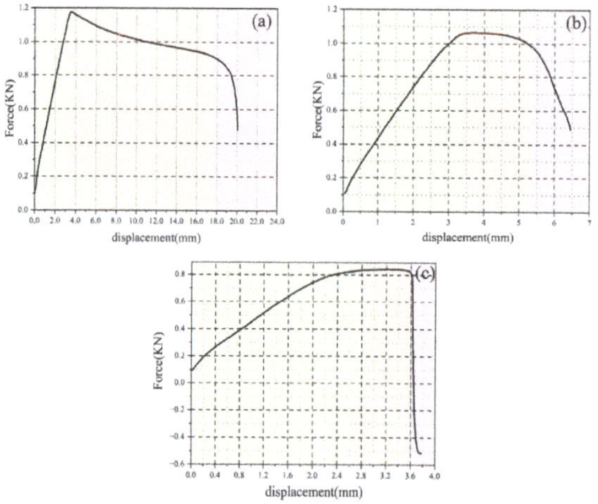

Figure 7. Tensile specimen curve forms. The green area represents the linear stage, the yellow-brown area represents the yield stage, and the pink area represents the fracture stage. (**a**) represents an inclination angle of 0°, (**b**) represents an inclination angle of 15°, and (**c**) represents an inclination angle of 30°.

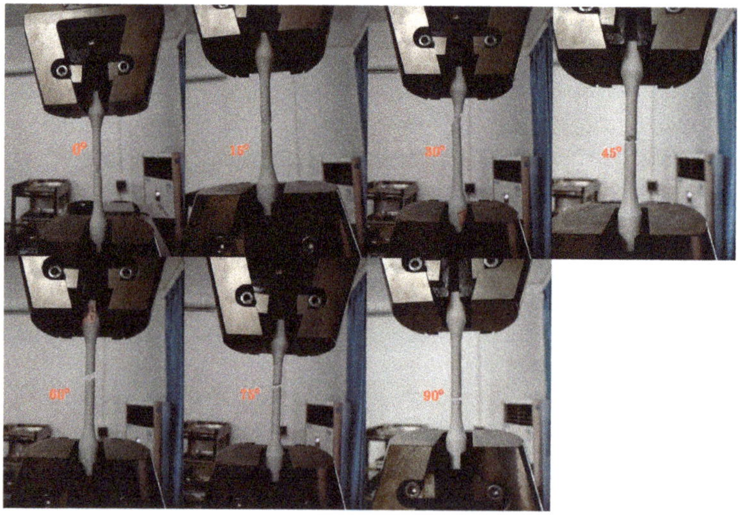

Figure 8. Tensile specimen fracture surface form.

From the loading stage to the specimen fracture process, the fracture time of the specimen inclination angle of 0° is much greater than that of the other inclination angles, followed by the fracture time of the 15° inclination angle, which is significantly greater than the 30° inclination angle; the fracture time of the 30° inclination angle does not differ significantly. Combined with the experimental phenomena and computer-generated image curves, we can initially conclude that the inclination angle of 0° allows the material to withstand the tension of the fiber; when the tension exceeds the range of the material, the fibers can withstand fracture. When a certain angle of inclination printing is used to produce a specimen, there is an interlayer bonding force between the printing layers; therefore, the force is much smaller than that of the PLA fiber, which can thus withstand the force. As such, the specimen has an angle of inclination along the print layer fracture.

4.3. Tensile Strength Data Analysis

The maximum tensile force that the specimen can withstand before tensile rupture, as a measure of the tensile strength of the specimen, is shown in Figure 9, below, with the conditions of different filling rates. The vertical axis represents force, and the horizontal axis represents the specimen's tilt angle. Figure 9a–d, respectively, show filling rates of 25%, 35%, 50%, and 75% under three kinds of filler structure under maximum force. For the tilt angle change, the black, solid-line nodes are rectangular, indicating that the filler structure is a honeycomb. The red dashed line nodes are round, indicating that the filler structure is a spiral. The last symbol indicates that the filler structure is a straight line. It is not difficult to see in Figure 9 that, regardless of the filling ratio, the maximum tensile stress that the component can withstand generally shows a decreasing trend as the printing inclination angle increases.

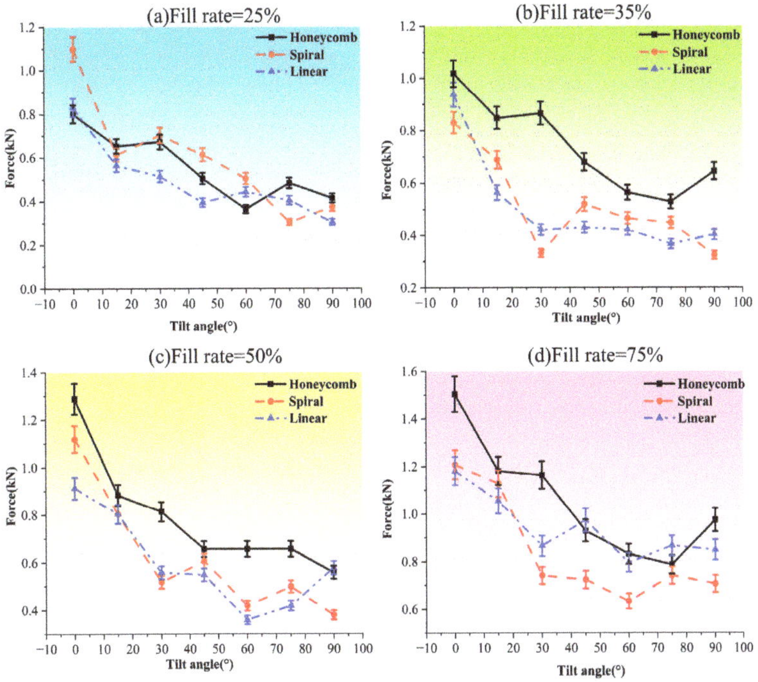

Figure 9. Variation in maximum load of tensile specimen with printing inclination angle. (**a**) represents a fill rate of 25%, (**b**) represents a fill rate of 35%, (**c**) represents a fill rate of 50%, and (**d**) represents a fill rate of 75%.

Removing the individual accidental anomalies from the graphs, the specimens of the three filling structures have a similar trend with the changes in specimen inclination angle at different filling rates. The trend is that the maximum tensile stress that the specimen can withstand decreases from 0° to 30°, increases from 30° to 45°, and decreases from 45° onward. Under the four filling rates, the curves of the honeycomb filling structure are, in most cases, located on the curves of the other two filling structures; from this, we can conclude that the tensile properties of the honeycomb filling structure are greater than those of the other two. At 35% and 50% filling rates, the curves of the spiral and linear filling structures are close to each other; at a 25% filling rate, the tensile properties of the spiral filling structures are better than those of the linear filling structures at inclination angles from 15° to 60°; and at the 75% filling rate, the tensile properties of the linear filling structures are better than those of the spiral filling structures.

The data images in Figure 10a–c, below, show the fill rate as the horizontal axis and the force size as the vertical axis; data points in the longitudinal axis coordinates show the specimen before the fracture of the maximum load appears in the transverse coordinates of the specimen's fill rate. From 10a–c, it can be seen that when the filling rate is greater than 50%, the maximum tensile load that the specimen can withstand at any inclination angle and under any filling structure increases with the increase in filling rate. The maximum tensile load that the specimen can withstand under each parameter when the filling rate is in an interval of 25% to 50% is uncertain, with the increase in the filling rate, so this interval is unstable. It can also be seen from the Figure 10 that when the specimen tilt angle is 0°, any filling rate and any filling structure of the specimen can withstand a large maximum load better than the other inclination angles; this again proves that an inclination angle of 0° for the tensile properties of the specimen is better than the other inclination angles. Secondly, the tensile properties of specimens with an inclination angle of 15° are also better than those with an inclination angle of more than 15° in general. It is not difficult to see from Figure 10 that, for the three different infill patterns, the maximum tensile stress that the component can withstand increases with the rise in the infill density ratio. This indicates that the infill ratio is one of the key factors determining the tensile capacity of the component.

Given the comprehensive experimental conclusions drawn in this section, in the manufacture of ships and marine engineering structures, we should make the inclination angle 0° when 3D printing structures that are subjected to tensile force; i.e., the nozzle-extruded material fiber filaments should lie flat in the printing layer, and the direction of tension should be the same as the longitudinal direction of the fiber filaments. Andrea Manes [37] pointed out that in fishery-rich areas where submarine pipelines are subjected to impacts, after removing the load, these pipelines are straightened and deformed due to rebound and applied axial tensile force; the deformation zone of the material is very easy to fracture, and leakage occurs. Dapeng Zhang [38] and others provided the theoretical basis and design method for the design of marine pipelines to cope with the complex marine environment. Combining the problems raised above and the design method of the reference pipeline, for the application of 3D printing technology for the manufacture of marine pipelines, the axial direction of the pipeline should be parallel to the longitudinal direction of the fiber filaments of the printing material; i.e., the specimen is printed with an inclination angle of 0°, and at the same time, the cellular structure is chosen. In order to satisfy the requirements of engineering safety, the filling rate should be set at more than 50%, so as to keep the tensile properties in a stable interval. For other marine structures subjected to tensile loads, 3D printing parameters and marine pipeline setup parameters should be the same, and the requirements of the surface quality can be appropriately adjusted to print the tilt angle while increasing the filling rate to compensate for the change in the tilt angle brought about by the change in compressive properties. For the requirements of very high elasticity and deflection changes in special pipelines, 3D-printed soft materials can be used [39] and then combined with rigid materials in the design of the parameters, hopefully achieving a number of excellent properties in one place.

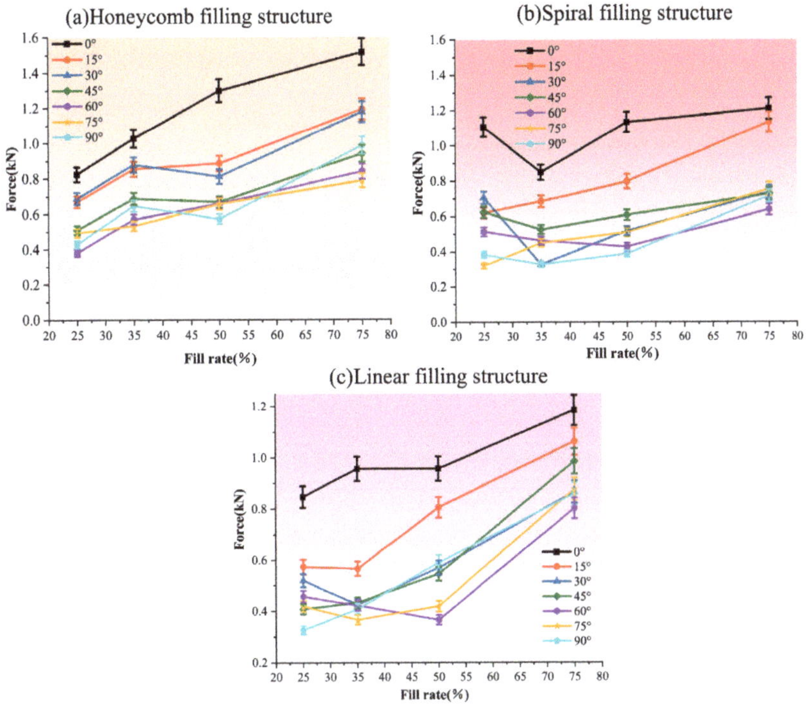

Figure 10. Variation in maximum load of tensile specimen with filling rate. (**a**) represents honeycomb filling structure, (**b**) represents spiral filling structure, and (**c**) represents linear filling structure.

5. Compression Experiment

5.1. Compression Specimen Load Changes with the Specimen Length in the Form of a Curve

The compression test machine we used is an ETM series electronic universal testing machine that is integrated with a tensile test machine. Before starting the experiment, the lower platform needs to be leveled so that it is in a horizontal position; the upper platform compresses the experimental specimen downward at 3 mm per minute. The curve of the force and compression amount has only one form (as shown in Figure 11); the initial stage involves force, increasing the amount of compression in the specimen, so it can bear the rapid increase in pressure after reaching a certain value at a slow increase. Up and down the two end surfaces of the platform's sassafras force, the specimen's constraints cannot produce deformation; the middle of the specimen with the section with increased pressure will continue to become thicker; and the bearing capacity increases. When the amount of deformation is close to 5 mm, the experiment is stopped.

5.2. Compression of Experimental Data Analysis

As shown in Figure 12a–d, below, the print tilt angle of the specimen is used as the horizontal axis, and the pressure is used as the vertical axis; the coordinates of the vertical axis of the data points are the pressures on the specimen when the deformation of the printed specimen is close to 5 mm. Figure 12a–d represent the variations in the maximum force of the three infill structures with the tilt angle under infill rates of 25%, 35%, 50%, and 75%, respectively. The black solid line nodes are rectangular, indicating that the filling structure is a honeycomb; the red dotted line nodes are circular, indicating that the filling structure is a spiral; and the last symbol indicates that the filling structure is straight.

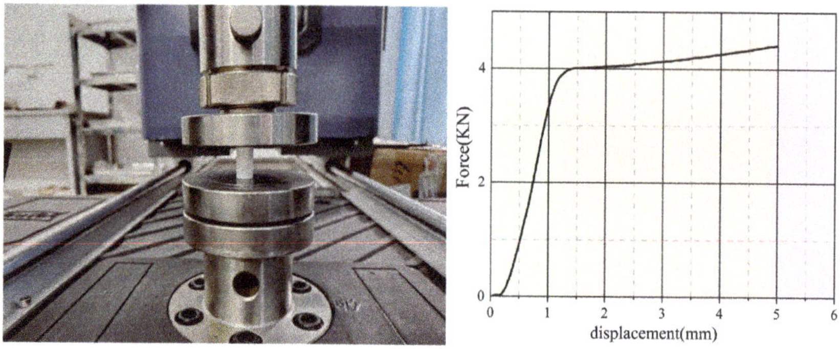

Figure 11. Compression specimen and force versus compression curve form. The green area represents the linear stage, the yellow-brown area represents the yield stage, and the pink area represents the fracture stage.

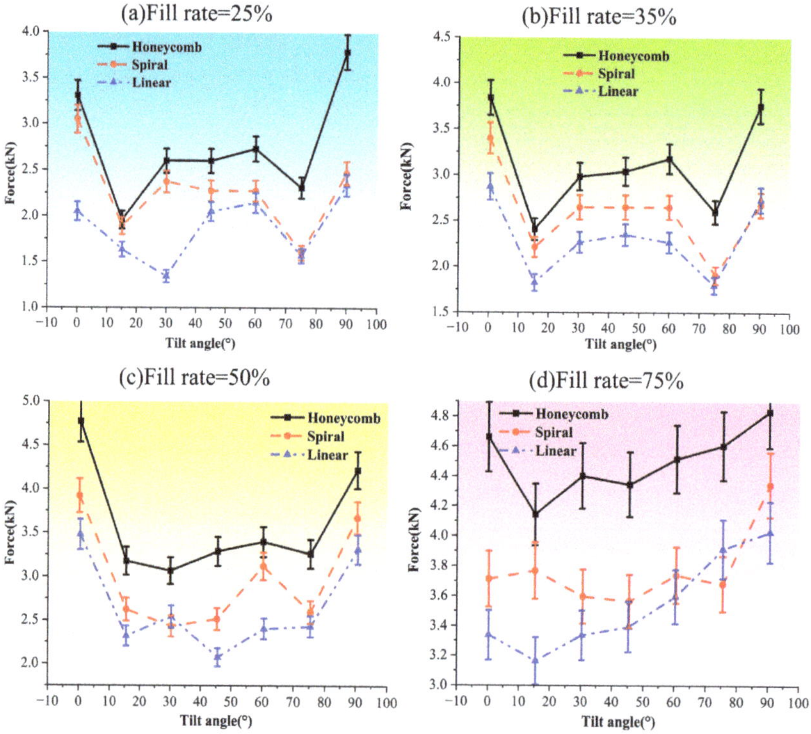

Figure 12. Variation in the maximum load that can be carried by a compression specimen with print tilt angle. (**a**) represents a fill rate of 25%, (**b**) represents a fill rate of 35%, (**c**) represents a fill rate of 50%, and (**d**) represents a fill rate of 75%.

When the filling rate is lower than 50%, the change curve of the loads of the three types of filling structures at a specimen deformation of 5 mm, with the print inclination angle of the specimen, is in the overall shape of a W. It decreases in an interval of 0° to 15°, rises in an interval of 15° to 45°, decreases in an interval of 45° to 75°, and rises in an interval of 75° to 90°; the trend of this curve is determined based on the statistical data of all the data, but there is bias in the individual data. At a 75% fill rate, the curves for the three fill structures show a decreasing trend in the 0° to 15° interval and an increasing trend in the 15° to 90° zone. The curve of the specimen with the honeycomb filling structure is located above the curves of the other two filling structures in all four filling rates, followed by the spiral filling curve above the curve of the linear filling structure. In summary, it can be concluded that the compressive performance of the specimens with honeycomb filling structure is better than that of the spiral one, and the compressive performance of the spiral filling structure is better than that of the linear filling structure.

As shown in Figure 13a–c, with the magnitude of the force as the vertical axis and the size of the filling rate as the horizontal axis, the vertical coordinates of the data points in the Figure 13 represent the magnitude of the compressive deformation of the specimen under pressure at the respective printing inclinations. The amount of deformation reaches 5 mm when subjected to pressure. The horizontal coordinates are the filling rate of the specimen. Figure 13a–c in the Figure 13 denote the honeycomb filling structure, the spiral filling structure, and the linear filling structure, respectively.

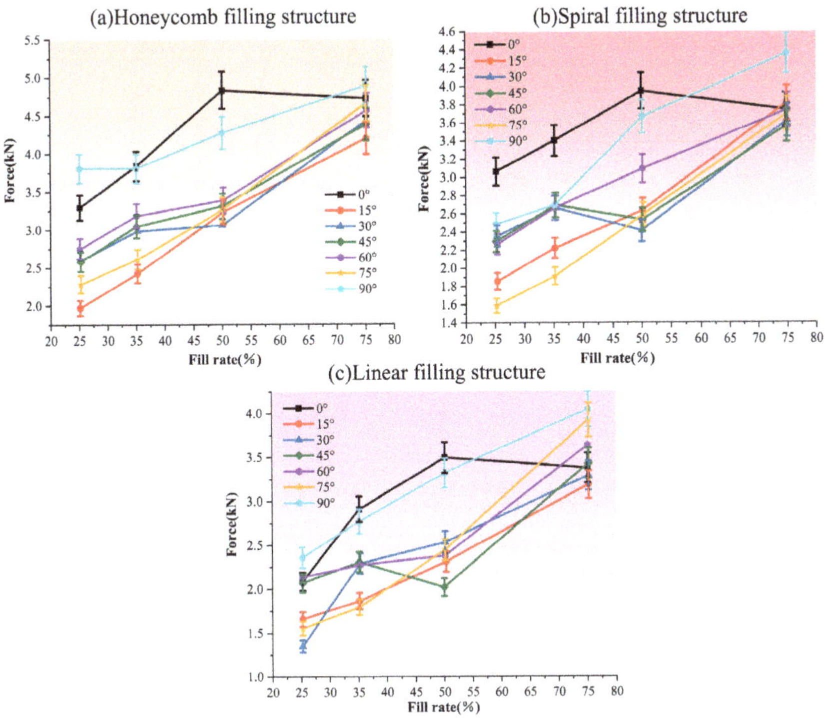

Figure 13. Variation in the maximum load that can be carried by a compression specimen with filling rates. (**a**) represents the honeycomb filling structure, (**b**) represents the spiral filling structure, and (**c**) represents the linear filling structure.

From a large number of data points, we can see that the size of the pressure on the specimen increases with the increase in the filling rate, and the difference between the values of each specimen becomes smaller and smaller, which leads to the conclusion that the compressive resistance of the specimen increases with the increase in the filling rate. It can be inferred that the compressive resistance grows closer and closer to the filling rate when it reaches a certain value. Individually, the compressive properties of the specimens with printing inclination angles of 0° and 15° are better than the others, while the line with a printing inclination angle of 0° shows an increasing trend in the filling rate range of 25% to 50% and a decreasing trend in the range of 50% to 75%.

Columns [40] are a typical solution for designing wind turbine towers and supporting foundations (e.g., monopiles and buoyancy), and the study of the structural stability and strength of column structures when subjected to axial pressures is of great significance for the wind energy industry [41]. Combined with the conclusions drawn from the compression experiments, in the use of 3D printing technology to manufacture towers, monopiles, offshore platforms, steel guides, and other such pressure-bearing structures, we can see that, regarding the compressive performance of the three infill structures under high infill rates of compression resistance, the honeycomb infill structure is more stable. The tilt angle of each of the compression resistance values also tends to be similar to the printing tilt angle. The two choices are 0° and 90°. If the overall printing structures need to be adjusted for the parameters or reasonable design of the local angle of the print bed components to obtain sufficient strength, we should choose high-fill-rate printing to achieve more than 80% success. However, then, the lightweight requirements will be reduced; therefore, this is a matter of trade-offs.

6. Torsion Experiment

6.1. Curve Form

This torsion experiment uses the ETT series torsion testing machine, as shown in Figure 14, below; it uses a computer through a sensor generated by the torque of the torsion angle of the curve. All the torsion specimens for the experiment generated three forms of curves, as shown in Figure 14a–c, which correspond to print inclination angles of 0°, 15°, and greater than 30° for the specimen. The three forms of the curve can be divided into three stages. The first stage is a rapid climb stage; that is, the specimen withstands the torque with a rapid increase in the torsion angle. In the second stage, there are three forms of the 0° curve, with the increase in the torsion angle of the torque withstanding a rate of increase, which is slightly slower than the first stage of the torque. The torque then increases to a certain value with the torsion angle, which continues to increase in a small range of change, that is, 15° for the specimen in this stage of the curve, within this range of the torque. The specimen in this stage increases its ability to withstand the torque with the increase in torsion angle, but there is a faster rate of decline. At greater than 30°, the specimen's ability to withstand the tilt angle with the torque increase will last longer within a small range of change. However, the 30° specimen, with the exception of the tilt angle of the specimen at the second stage, will endure for a relatively short period of time. In the third stage, the specimen is unable to withstand the torque falling sharply, proving that the specimen has been completely destroyed. From the curve of the second stage, it can be initially judged that the specimen's torsional performance is stronger at an inclination of 0°.

6.2. Torsional Experimental Fracture Forms

As shown in Figure 15, two fractures form in the torsion specimen. The first is for a specimen with a print inclination angle of 0°; with an increase in the torsion angle of the threaded specimen, the specimen surface will protrude from the marking distance, which is of a cyclic nature; this fracture and a tensile fracture in similar fiber filament material occur from the outside to the inside layer. The second form of fracture is along the print layer, with fractures and tensile fractures consistent with the larger angle of the cross-section, which is flatter. The 0° torsion specimen with the large number of threads on the surface is positively correlated with the torsion angle of the test piece before fracturing, which usually reaches 800° or more, while other specimen angles are usually within 500°, which further illustrates that the 0° print inclination angle of the specimen is better for its torsional performance than other inclination angles. This further indicates that the torsional properties of specimens printed with an inclination of 0° are better than those of specimens printed with other inclinations.

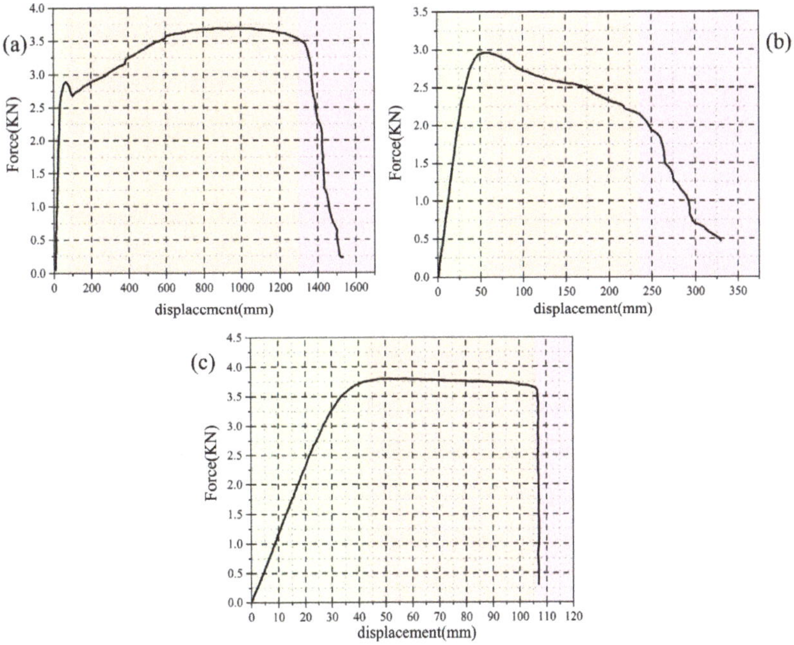

Figure 14. The maximum torque that can be carried by a torsion specimen as a function of the torsion angle. The green area represents the linear stage, the yellow-brown area represents the yield stage, and the pink area represents the fracture stage. (**a**) represents an inclination angle of 0°, (**b**) represents an inclination angle of 15°, and (**c**) represents an inclination angle of 30°.

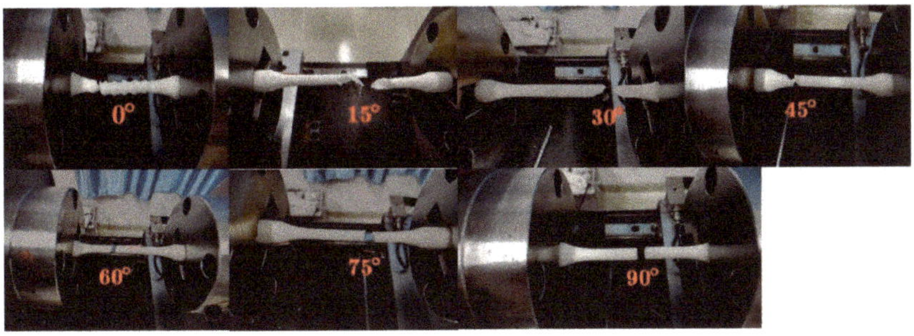

Figure 15. Sectioning of torsion specimens.

As shown by a,b,c,d in Figure 16 below, the horizontal axis of the coordinate system represents the print inclination of the specimen, and the vertical axis represents the torque that the specimen is subjected to. From Figure 16a–d, it can be seen that the fold line of the honeycomb infill structure is located above the other two infill structures in most cases, which shows that the torsional performance of the honeycomb infill structure is better than that of the other two infill structures. The variation curves of the three filler structures tend to decrease in an interval of 0° to 15°; in an interval of 15° to 30°, there is no obvious pattern in an unstable region, and there is a decrease in the interval from 45° to 90°. The maximum values of the data points of the three infill structures in the figure mostly appear at printing inclination angles of 0° and 45°, which indicates that the torsional performance of the specimens with these two printing inclinations is better. The torsional properties of the three infill structures are similar at low infill rates, and the honeycomb infill structure has a stronger variation pattern than the other two infill structures, indicating that it is more predictable.

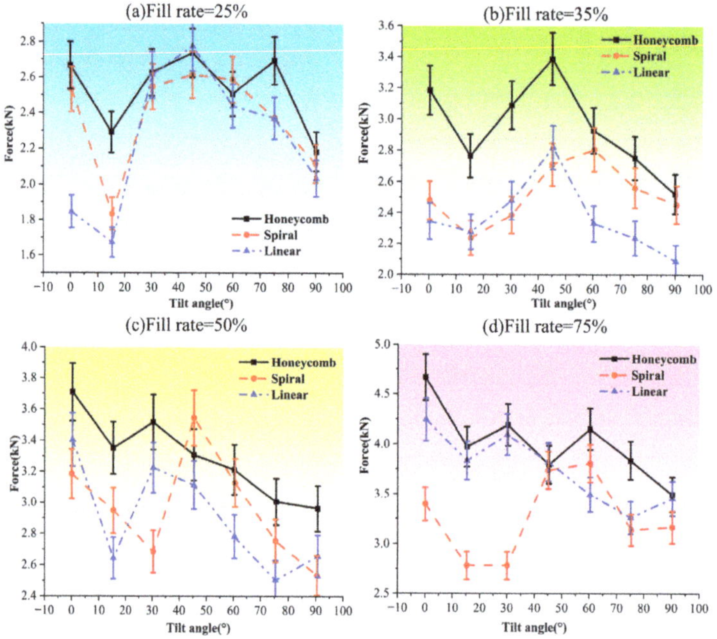

Figure 16. Variation in the maximum torque print with inclination angle that the torsion specimen can carry. (**a**) represents a fill rate of 25%, (**b**) represents a fill rate of 35%, (**c**) represents a fill rate of 50%, and (**d**) represents a fill rate of 75%.

As shown in Figure 17a–c, the horizontal axis represents the filling rate, and the vertical axis represents the torque; each folded line represents a particular print inclination and a filling structure of the torsional specimen and its maximum torque, along with its change in filling rate. From Figure 17, it can be seen that, for the filling rate interval of 25% to 50%, the majority of the data points carrying the torque are increased by the increase in filling rate, but there are still a few parts that do not comply with this rule; furthermore, the filling rate of more than 50% also increased the compressive properties of specimen. For specimens with a spiral filling structure, inclination angles of 0° and 30° and a filling rate greater than 50% torsional performance are better; this is also true of 45°. From Figure 17, we can observe that under the three printing methods, the changing trend of the component's torsional resistance mainly shows an upward pattern as the infill ratio increases.

Marine structures subjected to torsion are very important for the study of their failure, and determining the torsional performance of traditional marine structures is mostly based on the material itself [42]. The use of 3D printing technology to manufacture marine structures requires us to determine the torsional performance of not only the material itself but also the bonding force between the printed layers. A tilt angle of 0° for print specimens is more beneficial to the material's torsional capacity, and the design of other parameters can improve the torsional performance of a material. In summary, it can be seen that the choice of the honeycomb filling structure fulfils the principle of a higher filler rate being better. A print tilt angle of 0° is preferred, followed by 30°.

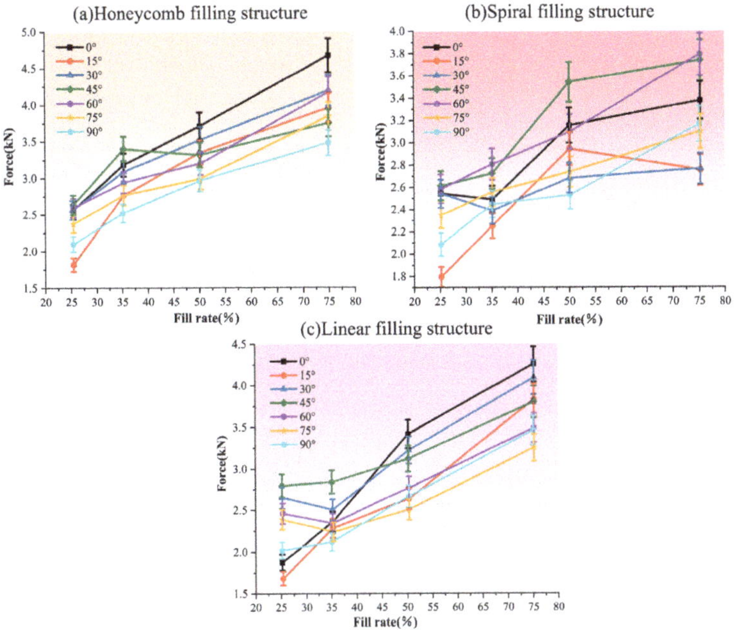

Figure 17. Variation in maximum torque that can be carried by a torsion specimen with filling rates. (**a**) represents the honeycomb filling structure, (**b**) represents the spiral filling structure, and (**c**) represents the linear filling structure.

7. Conclusions

Through tensile experiments, it can be concluded that the tensile properties of the honeycomb infill structure are better than those of the other two infill structures, and the tensile properties of the specimens with a printing inclination angle of 0° are determined by the strength of the fiber filaments of the material itself; the tensile specimen inclination angle of the specimen breaks along the printing layer between 15° and 90°, which indicates that the inter-layer bonding force is also one of the indicators for determining the tensile properties of 3D-printed specimens. The tensile properties of specimens with inclination angles of 0° and 15° are significantly greater than those with other inclination angles. A filling rate above 50% is in a regular and stable interval, in which the tensile properties of the specimens are positively correlated with the filling rate. The statistical data are shown in Table 2.

Through the compression experiment, it can be concluded that the quality of the compressive properties of the three kinds of filling structures is in the order of honeycomb filling structure, then spiral filling structure, and then linear filling structure. The compressive properties of the specimens with inclination angles of 0° and 90° are better than those of the other inclinations, but the specimens with an inclination angle of 90° have the best print surface quality (which does not have any effect on the mechanical properties); a filling rate of 50% is positively correlated with the filling rate when the compressive properties are positively correlated with the filling rate. The compressive properties are positively correlated with the filling rate when the filling rate is above 50%. The statistical data are shown in Table 3.

Table 2. The statistical data of the tensile experiments.

Tensile Experiments			Fill Rate			
			25%	35%	50%	75%
Honeycomb filling structure	Maximum load (kN)	0°	0.80	1.02	1.28	1.50
		15°	0.65	0.85	0.88	1.18
		30°	0.68	0.87	0.81	1.16
		45°	0.50	0.69	0.65	0.93
		60°	0.35	0.58	0.65	0.83
		75°	0.47	0.53	0.65	0.78
		90°	0.40	0.64	0.55	0.97
Spiral filling structure		0°	1.10	0.83	1.12	1.20
		15°	0.60	0.69	0.80	1.12
		30°	0.69	0.3	0.51	0.74
		45°	0.62	0.52	0.60	0.72
		60°	0.51	0.46	0.41	0.63
		75°	0.29	0.44	0.49	0.74
		90°	0.39	0.32	0.37	0.70
Linear filling structure		0°	0.84	0.93	0.91	1.18
		15°	0.56	0.56	0.80	1.05
		30°	0.51	0.42	0.55	0.86
		45°	0.38	0.43	0.54	0.97
		60°	0.42	0.42	0.36	0.79
		75°	0.40	0.36	0.41	0.86
		90°	0.29	0.40	0.57	0.84

Table 3. The statistical data of the compression experiment.

Compression Experiment			Fill Rate			
			25%	35%	50%	75%
Honeycomb filling structure	Maximum load (kN)	0°	3.30	3.83	4.77	4.66
		15°	1.95	2.40	3.17	4.14
		30°	2.59	2.98	3.06	4.40
		45°	2.59	3.03	3.28	4.34
		60°	2.72	3.17	3.39	4.51
		75°	2.30	2.59	3.25	4.60
		90°	3.78	3.75	4.22	4.83
Spiral filling structure		0°	3.04	3.39	3.91	3.71
		15°	1.88	2.21	2.62	3.76
		30°	2.37	2.65	2.43	3.59
		45°	2.27	2.65	2.52	3.56
		60°	2.27	2.65	3.11	3.74
		75°	1.60	1.90	2.59	3.68
		90°	2.46	2.67	3.67	4.34
Linear filling structure		0°	2.05	2.87	3.47	3.33
		15°	1.63	1.82	2.32	3.16
		30°	1.34	2.26	2.54	3.33
		45°	2.05	2.35	2.07	3.39
		60°	2.14	2.26	2.40	3.59
		75°	1.56	1.79	2.43	3.91
		90°	2.34	2.73	3.31	4.02

The torsion experiment shows that the specimen with honeycomb filling structure has better torsional performance; the specimen with 0° inclination angle breaks in torsion from the outer layer to the inner neck shrinkage, and then, one layer breaks; the specimens with other inclination angles break along the printed layer, similar to the tensile specimen; and the larger the angle, the flatter the section. This proves that the bonding force between the layers is one of the most important indices for determining torsional performance; the specimen printed with 0° is able to withstand larger torque and not fail at larger torsion angles, while the specimen with a 15° inclination angle has poorer torsional performance. The print specimens at 0° are able to withstand large torsion angles without failing and can withstand large torque, while torsional performance at a 15° inclination angle is poor; the positive correlation between torsional performance and the filling rate increases when the filling rate is higher than 50%. The statistical data are shown in Table 4.

Table 4. The statistical data of the torsion experiment.

Torsion Experiment			Fill Rate			
			25%	**35%**	**50%**	**75%**
Honeycomb filling structure	Maximum load (kN)	0°	2.66	3.18	3.70	4.66
		15°	2.29	2.76	3.34	3.97
		30°	2.62	3.08	3.51	4.19
		45°	2.73	3.38	3.30	3.79
		60°	2.50	2.92	3.21	4.15
		75°	2.69	2.75	3.00	3.83
		90°	2.18	2.52	2.96	3.49
Spiral filling structure		0°	2.53	2.47	3.18	3.39
		15°	1.83	2.23	2.94	2.78
		30°	2.54	2.38	2.68	2.78
		45°	2.61	2.71	3.54	3.73
		60°	2.58	2.80	3.12	3.80
		75°	2.37	2.56	2.75	3.13
		90°	2.11	2.45	2.53	3.16
Linear filling structure		0°	1.84	2.34	3.40	4.24
		15°	1.67	2.27	2.64	3.83
		30°	2.61	2.47	3.22	4.09
		45°	2.77	2.81	3.11	3.82
		60°	2.43	2.33	2.78	3.49
		75°	2.37	2.23	2.50	3.26
		90°	2.03	2.08	2.65	3.45

The above situation is shown in Table 5. In the table below, we have compared the aforementioned experiments and identified the advantages of and variations in different printing methods under various experimental operations. In the above experiments, we used the maximum stress that the components can withstand as the reference standard. This is also our basic reference method for evaluating a component. The ability to withstand higher ultimate stress means that the component possesses better performance parameters.

In summary, for the three properties, the best performance parameters are a honeycomb filling structure, an inclination angle of 0° specimen, and 15° specimen tensile performance. These are good for good compressive performance. For the 90° specimen, compressive performance is good, and torsional and tensile performance are poor, but it has the best surface print quality. The filling rate applies a full range of enhancements to the overall performance, but although increased fill density delivers superior performance, higher fill density inevitably leads to increased weight. The objective of this study is to reduce the weight of turbines and hydroelectric generators using 3D-printed materials while maintaining strength, thereby enhancing efficiency. The definition of lightweight is the ratio of strength to weight. When the strength requirements of turbines and hydrogenerators are met, appropriately increasing the lightweight ratio can reduce the weight of components, which can effectively improve the mechanical motion efficiency of the machines and thus enhance energy generation efficiency. However, since PLC material is not a traditional high-strength material, its mechanical performance under severe sea conditions remains to be verified. Additionally, its lightweight nature means that when used in large marine structures, it may exhibit excessive buoyancy, resulting in poor wave resistance. These are areas requiring future optimization. Future research may explore the varying structural impacts of different printing methods at equivalent mass.

Table 5. Summary of different performance characteristics.

	Torsional Strength	Stress Strength	Tensile Strength
Filling method	Honeycomb filling outperforms the other two continuous filling methods.		
Filling ratio	When the filling ratio exceeds 50%, all three properties of the component begin to increase significantly.		
Print tilt angle	Specimens printed at 0° can withstand greater torque and do not fail at larger twist angles, whereas specimens with a 15° tilt exhibit poorer torsional performance.	Specimens with inclination angles of 0° and 90° exhibit superior compressive strength compared to those with other inclination angles.	The tensile properties of specimens with inclination angles of 0° and 15° are significantly greater than those of specimens with other inclination angles.

Author Contributions: Conceptualization, H.J. and D.Z.; methodology, Y.X.; software, S.Z.; validation, Z.L. and S.G.; formal analysis, Y.X.; investigation, D.Z.; resources, D.Z.; data curation, S.Z.; writing—original draft preparation, Z.L.; writing—review and editing, D.Z. and Z.C.; visualization, D.Z.; supervision, D.Z.; project administration, D.Z.; funding acquisition, H.J. All authors have read and agreed to the published version of the manuscript.

Funding: This research was funded by the National Natural Science Foundation of China, grant number 62272109.

Data Availability Statement: The original contributions presented in this study are included in the article. Further inquiries can be directed to the corresponding author.

Conflicts of Interest: The authors declare that they have no known competing financial interests or personal relationships that could have appeared to influence the work reported in this paper.

References

1. Tan, D.-P.; Li, P.-Y.; Ji, Y.-X.; Wen, D.-H.; Li, C. SA-ANN-based slag carry-over detection method and the embedded WME platform. *IEEE Trans. Ind. Electron.* **2012**, *60*, 4702–4713. [CrossRef]
2. Wu, J.; Li, L.; Yin, Z.; Li, Z.; Wang, T.; Tan, Y.; Tan, D. Mass transfer mechanism of multiphase shear flows and interphase optimization solving method. *Energy* **2024**, *292*, 130475. [CrossRef]
3. Soto-Rivas, K.; Richter, D.; Escauriaza, C. Flow effects of finite-sized tidal turbine arrays in the Chacao Channel, Southern Chile. *Renew. Energy* **2022**, *195*, 637–647. [CrossRef]
4. Guo, X.; Wang, H.; Xiao, Y.; Liu, G.; Wang, W. Reassessing the cost of hydropower for multi-function hydropower bases: A case study from Sichuan. *J. Clean. Prod.* **2023**, *413*, 137422. [CrossRef]
5. Comoglio, C.; Castelluccio, S.; Fiore, S. Environmental reporting in the hydropower sector: Analysis of EMAS registered hydropower companies in Italy. *Front. Environ. Sci.* **2023**, *11*, 1178037. [CrossRef]
6. Wagner, B.; Hauer, C.; Habersack, H. Current hydropower developments in Europe. *Curr. Opin. Environ. Sustain.* **2019**, *37*, 41–49. [CrossRef]
7. Guillou, N.; Chapalain, G.; Neill, S.P. The influence of waves on the tidal kinetic energy resource at a tidal stream energy site. *Appl. Energy* **2016**, *180*, 402–415. [CrossRef]
8. Milford, C.; Martí, V.; Neill, S.P. Influence of the phase relationship between tidal currents and elevations on tidal energy conversion. *Renew. Energy* **2025**, *247*, 122774. [CrossRef]

9. Rome, L.C.; Flynn, L.; Goldman, E.M.; Yoo, T.D. Generating electricity while walking with loads. *Science* **2005**, *309*, 1725–1728. [CrossRef]
10. Donelan, J.M.; Li, Q.; Naing, V.; Hoffer, J.A.; Weber, D.J.; Kuo, A.D. Biomechanical energy harvesting: Generating electricity during walking with minimal user effort. *Science* **2008**, *319*, 807–810. [CrossRef]
11. Cobb, B.R.; Sharp, K.V. Impulse (Turgo and Pelton) turbine performance characteristics and their impact on pico-hydro installations. *Renew. Energy* **2013**, *50*, 959–964. [CrossRef]
12. Qi, Y.; Kim, J.; Nguyen, T.D.; Lisko, B.; Purohit, P.K.; McAlpine, M.C. Enhanced piezoelectricity and stretchability in energy harvesting devices fabricated from buckled pzt ribbons. *Nano Lett.* **2011**, *11*, 1331–1336. [CrossRef] [PubMed]
13. Cottone, F.; Vocca, H.; Gammaitoni, L. Nonlinear energy harvesting. *Phys. Rev. Lett.* **2009**, *102*, 080601. [CrossRef]
14. Qi, Y.; McAlpine, M.C. Nanotechnology-enabled flexible and biocompatible energy harvesting. *Energy Environ. Sci.* **2010**, *3*, 1275–1285. [CrossRef]
15. Platt, S.; Farritor, S.; Garvin, K.; Haider, H. The use of piezoelectric ceramics for electric power generation within orthopedic implants. *IEEE/ASME Trans. Mechatron.* **2005**, *10*, 455–461. [CrossRef]
16. Chen, X.; Xu, S.; Yao, N.; Shi, Y. 1.6 V nanogenerator for mechanical energy harvesting using PZT nanofibers. *Nano Lett.* **2010**, *10*, 2133–2137.
17. Jeon, Y.; Sood, R.; Jeong, J.-H.; Kim, S.-G. MEMS power generator with transverse mode thin film PZT. *Sens. Actuators A Phys.* **2005**, *122*, 16–22. [CrossRef]
18. Erturk, A.; Hoffmann, J.; Inman, D.J. A piezomagnetoelastic structure for broadband vibration energy harvesting. *Appl. Phys. Lett.* **2009**, *94*, 254102. [CrossRef]
19. Wu, H.; Zheng, H.; Qin, X.; Jin, Y.; Li, Y.; Yang, S.; Yi, Z.; Gao, S.; Wang, S.; Wang, Z. Drinking-Bird-Enabled Triboelectric Hydrovoltaic Generator. *Device* **2024**, *2*, 100318. [CrossRef]
20. Wei, L.; Nakamura, T.; Imai, K. Development and optimization of low-speed and high-efficiency permanent magnet generator for micro hydro-electrical generation system. *Renew. Energy* **2020**, *147*, 1653–1662. [CrossRef]
21. Eick, B.; Brown, A.; Wilcoski, J.; Thurmer, C. Localization of flow-induced vibrations from wicket gates in hydropower generating units. *J. Civ. Struct. Health Monit.* **2023**, *13*, 811–825. [CrossRef]
22. Yadav, G.; Chauhan, A.K. Design and development of pico micro hydro system by using house hold water supply. *Int. J. Res. Eng. Technol.* **2014**, *3*, 114–119.
23. Ridzuan, M.J.M.; Hafis, S.; Azduwin, K.; Firdaus, K.; Zarina, Z. Development of pico-hydro turbine for domestic use. *Appl. Mech. Mater.* **2015**, *695*, 408–412. [CrossRef]
24. Zainuddin, H.; Khamis, A.; Yahaya, M.S.; Basar, M.F.M.; Lazi, J.M.; Ibrahim, Z. Investigation on the performance of pico-hydro generation system using consuming water distributed to houses. In Proceedings of the 2009 1st International Conference on the Developements in Renewable Energy Technology (ICDRET), Dhaka, Bangladesh, 17–19 December 2009.
25. Wang, C.X. Application of 3D printing technology in product styling design. *Heilongjiang Sci.* **2022**, *13*, 31–33. (In Chinese)
26. Wu, Y.R. 3D Printing in aero engine manufacturing. *China Sci. Technol. Inf.* **2022**, *11*, 46–48. (In Chinese)
27. Parwani, R.; Curto, M.; Kao, A.P.; Rowley, P.J.; Pani, M.; Tozzi, G.; Barber, A.H. Morphological and mechanical biomimetic bone structures. *ACS Biomater. Sci. Eng.* **2017**, *3*, 2761–2767. [CrossRef]
28. Wong, K.V.; Hernandez, A. A review of additive manufacturing. *Int. Sch. Res. Not.* **2012**, *2012*, 208760. [CrossRef]
29. Attaran, M. The rise of 3-D printing: The advantages of additive manufacturing over traditional manufacturing. *Bus. Horiz.* **2017**, *60*, 677–688. [CrossRef]
30. Feng, J.; Fu, J.; Lin, Z.; Shang, C.; Li, B. A review of the design methods of complex topology structures for 3D printing. *Vis. Comput. Ind. Biomed. Art* **2018**, *1*, 5. [CrossRef]
31. Olsson, A.; Hellsing, M.S.; Rennie, A.R. New possibilities using additive manufacturing with materials that are difficult to process and with complex structures. *Phys. Scr.* **2017**, *92*, 053002. [CrossRef]
32. Ngo, T.D.; Kashani, A.; Imbalzano, G.; Nguyen, K.T.Q.; Hui, D. Additive manufacturing (3D printing): A review of materials, methods, applications and challenges. *Compos. Part B Eng.* **2018**, *143*, 172–196. [CrossRef]
33. *ISO 527-2-2012*; Plastics–Determination of Tensile Properties–Part 2: Test Conditions for Moulding and Extrusion Plastics. The State Administration for Market Regulation of the People's Republic of China, the Standardization Administration of the People's Republic of China: Shanghai, China, 2013.
34. *ASTM D695-15*; Standard Test Method for Compressive Properties of Rigid Plastics. ASTM International: West Conshohocken, PA, USA, 2010.
35. Liang, Z.; Xie, Y.; Zhang, Y.; Zhao, G.; Lin, Q.; Zhang, D.; Jiang, H.; Yang, K.; Si, K.; Yang, K.; et al. The Effect of the Support Mass in 3D Printing on the Forming Time of the Model. *Eng. Solut. Mech. Mar. Struct. Infrastruct.* **2024**, *1*, 7. [CrossRef]
36. Yao, T.; Deng, Z.; Zhang, K.; Li, S. A method to predict the ultimate tensile strength of 3D printing polylactic acid (PLA) materials with different printing orientations. *Compos. Part B Eng.* **2019**, *163*, 393–402. [CrossRef]

37. Manes, A.; Porcaro, R.; Ilstad, H.; Levold, E.; Langseth, M.; Børvik, T. The behaviour of an offshore steel pipeline material subjected to bending and stretching. *Ships Offshore Struct.* **2012**, *7*, 371–387. [CrossRef]
38. Zhang, D.; Zhang, Y.; Zhao, B.; Ma, Y.; Si, K. Exploring subsea dynamics: A comprehensive review of underwater pipelines and cables. *Phys. Fluids* **2024**, *36*, 101304. [CrossRef]
39. Wallin, T.J.; Pikul, J.; Shepherd, R.F. 3D printing of soft robotic systems. *Nat. Rev. Mater.* **2018**, *3*, 84–100. [CrossRef]
40. Li, S.; Kim, D.K. Ultimate strength characteristics of unstiffened cylindrical shell in axial compression. *Ocean Eng.* **2022**, *243*, 110253. [CrossRef]
41. Zhang, Y.; Zhang, D.; Jiang, H. A review of offshore wind and wave installations in some areas with an eye towards generating economic benefits and offering commercial inspiration. *Sustainability* **2023**, *15*, 8429. [CrossRef]
42. Buccino, F.; Martinoia, G.; Vergani, L.M. Torsion—Resistant structures: A nature addressed solution. *Materials* **2021**, *14*, 5368. [CrossRef] [PubMed]

Journal of
Marine Science and Engineering

MDPI

Review

From Multi-Field Coupling Behaviors to Self-Powered Monitoring: Triboelectric Nanogenerator Arrays for Deep-Sea Large-Scale Cages

Kefan Yang [1], Shengqing Zeng [1], Keqi Yang [2], Dapeng Zhang [1,3,*] and Yi Zhang [1]

1 Ship and Maritime College, Guangdong Ocean University, Zhanjiang 524005, China
2 School of Electronics and Information Engineering, Guangdong Ocean University, Zhanjiang 524088, China
3 Guangdong Provincial Key Laboratory of Intelligent Equipment for South China Sea Marine Ranching, Guangdong Ocean University, Zhanjiang 524088, China
* Correspondence: zhangdapeng@gdou.edu.cn

Abstract

As global Marine resource development continues to expand into deep-sea and ultra-deep-sea domains, the intelligent and green transformation of deep-sea aquaculture equipment has become a key direction for high-quality development of the Marine economy. Large deep-sea cages are considered essential equipment for deep-sea aquaculture. However, there are significant challenges associated with ensuring their structural integrity and long-term monitoring capabilities in the complex Marine environments characteristic of deep-sea aquaculture. The present study focuses on large deep-sea cages, addressing their dynamic response challenges and long-term monitoring power supply needs in complex Marine environments. The present study investigates the nonlinear vibration characteristics of flexible net structures under complex fluid loads. To this end, a multi-field coupled dynamic model is constructed to reveal vibration response patterns and instability mechanisms. A self-powered sensing system based on triboelectric nanogenerator (TENG) technology has been developed, featuring a curved surface adaptive TENG array for the real-time monitoring of net vibration states. This review aims to focus on the research of optimizing the design of curved surface adaptive TENG arrays and deep-sea cage monitoring. The present study will investigate the mechanisms of energy transfer and cooperative capture within multi-body coupled cage systems. In addition, the biomechanics of fish–cage flow field interactions and micro-energy capture technologies will be examined. By integrating different disciplinary perspectives and adopting innovative approaches, this work aims to break through key technical bottlenecks, thereby laying the necessary theoretical and technical foundations for optimizing the design and safe operation of large deep-sea cages.

Keywords: deep-sea large-scale net pens; multi-field coupled dynamic mechanisms; triboelectric nanogenerators; self-powered sensing systems

1. Introduction

With the global demand for protein resources constantly rising and the gradual depletion of near-shore aquatic ecosystems, the exploration and utilization of Marine resources have been expanding into deep-sea areas. As a key infrastructure for offshore Marine aquaculture, large deep-sea cages are an important support for the sustainable development of the "Blue Granary" initiative, which has been explicitly included in China's 14th Five-Year Plan and has become a core strategy for promoting high-quality development of the Marine economy. However, the operational reliability and long-term service performance of these cages are severely constrained by the complex and harsh deep-sea environment, which features multi-field coupling effects and encompasses aspects such as wave dynamics, ocean current fluid mechanics, biofouling, and sedimentary geomechanics. This complex interaction poses a severe challenge to the structural integrity of the net cages, often leading to problems such as deformation of the net coating, failure of the mooring system, and decline in aquaculture capacity, thereby hindering the large-scale commercialization process of deep-sea aquaculture [1–5]. Therefore, studying the multi-field coupling dynamic mechanism of large deep-sea cages is not only a fundamental scientific task but also a key engineering requirement for optimizing structural design, enhancing disaster resistance capacity, and supporting the construction of the national Marine ranch system.

Precise and continuous environmental monitoring is crucial for ensuring the safe operation of deep-sea cages, as it can assess the structural health status, aquatic ecological conditions, and aquaculture productivity in real time. The traditional power supply solutions for underwater sensor networks, such as wired cables and chemical batteries, have inherent limitations: the deployment cost of cables is high and they are prone to damage in turbulent currents; however, batteries need to be replaced frequently. In deep-sea environments, this process is not only labor-intensive but also poses environmental risks [6–10]. The emergence of triboelectric nanogenerator (TENG) technology has brought about a revolution in the field of low-frequency environmental energy harvesting and provided a highly promising solution to the energy supply bottleneck of deep-sea equipment [11–15]. TENG performs relatively well in converting irregular and low-amplitude mechanical energy into electrical energy, and its compatibility with flexible and corrosion-resistant materials makes it highly suitable for harsh Marine environments [16–20]. Therefore, the combination of TENG technology with self-powered sensing systems is expected to achieve autonomous and long-term monitoring of deep-sea cages, paving the way for the engineering application of ocean energy-driven sensing technology and establishing a new paradigm for the intelligence of deep-sea equipment.

Although significant progress has been made in deep-sea cage design and energy harvesting based on TENG, there are still key knowledge gaps and technical bottlenecks. In the field of cage dynamics, existing research mainly focuses on single-factor (such as waves or water flow) or dual-field (such as fluid–solid) interactions, with limited attention paid to the synergistic effects of multi-field coupling, especially the biomechanical interactions among fish schools, cage structures, and the surrounding flow fields. This oversight hinders the accurate prediction of cage deformation, load distribution, and potential instability mechanisms under actual sea conditions. In terms of TENG applications, although the progress in material modification and structural optimization has improved energy conversion efficiency and durability, the research on integrating TENG arrays with the flexible net jacket structure of deep-sea cages to achieve synchronous vibration monitoring and energy harvesting still needs to be further deepened. In addition, the lack of systematic research on the energy transfer mechanism within the multi-body coupling system of the grid box has restricted the rational design of the self-powered sensing system.

To address these challenges, this study adopts a multidisciplinary approach, integrating Marine fluid mechanics, structural mechanics, triboelectric energy harvesting, and aquatic biomechanics. In response to the dynamic response challenges of deep-sea aquaculture cages in complex Marine environments and the long-term energy supply requirements for monitoring, in this study, a three-level multi-field coupled dynamic model of "mesh clothes-fluids-fish school" was established, a surface adaptive TENG array and a multimodal energy management system were developed, and an integrated scheme of "self-powered monitoring-fish school state perception" was proposed. The study focuses on the three-level multi-field coupling mechanism of "net jacket fluid-structure coupling—multi-body system energy transfer—fish population flow field" [21–25], explores the self-powered sensing system based on TENG, and breaks through the key technical bottlenecks in the safety assessment and sustainable monitoring of deep-sea cages, providing theoretical and technical support for the intelligent upgrade of deep-sea and far-sea green aquaculture equipment in our country. This study aims to systematically explore the nonlinear vibration characteristics of the flexible structure of deep-sea large-scale net cages under complex fluid loads, establish a multi-field coupling dynamic model, and reveal its vibration response law and instability mechanism; develop a self-powered sensing system based on frictional nanopower generation technology to achieve real-time monitoring of the vibration state of the net garment; provide a theoretical basis and technical support for the optimal design and safe operation of large-scale deep-sea cages; and to enhance the adaptability and reliability of the cages in complex Marine environments. Figure 1 shows the main analysis content of this study.

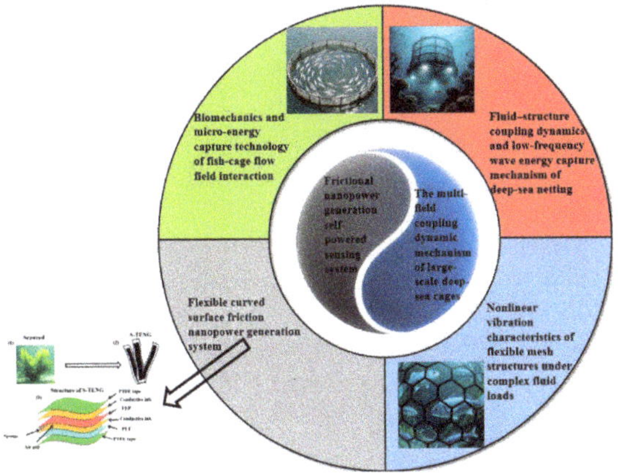

Figure 1. Schematic diagram of the technical field related to multi-field coupling dynamic mechanism and self-power supply system of deep-sea cages. (Among them, in the outer circle, the green area focuses on the biomechanics and micro-energy capture technology of flow field interactions in fish cages, the red area involves fluid–structure coupling dynamics and the low-frequency wave energy capture mechanism of deep-sea cages, the blue area discusses the nonlinear vibration characteristics of flexible net structures under complex fluid loads, and the gray area elaborates on the frictional nanopower generation system of flexible curved surfaces).

Unlike traditional review articles that systematically summarize and integrate all of the existing literature to provide a comprehensive overview of a certain research field, this paper adopts the perspective/positioning review method. Its core objective is not to enumerate all previous studies in detail but to focus on the key technical bottlenecks and unsolved scientific issues in the multi-domain coupling dynamics of large deep-sea cages and their self-powered monitoring systems. By clarifying this positioning, this review aims to provide targeted research directions for scholars in related fields and promote the transformation from theoretical research on deep-sea cage systems to engineering applications.

2. Flow–Solid Coupling Dynamics of Deep-Sea Cage Nets and Low-Frequency Wave Energy Capture Mechanism

Deep-sea large-scale cages, as core equipment for deep-sea and far-sea aquaculture and resource development, face extreme fluid–solid coupling loads in complex Marine environments due to their flexible net jacket structure [26–30]. The efficient capture of low-frequency wave energy is the key bottleneck for realizing the self-powered sensing system of the cages. Current research mostly focuses on the high-frequency vibration response and intensity verification of the mesh structure but lacks a systematic understanding of the large nonlinear motion induced by low-frequency waves and its energy capture mechanism, which is specifically manifested as shown in Figure 2:

1. The multi-scale evolution mechanism of fluid–structure coupling nonlinear behavior is unclear: the nonlinear vibrations such as local wrinkles and large swings caused by the porosity and viscoelastic properties of the mesh fabric under the combined action of waves and ocean currents are significantly affected by the scale effect in terms of the mapping relationship between the amplitude, frequency, and fluid load. The traditional rigid structure dynamics theory is difficult to accurately describe the coupling process of "fluid-induced deformation and load redistribution" of flexible mesh fabrics.
2. The spatio-temporal matching mechanism for low-frequency energy capture is missing: the energy conversion rate of wave energy in the vibration of the mesh jacket is less than 15%, and the distribution of energy along the curved surface shows significant non-uniformity. The coupling effect law between the high-frequency resonance region (such as the edge of the mesh jacket) and the low-frequency energy-rich region (such as the center of the mesh jacket) is not yet clear, resulting in a lack of theoretical guidance for the layout of the energy capture device.
3. The adaptive bottleneck of energy conversion under surface deformation: the existing triboelectric nanogenerators are mostly designed for planar structures. Under the significant deformation of the mesh coating surface, problems such as electrode fracture and unit coupling failure are prone to occur. Moreover, the seawater corrosion resistance life of the TENG materials is mostly between 6 and 12 months. There is an urgent need to break through the collaborative design theory of flexible arrays and surface deformation.

Therefore, the research content of the second part also revolves around these three manifestations.

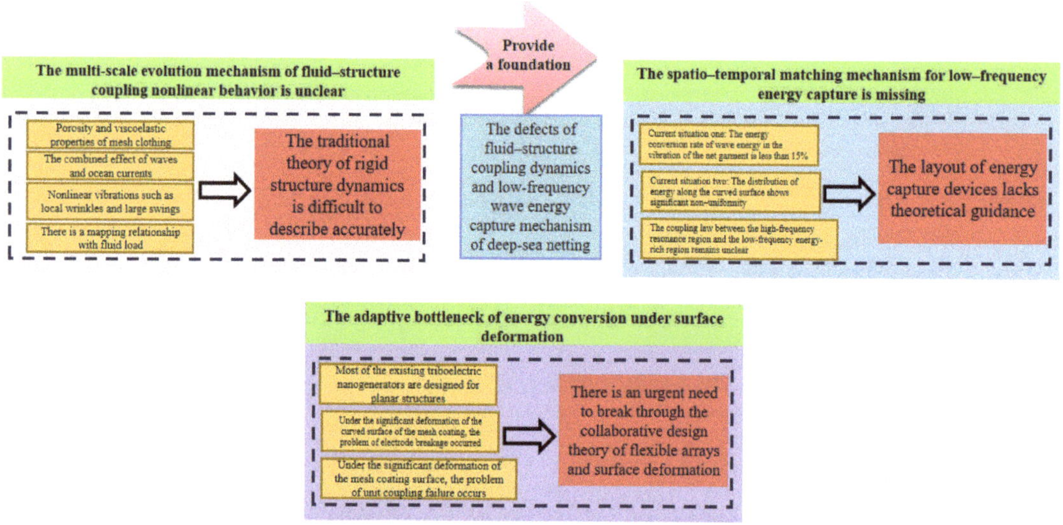

Figure 2. Logical representation of the fluid–structure coupling dynamics of deep-sea net clothing and the defect of the low-frequency wave energy capture mechanism.

2.1. Nonlinear Vibration Characteristics of Net Flexible Structures Under Complex Fluid Loads

The flexible net of deep-sea cages is constantly exposed to complex fluid environments. Its nonlinear vibration characteristics under the influence of waves and ocean currents are of vital importance to the structural safety of the cages. Scholars have carried out numerical simulation studies on the force characteristics of cages under complex sea conditions. For instance, the fluid–structure interaction analysis of cylindrical sunken and floating cages, hydrodynamic calculation of disk-shaped cages, etc. However, the verification of actual sea conditions still faces high costs and technical bottlenecks [31,32]. Zhang et al. [33] conducted a dynamic coupling analysis on a wave energy converter equipped with a floating-point absorber and catchain mooring. Based on potential flow theory and rigid body dynamics, they consider the six-degrees-of-freedom motion of a floating body under wave excitation, focusing on depicting the energy exchange process between the floating body and the waves, as well as the driving characteristics of the floating body's motion on the energy capture device. And they adopted the Catenary Theory to model an elastic and quasi-static catenary. In the local Cartesian coordinate system, the shape of the catchain can be explained by the parametric equation:

$$x = au + bsinh(u) + \alpha(12) \tag{1}$$

$$y = acosh(u) + \frac{b}{2}sinh^2(u) + \beta(13) \tag{2}$$

where $a = \dfrac{c}{\lambda_0 g}$, $b = \dfrac{ca}{DL_{eq}}$, $c = \dfrac{\lambda_0 L_{eq} g}{\sinh(u_2) - \sinh(u_1)}$.

By conducting a simulation analysis on this model, they were able to determine the influence of the mass, stiffness of the mooring chain, and the Marine environment on its forces. As shown in Figure 3, the sagging and undulating movements of the floating body under the action of waves will cause changes in the length and sag of the mooring chain, which, in turn, will trigger fluctuations in mooring tension. The change in mooring tension will generate a counterforce constraining the excessive movement of the floating body and forming a closed-loop coupling of "motion—tension—motion". These provide new ideas and methods for the dynamic study of cage systems in wave environments, which is conducive to a deeper understanding of the response characteristics of cages in similar complex, dynamic Marine environments. Not only that, they also conducted a bibliometric analysis on the progress and trends of global Marine ranching research [34], providing a macroscopic research direction reference for the study of multi-field coupling dynamic mechanisms of large-scale deep-sea cages. This can help researchers grasp the overall development context of this field, understand research hotspots and blind spots, and thus carry out related research more specifically. In addition, the team [35] studied the dynamic characteristics and fracture failure of the rigid truss trawl system during the towing process. Their research methods and analysis ideas for the mechanical properties of the structure can provide references for the study of the structural strength and stability of large deep-sea cages under complex sea conditions.

In deep-sea environments, the net garment, as a key structural component, is constantly subjected to the action of complex fluid loads. The periodic undulations of waves, the directional flow of ocean currents, and their combined effects make the vibration behavior of the net garment extremely complex [36–40], as shown in Figure 4. Therefore, it is a necessary project to study the nonlinear vibration characteristics of flexible mesh structures under complex fluid loads.

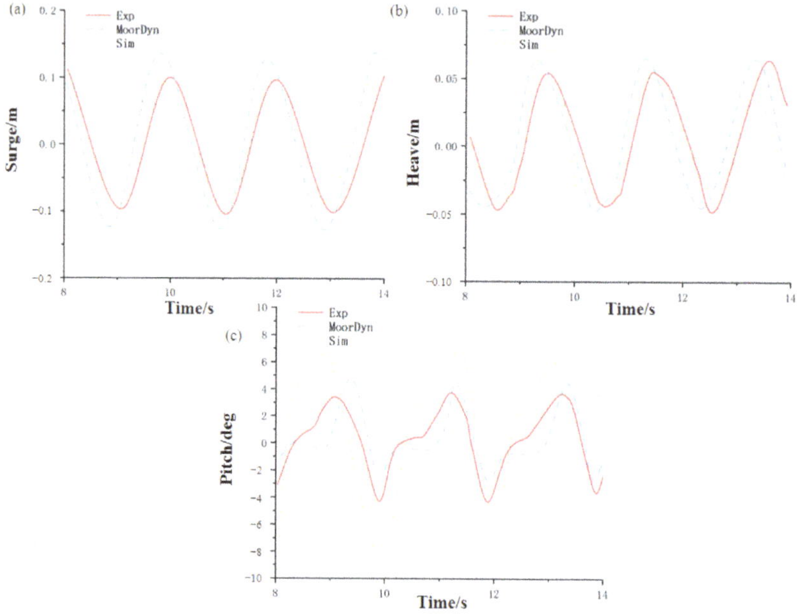

Figure 3. The movement of the floating box: (**a**) surge; (**b**) floating and sinking; and (**c**) swaying. The data is sourced from [33].

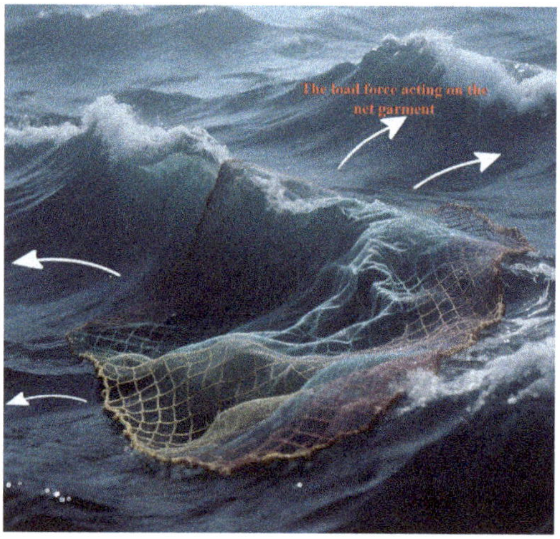

Figure 4. Deformation diagram of the net garment under load in complex Marine environments.

2.1.1. Construction of Multi-Factor Fluid–Structure Coupling Dynamics Model

Based on the fluid dynamics theory of porous media, a bidirectional coupling model of pore seepage and structural deformation is established. Focus on breaking through the following key technologies:

1. A dynamic porosity model with spatial heterogeneity was constructed using the Brinkman–Darcy equation:

$$\mu \nabla^2 \vec{u} - \frac{\mu}{K} \vec{u} + \rho (\vec{u} \cdot \nabla) \vec{u} = -\nabla p + \rho \vec{g} \quad (3)$$

Among them, $\vec{u}$ is the seepage velocity vector of the fluid, μ is the dynamic viscosity of the fluid, K is the permeability of the porous medium, ρ is the density of the fluid, P is the fluid pressure, $\vec{g}$ is the gravitational acceleration vector, ∇^2 is the Laplacian operator, and ∇ is the gradient operator.

The influence of the porosity gradient distribution on fluid permeation resistance is characterized by this equation, and the quantitative relationship between the time-varying characteristics of porosity and the frequency of vortex shedding is established.

2. Introduce the fractional-order Burgers viscoelastic constitutive model, describe the frequency-dependent characteristics of the material through Caputo fractional-order derivatives, and construct a nonlinear constitutive equation considering the historical effects of strain:

$$\sigma(t) = \eta\int_{-\infty}^{t} K(t-\tau)\ddot{\varepsilon}(\tau)d\tau + E\int_{-\infty}^{t} K(t-\tau)\dot{\varepsilon}(\tau)d\tau + \eta_f D_t^{\alpha}\left[\int_{-\infty}^{t} K(t-\tau)\dot{\varepsilon}(\tau)d\tau\right] \quad (4)$$

Here, $\sigma(t)$ represents the stress the material is subjected to at time t, η is the viscosity coefficient, E is the elastic modulus, and η_f is the coefficient related to fractional viscosity.

3. Develop a fluid–structure coupling solver based on the immersion boundary method to achieve a precise solution of the bidirectional coupling of flow field and structure under large deformation conditions.

2.1.2. Research on Nonlinear Vibration Evolution Mechanism in Complex Flow Fields

Establish a wave–current combined action test system, integrating PIV flow field measurement and DIC full-field strain measurement technology, and conduct systematic research:

1. Multimodal coupling vibration characteristics: Analyze the nonlinear phenomena such as parametric resonance and internal resonance of the mesh structure under the coupling action of regular waves, irregular waves, and shear flow;
2. Instability critical criterion: A periodic disturbance stability analysis framework is established through Floquet theory to determine the critical conditions for the transformation of the structure from steady-state vibration to chaotic motion under different flow velocity ratios. The vibration system of the net garment under the combined action of waves and the current is regarded as a system of periodic linear differential equations, and its dynamic equation can be expressed as follows:

$$\dot{x}(t) = A(t)x(t) \quad (5)$$

Among them, $x(t)$ represents the vibration state vector of the net garment (including key parameters such as displacement and velocity), $A(t)$ is the period coefficient matrix, and period T corresponds to the characteristic period of waves or ocean currents.

According to Floquet's fundamental theorem, the fundamental solution matrix $\phi(t)$ of the net garment vibration system is constructed:

$$\phi(t+T) = \phi(t)\mathbf{C} \quad (6)$$

The Floquet multiplier $\mu_1, \mu_2, \ldots, \mu_n$ is obtained by solving the eigenvalues of the matrix. As shown in Table 1, this demonstrates the relationship between Floquet multiplier modulus values and the vibration state of the net.

Table 1. The modulus value $|\mu_i|$ of the Floquet multiplier is directly related to the vibration state of the net garment.

Modulus Value 0	Status	Vibration Amplitude
All $|\mu_i| < 1$	The net garment is in a steady-state vibration	The vibration amplitude shows no increasing trend over time, and the dynamic behavior of the structure is controllable
When there is at least one $|\mu_i| > 1$	The system has entered the initial stage of instability	The vibration amplitude increases exponentially with the periodic iteration
When $|\mu_i| = 1$	The system is in a critical stable state	The vibration amplitude remains constant but is prone to instability triggered by flow field disturbances

3. Local fold evolution law: Based on the curvature mode decomposition method, the central difference method is used to calculate the curvature modes of each measurement point of the net garment:

$$\kappa(x_j) \approx \frac{\phi(x_{j+1}) - 2\phi(x_j) + \phi(x_{j-1})}{h^2} \tag{7}$$

Here, $\phi(x)$ represents the displacement mode. Then, obtain the curvature mode through the second derivative:

$$\kappa(x) = \frac{d^2S(x)}{dx^2} \tag{8}$$

Thus, a correlation model between the two is established to quantify the associated parameters, such as the offset and distance between the initial position of the fold and the direction to the vortex core, and to reveal the influential mechanism of fold expansion on the overall damping characteristics of the structure.

2.1.3. Intelligent Mapping Model of Vibration Response and Load Characteristics

Build a multi-parameter coupled database and develop an intelligent analysis method driven by data and models:

1. The random forest algorithm is adopted to rank the importance of features, and a vibration amplitude prediction model based on XGBoost is constructed to achieve the prediction of the probability distribution of response extremum. The vibration response of the net garment under the combined action of waves and the current shows significant nonlinearity (such as local wrinkles and large swings), and there is a complex coupling relationship between its vibration amplitude and fluid load as well as structural parameters. XGBoost is based on the iterative optimization logic of gradient boosting trees (GBDT). By constructing regression trees one by one to minimize prediction errors, it can efficiently fit the nonlinear mapping relationship between multiple factors and vibration amplitudes. Compared with traditional linear models that cannot capture nonlinear features such as a "sudden increase in flow velocity causing sudden changes in vibration amplitude", XGBoost uses a "residual compensation" mechanism. Each new tree corrects the prediction error of the previous model in each round, and the final integrated strong learner can accurately depict the dynamic evolution law of the net garment vibration.
2. Develop a convolutional long short-term memory network, integrate flow field time series data with structural response signals, and establish an end-to-end vibration state prediction framework.

2.2. Spatiotemporal Distribution Characteristics of Wave-Induced Net Motion Energy

An in-depth study of the energy spatio-temporal distribution characteristics of wave-induced net garment movement, on the one hand, can accurately quantify the energy conversion rate of wave energy in the vibration of the net garment, which is conducive to evaluating the feasibility and efficiency of the net garment as an energy capture carrier. On the other hand, it can clearly analyze the spatio-temporal distribution law of energy along the surface of the net coating, providing a scientific basis for the reasonable layout of energy capture devices, thereby enhancing the energy collection capacity of the entire system and promoting the development of deep-sea self-energy supply technology.

2.2.1. Quantitative Model of Multimodal Energy Conversion Efficiency

By comprehensively applying the linear wave theory, Hamilton's principle, and the knowledge of fluid–structure interaction dynamics, an energy conservation equation considering multiple factors is established. Innovatively construct a time-varying energy conservation equation based on the improved Hamilton principle:

$$\frac{\partial E}{\partial t} + \nabla \cdot S = \sum_{i=1}^{n} Q_i \tag{9}$$

Here, E represents the total energy density of the system, S is the energy flow vector, and Q_i characterizes pore dissipation.

2.2.2. Analysis of Topological Characteristics of Energy Spatio-Temporal Distribution

By adopting advanced energy analysis methods, the net garment is divided into multiple characteristic regions, and the process of energy transfer and conversion between different regions is deeply studied. Based on the energy field reconstruction method of eigen-orthogonal decomposition, the energy density distribution function is established:

$$\Psi(x, y, t) = \sum_{k=1}^{m} a_k(t)\Phi_k(x, y) \tag{10}$$

A three-dimensional energy cloud map database is constructed through high spatiotemporal resolution measurement.

2.2.3. The Correlation Mechanism Between Resonance Modes and Energy Accumulation

As shown in Figure 5, a comprehensive analysis of the vibration response of the net garment under the action of waves was conducted by using technical means such as spectral analysis and modal analysis. Through precise calculation and experimental measurement, the high-frequency resonance areas and low-frequency energy-rich areas on the surface of the net garment are identified. For the high-frequency resonance region, study the matching relationship between its resonance frequency and wave frequency, and analyze the influence of resonance on the structural stability and energy conversion of the mesh jacket. For low-frequency energy-rich regions, an in-depth exploration of the accumulation mechanism of low-frequency energy in this area is conducted, including fluid dynamic effects, structural dynamic characteristics, and the coupling effect between the two.

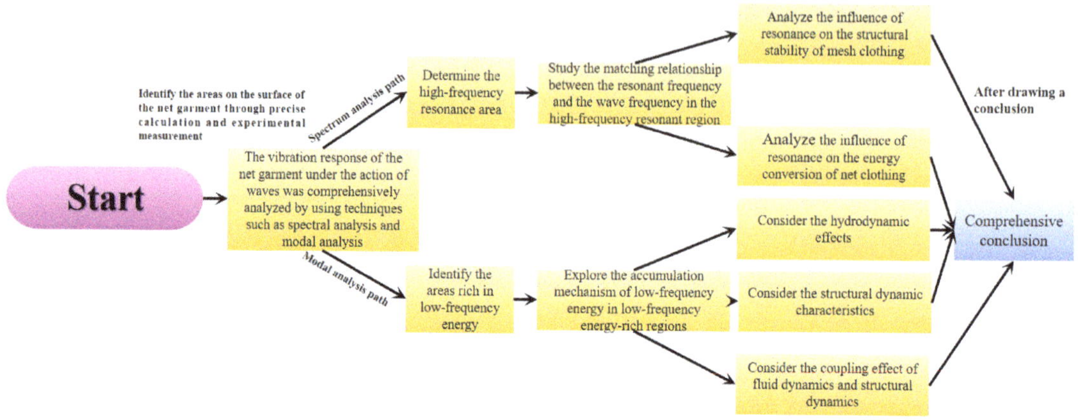

Figure 5. The technical route for studying the correlation mechanism between resonance modes and energy accumulation.

2.3. Design of Curved Surface-Adaptive TENG Arrays

As a new type of energy conversion device, TENG has broad application prospects in the field of deep-sea energy collection. A team from Chongqing University [41] has developed a TENG with an energy conversion efficiency of 48% and a durability of 500,000 cycles through interface lubrication and voltage dispersion design, which can be applied to Marine environment monitoring and data acquisition [42,43]. Shanghai University [44] proposed an electromagnetic–triboelectric hybrid generator, achieving wideband fluid energy harvesting and wireless sensing integration. The team from Beijing Institute of Technology [45] integrated TENG into the bearings of wind turbines to achieve self-power supply fault diagnosis (with an accuracy rate of 95.6%), providing a reference for the mechanical structure health monitoring of deep-sea cages. Not only that, in the latest research in 2024, the research article proposed a self-powered wireless sensing system based on human body antennas, supporting 8 m signal transmission, which is suitable for extreme environment monitoring [46]. In addition, Wang Wei et al. studied the performance of liquid–solid triboelectric nanogenerators based on multiphase liquids [47], the research progress of triboelectric nanogenerator technology in wave energy collection [48], and the research progress of fluid energy collection based on triboelectric nanogenerators [49]. However, as mentioned above, the development of TENG was not perfect. If TENG is to be widely applied in fields such as ocean energy collection (such as wave energy) and power supply for underwater sensors, it is necessary to enhance its stability in seawater environments by modifying the materials and designing the structure of TENG.

In deep-sea cage systems, achieving the efficient capture of low-frequency wave energy and conversion of electrical energy is crucial for solving the problem of self-power supply. TENG has become a research hotspot in this field due to its advantages such as simple structure, ability to convert low-frequency energy, and strong adaptability [50,51]. The design of a surface-adaptive TENG array enables the TENG to perfectly fit the surface of the net jacket, maintaining a stable and efficient energy conversion performance even when subjected to tensile, bending, and other deformations. This is of crucial significance for enhancing the reliability and practicality of the self-energy supply system in deep-sea cages and is expected to break through the bottlenecks of existing self-energy supply technologies. It also promotes the sustainable development of deep-sea aquaculture and related Marine industries. For instance, in order to reduce the manufacturing cost of TENG while significantly enhancing the stability and energy conversion efficiency of the equipment in the Marine environment, Kefan Yang [48] designed a flexible structure TENG inspired by seaweed in nature (as shown in Figure 6). The "flexibility" of this device is not only reflected in the physical ductility of the material itself but also in the design logic that adapts to the environment. In this way, the equipment can maintain stable operation in complex Marine scenarios.

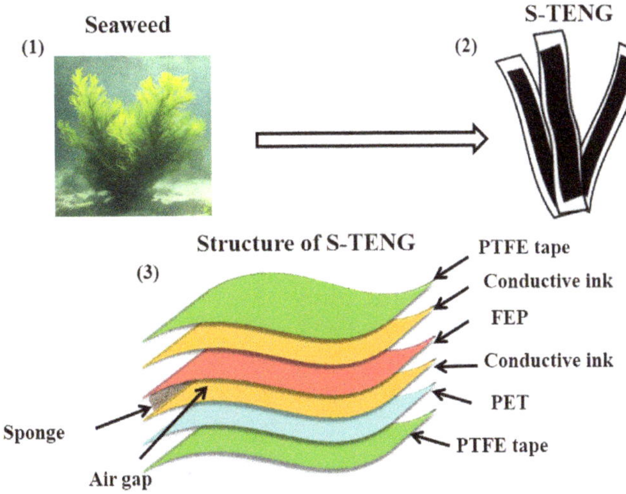

Figure 6. Schematic diagram of TENG's flexible structure. ((1) Schematic diagram of seagrass swaying, (2) Schematic diagram of S-TENG, (3) structural schematic diagram of S-TENG). The image is sourced from [48].

In this paper, a dynamic, adaptive, flexible TENG array is designed for the curved surface deformation of mesh clothing to break through the limitations of traditional planar structures, as shown in Figure 7.

This design selects PDMS/AgNWs composite film as the electrode material and optimizes the conductivity and flexibility by changing the concentration of AgNWs. Meanwhile, polytetrafluoroethylene (PTFE) film is coated on the surface. From the perspective of material properties, PDMS itself possesses excellent flexibility and chemical stability, while AgNWs (silver nanowires) have outstanding electrical conductivity. The PDMS/AgNWs composite film formed by the combination of the two can maintain good electrical conductivity while meeting the flexibility requirements for the deformation of the mesh surface. Lay a material foundation for the stable operation of the array in a dynamic curved surface environment. Ji Chao [52] demonstrated through research that, when preparing the flexible acoustic sensor based on PDMS-AgNWs, the material prepared by combining the two exhibited high electrical conductivity, high flexibility, and good stability and could respond quickly to acoustic wave signals. As shown in Figure 8, the volt–ampere curve of the flexible stress sensor of PDMS-AgNWs is presented.

The coating design of PTFE film specifically addresses the issue of material loss in humid and corrosive environments such as the ocean. PTFE material, due to its extremely strong chemical inertness and corrosion resistance, can effectively prevent seawater from eroding electrode materials. It is often used in the chemical industry for anti-corrosion linings and sealing gaskets, effectively protecting equipment from the damage of highly corrosive media [53]. For the areas prone to high strain when the mesh fabric deforms (such as the edges of the mesh fabric), a serpentine electrode structure is designed. During the stretching process, the serpentine structure can buffer the strain force through the expansion of its own shape, preventing the electrode from breaking or the decline of electrical conductivity due to direct stretching. This significantly enhances the structural stability and operational reliability of the array in the dynamic deformation scenarios of the mesh jacket, meeting the design requirements of "improving stretching reliability and adapting to the curved surface deformation of the mesh jacket". In 2023, among the related electrodes developed by the Printed Electronics team of the Suzhou Institute of Nano-Tech and Nano-Bionics, Chinese Academy of Sciences [54], as shown in Figure 9, the horseshoe-shaped structure with an Angle of 216° can be cycled and stretched 230 times at a tensile strength of 30%. When stretched simultaneously in four different directions, it also exhibits excellent electrical conductivity. At a tensile strength of 50%, the resistance in all directions remains almost constant (as shown in Figure 9). The serpentine electrode structure of this design can also effectively cope with tensile conditions and can guarantee the performance of the equipment in high-strain environments.

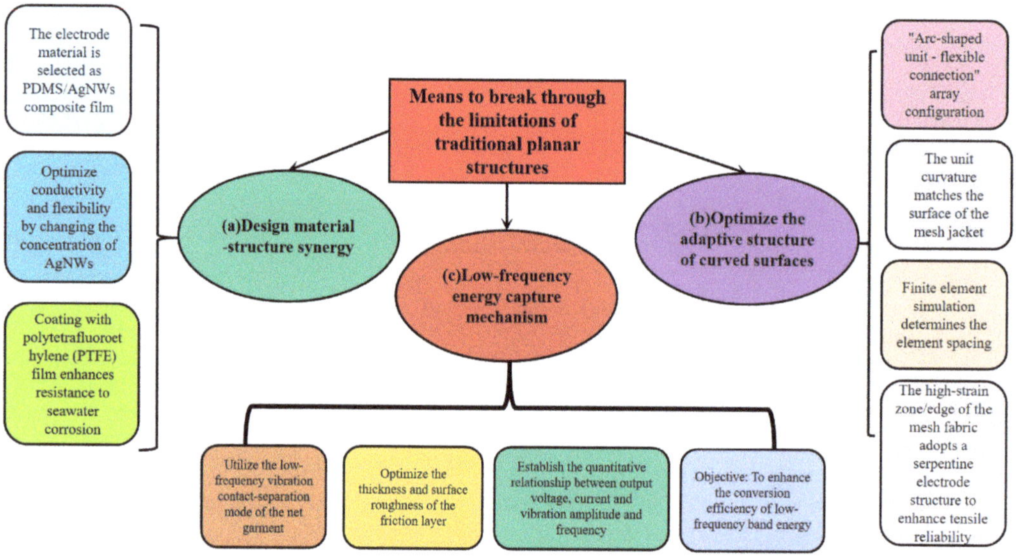

Figure 7. The logical presentation of designing a dynamically adaptable flexible TENG array: (**a**) Material–structure co-design: PDMS/AgNWs composite film is selected as the electrode material. The conductivity and flexibility are optimized by changing the concentration of AgNWs, and the surface is coated with polytetrafluoroethylene (PTFE) film to enhance the resistance to seawater corrosion. (**b**) Surface-adaptive structure optimization: Based on the "arc-shaped element—flexible connection" array configuration, the element curvature is matched with the surface of the mesh jacket. The element spacing is determined through finite element simulation. In the high-strain zone/edge of the mesh jacket, a serpentine electrode structure is adopted to enhance tensile reliability. (**c**) Low-frequency energy capture mechanism: By utilizing the contact–separation mode triggered by the low-frequency vibration of the mesh garment, the thickness and surface roughness of the friction layer are optimized, and the quantitative relationship between the output voltage, current, vibration amplitude, and frequency is established. The objective is to enhance the conversion efficiency of energy in the low-frequency band.

In addition to the curved surface-adaptive TENG array, electromagnetic power generation devices, piezoelectric energy collectors, and bionic integrated power generation systems have also shown application potential in the self-energy supply monitoring of deep-sea cages due to their excellent Marine environment tolerance and multimodal energy capture capabilities. To clarify the optimal application scenarios of various types of equipment in deep-sea cages, based on indicators such as energy capture frequency, power density, and environmental tolerance, this paper establishes a comparison model, as shown in Table 2, through the extensive literature comparison.

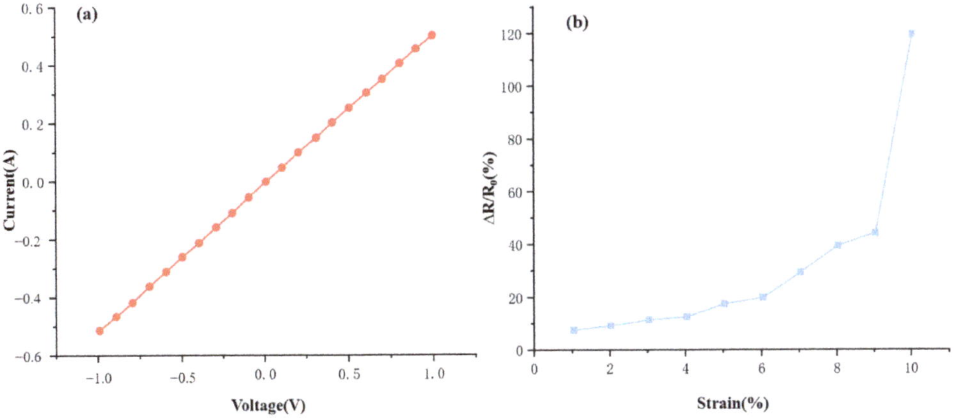

Figure 8. (**a**) Volt–ampere characteristic curve, (**b**) tensile characteristic curve. The data is sourced from [52].

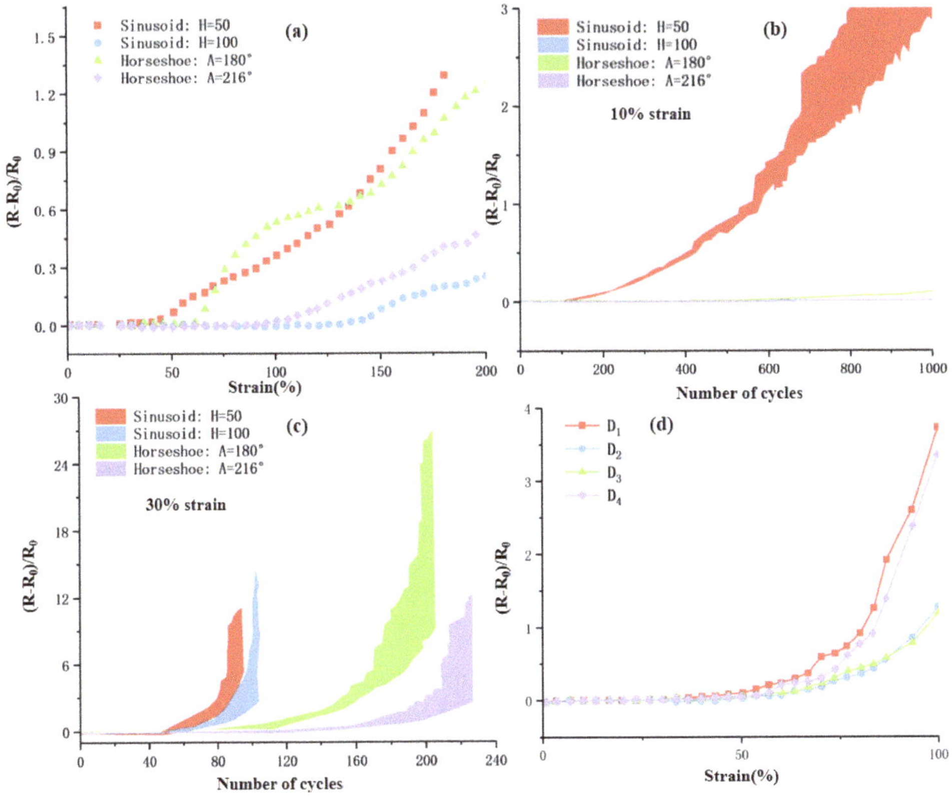

Figure 9. (**a–c**) Tensile cycling stability of the stretchable transparent electrode, (**d**) the multi-directional tensile performance of the stretchable transparent electrode. The data is sourced from [54].

Table 2. Comparison of the adaptability of energy harvesting equipment in complex Marine environments.

Equipment Type	Capture Frequency Range	Power Density	Deep-Sea Pressure Adaptation (Maximum Water Depth)	Corrosion-Resistant Life	The Adaptation Area of the Net Box	Core Advantage
Surface-adaptive TENG array	0.1–10 HZ	0.1–0.5 mW/cm^2	1500 m	6 to 12 months	The curved surface and floating body of the net garment	Flexible conformal, low-frequency adaptation
Deep-sea compatible EMG	0.5–50 HZ	5–15 mW/cm^2	2000 m	5 years and more	Frame, anchor chain	High-power and high-voltage tolerance
Flexible PEH	50–100 HZ	0.5–12 mW/cm^2	1000 m	2 years and more	Local areas of the net garment and turbulent zones	High-frequency capture, flexible integration
Bionic multimodal integrated system	0.1–500 HZ	2–8 mW/cm^2	1800 m	3 years and more	Full net box coverage	Full-band capture and strong environmental adaptability

2.4. Quantitative Analysis of Power Demand for Core Equipment in Deep-Sea Aquaculture Systems

To verify the adaptability of the TENG array as a self-powered solution, it is necessary to first clarify the power consumption characteristics of each key device in the deep-sea aquaculture system. Based on the existing deep-sea aquaculture engineering cases and technical sensor parameters, a database of the equipment power demand (as shown in Table 3) was constructed to provide a quantitative basis for the subsequent selection of energy capture technology.

From this, it is not difficult to find that, in Table 3, the power consumption characteristics and power supply requirements of different devices vary significantly, specifically manifested in three types of patterns. Low-power devices dominate the core monitoring function. The total average power consumption of their structure and environmental sensors is approximately 1.4–2.2 Wh/day, accounting for only 8–12% of the system's total power consumption. However, as 30–50 sensors need to be networked, distributed power supply must be adopted. High-power consumption devices rely on instantaneous power supply. The instantaneous power of actuating devices such as automatic feeders and water quality regulating pumps can reach 10–30 W. Although they work for only 0.5–1 h per day, their average power consumption still accounts for over 80%, and they need to be used in conjunction with energy storage modules. Meanwhile, over 90% of the equipment within the system operates in an intermittent mode of "sampling/work—sleep", with a distinct "low power period", which provides a necessary time window for the energy collection and storage of the TENG array.

Based on the power demand data in Table 3, this paper quantitatively compares the surface-adaptive TENG array with mainstream energy capture technologies such as solar energy, piezoelectric energy, and electromagnetic energy from three dimensions: "power matching degree—environmental adaptability—cost-effectiveness", thereby clarifying the core advantages of TENG.

Based on the multi-dimensional data analysis and comprehensive consideration in Tables 3 and 4, it is found that TENG technology has three core adaptation advantages in deep-sea net clothing applications, which can effectively solve the pain points of existing power supply solutions. In terms of precise power demand matching, the energy conversion efficiency of its array in the 0.1–10 Hz frequency band reaches 15–20%, which can precisely cover the main vibration frequency band of 0.5–5 Hz for deep-sea net clothing. The daily output energy of a single array (with an area of 100 cm^2) is approximately 1.44–3.6 Wh, which can meet the continuous power supply requirements of 15–20 sensors.

When paired with a supercapacitor of 1 F capacity, it can also store energy to supply power to instantaneous high-power equipment such as automatic feeders. In terms of environmental adaptability, the design of the PDMS/AgNWs composite membrane and PTFE coating can withstand pressure at a water depth of 1500 m and seawater corrosion for 6 to 12 months without the need for frequent maintenance, thus solving the problem of the "difficult maintenance and high cost" of deep-sea equipment. In terms of power supply layout, the TENG array supports modular deployment in different areas of the mesh jacket and can directly supply power to nearby sensors, avoiding the problems of "difficult wiring and easy damage" in traditional wired power supply and significantly reducing system complexity.

Table 3. Comparison table of power and energy consumption characteristics of core equipment for deep-sea aquaculture.

Equipment Category	Specific Equipment	Power Range	Working Mode	Average Daily Power Consumption (Wh)
Structural monitoring sensor	Strain sensor	5–15 μW	Intermittent sampling (once per minute)	0.012–0.036
	Vibration sensor	10–20 μW	Intermittent sampling (once per 30 s)	0.024–0.048
	Acceleration sensor	8–12 μW	Intermittent sampling (once/2 min)	0.0096–0.0144
Environmental monitoring sensor	Temperature sensor	3–8 μW	Continuous monitoring (once per 10 s)	0.0216–0.0576
	Salinity sensor	15–25 μW	Intermittent sampling (once for 5 min)	0.0432–0.072
	Dissolved oxygen sensor	30–50 μW	Continuous monitoring (once per 5 s)	0.432–0.72
	pH sensor	20–35 μW	Intermittent sampling (once/2 min)	0.0576–0.1008
Execution device	Automatic feeding machine	10–30 W	Intermittent work (three times a day, 10 min each time)	5–15
	Underwater camera	50–100 mW	Intermittent shooting (1 h each day)	0.05–0.1
Data transmission system	Short-range wireless transmission module	50–100 mW	Intermittent transmission (once per 10 min)	0.072–0.144
	Long-range wireless transmission module	1–5 W	Intermittent transmission (once per hour)	0.04–0.2
Auxiliary system	Water quality regulating pump	5–15 W	Intermittent work (twice a day, 30 min each time)	5–15
	Biological cleaning device	8–20 W	Intermittent work (once a day, 20 min each time)	2.67–6.67

Table 4. A comparison table of the adaptability of different energy capture technologies in deep-sea aquaculture scenarios.

Dimensions	Surface-Adaptive TENG Array	Underwater Solar Energy	Electromagnetic EMG
Power density	0.1–0.5 mW/cm^2	<0.01 mW/cm^2	5–15 mW/cm^2
Adaptation frequency	0.1–10 Hz (low)	The deep sea is ineffective	0.5–50 Hz (medium to high)

Table 4. *Cont.*

Dimensions	Surface-Adaptive TENG Array	Underwater Solar Energy	Electromagnetic EMG
Environmental tolerance	1500 m water pressure, corrosion-resistant for 6 to 12 months	Easy to adhere, with a 50% reduction in effectiveness in March	2000 m water pressure, corrosion resistance for over 5 years
Conversion efficiency	15–20% (low frequency)	<1% (deep sea)	40–50% (mid and high frequencies)
Adaptability	Flexible mesh fabric	It takes up space and is not usable in the deep sea	A fixed frame is required
Unit cost	0.02 dollar/Wh	0.8 dollar/Wh	0.08 dollar/Wh

3. Energy Transfer and Cooperative Capture of Multi-Body Coupling Cage Systems Under Combined Wind–Wave–Current Actions

As shown in Figure 10, based on the energy flow logic of the multi-body coupling system of the cage, this paper systematically reviews the research progress in recent years from three dimensions, energy dissipation, frequency response, and energy capture, analyzes the advantages and limitations of different research methods, and finally proposes future research directions, providing theoretical support for the optimal design and engineering application of deep-sea cage systems.

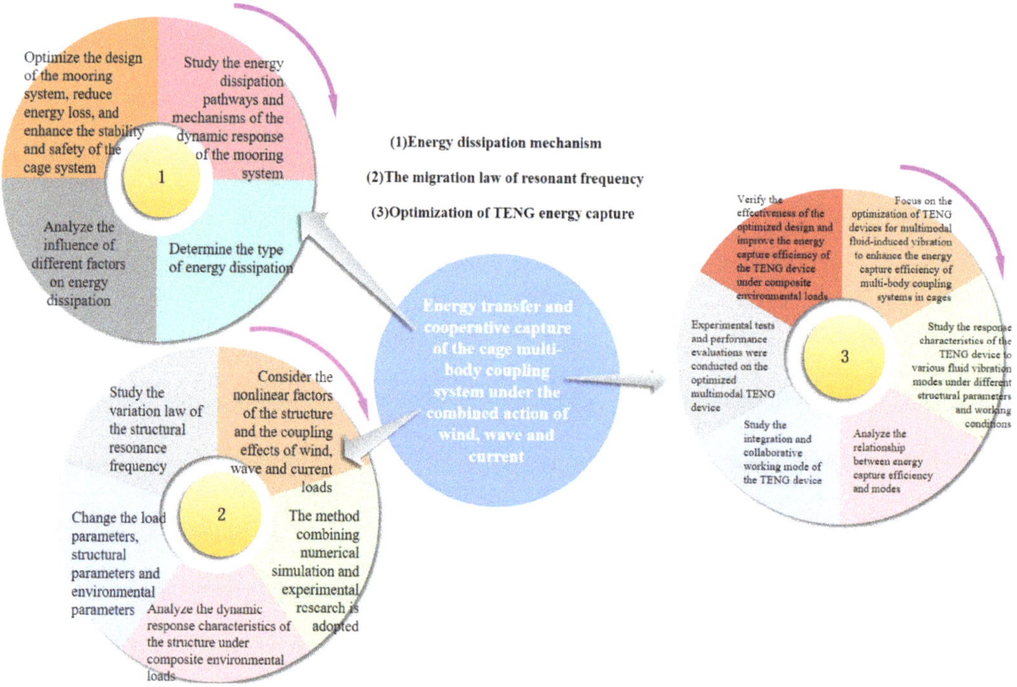

Figure 10. The logical presentation of the research content on energy transfer and cooperative capture of the multi-body cage coupling system under the combined action of wind, waves, and current.

3.1. Energy Dissipation Mechanism of Dynamic Response of Multi-Degrees-of-Freedom Mooring Systems

This paper studies the energy dissipation pathways and mechanisms in the dynamic response process of the mooring system through energy analysis methods. Energy dissipation mainly includes internal friction dissipation of the mooring line material, viscous friction dissipation with seawater, and collision dissipation with the seabed, etc. [55–59]. Analyze the influence of different factors on energy dissipation, such as the material properties, length, diameter, and pretension of the mooring line, as well as sea conditions, etc. Through research on the energy dissipation mechanism, theoretical guidance is provided for optimizing the design of the mooring system, reducing unnecessary energy loss, and improving the stability and safety of the cage system.

3.1.1. Internal Friction Dissipation of the Mooring Line Material

The viscoelastic properties of mooring line materials are the core source of internal friction dissipation, and the degree of energy loss can be quantitatively described by the loss factor (tanδ). Consider polyester (PET), a commonly used synthetic fiber material, which has become one of the mainstream materials for deep-sea cage mooring systems due to its excellent tensile strength, low elongation, and resistance to seawater corrosion. This paper will start from the viscoelastic characteristics of polyester materials [60] and, in combination with relevant research, systematically expound the internal friction dissipation mechanism of anchor line materials under dynamic loads.

In the field of Marine engineering, the application of polyester materials can be traced back to earlier mooring system practices. For example, as Michel Francois and Peter Davies [61] pointed out at the 27th ASME International Conference on Marine Mechanics and Arctic Engineering (OMAE2008) in 2008, polyester has been maturely applied in floating platform mooring systems in the offshore areas of Brazil (for over 10 years), West Africa, the Gulf of Mexico, and other regions, and its stability under long-term loads has been verified by engineering. From the perspective of material structure, the molecular structure of polyester fibers contains a large number of ester functional groups, which endows the material with a relatively high modulus and strength but also makes it exhibit certain viscoelastic characteristics. The specific manifestation of this viscoelastic characteristic under dynamic loading can be further explained by the polyester rope load-elongation model proposed by Francois and Davies.

1. Through three-parameter fitting, they wrote the elongation $L(t)$ on each creep (or recovery) platform as follows:

$$L(t) = L(t_p) + a_c \cdot log\left[\frac{1 + (t - t_p)}{t_a}\right] \tag{11}$$

This fitting (for a_c, t_a, and $L(t_p)$) is independent of the time unit and the starting point, and its result does not depend on the selection of t_p (i.e., the time at any point on the load platform).

2. In $L(t)$, they use

$$L\tau \approx L(t_p) + a_c \cdot log[\tau / t_a] \tag{12}$$

It is used to represent the elongation $L\tau$ at the end of each $\frac{1}{2}$ period of duration τ.

3. Then, the linearized quasi-stiffness is regarded as the secant stiffness between the last consecutive $\frac{1}{2}$ period (duration τ) endpoints. The load is normalized through the rope MBS to obtain the dimensionless quasi-static stiffness KrS.

This model decomposes the response of polyester rope into average elongation, quasi-static response, and dynamic response. Among them, the dynamic response corresponds to the cyclic loads (low frequency and wave frequency) caused by the wind, waves and currents, and its mechanical behavior needs to be quantified by dynamic stiffness (KrD).

The essence of dynamic stiffness is precisely the macroscopic manifestation of the viscoelasticity of materials under cyclic loading—this is highly consistent with the physical mechanism of internal friction dissipation; that is, energy loss occurs in materials during dynamic deformation due to molecular chain movement, interface friction, etc. In Marine environments, polyester mooring lines are constantly subjected to cyclic loads, and their dynamic mechanical properties have a significant impact on the stability and service life of the mooring system. As shown in Figure 11, Francois and Davies experimentally verified that the dynamic stiffness (KrD) of the polyester rope is mainly controlled by the average load, and under actual random loads (such as wideband signals simulating Marine environments), its load-elongation relationship is close to linear. Meanwhile, the correlation between energy loss and load frequency is relatively weak (stiffness variation is only 1.4% within a period of 12.5 s to 500 s). This conclusion provides engineering background support for subsequent experimental research; that is, the internal friction dissipation characteristics of polyester mooring lines need to be analyzed in combination with actual load conditions.

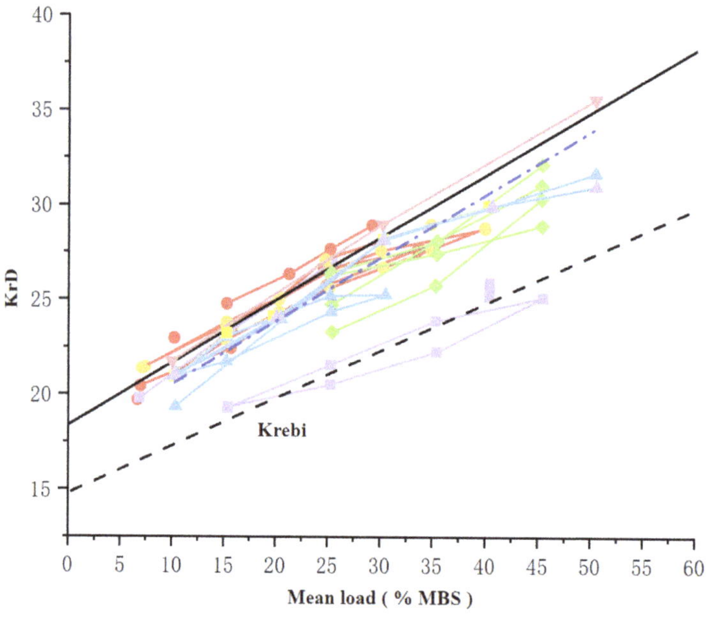

Figure 11. Dynamic stiffness as a function of mean load. The data is sourced from [61].

From a further perspective of the mesoscopic characteristics of dynamic mechanical behavior, polyester materials exhibit obvious frequency and temperature dependence. This phenomenon is widespread in different types of polyester materials. For instance, Liu Xiu et al. [62] found in their 2024 research that in the frequency scanning test of PET materials, at the same temperature, the energy storage modulus gradually increased with the increase in the test frequency, showing a trend of enhanced rigidity. The loss factor, near the glass transition temperature, first increases and then decreases with the increase in frequency, showing a distinct peak. This result complements the "linear increase of dynamic stiffness with average load" observed by Francois and Davies: the former reveals the influence of frequency and temperature on viscoelasticity from the molecular motion level, while the latter quantifies the regulatory effect of load conditions on dynamic mechanical properties from the engineering application level. Together, they construct a "microscopy-macro" analysis framework for the internal friction dissipation characteristics of polyester mooring lines, providing a complete mechanical basis for the material selection and life assessment of deep-sea cage mooring systems.

3.1.2. Viscous Frictional Dissipation of Seawater

The viscous friction between the mooring line and seawater is one of the main paths for energy dissipation, and its power loss can be calculated by the Morison equation:

$$F_D = \frac{1}{2} C_D \rho D |u - v|(u - v) \tag{13}$$

Among them, F_D represents the viscous frictional force per unit length, N/m; C_D is the drag coefficient, dimensionless; ρ is the density of seawater, kg/m^3; D is the diameter of the anchor line; u is the velocity of seawater, m/s; and v is the velocity of the anchor line, m/s.

Leandro S. P. da Silva [63], in the article "Statistical linearization of the Morison's equation applied to wave energy converters", points out that viscous drag has a significant effect on the dynamics of wave energy conversion devices in high-energy sea states and large motion caused by resonance. In this paper, the viscous resistance is derived by using the statistical linearization technique, and its equivalent linear term is decomposed into the excitation term and the damping term, providing a reliable and low-cost method for the estimation of system dynamics. This indicates that the Morrison equation has significant application value in studying the energy loss and dynamic response of Marine energy-related structures and also reflects from the side the key significance of calculating the viscous friction dissipation of seawater through this equation for understanding the performance of Marine engineering structures. According to S. Beji's research [64], the drag force stems from convective acceleration and surface frictional resistance. Although surface frictional resistance is not explicitly expressed in the Morrison equation, it is part of the drag force. This literature derives the momentum equation expression of one-dimensional flow from the one-dimensional momentum equation:

$$\rho \frac{\partial u}{\partial t} + \rho u \frac{\partial u}{\partial x} = -\frac{\partial p}{\partial x} + \frac{\partial \tau_{yx}}{\partial y} \tag{14}$$

It also indicates that the Morison equation is a special case, providing a solid theoretical foundation for the Morison equation. This further indicates that when calculating the viscous frictional dissipation of seawater using the Morrison equation, the accurate understanding and calculation of the drag force are crucial for precisely assessing energy dissipation. Liu et al. [65] discovered that the key to calculating the Morrison equation lies in determining the drag force coefficient C_D and the inertial force coefficient C_M. Therefore, they weighted and averaged the pure wave quantity value and the pure flow value to obtain the hydrodynamic coefficient in the wave–current coexistence field:

$$C_D = \frac{C_{Dw} u_o + C_{Dc} U \alpha}{u_o + U \alpha} = \frac{C_{Dw} KC_w + C_{Dc} KC_c \alpha}{KC_w + KC_c \alpha} = \frac{C_{Dw} + C_{Dc} V_T \alpha}{1 + V_T \alpha} \tag{15}$$

$$C_M = \frac{C_{Mw} u_o + C_{Mc} U \alpha}{u_o + U \alpha} = \frac{C_{Mw} KC_w + C_{Mc} KC_c \alpha}{KC_w + KC_c \alpha} = \frac{C_{Mw} + C_{Mc} V_T \alpha}{1 + V_T \alpha} \tag{16}$$

In the formula, α is the weighting coefficient; C_{Dw} and C_{Dc} are determined by KC_w and Re_w (Re_w refers to the wave Reynolds number); and the subscript w represents the pure wave quantity. C_{Mw} and C_{Mc} are determined by Re_c (Re_c refers to the Reynolds number of water flow), and the subscript c indicates the pure flow rate.

As shown in Figure 12, the drag force coefficient is affected by multiple factors such as the shape of the pile column, its placement state, the number of KC, the number of Re, and the wave and flow conditions. Liu Guijie and others summarized the research results under different conditions, such as the variation laws of coefficients when pure waves and wave–current combined effects occur, which enables more precise calculation of the viscous frictional dissipation of seawater using the Morison equation.

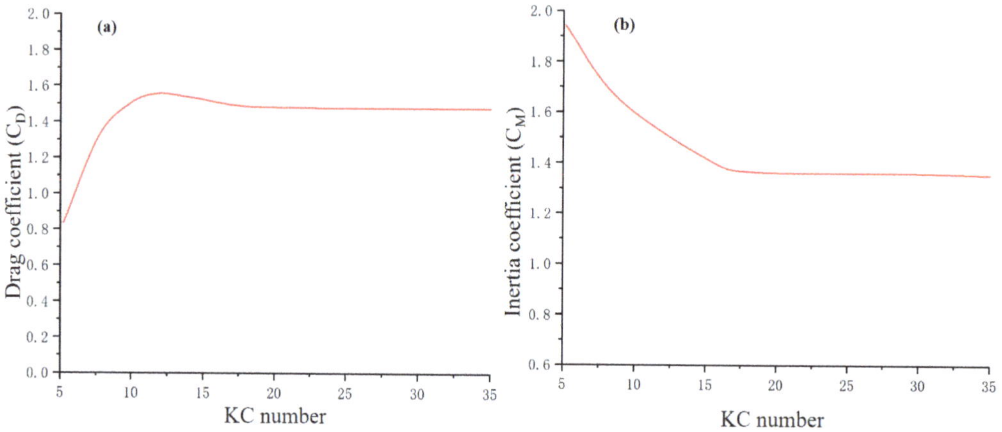

Figure 12. (**a**) The relationship between the numbers C_D and KC in pure wave fields and wave–current coexistence fields. (**b**) The relationship between the numbers C_M and KC in pure wave fields and wave–current coexistence fields. The data is sourced from [65].

3.2. Migration Law of Structural Resonance Frequency Under Composite Environmental Loads

The resonance response of the cage multi-body coupling system is the main cause of its structural fatigue failure. In practical application scenarios, the combined effect of wind, waves, and the current will introduce various nonlinear factors such as geometric nonlinearity, material nonlinearity, and fluid–structure coupling nonlinearity, which will cause a significant shift in the original resonance frequency of the system [66–70]. Therefore, clarifying the migration law of this resonant frequency has become the core key to avoiding resonant disasters. To achieve this goal, the research should be based on the dynamic model of the multi-body coupling system of the cage. In the model, the nonlinear factors of the structure itself and the coupling effects of the wind, waves, and current loads should be fully considered. Then, a method combining numerical simulation and experimental research should be adopted to systematically analyze the dynamic response characteristics of the structure under composite environmental loads. During this process, by specifically changing the load parameters, structural parameters, and environmental parameters, the variation law of the structural resonance frequency is ultimately explored and clarified, providing theoretical and data support for resonance prevention and control.

3.2.1. The Influence of Load Parameters on Resonant Frequency

The period and wave height of the wave load are the core parameters for regulating the resonant frequency [71–75]. In the multi-body coupling system of the cage, the regulation effect of the wave load period and wave height on the resonant frequency needs to be analyzed in combination with the dynamic response characteristics of the system. Relevant numerical simulation studies have provided solid technical support and data verification for this regulation mechanism. From the perspective of numerical modeling of the multi-body coupling system of the cage, Shuo Mi et al. [76] proposed a Euler–Lagrangian implicit coupling model based on the OpenFOAM platform. The immersion boundary method was adopted to achieve precise coupling between the flow field and the mesh structure. Meanwhile, the mesh element was constructed through the mass-spring model, and the resistance coefficient was defined by piecewise functions. The multi-body collaborative mechanism of "net jacket—floating body—mooring" in the cage has been completely restored. When simulating different wave conditions, this study found that when the wave period approached the inherent vibration period of the cage system, the dynamic response of the system (such as the sag displacement of the floating body and the tension of the mooring rope) showed a significant increase, and the increase showed a nonlinear upward trend with the increase in wave height—this result directly confirmed the "matching trigger" effect of the wave period on the resonant frequency, and the "amplification regulation" effect of wave height on the intensity of resonance response provides a quantitative basis for the correlation law between load parameters and resonance frequency.

At the level of precise application and verification of wave load parameters, Cao and Wan [77] constructed a viscous numerical wave pool suitable for Marine engineering structure analysis based on the OpenFOAM platform by improving the wave generation algorithm and the design of the wave dissipation boundary:

$$X(t) = S_e\left[\sin(\omega t) - \frac{H}{4d_T}\left(\frac{3}{\sinh^2 kd} - \frac{n_1}{2}\right)\sin 2\omega t\right] \tag{17}$$

Among them, S_e represents the stroke calculated based on the theory of linear wave plates; $n_1 = \frac{1}{2}\left(1 + \frac{2kd}{\sin 2kd}\right)$.

$$\mu_s(x) = \begin{cases} \alpha_s[(x - x_0)/L_s]^2 & , x > x_0 \\ 0 & , x \le x_0 \end{cases} \tag{18}$$

For this formula, x_0 represents the starting position of the cancelation zone; L_s represents the length of the wave cancelation zone; and α_s is a dimensionless artificial viscosity coefficient used to control the wave cancelation intensity. The larger the value of α_s, the greater the corresponding wave cancelation intensity. Unlike the traditional sponge layer wave elimination methods, this paper introduces a correction velocity in the wave elimination source term f_s. Its main function is to perform quality correction and ensure the conservation of mass during the calculation process.

Thus, the stable generation and precise control of regular and irregular waves have been achieved. As shown in Figure 13, the research takes the cage multi-body system as the analysis object. By comparing the resonance response characteristics of the system under different wave parameters, it is found that the numerical simulation results of waves A and B demonstrate the good computational accuracy of the Stokes second-order wave theory solution. The wave-forming boundary is given a linear wave condition but it can be seen that when the wave steepness is large, the regular wave will still undergo nonlinear evolution during propagation, and the resulting wave surface has sharp crests and relatively flat troughs.

Meanwhile, this study verified the reliability of the established OpenFOAM model in wave load transmission and system resonance response prediction by comparing the numerical simulation results with the physical model test data, further consolidating the technical foundation of the "wave period—wave height—resonance frequency" correlation analysis. It provides a numerical simulation method and data support for the resonance frequency regulation of the multi-body coupling system of the cage.

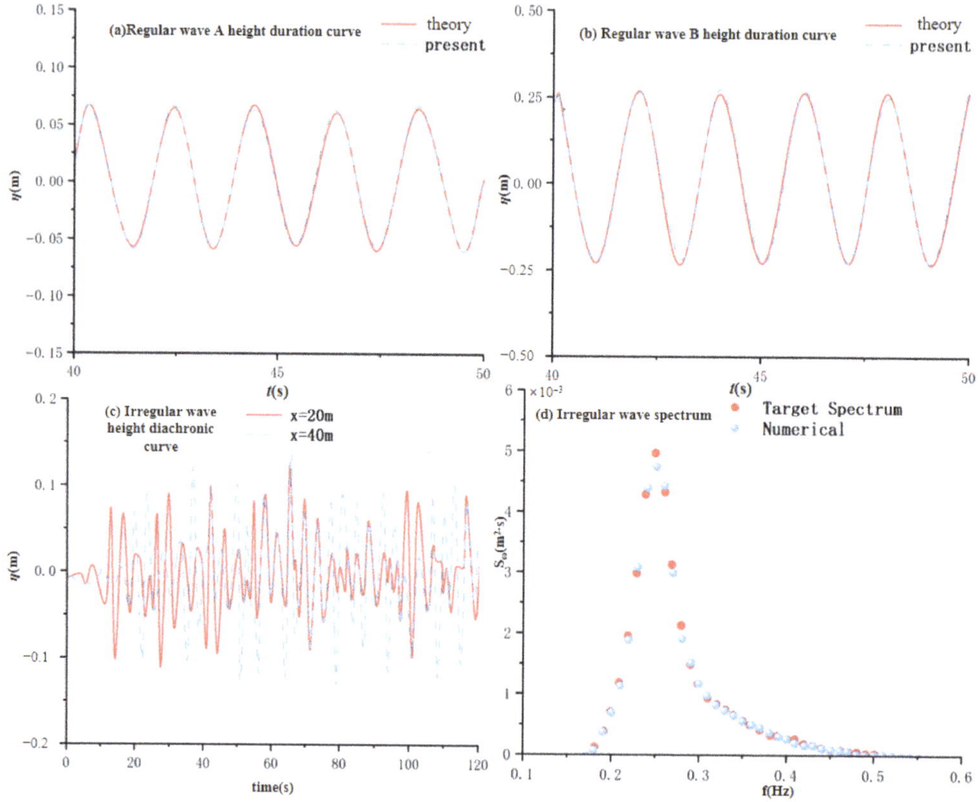

Figure 13. Simulation results of regular waves and irregular waves. (**a**) Represents the diachronic curve of the height of the regular wave A, (**b**) represents the diachronic curve of the height of the regular wave B, (**c**) represents the irregular wave height diachronic curve, and (**d**) represents the spectrum of irregular waves. The data is sourced from [77].

3.2.2. The Regulatory Role of Structural Parameters

Structural parameters, such as the size of the cage, the material of the frame, and the diameter of the mesh, play an important regulatory role in the resonant frequency of the multi-body coupled cage system. The mesh size and wire diameter of the net garment directly affect the stiffness characteristics of the system and thereby change the resonant frequency. This influential mechanism has been verified in the research of various engineering scenarios related to net garments. Specifically, it can be further clarified through the following research conclusions:

From the perspective of the basic mechanical properties of mesh clothing, Chen et al. [78] quantitatively analyzed the influence laws of mesh size and wire diameter on stiffness-related indicators through numerical simulation experiments. The research finds that, under the condition that other parameters remain unchanged, when the mesh size of the net garment increases, the force-bearing capacity of the net garment significantly decreases, and the corresponding equivalent stiffness also reduces accordingly. An increase in the wire diameter will enhance the structural load-bearing capacity of the mesh garment, causing the equivalent stiffness to show an upward trend. As shown in Figure 14, the quantitative relationship between the variation range of different mesh pins/wire diameters and the force and flow rate is presented.

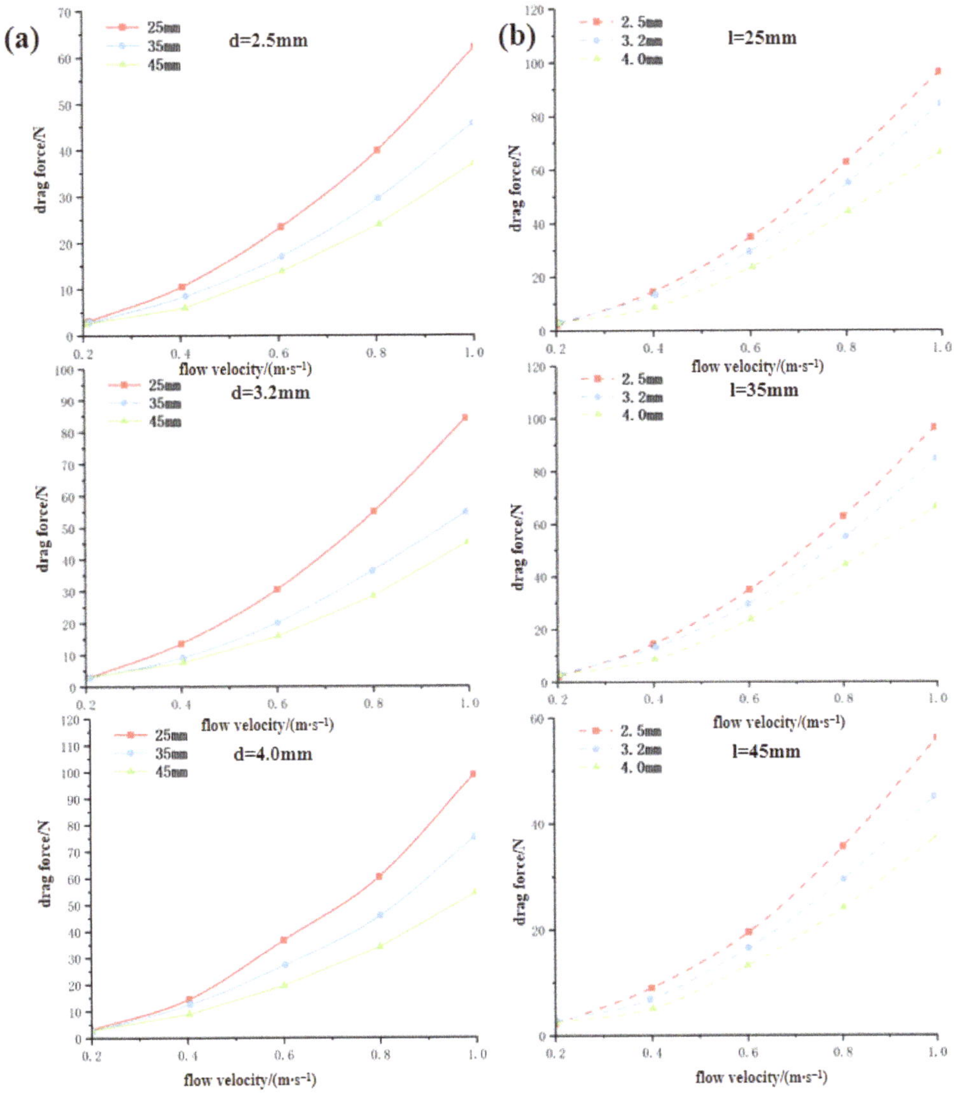

Figure 14. (**a**) Comparison of the force on the mesh jacket under different mesh size conditions, (**b**) comparison of the force on the mesh jacket under different wire diameter conditions. The data is sourced from [78].

In the application scenarios of multi-body cage systems, the research by Huang et al. [79] further confirmed this correlation. This study, focusing on the typical multi-body coupling system of the HDPE circular gravity cage, analyzed the influence of mesh size changes on the overall mechanical response of the system. It was found that an increase in mesh size can reduce the overall deformation of the cage during the force application process, indirectly reflecting the improvement of the equivalent stiffness of the system. Moreover, the quantitative change trend between mesh size and deformation was clarified. It is indicated that, in the multi-body coupling cage system, the stiffness can be regulated by optimizing the mesh size, thereby achieving the adjustment of the resonant frequency.

From the perspective of engineering practice application, the research by Chen et al. [80] emphasizes the crucial role of mesh parameters in multi-body systems. This study points out that the mesh jacket is the core component affecting the overall stiffness of the self-elevating truss box. After taking into account the action of the mesh jacket, the reaction force of the system's pile legs and the utilization coefficient of the truss increase significantly, demonstrating the constraint effect of the mesh jacket on the overall stiffness of the system. Meanwhile, the paper provides the cluster modeling parameters corresponding to the specific mesh size. In the design of multi-body coupled cage systems, by determining parameters such as mesh size and wire diameter to precisely control stiffness and resonant frequency, it offers a reference for engineering practice.

3.2.3. Coupling Effects of Nonlinear Factors

The nonlinear factors in the multi-body coupling cage system, including geometric nonlinearity, material nonlinearity, and fluid–structure coupling nonlinearity, do not act independently but have complex coupling effects, further influencing the migration of the system's resonant frequency. Geometric nonlinearity mainly occurs when the structure undergoes significant deformation [81]. At this point, the relationship between the stress and strain of the structure is no longer linear, thereby altering the stiffness of the structure. Material nonlinearity is caused by the inelastic properties of the material itself, such as plastic deformation and creep, and also affects the stiffness and mass distribution of the structure [82]. Fluid–structure coupling nonlinearity is due to the interaction between the fluid and the structure. The fluid load acting on the structure varies with the deformation of the structure, and the deformation of the structure is also affected by the fluid load, forming a complex coupling relationship, which cannot be ignored in the research [83]. The influence of the above-mentioned nonlinear fluid–structure coupling on the vibration frequency, strain distribution, and energy density of the net garment directly determines the structural adaptability, layout rationality, and upper limit of energy output of the TENG array, and the design parameters need to be optimized specifically.

Nonlinear fluid–structure coupling can alter the vibration characteristics of the net garment, thereby affecting the energy capture conditions of the TENG. The specific manifestations and corresponding design optimizations are as follows: large deformation of the mesh jacket can cause "contact misalignment" in traditional planar TENG cells. Finite element simulation verification shows that when the mesh jacket stretching rate reaches 15–20%, the contact area of the planar cells decreases by 30–40%, and the output power drops by more than 50%. To address this issue, the original "serpentine electrode" needs to be upgraded to a "stretchable wavy electrode". And the matching relationship between the peak spacing of the electrode and the wrinkling wavelength of the mesh jacket was determined through COMSOL (6.2) simulation to ensure the retention rate of the contact area after deformation. The viscoelasticity of the mesh fabric material can cause a "drift" in vibration frequency. Based on the fractional Burgers model calculation, it is known that under the wave excitation of 5–10 Hz, the material hysteresis effect will reduce the actual vibration frequency by 10–15%, resulting in a mismatch with the original design frequency of TENG (0.1–10 Hz). To address this issue, a "frequency adaptive module" needs to be added to the TENG array. By increasing the thickness of the friction layer from 50 μm to 80 μm, the resonant frequency of the TENG is shifted by 12–18% towards the low frequency, thereby complementing the frequency drift caused by the nonlinearity of the material. In addition, fluid nonlinearity such as vortex-induced vibration and wave–flow coupling can cause uneven energy distribution on the surface of the mesh jacket. Through PIV experiments, it was measured that the energy density in the vortex zone at the edge of the mesh jacket (0.3–0.5 mW/cm^2) is 2–3 times that in the central zone (0.1–0.2 mW/cm^2).

Moreover, there are instantaneous energy pulses (with a peak of up to 1.2 mW/cm^2), so the layout of the TENG array needs to be optimized, dividing it into "high-energy zone units" and "conventional zone units". The edge zone adopts a "double-layer friction structure" (PTFE + FEP) to enhance the upper limit of energy capture. The central area adopts a "lightweight unit" (PET substrate) to reduce costs. Meanwhile, an energy distribution model is established through MATLAB (R2023a), which can ultimately increase the total output of the array by 40% to 60%.

3.3. Optimization of TENG Devices for Multi-Modal Fluid-Induced Vibration

As a new type of energy harvesting device, TENG can convert the mechanical energy of fluid-induced vibration into electrical energy, providing a solution for the self-power supply of deep-sea cages [84–88]. Multimodal TENG can achieve significant improvements by adapting to multiple fluid vibration modes, such as the up-and-down vibration of waves and the horizontal vibration of ocean currents, including the efficiency of energy capture. Structural parameter optimization and modal adaptation are the main aspects of optimizing its design.

The structural form of TENG directly determines its response capability to different fluid vibration modes. The structural parameters of multimodal TENG have a significant impact on its modal adaptation and energy capture efficiency [89,90]. The magnitude of the vibration mass determines the natural frequency of the TENG device. Inspired by the classic Archimedes helix, Zhou et al. [91] prepared composite fibers (SMFs) with helical element structures by rolling and cold stretching a double-layer film of MXene and solid polymer electrolyte (as shown in Figure 15). This unique helical structure demonstrates significant advantages in multimodal energy harvesting. From the perspective of structural parameters, parameters such as the helical period, number of layers, and MXene content have a significant impact on its performance. When dealing with mechanical vibration modes, a reasonable helical period design can effectively disperse external stress, enabling fibers to evenly distribute stress when subjected to vibration, avoiding structural damage caused by local stress concentration, and thus ensuring the stability of energy capture. Meanwhile, the variation in MXene content will affect the electrical properties of the material, thereby altering the generation and transmission efficiency of charges during vibration, and ultimately influencing the energy capture efficiency. For humidity gradients and temperature difference modes, the ion–electron coupling device formed by the alternating layers of MXene and polymer electrolytes plays a role. An appropriate number of layer and material ratios can optimize the transport path of ions and electrons, enhance the response to humidity and temperature changes, and improve energy capture efficiency. This study, through experimental and theoretical analysis, clarified the relationship between these structural parameters and multimodal energy harvesting, providing an important basis for designing efficient multimodal energy harvesting fibers.

Jin Sun Yangyang [92] focuses on the application of drum-type TENG devices in vibration energy collection. The structural parameters of the drum TENG, such as the radius of the drum surface, the number of electrode pairs, and the electrode spacing, have a direct impact on its energy capture efficiency in the vibration mode. In the vertical contact–separation vibration mode, an increase in the number of electrode pairs can significantly enhance the output power. He experimentally measured that the output power of a single electrode pair could reach 1.32×10^{-6} W within a single cycle. When an eight-electrode pair drum TENG device was used, the output power increased to 1.06×10^{-5} W. This is because more electrode pairs increase the area of action of the triboelectric effect, enabling the generation of more charges under the same vibration conditions and thereby enhancing the energy capture efficiency. The variation in the drum surface radius and the electrode spacing will affect the natural frequency and electric field distribution of the device. If the radius of the drum surface is too large or too small, it may cause the natural frequency of the device to not match the external vibration frequency, reducing the energy capture efficiency. If the electrode spacing is inappropriate, it will affect the transfer and collection of charges and thus impact the output performance. Through finite element analysis and experimental verification, this study revealed the intrinsic connection between the structural parameters of the drum TENG and the vibration mode adaptation as well as the energy capture efficiency. Not only that, but Yang Hongbo [93] also designed a triboelectric nanogenerator with a specific structure for the low-frequency environmental energy of water waves. This further fully demonstrates the significant influence of structural parameters on the multimodal TENG in the specific mode of water wave energy harvesting.

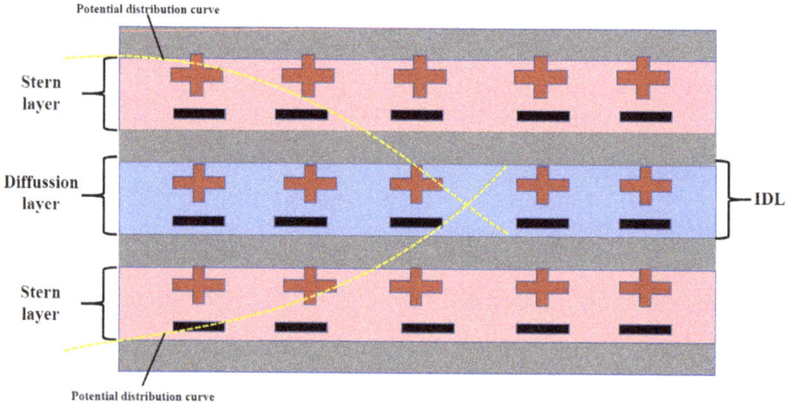

Figure 15. Schematic diagram of mechanical vibration modes.

4. Biomechanics of Interactions Between Fish Schools and Cage Flow Fields and Micro-Energy Harvesting Technology

The behavior of fish schools can serve as a "natural indicator" of the dynamic load on the cage structure, which is specifically manifested in two aspects: first, the net jacket in the area where fish schools gather will generate additional local loads due to the water flow blocking effect; second, the vortices caused by the rapid swimming of fish schools will intensify the vibration of the net jacket. All these characteristics of fish school behavior can be directly captured by fish-on-board sensors.

In the fish–cage flow field interaction system, the tail swing movement of fish is a key factor affecting the characteristics of the flow field. The periodic tail wagging of fish induces the formation of a complex vortex energy field around them. The distribution and evolution of this vortex energy field directly affect the energy exchange between fish stocks and the flow field, as well as the energy scale available for micro-energy capture [94–97]. In addition, the research and development of high-performance micro-energy sensors and management systems is an important guarantee for achieving the effective capture and utilization of micro-energy. Bionic flexible TENG sensors, with their advantages of high sensitivity, low cost, and self-driving capabilities, have demonstrated great application potential in the field of Marine micro-energy monitoring and capture. However, there are still many unsolved problems in the current research, such as the precise description of the vortex energy field induced by fish tail swinging, the improvement of the performance of bionic flexible TENG sensors in complex Marine environments, and the efficient management of multi-physics field coupled micro-energy, etc. This paper will systematically review the research progress in the above-mentioned fields, analyze the existing problems and challenges, and look forward to future research directions with the aim of promoting the in-depth development of biomechanics and micro-energy capture technology of fish–cage flow field interaction. Figure 16 shows the main content of this part of the analysis.

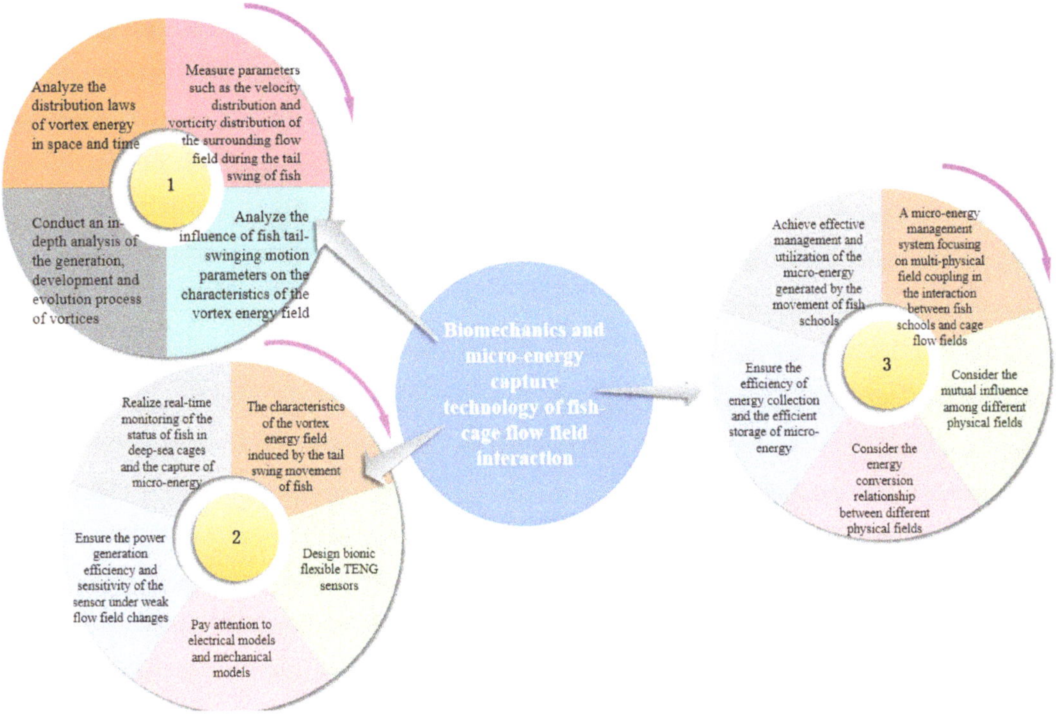

Figure 16. Logical presentation of the research content on biomechanics and micro-energy capture technology of the interaction between fish schools and cage flow fields.

4.1. Characteristics of the Vortex Energy Field Induced by Fish Tail Swing Motion

The tail swing movement of fish is a typical biological movement and the main means for fish to generate propulsion and achieve movement in water. During the tail swing process, the fish tail pushes the surrounding water to form a series of vortices. These vortices carry certain energy and constitute a unique vortex energy field [98]. Studying the characteristics of this vortex energy field is the basis for understanding the interaction between fish schools and cage flow fields and developing micro-energy capture technology.

4.1.1. Experimental Study on the Characteristics of Vortex Energy Fields

In the experimental study of the vortex energy field induced by fish tail wagging movement, it is of vital importance to construct a high-precision fish swimming experimental platform. This platform needs to meet the requirements of synchronous and high-precision capture of fish swimming postures, tail fin movement parameters, and flow field vortex characteristics. The existing related experimental research has provided key references for the core technical direction and performance verification of the platform construction. Specifically, the following two representative research results can be combined for supplementary explanation:

1. The reference basis for the core measurement technology of the experimental platform:

The analysis of the energy field of the tailing-induced vortex in fish relies on advanced flow field visualization and data analysis techniques. Guo Chunyu et al. [99] achieved dynamic capture of the shedding process of the tailing-induced wake vortex pair during the linear acceleration of zebrafish based on particle image velocimetry (TR-PIV) technology and the biorthogonal decomposition (BOD) method. The evolution law of the dominant vortex structure in the flow field was precisely characterized through low-order modes. This technical approach can be directly applied to the construction of the flow field measurement module on the experimental platform, clarifying that the spatial resolution and sampling frequency of the PIV system need to match the tail swing frequency of fish (for example, the tail swing frequency of zebrafish during acceleration is approximately 5–8 Hz). Meanwhile, the effectiveness of the BOD method in extracting key features of the vortex energy field (such as vorticity intensity and vortex core position) was verified, providing experimental support for the selection of data analysis algorithms on the platform.

2. Comparison indicators for performance verification of the experimental platform:

After the experimental platform is constructed, its reliability needs to be verified through tests in specific fish swimming scenarios. Yang Guodang et al. [100] provided quantitative comparison indicators under multiple swimming conditions. This study focuses on three typical postures of grass carp: straight-line swimming, turning swimming, and backward swimming. As shown in Figure 17, they measured the thrust and lateral force around the fish body under different swimming conditions through the PIV system, thereby obtaining the vortex characteristics induced by tail swing; among them, the positive vortex at the tail dominated the generation of thrust when swimming in a straight line, and when the head deflection Angle was controlled at 9–10°, the energy conversion efficiency reached the highest (83%). This result can serve as a benchmark for verifying the performance of the experimental platform. After the platform is built, herbivorous fish of the same body type (such as grass carp and crucian carp) can be selected for linear swimming tests to compare and measure the vortex distribution, thrust coefficient, and energy efficiency of the tail-induced vortex.

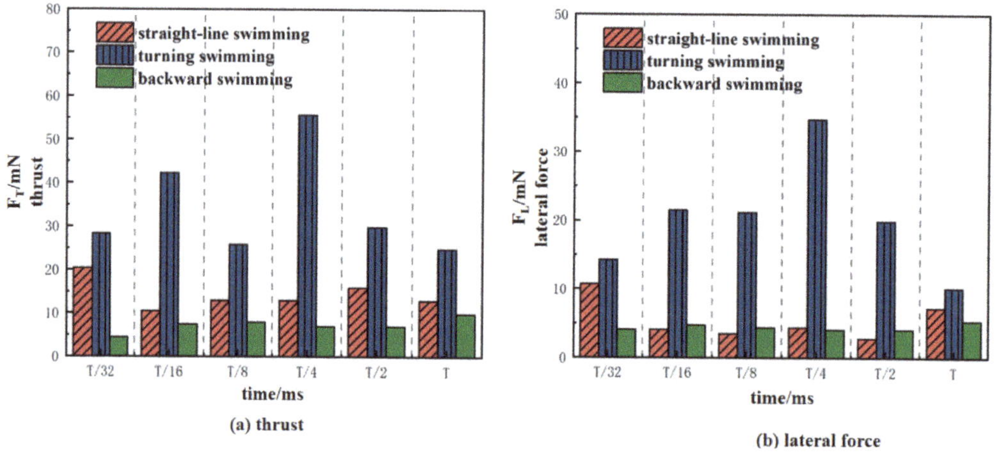

Figure 17. The thrust and lateral force throughout the fish's body under different swimming conditions. The data is sourced from [101].

4.1.2. Numerical Simulation of Vortex Energy Field Characteristics

Numerical simulation is an important means for in-depth analysis of the generation, development, and evolution process of the vortex energy field induced by fish tail wagging movement. At present, the academic circle has formed multi-dimensional research results in this direction. The literature with different numerical methods and research perspectives provides key support for the complete analysis of the "generation mechanism—evolution law—energy transfer" of the vortex energy field. Specifically, the following three representative studies can be supplemented and explained:

1. The revelation of the vortex energy evolution mechanism by two-dimensional unsteady models:

Hu Wenrong et al. [101], in their early classical studies, adopted the two-dimensional unsteady NS equation to focus on the vortex energy evolution process in the rapid start (C-start) stage of fish. The simulation results show that there is a significant difference in energy evolution in the "burst—deceleration" stage of C-start tail swing—during the burst period (0–0.1 s), the tail edge vortex is rapidly generated, the energy is mainly radial diffusion, and the instantaneous energy growth rate reaches 500%/s; during the deceleration period (0.1 to 0.3 s), the vortex begins to fall off and fuse with the forward vortex system. The energy gradually converts into propulsion kinetic energy, and the contribution of the additional inertial force to the energy conversion efficiency accounts for 35%. This study, for the first time, distinguished the temporal correspondence between the tail swing motion stage and the evolution of vortex energy.

2. Analysis of the vortex energy development law under the dynamic mesh RANS method:

In light of the dynamic variation law of the vortex energy field with the tail swing parameters, Su Boyue et al. [102] systematically explored the regulatory effect of the Strouhal number (St) on the development of vortex energy by using a dynamic grid combined with the RNG k-ε turbulence model. As shown in Figure 18 and Table 5, in order to study the influence of the St number on the hydrodynamic coefficient, they achieved the change in the St number by regularly varying the flow velocity, swing frequency, and lateral movement amplitude. It was found that as the St increased, the mean hydrodynamic coefficient Cxm gradually decreased; that is, the average thrust coefficient Kt gradually increased. This achievement also further clarifies the quantitative correlation between the tailing frequency and the development of vortex energy.

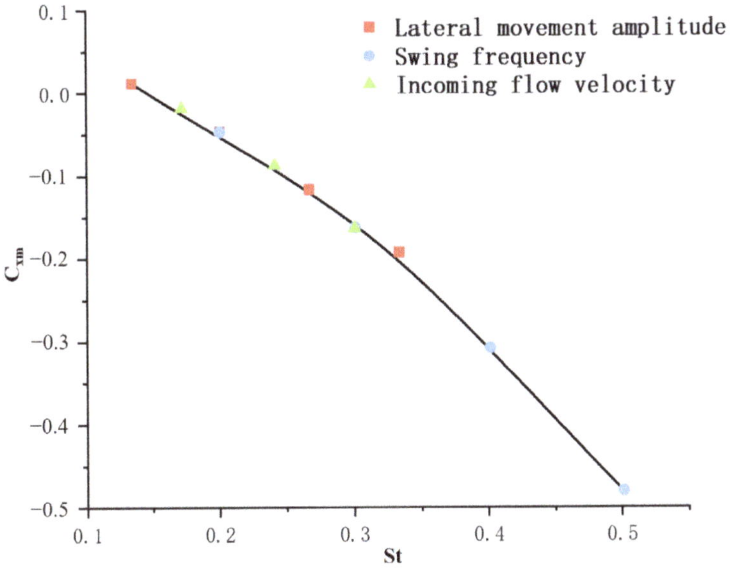

Figure 18. Calculated results of Cxm with different St. The data is sourced from [103].

Table 5. Parameters for calculation. The data is sourced from [103].

Different Incoming Flow Velocity Operating Conditions					Operating Conditions with Different Swing Frequencies					Working Conditions with Different Lateral Movement Amplitudes				
Serial Number	V/ $(m{\cdot}s^{-1})$	A/C_0	f/Hz	St	Serial Number	V/ $(m{\cdot}s^{-1})$	A/C_0	f/Hz	St	Serial Number	V/ $(m{\cdot}s^{-1})$	A/C_0	f/Hz	St
1	0.8	0.6	0.5	0.30	5	1.2	0.6	0.25	0.10	9	1.2	0.4	0.5	0.13
2	1.0	0.6	0.5	0.24	6	1.2	0.6	0.5	0.20	10	1.2	0.6	0.5	0.20
3	1.2	0.6	0.5	0.20	7	1.2	0.6	0.75	0.30	11	1.2	0.8	0.5	0.27
4	1.4	0.6	0.5	0.17	8	1.2	0.6	1	0.40	12	1.2	1	0.5	0.33

4.2. Self-Driving Principle of Bionic Flexible TENG Sensors

Based on the characteristics of the vortex energy field induced by fish tail wagging movement [103–106], the development of self-driven sensors that can capture micro-energy from the vortex field and monitor the state of fish schools is an important research direction in this field. The bionic flexible TENG sensor is designed by imitating the structure and function of biological organs. It features high flexibility, high sensitivity, self-driving capabilities, and other advantages and is suitable for complex Marine environments.

From the current research status of technical realization and application implementation, there are already many research achievements that have provided key support for the application of bionic flexible TENG sensors in deep-sea cage aquaculture scenarios. In terms of the integration of bionic structural design and underwater perception functions, the bionic self-driven triboelectric tactile sensor proposed by Peng Xu et al. [107] achieves high-precision perception of the underwater environment by simulating the structural characteristics of biological perception organs. The self-driven characteristics and high-sensitivity design of this sensor can be directly used for real-time capture of the swimming trajectories of fish in deep-sea cages. It provides an accurate perception tool for analyzing the distribution of vortex energy fields induced by fish tail wagging movement. In response to the strong corrosiveness and sealing protection requirements of deep-sea environments, Zhang Yanrong et al. [108] developed the sealed bionic fishtail structure TENG, which is more adaptable. This sensor takes the bionic fishtail as its core structure and combines anti-corrosion coating technology. While ensuring structural stability, it can efficiently capture mechanical energy during underwater movement (such as the vortex energy generated by fish tail wagging) and convert it into electrical energy. Its energy capture mechanism is highly consistent with the research direction of "vortex field micro-energy capture" focused on in this paper. It can be used as the core component of self-driven monitoring equipment in deep-sea cages to continuously monitor the activity status of fish and environmental parameters. From the perspective of a systematic review of technical application scenarios, Li Daoliang et al. [109] pointed out that TENG technology, with its self-driven and low-power consumption characteristics, has become one of the core technologies for the wearable monitoring of fish and in situ monitoring of aquaculture environments. This review further verifies the feasibility of bionic flexible TENG sensors in fishery scenarios, as shown in Figure 19. The "sensor-energy capture integration" design concept summarized in it also provides theoretical and application references for the development of the "vortex energy capture—fish population status monitoring" dual-function sensor studied in this paper. It is conducive to promoting the transformation of technology from laboratory research to practical application in deep-sea cage aquaculture.

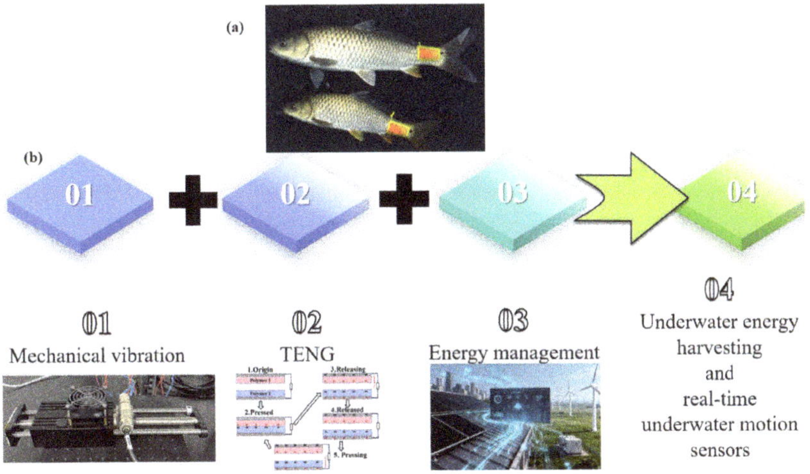

Figure 19. (**a**) Schematic diagram of sensor wearing, (**b**) flowchart of sensor system operation.

4.3. Multi-Physical Field Coupling Micro-Energy Management System

During the interaction between fish schools and net cage flow fields, the micro-energy generated by the tail wagging movement of fish involves various physical fields such as flow fields, electric fields, and mechanical fields [110]. The coupling effect among these physical fields makes the process of energy conversion and transfer more complex. Therefore, the development of a multi-physics field coupled micro-energy management system is the key to achieving the effective collection, storage, and utilization of micro-energy.

The main objective of this chapter is to establish a multi-physics field coupled micro-energy management system to simulate the hydrodynamic characteristics and motion behavior of deep-sea cages and to provide a new numerical method for the design and optimization of deep-sea cages.

The specific research methods are as follows:

1. CFD model establishment:

Utilize the CFD model to calculate the flow field around the cage. The governing equation of fluid flow is the Navier–Stokes equation, and the k-ε turbulence model is adopted to seal this equation. The cage is simplified as a porous medium, and the resistance of the cage to water flow is simulated through the porous medium model.

2. DEM model establishment:

Utilize the DEM model to simulate the movement of the cage structure. The cage structure is discretized into a series of discrete units, and the contact force between the units is calculated through the Hertz–Mindlin contact model. The explicit central difference method is adopted to solve the motion equation of the element.

3. Coupling of CFD and DEM models:

Coupling between CFD and DEM models is achieved through the exchange of force and displacement. The force exerted by the fluid on the cage structure is calculated by the CFD model and passed to the DEM model. The displacement of the cage structure calculated by the DEM model is then passed to the CFD model to update the position of the porous medium.

4. Model verification:

By comparing the simulation results with the experimental data, the established CFD-DEM coupling model is verified. The experimental data are derived from the literature or self-designed experiments.

5. Case study:

Conduct a case study on deep-sea cages using the validated CFD-DEM coupling model. Analyze the influence of different factors such as flow velocity, cage structure, and mooring system on the hydrodynamic characteristics and movement behavior of the cage.

4.3.1. CFD Model

The governing equations of fluid flow in the CFD model are the Navier–Stokes equations, which can be expressed in the tensor form as follows:

$$\frac{\partial \rho}{\partial t} + \frac{\partial(\rho u_i)}{\partial x_i} = 0 \tag{19}$$

$$\frac{\partial(\rho u_i)}{\partial t} + \frac{\partial(\rho u_i u_j)}{\partial x_j} = -\frac{\partial p}{\partial x_i} + \frac{\partial}{\partial x_j}\left(\mu\left(\frac{\partial u_i}{\partial x_j} + \frac{\partial u_j}{\partial x_i}\right)\right) + S_i \tag{20}$$

where ρ is the fluid density, t is time, u_i and u_j are the velocity components in the x_i and x_j directions, respectively, p is the pressure, μ is the dynamic viscosity, and S_i is the source term.

In this study, the k-ε turbulence model is used to close the Navier–Stokes equations. The transport equations of the turbulent kinetic energy k and the dissipation rate ε are as follows:

$$\frac{\partial(\rho k)}{\partial t} + \frac{\partial(\rho u_i k)}{\partial x_i} = \frac{\partial}{\partial x_j}\left(\frac{\mu_t}{\sigma_k}\frac{\partial k}{\partial x_j}\right) + G_k - \rho\varepsilon \tag{21}$$

$$\frac{\partial(\rho \varepsilon)}{\partial t} + \frac{\partial(\rho u_i \varepsilon)}{\partial x_i} = \frac{\partial}{\partial x_j}\left(\frac{\mu_t}{\sigma_\varepsilon}\frac{\partial \varepsilon}{\partial x_j}\right) + C_{1\varepsilon}\frac{\varepsilon}{k}G_k - C_{2\varepsilon}\rho\frac{\varepsilon^2}{k} \tag{22}$$

where μ_t is the turbulent viscosity, σ_k and σ_ε are the turbulent Prandtl numbers for k and ε, respectively, G_k is the generation of turbulent kinetic energy due to mean velocity gradients, and $C_{1\varepsilon}$ and $C_{2\varepsilon}$ are model constants.

The deep-sea cage is simplified as a porous medium in the CFD model. The porous medium model is used to simulate the resistance of the cage to the water flow. The momentum equation of the porous medium is modified as follows:

$$\frac{\partial(\rho u_i)}{\partial t} + \frac{\partial(\rho u_i u_j)}{\partial x_j} = -\frac{\partial p}{\partial x_i} + \frac{\partial}{\partial x_j}\left(\mu\left(\frac{\partial u_i}{\partial x_j} + \frac{\partial u_j}{\partial x_i}\right)\right) - \frac{\mu}{K}u_i - C_2\frac{\rho}{2}|u|u_i + S_i \tag{23}$$

where K is the permeability of the porous medium, C_2 is the inertial resistance coefficient, and $|u| = \sqrt{u_1^2 + u_2^2 + u_3^2}$.

The permeability K and inertial resistance coefficient C_2 of porous media can be determined by experimental data or empirical formulas. This study employs the empirical formula used by E. C. Clukey et al. [111] to calculate the permeability and inertial resistance coefficient of deep-sea cages:

$$K = \frac{d^2}{150}\frac{\varepsilon^3}{(1-\varepsilon)^2} \tag{24}$$

$$C_2 = \frac{3.5}{d}\frac{1-\varepsilon}{\varepsilon^3} \tag{25}$$

where d is the diameter of the cage wire, and ε is the porosity of the cage.

4.3.2. DEM Model

The cage structure is discretized into a series of discrete elements in the DEM model. In this study, the cage structure is composed of rigid rods and flexible nets. The rigid rods are discretized into beam elements, and the flexible nets are discretized into truss elements.

The beam element is used to simulate the rigid rod, and its motion equation can be expressed as follows:

$$M\ddot{u} + C\dot{u} + Ku = F \tag{26}$$

where M is the mass matrix, C is the damping matrix, K is the stiffness matrix, u is the displacement vector, $\dot{u}$ is the velocity vector, $\ddot{u}$ is the acceleration vector, and F is the force vector.

The truss element is used to simulate the flexible net, and its motion equation is also expressed by Formula (26).

The contact force between elements is calculated by the Hertz–Mindlin contact model in the DEM model. The Hertz–Mindlin contact model considers the normal and tangential contact forces between elements and has been widely used in the simulation of granular materials.

The normal contact force between two elements can be calculated as follows:

$$F_n = \frac{4}{3}E^*\sqrt{R^*}\delta_n^{\frac{3}{2}} \tag{27}$$

where E^* is the equivalent Young's modulus, R^* is the equivalent radius, and δ_n is the normal overlap between the two elements.

The tangential contact force between two elements can be calculated as follows:

$$F_t = -G^*\sqrt{R^*}\delta_t \tag{28}$$

where G^* is the equivalent shear modulus, and δ_t is the tangential displacement between the two elements.

The motion equations of the elements are solved by the explicit central difference method in the DEM model. The explicit central difference method is a simple and efficient numerical integration method, which has been widely used in the simulation of dynamic systems.

The acceleration, velocity, and displacement of the elements at time $t + \Delta t$ can be calculated as follows:

$$\begin{cases} \ddot{u}_{t+\Delta t} = \dfrac{F_{t+\Delta\frac{t}{2}} - C\dot{u}_{t+\Delta\frac{t}{2}} - Ku_{t+\Delta\frac{t}{2}}}{M} \\ \dot{u}_{t+\Delta t} = \dot{u}_t + \dfrac{1}{2}(\ddot{u}_t + \ddot{u}_{t+\Delta t})\Delta t \\ u_{t+\Delta t} = u_t + \dot{u}_t\Delta t + \dfrac{1}{2}\ddot{u}_t(\Delta t)^2 \end{cases} \tag{29}$$

where Δt is the time step.

4.3.3. Coupling of CFD and DEM Models

Coupling between the CFD and DEM models is achieved through the exchange of forces and displacements. The force exerted by the fluid on the cage structure is calculated by the CFD model and transferred to the DEM model, and the displacement of the cage structure calculated by the DEM model is transferred to the CFD model to update the position of the porous medium.

The force exerted by the fluid on the cage structure can be calculated by integrating the pressure and shear stress on the surface of the cage structure:

$$F_f = \int_S (p\vec{n} + \tau\vec{t})dS \tag{30}$$

where F_f is the force exerted by the fluid on the cage structure, S is the surface area of the cage structure, $\vec{n}$ is the unit normal vector of the surface, $\vec{t}$ is the unit tangent vector of the surface, p is the pressure, and τ is the shear stress.

The displacement of the cage structure calculated by the DEM model is transferred to the CFD model to update the position of the porous medium. The updated position of the porous medium is used to recalculate the permeability and inertial resistance coefficient of the porous medium, and then the flow field around the cage structure is recalculated by the CFD model.

4.4. Model Verification and Convergence Analysis

To ensure the accuracy and reliability of the CFD-DEM coupling model established in this paper, the grid sensitivity analysis and the verification of the convergence of time steps were carried out successively in this section, and the numerical simulation results were compared with the published experimental data.

4.4.1. Grid Sensitivity Analysis

To determine the optimal grid scheme, three groups of structured grids of different sizes were designed (as shown in Table 6). The core computing domain (cage area) was encrypted with grids, while the non-core domain (five times the diameter of the cage) adopted a gradient transition. The influence of the grid size on the results was evaluated by taking the total resistance of the net coating and the average flow velocity inside the net box as indicators.

Table 6. Comparison of calculation results from different grid schemes.

Grid Scheme	Core Domain Size	Total Resistance (N)	Internal Flow Velocity (m/s)	Fine Grid Error
Coarse grid	5 cm	185	0.28	12.3%
Middle grid	3 cm	205	0.32	4.7%
Fine grid	1 cm	215	0.33	0%

The results show that when the grid size is reduced from 5 cm (thick grid) to 3 cm (medium grid), the calculated resistance value changes by 10.8% and the flow velocity changes by 11.5%. When further reduced to 1 cm (fine grid), the resistance change was only 4.7%, and the flow velocity change was 3.1%. According to the grid independence determination criterion (change rate $\leq$ 5%), the middle grid (3 cm of the cage area) is selected as the benchmark scheme, taking into account both calculation accuracy and efficiency.

4.4.2. Verification of Convergence of Time Steps

The selection of time steps is crucial for the stability of explicit algorithms. This paper establishes three different time steps: 0.01 s, 0.005 s (reference), and 0.001 s. The total kinetic energy of the system and the tension of the mooring line are taken as the convergence criteria. The results show that when the time step is reduced from 0.005 s to 0.001 s, the variation range of the key physical quantity is less than 1%. Therefore, choosing 0.005 s as the calculation time step can ensure the accuracy of the results while achieving efficient calculation.

4.4.3. Verification by Comparison with Experimental Data

To further verify the model, we compared the simulation results with the deformation experimental data of HDPE circular gravity cages under constant flow reported in Huang [79]. Under the same flow velocity (0.4 m/s) and cage structure parameters, the maximum horizontal displacement of the net garment calculated by the model in this paper is 2.85 m. Compared with the experimental value of 2.78 m in the literature, the relative error is 2.5%. The overall deformation contour of the net garment is also consistent with the height observed in the experiment.

The verification work of the above series has fully demonstrated that the CFD-DEM coupling model established in this paper has sufficient accuracy and reliability in predicting the hydrodynamic response and structural deformation of deep-water cages, laying a solid numerical foundation for subsequent case studies and analyses.

5. Conclusions

5.1. Limitations

5.1.1. Limitations in Multi-Field Coupling Dynamic Mechanism Research

The multi-field coupling dynamic model constructed in this study focuses on the "net jacket-fluid-structure coupling—multi-body system energy transfer—fish population flow field" three-level mechanism, but there are still deficiencies in the integration of spatiotemporal scales and the characterization of nonlinear factors. In terms of scale integration, the current model separately describes the microscale porous media seepage behavior of the net jacket (based on the Brinkman–Darcy equation) and the macroscale mooring system dynamics (based on Catenary Theory) but lacks a unified scale transition framework. This leads to inaccuracies in the simulation of energy transfer between microscale vortex shedding and macroscale structural vibration, especially when the net jacket undergoes large nonlinear deformation (such as local folding with curvature changes greater than 0.5 m^{-1}).

In the characterization of nonlinear factors, the model considers the viscoelasticity of the net material (fractional-order Burgers model) and fluid–structure coupling nonlinearity but ignores the time-varying characteristics of biological factors. For example, the dynamic feedback of fish school density and swimming speed on the flow field is treated as a static boundary condition, failing to capture the real-time interaction between the "fish-induced vortex field" and the "net structure vibration field". Experimental verification also has limitations: the wave–current combined action test system adopts regular waves and steady currents, which cannot fully replicate the randomness of extreme sea conditions (such as freak waves with a wave steepness exceeding 0.14) and the unsteadiness of upwelling currents, resulting in insufficient verification of the model's reliability under extreme loads.

5.1.2. Deficiencies in Self-Powered Sensing System Development

The curved surface adaptive TENG array designed for the net jacket structure still faces bottlenecks in energy conversion efficiency and long-term reliability. In terms of energy conversion, the current design achieves a low-frequency energy conversion rate of approximately 15–20% under laboratory conditions, but in actual sea trials, the conversion efficiency drops by more than 40%. This is mainly due to the failure of the array to adapt to the multimodal vibration characteristics of the net jacket—under the combined action of wind, waves, and the current, the net jacket exhibits both low-frequency (0.1–1 Hz) overall swing and high-frequency (5–10 Hz) local vibration, while the current TENG unit only optimizes the contact–separation mode for low-frequency motion, resulting in inefficient utilization of high-frequency vibration energy [48]. The existing TENG arrays face three core limitations in Marine environments: Firstly, the NaCl solution leads to charge dissipation in the friction layer and electrode corrosion, resulting in an output decline of nearly 50% within three months [112]. Second, traditional packaging cannot withstand the water pressure above 1000 m and is prone to cracking and failure [113]. Thirdly, in a humid environment, the leakage current increases, and the output stability drops by more than 30%, making it difficult to meet the one-year maintenance-free requirement of deep-sea cages [114].

In terms of long-term reliability, although the PDMS/AgNWs composite film coated with PTFE has passed 500,000 cycles of durability testing in simulated seawater [109], the actual Marine environment contains suspended sediments (particle size 1–100 μm) and biofouling organisms (such as barnacles and algae) [53]. The abrasion of sediments on the friction layer and the adhesion of biofouling lead to a 30–50% decrease in the output voltage of the TENG array after 3 months of immersion, which is far from meeting the 1-year maintenance-free requirement of deep-sea aquaculture equipment. In addition, the energy management module of the sensing system lacks adaptive adjustment capabilities—when the vibration amplitude of the net jacket is less than 5 mm, the output power of the TENG is insufficient to drive the wireless transmission module, resulting in data loss.

5.1.3. Gaps in Fish–Cage Flow Field Interactions and Micro-Energy Capture

The research on the vortex energy field induced by fish tail swing mainly focuses on single fish or small-scale fish schools ($\leq$50 individuals), and there is a lack of exploration of the "collective vortex effect" of large-scale fish schools ($\geq$1000 individuals) in deep-sea cages. Experimental observations show that large-scale fish schools form a "rotational flow field" with a tangential velocity of 0.2–0.5 m/s inside the cage, which superimposes with the external wave–current flow field to generate complex turbulent structures. However, the current numerical simulation cannot accurately characterize the energy distribution of this superimposed flow field, leading to errors in the design of micro-energy capture points.

For the bionic flexible TENG sensor, its sensitivity to low-amplitude vortex signals (pressure fluctuation amplitude < 1 Pa) is insufficient. The sensor's current detection limit is 2 Pa, which cannot capture the weak vortex disturbance generated by juvenile fish (body length < 10 cm) tail swing. In addition, the multi-physics field coupling micro-energy management system based on CFD-DEM coupling only realizes the one-way transmission of force and displacement between the flow field and the structure and does not consider the reverse influence of the TENG array's energy capture behavior on the net jacket vibration—when the TENG unit captures energy, it generates additional damping (approximately 0.05 N·s/m), which may change the natural frequency of the net jacket and affect the structural stability assessment results.

5.2. Prospects

5.2.1. Optimization of Multi-Field Coupling Dynamic Model

In the future, researchers can build an integrated multi-scale coupling model of "micro-meso-macro" based on homogenization methods. At the microscopic scale, the lattice Boltzmann method is used to simulate the infiltration behavior of seawater in the mesh (pore diameter 2–5 cm) and the interaction between suspended particles and the mesh surface. At the mesoscale scale, the deformation and contact force transfer of the net nodes will be characterized by the discrete element method. At the macroscale, the finite volume method will be used to calculate the overall dynamic response of the cage mooring system. The three scales are coupled by the volume averaging method to achieve a seamless connection of energy transfer between different scales.

In terms of biological factor integration, a dynamic fish school model based on the Reynolds-Averaged Navier–Stokes (RANS) equation was established. The model will consider the collective behavior of fish schools (such as aggregation and dispersion) and the time-varying characteristics of tail swing frequency (1–5 Hz) and amplitude (2–10 cm) and couple it with the net sleeve flow field model to realize the real-time simulation of the "fish-structure-fluid" interaction. In addition, a simulation platform can be built to generate strange waves and irregular upwelling, and the model can be verified by measuring the net sheath deformation and mooring tension under extreme loads so as to improve the applicability of the model under complex sea conditions.

5.2.2. Upgrade of Self-Powered Sensing System

Researchers can develop a multimodal adaptive TENG array to achieve the efficient capture of low- and high-frequency vibrational energy. The array will have a "double-layer structure": the bottom layer will use a serpentine electrode-based contact–separation mode TENG to capture low-frequency oscillation energy, and the top layer will integrate a piezoelectric–TENG composite unit to collect high-frequency local vibration energy. The two layers will share the same PDMS/AgNWs substrate, and the output energy will be integrated through a multi-port rectifier circuit; the overall energy conversion efficiency is expected to increase to more than 35%.

For long-term reliability, a "self-cleaning-antifouling" composite coating can be developed. Modified titanium dioxide nanoparticles were added to achieve photocatalytic self-cleaning under low-light conditions in deep-sea low-light conditions. At the same time, the surface will be modified with fluorosilane to reduce the adhesion of biofouling organisms, extending the service life of the TENG array to more than 18 months. In addition, an adaptive energy management module based on a boost converter with maximum power point tracking can be designed. When the vibration amplitude is less than 5 mm, the module stores the energy in the supercapacitor and triggers data transmission in batches; when the amplitude is greater than 5 mm, real-time data transmission will be realized to avoid data loss. Not only that, considering the limitations discussed above, the Marine adaptability of TENG needs to be further optimized in the future: First, develop an integrated friction layer material of "superhydrophobic—anti-corrosion", such as PFOTS-modified PTFE/Al_2O_3 composite film, with the goal of extending the resistance to seawater corrosion life to more than two years; the second is to develop lightweight and high-voltage-resistant packaging, achieving no leakage for two years at a water depth of more than 1000 m; and thirdly, it integrates an intelligent humidity control module, which monitors the internal humidity in real time through micro-sensors and triggers the dehumidification function to ensure the stability of output throughout the entire life cycle.

5.2.3. Deepening of Fish–Cage Flow Field Interaction and Micro-Energy Capture Research

Researchers can expand the study of eddy energy fields in large schools of fish. A large circulating water tank was used to simulate the collective movement of fish, and the three-dimensional distribution of the eddy current energy field was captured by particle image velocimetry (PIV) technology with a spatial resolution of 0.1 mm and a temporal resolution of 1000 fps. A numerical model based on the large eddy simulation (LES) method was established to reveal the energy transfer mechanism between the collective vortex of fish schools and the vibration of the net sleeve and to determine the optimal layout of micro-energy capture points.

By imitating the lateral line structure of fish, a highly sensitive biomimetic flexible TENG sensor was developed. The sensor will use a hollow PDMS microchannel filled with ionic liquid as a sensitive unit to improve sensitivity by optimizing electrode spacing and ionic liquid concentration. The sensor will be integrated with the TENG array to realize the simultaneous monitoring of fish activity status and net structure status. In addition, a bidirectionally coupled CFD-DEM-TENG model can be established to consider the damping effect of energy capture on net sheath vibration, and the model can be used to optimize the structural parameters of the TENG element, such as mass and stiffness, to minimize the impact on the structural stability evaluation.

5.2.4. Broader Application Prospects

The theoretical system and technical achievements of this study can be extended to other deep-sea equipment, such as deep-sea observatories and underwater robots. The multi-field coupling dynamic model can provide a reference for the structural design of flexible components (such as observation platform mooring systems and robot manipulators) under complex Marine environments. The self-powered sensing system based on TENG can be adapted to the energy supply needs of underwater sensor networks in deep-sea observatories, reducing reliance on chemical batteries. In addition, the research on fish–cage flow field interactions can provide theoretical support for the design of eco-friendly deep-sea aquaculture systems, optimizing the flow field inside the cage to improve aquaculture efficiency and reduce environmental impacts.

In the long term, the integration of "structural health monitoring—aquaculture environment perception—energy self-supply" will be realized. By combining the multi-field coupling dynamic model, self-powered sensing system, and fish school behavior monitoring technology, an intelligent management platform for deep-sea aquaculture will be built. The platform can predict structural failures of cages 24–48 h in advance, adjust aquaculture density based on flow field changes, and achieve fully autonomous operation of deep-sea aquaculture, providing key technical support for the construction of the "Blue Granary" and the high-quality development of the Marine economy.

Author Contributions: Conceptualization, K.Y. (Kefan Yang) and D.Z.; methodology, K.Y. (Kefan Yang); software, S.Z.; validation, K.Y. (Keqi Yang) and Y.Z.; formal analysis, K.Y. (Kefan Yang); investigation, D.Z.; resources, D.Z.; data curation, K.Y. (Keqi Yang); writing—original draft preparation, K.Y. (Kefan Yang); writing—review and editing, D.Z.; visualization, D.Z.; supervision, D.Z.; project administration, D.Z.; funding acquisition, D.Z. All authors have read and agreed to the published version of the manuscript.

Funding: This study is supported by the Fund of Guangdong Provincial Key Laboratory of Intelligent Equipment for South China Sea Marine Ranching (Grant No. 2023B1212030003).

Data Availability Statement: Data is contained within the article.

Conflicts of Interest: The authors declare no conflict of interest.

References

1. Yan, G.; Ni, X.; Mo, J. Research status and development trend of profound Marine aquaculture equipment technology. *J. Dalian Ocean. Univ.* **2018**, *33*, 123–129. (In Chinese) [CrossRef]
2. Chen, Y.; Fang, R.; Qiao, J. China's Marine aquaculture development research. *J. Agric. Econ.* **2012**, 72–77+112. (In Chinese) [CrossRef]
3. Zhu, Y.; Ju, X.; Chen, Y. Current Situation, Problems and Countermeasures of Deep-sea Cage Aquaculture in China. *China Fish. Econ.* **2017**, *35*, 72–78. (In Chinese)
4. Hou, H.; Ju, X.; Chen, Y. The development trends of Deep-sea cage aquaculture abroad and its implications for China. *World Agric.* **2017**, 162–166. (In Chinese) [CrossRef]
5. Song, R.; Zhou, M.; Li, Y.; Zhang, H.; Lin, Z. Research Progress on Key Technologies of Deep-sea Cage Aquaculture Equipment. *Mech. Eng.* **2015**, 134–138. (In Chinese)
6. Hou, Y.; Li, Y.; Li, G.; Tong, X.; Wang, Y. Oil-spill detection sensor using ultraviolet-induced fluorescence for routine surveillance in coastal environments. *Appl. Phys. B* **2022**, *128*, 41. [CrossRef]
7. Gonzaga, B.A.; Alves, D.L.; Albuquerque, M.D.G.; Espinoza, J.M.D.A.; Almeida, L.P.; Weschenfelder, J. Development of a Low-cost Ultrasonic Sensor for Groundwater Monitoring in Coastal Environments: Validation using Field and Laboratory Observations. *J. Coast. Res.* **2020**, *95* (Suppl. 1), 1001–1005. [CrossRef]

3. Xie, Y. Based on the Internet of Things of Marine Water Quality Wise Observation Solution Research. Master's Thesis, Shanghai Ocean University, Shanghai, China, 2020. (In Chinese)
). Diaz, K.; Teh, Y. Design and Power Management of a Secured Wireless Sensor System for Salton Sea Environmental Monitoring. *Electronics* **2020**, *9*, 544. [CrossRef]
10. Zhang, Y. Review on the Development of Marine Environment Monitoring Equipment Technology. *Digit. Ocean. Underw. Attack Def.* **2018**, *1*, 7–14. (In Chinese)
11. Rodrigues, C.; Nunes, D.; Clemente, D.; Mathias, N.; Correia, J.M.; Rosa-Santos, P.; Taveira-Pinto, F.; Morais, T.; Pereira, A.; Ventura, J. Emerging triboelectric nanogenerators for ocean wave energy harvesting: State of the art and future perspectives. *Energy Environ. Sci.* **2020**, *13*, 2657–2683. [CrossRef]
12. Liu, L.; Hu, T.; Zhao, X.; Lee, C. Recent progress in blue energy harvesting based on triboelectric nanogenerators. *Nanoenergy Adv.* **2024**, *4*, 156–173. [CrossRef]
13. Jiang, Y.; Liang, X.; Jiang, T.; Wang, Z.L. Advances in triboelectric nanogenerators for blue energy harvesting and marine environmental monitoring. *Engineering* **2024**, *33*, 204–224. [CrossRef]
14. Isabel, G.; Rodrigues, C.; Ventura, J. Sea state adaptation enhances power output of triboelectric nanogenerators for tailored ocean wave energy harvesting. *Adv. Energy Mater.* **2024**, *14*, 2302627.
15. Li, X.; Xu, L.; Wang, Z.L. Networking strategies of triboelectric nanogenerators for harvesting ocean blue energy. *Nanoenergy Adv.* **2024**, *4*, 70–96. [CrossRef]
16. Tan, D.; Chi, S.; Ou, X.; Zhou, J.; Wang, K. Some Advances in nonlinear frictional nanopower generation Energy capture technology. *Acta Mech. Sin.* **2024**, *56*, 2495–2510. (In Chinese)
17. Gao, J.; Zhao, B.; Yang, C.; Rui, H.Y.; Gang, X.H. Development Trends and Prospects of Offshore Energy Storage Technology. *Adv. New Renew. Energy* **2020**, *8*. (In Chinese)
18. Wang, Y. Research on Friction Nano-Electromagnetic Composite Flow Energy Capture and Energy Supply for Observation Nodes. Ph.D. Thesis, Dalian Maritime University, Dalian, China, 2023. (In Chinese)
19. Dong, J. Research on Self-Powering Technology of Marine Buoys Based on Stackable Triboelectric Nanogenerators. Ph.D. Thesis, Dalian Maritime University, Dalian, China, 2021. (In Chinese)
20. Lin, Z.W. Triboelectric nanogenerators as new energy technology for self-powered systems and as active mechanical and chemical sensors. *ACS Nano* **2013**, *7*, 9533–9557. [CrossRef]
21. Li, P.; Tian, Y.; Qin, H.; Wei, B.; Deng, Z. Analysis of Flow Field Characteristics and Deformation of Nylon Mesh Based on Fluid-Structure Coupling. *China Offshore Platf.* **2023**, *38*, 48–56. (In Chinese)
22. Ma, J. Energy Transfer Analysis of Multi-Body Systems. Ph.D. Thesis, Dalian University of Technology, Dalian, China, 2008. (In Chinese)
23. Zhang, C. Fish Flow Field Interference Mechanism Research. Ph.D. Thesis, Nanjing University of Aeronautics and Astronautics, Nanjing, China, 2022. (In Chinese)
24. Lin, X. Research on the Hydrodynamic Mechanism of Fish and Bird Migration. Ph.D. Thesis, Nanchang Hangkong University, Nanchang, China, 2017. (In Chinese)
25. Tang, M. Flow Wave Net Structure Hydrodynamic Characteristics Research. Ph.D. Thesis, Dalian University of Technology, Dalian, China, 2020. (In Chinese)
26. Zhang, K.; Liu, B.; Chen, H.; Chen, J.; Yao, H.D. Solving fluid-structure interaction of flexible net structures in current and waves based on a coupled truss–porous medium model. *Ocean. Eng.* **2025**, *329*, 121115. [CrossRef]
27. Schäfer, F.; Kniesburges, S.; Uffinger, T.; Becker, S.; Grabinger, J.; Link, G.; Kaltenbacher, M. Numerical simulation of fluid-structure-and fluid-structure-acoustic interaction based on a partitioned coupling scheme. In *High Performance Computing in Science and Engineering, Garching/Munich 2007: Transactions of the Third Joint HLRB and KONWIHR Status and Result Workshop, Leibniz Supercomputing Centre, Garching/Munich, Germany 3–4 December 2007*; Springer: Berlin/Heidelberg, Germany, 2009.
28. Wang, M.; Li, G.; Xia, C.; Jing, X.; Wang, R.; Liu, Q.; Cai, X. Facile preparation of cyclodextrin polymer materials with rigid spherical structure and flexible network for sorption of organic contaminants in water. *Chem. Eng. J.* **2021**, *411*, 128489. [CrossRef]
29. Rosén, T.; Mittal, N.; Roth, S.V.; Zhang, P.; Lundell, F.; Söderberg, L.D. Flow fields control nanostructural organization in semiflexible networks. *Soft Matter* **2020**, *16*, 5439–5449. [CrossRef]
30. Hu, Z. Numerical Simulation Research on the Structural Safety of Net Jackets in Large Deep-Sea and Far-Sea Cages. Ph.D. Thesis, Shandong University, Jinan, China, 2024. (In Chinese)
31. Hu, B.; Yang, X. Current Situation and Development Trend of Deep-sea Aquaculture Cages at Home and Abroad. *Silicon Val.* **2008**, 28–29. (In Chinese)
32. Pan, Y.; Zhang, J.; Li, L.; Liu, C.; Sun, Z.; Hong, J. Numerical simulation of hydrodynamic characteristics of ultra-large diameter aquaculture cages under waves action. *Ocean. Eng.* **2025**, *338*, 121981. [CrossRef]

33. Zhang, D.; Zhao, B.; Jiang, H. Wave energy converter with floating-point absorber and catenary mooring: Dynamic coupling analysis. *Front. Mar. Sci.* **2024**, *11*, 1338330. [CrossRef]
34. Ma, Y.; Si, K.; Xie, Y.; Liang, Z.; Wu, J.; Zhang, D.; Zhang, Y.; Cai, R. Global Marine Ranching Research: Progress and Trends through Bibliometric Analysis. *Eng. Solut. Mech. Mar. Struct. Infrastruct.* **2024**, *1*, 1–23. [CrossRef]
35. Zhang, D.; Zhao, B.; Zhang, Y.; Zhu, K.; Yan, J. Study on Dynamic Characteristics and Fracture Failure of Rigid Truss Trawl System During Towing Process. *J. Mar. Sci. Eng.* **2025**, *13*, 586. [CrossRef]
36. Gui, F.; Zhu, H.; Feng, D. Research Progress on Hydrodynamic Characteristics of Marine Aquaculture Netting. *Fish. Mod.* **2019**, *46*, 9–14. (In Chinese)
37. Li, N.; Li, X.; Shi, W.; Liu, C. Numerical Simulation of Hydrodynamic Characteristics of Planar Mesh under Wave-Current Action. *J. Shanghai Ocean. Univ.* **2024**, *33*, 509–520. (In Chinese)
38. Wu, H.; Liu, Q.; Fan, W. Research on Hydrodynamic Loads of Cage Nets and Net Jackets in Aquaculture. *Mod. Fish.* **2023**, *50*, 43–51. (In Chinese)
39. Zhang, Y.M.; Bi, C.W.; Feng, L.; Huang, L.Y. Hydrodynamic response analysis of the aquaculture net under the combined action of irregular waves and currents. *Ships Offshore Struct.* **2024**, 1–17. [CrossRef]
40. Ling, A.; Liang, Y.; Zhao, D.; Fan, Z.; Zhao, Y. Marine aquaculture hydrodynamic characteristics is reviewed. The flexible net structure. *Fish. Sci. Prog.* **2024**, *45*, 223–236. (In Chinese) [CrossRef]
41. He, W.; Liu, W.; Fu, S.; Wu, H.; Shan, C.; Wang, Z.; Xi, Y.; Wang, X.; Guo, H.; Liu, H.; et al. Ultrahigh Performance Triboelectric Nanogenerator Enabled by Charge Transmission in Interfacial Lubrication and Potential Decentralization Design. *Research* **2022**, *2022*, 9812865. [CrossRef]
42. Zhang, C.; He, L.; Zhou, L.; Yang, O.; Yuan, W.; Wei, X.; Liu, Y.; Lu, L.; Wang, J.; Wang, Z.L. Active resonance triboelectric nanogenerator for harvesting omnidirectional water-wave energy. *Joule* **2021**, *5*, 1613–1623. [CrossRef]
43. Li, X.; Zhang, D.; Zhang, D.; Li, Z.; Wu, H.; Zhou, Y.; Wang, B.; Guo, H.; Peng, Y. Solid-liquid triboelectric nanogenerator based on vortex-induced resonance. *Nanomaterials* **2023**, *13*, 1036. [CrossRef]
44. Zhang, Q.; Li, L.; Wang, T.; Jiang, Y.; Tian, Y.; Jin, T.; Yue, T.; Lee, C. Self-sustainable flow-velocity detection via electromagnetic/triboelectric hybrid generator aiming at IoT-based environment monitoring. *Nano Energy* **2021**, *90*, 106501. [CrossRef]
45. Wang, S.; Zhang, X.; Ma, T.; Kong, Y.; Gao, S.; Han, Q. Symmetrical Triboelectric In Situ Self-Powered Sensing and Fault Diagnosis for Double-Row Tapered Roller Bearings in Wind Turbines: An Integrated and Real-Time Approach. *Adv. Sci.* **2025**, *12*, 2500981. [CrossRef] [PubMed]
46. Xu, L.; Wu, J.; Zhang, K.; Lu, J.; Li, J.; Hazarika, D.; Jin, H.; Luo, J. An instantaneous self-powered wireless sensing system based on a triboelectric nanogenerator and the human body. *Nanoscale* **2024**, *16*, 19355–19363. [CrossRef] [PubMed]
47. Wang, W.; Yan, J.; Wang, X.; Pang, H.; Sun, C.; Sun, Y.; Wang, L.; Zhang, D. Research on the Performance of a Liquid–Solid Triboelectric Nanogenerator Prototype Based on Multiphase Liquid. *Micromachines* **2025**, *16*, 78. [CrossRef]
48. Yang, K.F.; Xie, Y.; Si, K.; Yang, K.Q.; Zeng, S.; Zhang, D. Research Progress on Friction Nanogenerator Technology for Wave Energy Harvesting. *Eng. Solut. Mech. Mar. Struct. Infrastruct.* **2024**. [CrossRef]
49. Yan, J.; Sheng, Y.; Zhang, D.; Tang, Z. Research Progress in Fluid Energy Collection Based on Friction Nanogenerators. *Micromachines* **2023**, *15*, 40. [CrossRef]
50. Chen, P.; Zhu, X.; Tian, G.; Feng, X. Recent advances of TENGs for marine applications: Opportunities and challenges coexist. *Nano Energy* **2025**, *142*, 111212. [CrossRef]
51. Li, M.; Jin, S.; Meng, A.; Song, J.; Xu, Z.; Chen, Y.; Zeng, Y. A performance study on the drum-type energy harvesting device based on triboelectric nanogenerator. *J. Electrost.* **2025**, *135*, 104066. [CrossRef]
52. Ji, C. Research on Flexible Acoustic Sensor Based on PDMS-AgNWs. Ph.D. Thesis, Taiyuan University of Technology, Taiyuan, China, 2019. (In Chinese)
53. Deng, X.; Li, D.; Yang, J.; Wang, J. Research on Friction and Wear Properties of High-Performance High-Temperature Resistant Polymer Composites. *Bull. Polym. Sci.* **2008**, 41–46. (In Chinese)
54. Chen, X.; Chen, J.; Huang, L.; Nie, S.; Xu, W.; Yin, Y.; Zhang, S.; Pei, F.; Yu, K.; Su, W.; et al. Highly conductive omnidirectionally stretchable 2D transparent copper mesh electrodes and applications in optoelectronic devices. *Adv. Mater. Technol.* **2023**, *8*, 2201406. [CrossRef]
55. Sun, P.; Wang, L.; Wang, L. Deepwater semi-submersible platform anchoring auxiliary power positioning system power consumption research. *Ocean. Eng.* **2010**, *28*, 24–30. (In Chinese) [CrossRef]
56. Wen, X.; Gjerde, A.; Ong, M.C. Numerical investigation of pre-tension effects in mooring systems on global responses of gravity-type fish farms. *Ocean. Eng.* **2025**, *337*, 121832. [CrossRef]
57. Wang, B.; Wang, H.; Liu, Z.; Cao, Z.; Yang, D.; Wang, F. Research Progress on Structural Design and Performance Analysis of Mooring System for Aquaculture Cages. *J. Shanghai Ocean. Univ.* **2025**, *34*, 176–187. (In Chinese)

58. Liu, F.; Jiang, Z.; Cheng, T.; Xu, Y.; Zhu, H.; Wang, G.; Sun, G.; Zhang, Y. Study on the Dynamic Response of Mooring System of Multiple Fish Cages under the Combined Effects of Waves and Currents. *J. Mar. Sci. Eng.* **2024**, *12*, 1648. [CrossRef]
59. Wang, H.; Gu, S.; Ma, G.; Huang, W.; Sun, Y.; Du, Z. Dynamic performance analysis of grid mooring system for gravity cages. *Ships Offshore Struct.* **2025**, *20*, 1143–1160. [CrossRef]
60. Abdellah Mohammed, Y.; Sadek, M.G.; Alharthi, H.; Abdel-Jaber, G.T.; Backar, A.H. Characteristic properties of date-palm fibre/sheep wool reinforced polyester composites. *J. Bioresour. Bioprod.* **2023**, *8*, 430–443. [CrossRef]
61. Michel, F.; Davies, P. Characterization of polyester mooring lines. In Proceedings of the International Conference on Offshore Mechanics and Arctic Engineering, Estoril, Portugal, 15–20 June 2008; Volume 48180.
62. Liu, X.; Li, Z.; Zhu, D.; Gong, Q. Dynamic viscoelastic properties of polyethylene terephthalate. *Plastics* **2024**, *53*, 137–142. (In Chinese)
63. da Silva, L.S.; Cazzolato, B.S.; Sergiienko, N.Y.; Ding, B.; Morishita, H.M.; Pesce, C.P. Statistical linearization of the Morison's equation applied to wave energy converters. *J. Ocean. Eng. Mar. Energy* **2020**, *6*, 157–169. [CrossRef]
64. Beji, S. Formulation of wave and current forces acting on a body and resistance of ships. *Ocean. Eng.* **2020**, *218*, 108121. [CrossRef]
65. Liu, G.; Wang, Q.; Tian, X.; Xie, Y. The ocean structure small scale research and progress on hydrodynamic coefficients of pile. *Chin. J. Ocean. Univ. (Nat. Sci. Ed.)* **2020**, *50*, 136–144. (In Chinese) [CrossRef]
66. Wang, F.; Zhang, J.; Zhu, Y.; Geng, Z.; Wang, F. Research and Progress on Hydrodynamics of Boat-type Aquaculture cages. *Ships* **2025**, *36*, 27. (In Chinese)
67. Wang, G.; Tao, L.; Rong, B.; Rui, X. Research Progress of Dynamic Simulation Methods for Multi-field Coupled Systems. *Adv. Mech.* **2023**, *53*, 468–495. (In Chinese)
68. Lin, M.; Wu, Q.; Tian, H.; Huang, L.; Huang, G.; Liu, B.; Wang, X.; Wang, G. Tension of mooring ropes and safety of jacket Foundation structure in the integrated system of offshore wind turbines and aquaculture cages. *Acta Fish. Sin.* **2024**, *48*, 107–119. (In Chinese)
69. Kuan, J.; Čanić, S.; Muha, B. Fluid-poroviscoelastic structure interaction problem with nonlinear geometric coupling. *J. Math. Pures Appl.* **2024**, *188*, 345–445. [CrossRef]
70. Bungartz, H.J.; Lindner, F.; Mehl, M.; Uekermann, B. A plug-and-play coupling approach for parallel multi-field simulations. *Comput. Mech.* **2015**, *55*, 1119–1129. [CrossRef]
71. Gong, S.; Gao, J.; Yan, M.; Song, Z.; Shi, H. Effects of floater motion on wave loads during steady-state gap resonance occurring between two non-identical boxes. *Ocean. Eng.* **2025**, *323*, 120649. [CrossRef]
72. Zeng, X.; Shi, W.; Feng, X.; Shao, Y.; Li, X. Investigation of higher-harmonic wave loads and low-frequency resonance response of floating offshore wind turbine under extreme wave groups. *Mar. Struct.* **2023**, *89*, 103401. [CrossRef]
73. Ling, H.; Yang, P.; Zhang, Z.; Wang, Z. Complex wave loads and hydroelastic responses of a very large tourism floating platform 'Ocean Diamond'. *Ocean. Eng.* **2022**, *265*, 112608. [CrossRef]
74. Wang, Z.; Huang, D.; He, H.; Yang, F.; Li, W.; Chen, Y. Dynamic Response and Fatigue Analysis of a New Marine Gravitational Energy Storage System Under Wave Loads. *J. Mar. Sci. Eng.* **2024**, *12*, 2196. [CrossRef]
75. Fan, Y.; Zhao, Z.; Yu, J.; Wu, L. Net hydrodynamic load impact factor analysis. *Mar. Eng.* **2025**, *47*, 1–6+77. (In Chinese) [CrossRef]
76. Mi, S.; Wang, M.; Avital, E.J.; Williams, J.J.; Chatjigeorgiou, I.K. An implicit Eulerian–Lagrangian model for flow-net interaction using immersed boundary method in OpenFOAM. *Ocean. Eng.* **2022**, *264*, 112843. [CrossRef]
77. Cao, H.; Wan, D. Research on the Development and Application of Numerical Wave Water Pool Based on OpenFOAM. In *Strive to Be Strong and Realize the Dream of the Maritime Frontier—Proceedings of the 8th Plenary Session of the Ship Mechanics Academic Committee of the China Shipbuilding Society*; State Key Laboratory of Marine Engineering, School of Naval Architecture, Ocean and Civil Engineering, Shanghai Jiao Tong University: Shanghai, China, 2014; pp. 319–332. (In Chinese)
78. Chen, C.; Wang, W.; Zheng, J.; Zheng, Y.; Shi, X.; Liu, C. Water flow under the action of the flat metal net water resistance characteristic numerical simulation. *J. Dalian Ocean. Univ.* **2017**, *32*, 363–368. (In Chinese) [CrossRef]
79. Huang, X.; Guo, G.; Tao, Q.; Hu, Y. Numerical Simulation of Deformation Characteristics under Force of HDPE Circular Gravity Cage. *South. Fish. Sci.* **2013**, *9*, 126–131. (In Chinese)
80. Chen, Y.; Pang, G.; Huang, X.; Li, G.; Yuan, T.; Hu, Y.; Tao, Q. The influence of net coating on the structural response of self-elevating truss cages. *South. Fish. Sci.* **2024**, *20*, 43–53. (In Chinese)
81. Scherz, L.; Kriegesmann, B.; Pedersen, C.B. A condition number-based numerical stabilization method for geometrically nonlinear topology optimization. *Int. J. Numer. Methods Eng.* **2024**, *125*, e7574. [CrossRef]
82. Ceglie, M.; Violano, G.; Afferrante, L.; Menga, N. Hysteresis in Approach-Retraction Frictional Contacts with Geometric and Material Nonlinearity. *Mater. Sci. Forum* **2025**, *1151*, 109–115. [CrossRef]
83. Dai, H. Research on Fluid-Structure Coupled Nonlinear Vibration Mechanism and Its Driving Utilization. In *Abstracts of the 19th National Conference on Nonlinear Vibration and the 16th National Conference on Nonlinear Dynamics and Motion Stability*; Department of Engineering Mechanics, School of Aeronautics and Astronautics, Huazhong University of Science and Technology: Wuhan, China, 2023; Volume 4. (In Chinese) [CrossRef]

84. Pan, Y. Research on Ocean Energy Harvesting and Sensing Technology Based on Nanopower Generation. Master's Thesis, Beijing University of Technology, Beijing, China, 2024. (In Chinese)
85. Liang, N.; Miao, C.; Cui, H.; Wang, J.; Yang, X. Simulation Study on Coupling System of Raft Wave Energy Conversion and Triboelectric Nanogenerator Energy Output. *Adv. New Renew. Energy* **2022**, *10*, 265–270. (In Chinese) [CrossRef]
86. Pan, C.Y.; Dai, Z.H.; Ma, H.X.; Zheng, J.R.; Leng, J.; Xie, C.; Yuan, Y.P.; Yang, W.C.; Yaliun, Y.X.; Song, X.M.; et al. Self-powered and speed-adjustable sensor for abyssal ocean current measurements based on triboelectric nanogenerators. *Nat. Commun.* **2024**, *15*, 6133. (In Chinese) [CrossRef] [PubMed]
87. Zhao, Y.; Zhang, J.; Liu, Y.; Sun, K.; Zhang, C.; Wu, W.; Teng, F. Numerical assessment of the environmental impacts of deep sea cage culture in the Yellow Sea, China. *Sci. Total Environ.* **2020**, *706*, 135752. [CrossRef] [PubMed]
88. Zhang, C.; Hao, Y.; Yang, J.; Su, W.; Zhang, H.; Wang, J.; Wang, Z.; Li, X. Recent advances in triboelectric nanogenerators for marine exploitation. *Adv. Energy Mater.* **2023**, *13*, 2300387. [CrossRef]
89. Yu, J.; Guo, Y.; Yang, L. Solid-liquid Triboelectric Nanogenerator. *J. Mech. Eng.* **2021**, *57*, 160–181. (In Chinese)
90. Zhu, H.; Liang, J.; Wang, S.; Huang, J.; Chen, Z. A multimodal PDMS triboelectric nanogenerator sensor based on anodised aluminium oxide template preparation for object recognition. *J. Mater. Chem. A* **2023**, *11*, 24158–24168. [CrossRef]
91. Zhou, S.; Zhang, Y.; Li, X.; Xu, C.; Halim, J.; Cao, S.; Rosen, J.; Strömme, M. A mechanically robust spiral fiber with ionic–electronic coupling for multimodal energy harvesting. *Mater. Horiz.* **2024**, *11*, 3643–3650. [CrossRef]
92. Jin, S.Y. Based on Drum TENG Device of Energy Collection and Vibration Testing Research. Master's Thesis, Hangzhou University of Electronic Science and Technology, Hangzhou, China, 2024. (In Chinese)
93. Yang, H. Structural Design and Performance Research of a Triboelectric Nanogenerator for Water Wave Energy Harvesting. Master's Thesis, Zhejiang Normal University, Jinhua, China, 2024. (In Chinese)
94. Park, S.G.; Zhou, J.; Dong, S.; Li, Q.; Yoshida, T.; Kitazawa, D. Characteristics of the flow field inside and around a square fish cage considering the circular swimming pattern of a farmed fish school: Laboratory experiments and field observations. *Ocean. Eng.* **2022**, *261*, 112097. [CrossRef]
95. Gansel, L.C.; Rackebrandt, S.; Oppedal, F.; McClimans, T.A. Flow fields inside stocked fish cages and the near environment. *J. Offshore Mech. Arct. Eng.* **2014**, *136*, 031201. [CrossRef]
96. Dong, S.; Park, S.G.; Zhou, J.; Li, Q.; Yoshida, T.; Kitazawa, D. Flow Field Inside and Around a Square Fish Cage Considering Fish School Swimming Pattern. In Proceedings of the International Conference on Offshore Mechanics and Arctic Engineering, Virtual, 21–30 June 2021; Volume 85154.
97. Tang, M.F.; Xu, T.J.; Dong, G.H.; Zhao, Y.P.; Guo, W.J. Numerical simulation of the effects of fish behavior on flow dynamics around net cage. *Appl. Ocean. Res.* **2017**, *64*, 258–280. [CrossRef]
98. Borazjani, I.; Mohsen, D. The fish tail motion forms an attached leading edge vortex. *Proc. R. Soc. B Biol. Sci.* **2013**, *280*, 20122071. [CrossRef]
99. Guo, C.; Liang, Z.; Han, Y.; Kui, Y.; Xu, P.; Sun, L. Experimental Study on Linear Accelerated Swimming of Zebrafish. *Acta Mech. Sin.* **2022**, *54*, 2446–2459. (In Chinese)
100. Yang, G.; Hu, X.; Zhang, B.; Long, Z.; Zhang, Y.; Shi, X. Analysis of Dynamic Characteristics of Free-Swimming Grass Carp Based on Particle Image Velocimetry (PIV). *J. Dalian Ocean. Univ.* **2021**, *4*, 833–841. (In Chinese) [CrossRef]
101. Hu, W. Numerical and analytical study on a tail-flaping model for fish fast C-start. *Acta Mech. Sin.* **2004**, *20*, 16–23. (In Chinese) [PubMed]
102. Su, B.; Zhai, S.; Bai, Y.; Zhang, J. Analysis of hydrodynamic and vortex Structure Characteristics of Tail Fin Swing Propulsion. *J. Ship Mech.* **2024**, *28*, 220–228. (In Chinese)
103. Cui, Z.; Jiang, H.; He, J.; Tong, Z. Research Progress and Key Technology Analysis of Swimming Mechanism of BCF bionic fish. *J. Mech. Eng.* **2015**, *51*, 177–184. (In Chinese) [CrossRef]
104. Tong, B. The Propulsion Mechanism of Fish 'Wavy Swimming'. *Mech. Chin. Pract.* **2000**, *22*. (In Chinese)
105. Chu, W.; Yuan, Q.; Kong, X.; Zhang, Z.; Li, H.; Qian, W. Analysis of Tail fin Propulsion Mechanism and Influencing Parameters of Swimming Ability of Fish. *J. Shanghai Ocean. Univ.* **2022**, *31*, 1224–1234. (In Chinese)
106. Sun, W.; Liu, S.; Wu, G.; Zhou, Y.; Ren, X. Research Progress on the Swimming mechanism of fish. *J. Dalian Ocean. Univ.* **2024**, *39*, 698–707. (In Chinese)
107. Xu, P.; Liu, J.; Liu, X.; Wang, X.; Zheng, j.; Wang, S.; Chen, T.; Wang, H.; Wang, C.; Fu, X.; et al. A bio-inspired and self-powered triboelectric tactile sensor for underwater vehicle perception. *Npj Flex. Electron.* **2022**, *6*, 25. [CrossRef]
108. Zhang, Y.; Xia, C.; Zhong, L.W. The sealed bionic fishtail-structured TENG based on anticorrosive paint for ocean sensor systems. *Nano Energy* **2023**, *108*, 108210. [CrossRef]
109. Li, D.; Wang, S.; Wang, C. Application Progress of Flexible Wearable Sensing Technology in Smart Fishery. *Trans. Chin. Soc. Agric. Eng.* **2023**, *39*, 1–13. (In Chinese)
110. Zhang, B.; Hu, X.; Yang, G.; Long, Z.; Huang, H.; Shi, X. Study on the Dynamics of Juvenile Grass Carp Cruising Based on Pressure Field. *J. Hydroelectr. Power* **2021**, *40*, 79–88. (In Chinese)

11. Clukey, E.C.; Silva, A.J. Permeability of deep-sea clays: Northwestern Atlantic. *Mar. Georesources Geotechnol.* **1982**, *5*, 1–26. [CrossRef]
12. Du, J.; Hu, L.; Sun, J.; Song, Q.; Chen, M.; Xiao, J. Tribo-corrosion Performance of 7075-T6 Al-alloy in 3.5%NaCl Solution. *J. Chin. Soc. Corros. Prot.* **2025**, *45*, 803–811. [CrossRef]
13. Chen, Z.; Mei, Y.; Liu, S.; Li, H.; Liu, L.; Lei, X.; Zhou, Y.; Gao, X. Electronic Packaging reliability: Past, Present and future. *J. Mech. Eng.* **2021**, *57*, 248–268. (In Chinese)
14. Sun, Y.; Zheng, Y.; Wang, R.; Lei, T.; Liu, J.; Fan, J.; Shou, W.; Liu, Y. 3D micro-nanostructure based waterproof triboelectric nanogenerator as an outdoor adventure power source. *Nano Energy* **2022**, *100*, 107506. [CrossRef]

Journal of
Marine Science and Engineering

MDPI

Article

Design and Implementation of a LiDAR-Based Inspection Device for the Internal Surveying of Subsea Pipelines

Qingmiao Ma [1], Weige Liang [1,*], Haoming Chen [1], Qianshi Wang [1], Peiyi Zhou [1] and Qingshan Wang [2,*]

[1] Naval University of Engineering, Wuhan 430033, China; m24182607@nue.edu.cn (Q.M.)
[2] School of Mechatronic Engineering, Central South University, Changsha 410083, China
* Correspondence: 1312021010@nue.edu.cn (W.L.); qingshanwang@csu.edu.cn (Q.W.)

Abstract

Subsea pipelines are extensively utilized in transportation systems. Conducting regular and effective internal inspections of these pipelines to promptly identify internal defects and potential risks is of paramount importance to ensure safe operational practices. In response to the practical engineering requirements for the internal inspection of subsea pipelines, this paper presents the design of an inspection device capable of capturing point cloud data from pipelines with internal diameters of 100 mm and above and performing three-dimensional reconstruction through coding. This device clearly reveals internal pipeline defects and enables both qualitative and quantitative analyses. Upon designing the motion module, control system, and LiDAR-based detection module of the internal pipeline inspection device, the capacity to collect internal point cloud data and perform 3D reconstruction was achieved. An experimental prototype of the inspection device was manufactured and tested using a simulated pipeline constructed to replicate real-world conditions. An analysis of the inspection results demonstrates that the device can travel steadily inside the pipeline, and the collected point cloud data can be used for 3D reconstruction via coding, accurately and clearly displaying the internal 3D structure of the pipeline and its defects. This device provides a basis for the prediction of pipelines' service lives.

Keywords: subsea pipeline inspection; LiDAR; 3D reconstruction

1. Introduction

Subsea pipeline transportation offers significant advantages in the conveyance of gases and liquids, including a high capacity, reliability, speed, safety, and efficiency. However, external environmental factors, such as water currents and soil conditions, can subject these pipelines to corrosion, cracking, and depression. These issues not only compromise transport efficiency but also pose substantial safety hazards [1]. Consequently, it is imperative to employ effective methods for the regular inspection of pipelines to prevent accidental failures [2]. Figure 1 illustrates a cross-sectional view comparing the internal condition of a normal subsea pipeline against that of a corroded one, used to transport a mixture of petroleum, water, and certain gases. Existing pipeline inspection technologies are constrained by limitations in their practical engineering application, presenting significant challenges for survey operations. Manual inspection methods are not only time-consuming and labor-intensive but are also prone to oversight. While technology-assisted approaches offer alternatives, their effectiveness is often restricted by the realities of field deployment and the current state of technological development [3].

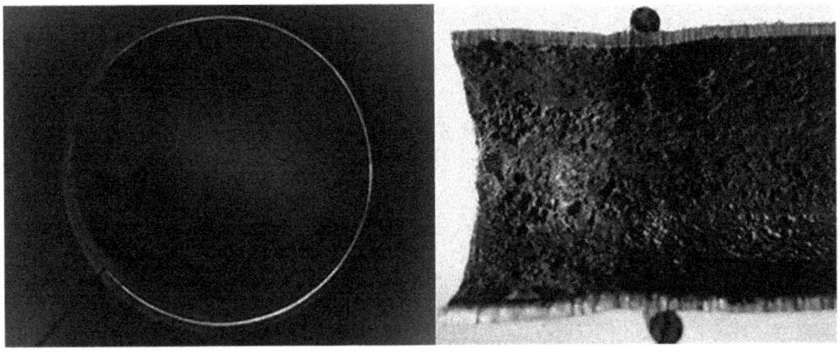

Figure 1. Internal view of pipeline: normal condition versus corrosion.

To address the aforementioned issues, current research focuses on utilizing various detection technologies, such as electromagnetic testing, acoustic testing, and machine vision inspection, to identify internal pipeline defects [4,5]. Electromagnetic testing offers high sensitivity for the detection of surface or near-surface defects in metallic pipelines; it boasts rapid inspection speeds and requires no coupling agent. However, its application is limited to metallic materials; it has a low detection rate for deep-seated defects; and its efficacy is easily affected by the pipe wall thickness and geometry [6]. Acoustic testing is capable of identifying deep-seated defects and leakage points; it is suitable for long-distance scanning and demonstrates good compatibility with non-metallic pipelines. However, it is susceptible to environmental noise, which can interfere with signals, and it involves complex data analysis, with poor real-time performance [7]. Machine vision inspection enables the intuitive visualization of defects. Nevertheless, it necessitates a clean and dry environment, fails under inadequate lighting conditions, and entails high equipment costs, as well as being sensitive to pipeline deformation [8]. Yun Liang [9] detected internal pipeline corrosion via in-line inspection using ultrasonic testing. Meanwhile, Zhe Liu [10] designed a magnetic flux leakage inspection device for subsea pipelines, with its measurement reliability and accuracy being experimentally verified. Meniconi, S. et al. [11,12] focused on the Trieste Submarine Pipeline. They first designed a test protocol for pressure wave generation using SDV; they then localized pipe wall degradation by integrating numerical and analytical models with on-site data—this approach was subsequently validated via diving inspections. Cawley, P. [13] employed ultrasonic-guided waves with frequencies below 100 kHz. By controlling wave modes and coherent noise, he achieved the inspection of pipelines, rails, and other structures and extended this technology's application to commercial pipeline systems. However, this method exhibits limitations in the inspection of complex structures and testing of high-attenuation systems. Davidson, R. [14] utilized various pig types, including mandrel-type and foam-type pigs. Driven by differential pressure, these pigs perform functions such as pipeline cleaning, diameter measurement, and fluid separation. Nevertheless, foam-type pigs suffer from rapid wear; intelligent pigs have poor real-time communication capabilities; and neither type is suitable for non-circular or multi-branch pipelines. Bohorquez, J. et al. [15] combined fluid transients with convolutional neural networks (CNNs). They incorporated stochastic resonance to optimize the noise intensity for model training, thereby realizing pipeline leak localization. This method, however, relies on the matching of numerical and experimental data, remains unverified in looped pipeline network scenarios, and shows reduced accuracy under large background pressure fluctuations. Covas, D. et al. [16] applied inverse transient analysis, coupled with a viscoelastic pipeline transient solver, for leak detection. However, this method requires precise knowledge of the pipeline's physical properties, is ineffective under slow valve operations, and struggles to detect small leaks (<0.12 L/s). Mohapatra, P.K. et al. [17] adopted the frequency response method (FRM; based on a transfer matrix) to analyze steady-state oscillatory flow. They used peak patterns to determine the location of a single blockage and mean values to quantify the blockage size. However, this approach does not enable the identification of individual positions of multiple blockages; actual friction impedes peak identification; and the frictionless assumption does not reflect real-world conditions. Brockhaus, S. et al. [18] reviewed in-line inspection (ILI) technologies—including magnetic flux leakage (MFL), ultrasonic testing (UT), and electromagnetic acoustic transducers (EMATs)—for corrosion detection in underground pipelines. Among these, MFL has low sensitivity to shallow corrosion; UT requires liquid coupling; EMATs are costly; and none are applicable to non-metallic pipelines. Mpesha, W. et al. [19] used the frequency response method, generating an oscillatory flow via periodic valve opening and closing, and they localized leaks using secondary pressure peaks. This method demands pressure and flow rate measurements at valves, is incompatible with closed pipeline networks, and fails to detect micro-leaks (<0.5% of the average flow rate). Kazeminasab, S. et al. [20] reviewed pipeline robots, classifying their mechanical systems and summarizing tasks such as localization, mapping, navigation, and inspection. Pipeline robots, face challenges in navigating elbows/reducing pipes, are vulnerable to localization interference, and have poor detection capabilities for non-metallic pipelines. Liou, C. [21] analyzed transient pressure signals via the impulse response extraction method to localize leaks. It was shown that impulses are susceptible to noise interference; signal attenuation in long pipelines leads to weak responses; and the method has not been validated in multi-leak scenarios. Gazis, D.C. [22] established a three-dimensional (3D) analytical foundation for wave propagation in hollow cylinders, providing theoretical support for the ultrasonic-guided wave testing of pipelines. This work, however, was limited to theoretical derivation, did not integrate practical inspection scenarios, and is difficult to adapt to complex cross-sections.

In summary, the aforementioned technologies fail to facilitate the qualitative and quantitative analyses of defects. Consequently, scholars have proposed the use of laser technology for the detection of internal pipeline defects. This technology offers high detection accuracy and has promising development prospects. It not only facilitates the qualitative and quantitative analyses of defects but also enables the 3D visualization of the pipeline interior with the aid of computer technology [23]. Liu Xuxiang [24] addressed the detection of partial wear on the inner walls of narrow, fine oil pipes with small internal diameters by proposing a method that employs a point laser sensor to perform continuous circumferential measurements at different positions on the pipe's inner wall. This approach involves reconstructing the 3D point cloud data of the pipeline's internal surface and allows for both qualitative analysis and quantitative calculation. Li Yi et al. [25] tackled the issues of missed and false judgements occurring during the manual assessment of deformation-type defects in pipelines by introducing an automated identification method based on laser point cloud data. This method enables the identification of pipeline deformation defects through the calculation of relevant information from fitted ellipses to cross-sections of the pipe point cloud. Mario Montoya et al. [26] established a fully functional laser system based on a freeform laser profiling method, characterized by a high resolution, excellent precision, and fast acquisition speeds. Tina Tian et al. [27] integrated a monocular camera, an inertial sensor, a ring-shaped laser profiler, and LiDAR to achieve the colored 3D reconstruction of the internal pipeline surface.

The current work presents the design of an internally navigable pipeline inspection device. The apparatus is equipped with a LiDAR sensor that acquires and transmits internal 3D point cloud data in real time during axial traversal. Subsequently, a 3D reconstruction algorithm is used to process the collected data to generate detailed internal models of pipelines with internal diameters $\geq$100 mm. In contrast to the aforementioned inspection techniques, the proposed method not only enables the comprehensive visualization of the internal pipeline condition for qualitative and quantitative defect analysis but also incorporates an integrated, modular design. This results in a compact, lightweight, and portable system, thereby enhancing the ease and practicality of routine pipeline operation and maintenance.

2. Overall Design of Subsea Pipeline Inspection Device

2.1. Overall Design Scheme

Considering the practical conditions and the inspection requirements for subsea pipelines, the proposed device must fulfil several criteria. Firstly, it must achieve stable locomotion within the complex internal pipeline environment. Secondly, it must ensure the stable operation of the onboard LiDAR system. Thirdly, it must be capable of collecting internal point cloud data for subsequent 3D reconstruction. Fourthly, it should be applicable to pipelines with an internal diameter of 100 mm and above. Finally, the device needs to be compact, lightweight, and portable.

To guarantee the stability of the inspection device and the reliability of the acquired data, the internal subsea pipeline inspection apparatus presented in this paper is divided into three modules: the locomotion module, the control system, and the detection module. The locomotion structure employs a combination of a hollow-shaft torque motor and Mecanum wheels. The control system utilizes an STM32 development board as its platform, implementing PID control algorithms to enable the autonomous movement of the inspection device. The detection module consists of a LiDAR sensor for point cloud data acquisition. The acquired point cloud data are transmitted to a data processing terminal for 3D reconstruction. The overall design scheme is illustrated in Figure 2.

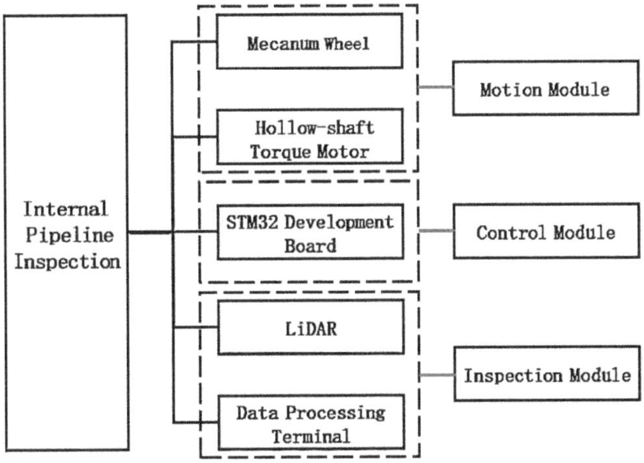

Figure 2. Schematic diagram of the internal pipeline inspection system.

The 3D model of the inspection device is shown in Figure 3. The device is broadly divided into two main parts: the locomotion module and the LiDAR unit. The locomotion module serves as the moving platform for the device, responsible for ensuring stable movement and maintaining central alignment within the pipe. The two Mecanum wheels are connected by a linking rod. The LiDAR unit is responsible for measuring the point cloud data of the internal bore. Furthermore, the STM32 development board is housed within the central cavity of the Mecanum wheel assembly. Wiring connects the various components through the through-hole of the hollow-shaft torque motor. Considering the constraints of the actual operating space, a custom-designed 12 V power supply is employed. This flat-shaped power system is mounted, along with the development board, within the central cavity of the Mecanum wheel assembly.

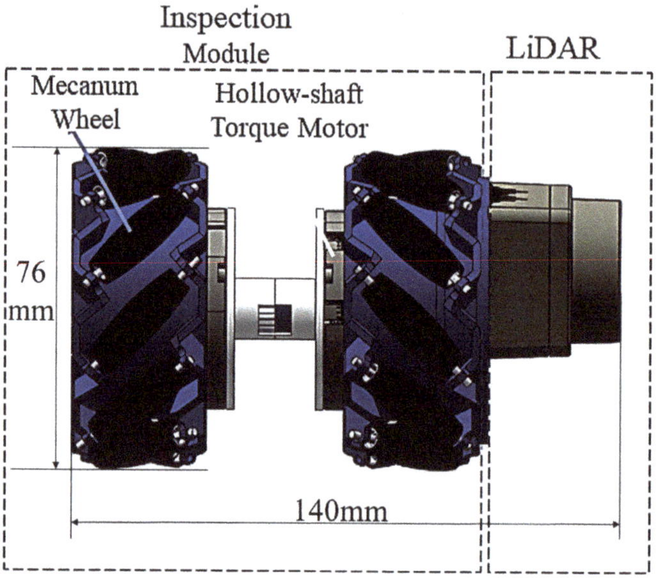

Figure 3. Model diagram of the internal pipeline inspection device.

2.2. Locomotion Module Design

Conventional architectures for in-pipe locomotion typically employ either wheeled or linkage-based mechanisms. Wheeled systems propel themselves through motor-driven wheels engaging the pipe inner wall, offering continuous and efficient motion. Their design is relatively simple and allows for high traversal speeds; however, they exhibit poor adaptability to irregular pipe geometries and obstacles, with their traction being highly dependent on the friction between the wheels and the pipe surface. In contrast, linkage-based mechanisms, often inspired by the biomechanics of organisms such as inchworms, achieve peristaltic crawling through the alternating extension, contraction, and locking of multiple body joints. This inchworm-like, step-by-step locomotion is characterized by relatively low speeds but demonstrates exceptional terrain adaptability and obstacle-crossing capabilities, making it particularly suitable for complex pipeline environments. To overcome the limitations of traditional locomotion systems [28], a novel structure integrating a hollow-shaft torque servo motor with Mecanum wheels [29] was designed. One hollow-shaft torque motor and one Mecanum wheel constitute a single locomotion module. The device incorporates two such modules connected in series at the front and rear. The structure of the locomotion module is depicted in Figure 4.

Within the module, (1) denotes the Mecanum wheel hub, which serves to fix the rollers, bear the load, and ensure the proper movement of the Mecanum wheel. Meanwhile, (2) refers to the Mecanum wheel rollers, which are the primary components responsible for the wheel's movement. They provide the driving force for the device and, crucially, ensure the central alignment of the inspection device within the pipeline. Next, (3) is the hollow-shaft torque motor, responsible for driving the Mecanum wheel, and (4) is the motor sleeve, which connects the rotor of the hollow-shaft torque motor to the Mecanum wheel hub, ensuring the effective transmission of torque from the motor. The synergistic operation of these four components guarantees the central alignment of the inspection device.

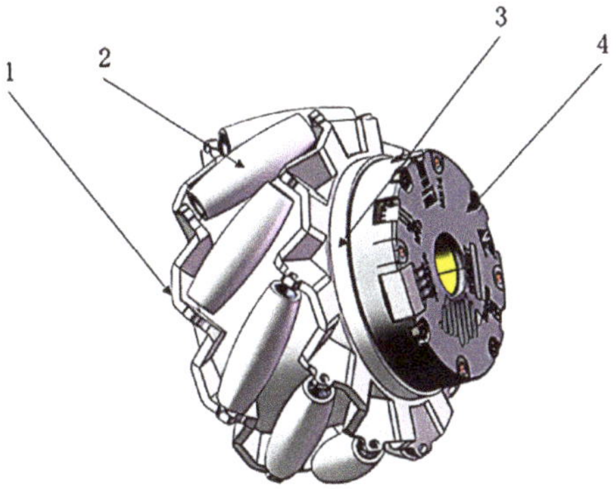

Figure 4. Structural diagram of the locomotion module: (1) Mecanum wheel hub; (2) Mecanum wheel roller; (3) motor sleeve; (4) hollow-shaft torque motor.

2.3. *LiDAR and Control Board Selection*

The front end of the internal pipeline inspection device is equipped with a LiDAR sensor, serving as the detection module to acquire internal point cloud data. The following key factors were considered during the component selection process: (1) the accuracy requirements for data acquisition; (2) the measurement range of the LiDAR; (3) the physical dimensions of the LiDAR unit; (4) its power supply requirements; and (5) the data transmission specifications.

After the comprehensive evaluation of these factors, the Orbbec MS200 LiDAR was selected, which is shown in Figure 5. This LiDAR operates based on the time-of-flight (ToF) ranging principle. It has a diameter not exceeding 40 mm, a weight of only 40 g, a measurement range of 0.03 m to 12 m, and a sampling frequency of 4500 points per second. These specifications ensure that sufficient point cloud data can be collected within a short time period. During operation, the LiDAR scans the pipeline interior to collect data. The acquired point cloud data are received by a data acquisition program on the processing terminal for subsequent analysis. Since the LiDAR's minimum measurement distance is 30 mm, it was installed with an offset (eccentric mounting), as illustrated in Figure 6, to avoid any blind spots in the measurement field.

Size	**37.7*37.5*33mm**
Sampling Frequency	**4500/s**
Scanning Frequency	**7Hz—15Hz**
Angle Accuracy	**0.8°**
Ranging Accuracy	**≤4mm(0.2m—2m) ≤15mm(2m—12m)**
Minimum Measurement Distance	**30mm**

Figure 5. Orbbec MS200 LiDAR.

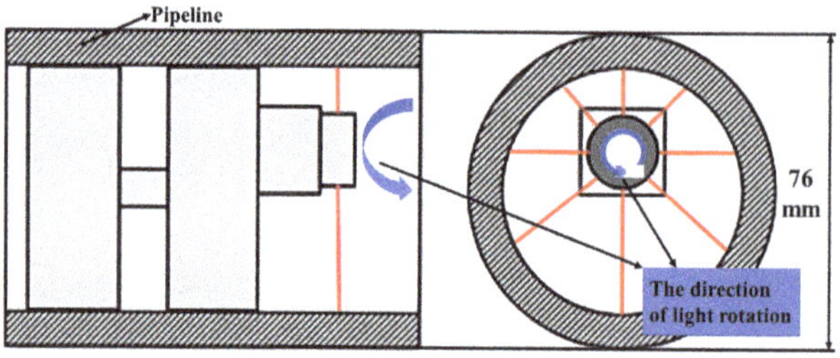

Figure 6. Schematic diagram of LiDAR mounting position.

Since the LiDAR was required to acquire internal point cloud data while the inspection device is in motion, it was necessary to regulate the device's step length and velocity. A development board was therefore adopted as the control center to manage both the motors and the LiDAR. Considering the confined internal space of the pipeline, an STM32 development board was selected for its high level of integration and compact dimensions. The STM32 board, which utilizes an ARM core, offers high performance, low power consumption, and excellent expandability, making it suitable for a variety of application scenarios. Power is supplied to the motors, and their operation is controlled via a booster module connected to the development board. The development board also interfaces with the LiDAR to manage data acquisition. The control algorithm implemented for the pipeline inspection device is a proportional–integral–derivative (PID) control algorithm.

3. Simulation Verification

Following the completion of the internal pipeline inspection device model, a kinematic simulation analysis was conducted to verify the feasibility of the device's movement within the pipeline and to analyze the stability of its motion. Using the ADAMS 2020 software, a pipeline internal model was established based on realistic operational conditions. The model incorporated multiple straight grooves, stepped grooves, and irregular recesses to simulate internal surface anomalies. The inspection device model was then positioned within this pipeline environment for simulation analysis. The overall assembly model is presented in Figure 7.

Figure 7. Kinematic simulation model of the internal pipeline inspection device.

According to the movement mode of the internal pipeline inspection device, as specified in the design scheme, torque is provided to the device's motor to drive it forward, causing the motor to rotate the Mecanum wheels. Rolling constraints are set between the rollers of the Mecanum wheels and the inner wall of the pipeline. After the motor operates, the inspection device travels inside the pipeline; thus, the kinematic simulation of the device's movement process could be successfully completed. The vibrational accelerations of the inspection device in the X, Y, and Z directions during travel were analyzed to evaluate its motion stability. The simulation results show that the internal pipeline inspection device can move at a constant speed inside the pipeline.

4. Experimental Validation

To validate the practical feasibility of the internal pipeline inspection device, practical traversal and data acquisition experiments were conducted. The inspection device was deployed inside a pipeline for testing, allowing the observation of its motion behavior and the operational status of the LiDAR. Point cloud data acquired by the LiDAR were collected for subsequent processing.

The inspection experiment required a section of a pipeline and a prototype of the inspection device. Considering the practical difficulties associated with measurement inside actual pipelines, an internal pipeline model was fabricated to simulate the complex internal conditions. This model, featuring a 100 mm internal diameter and incorporating linear grooves, a stepped groove, and irregular depressions, was produced using 3D printing and is shown in Figure 8. A photograph of the physical inspection device and the non-contact measurement data acquisition process during operation are presented in Figure 9.

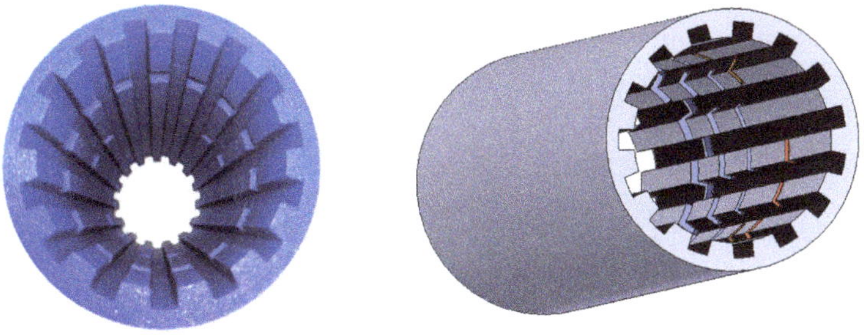

(a) (b)

Figure 8. Pipeline physical object and model diagrams: (**a**) physical object; (**b**) schematic diagram.

(a) (b)

Figure 9. Pipeline internal inspection device and data acquisition process: (**a**) physical device; (**b**) data acquisition process.

The specific data acquisition procedure is as follows:

(1) Device Advancement: The inspection device advances while the LiDAR remains in standby mode. The STM32 development board, utilizing a PID control algorithm, governs the device's movement, controlling its progression in precise increments of 0.5 mm according to preset parameters.

(2) Device Halt and Data Acquisition: The device halts, and the LiDAR acquires point cloud data. Following the cessation of the hollow-shaft torque motors, the host computer triggers the LiDAR to operate to capture the point cloud information.

(3) Data Transmission and Resumption: Upon completion of the scan, the LiDAR transmits the acquired point cloud data back to the host computer. The host computer then commands the LiDAR to cease operation, and the inspection device resumes its advancement based on the programmed instructions.
(4) Process Iteration: The cycle described above (steps 1–3) is repeated iteratively. This process accumulates point cloud data for a specific length of the pipeline's internal bore.
(5) Post-Processing and Reconstruction: After data acquisition is complete, the inspection device is retrieved from the pipeline. The collected data undergo preprocessing and subsequent 3D reconstruction through a dedicated data processing algorithm.

Through the coordinated operation of the device's locomotion and detection modules, the LiDAR successfully collects internal pipeline point cloud data and transmits them to the data processing terminal. The data collection page window is shown in Figure 10. A sample of the acquired data is presented in Table 1.

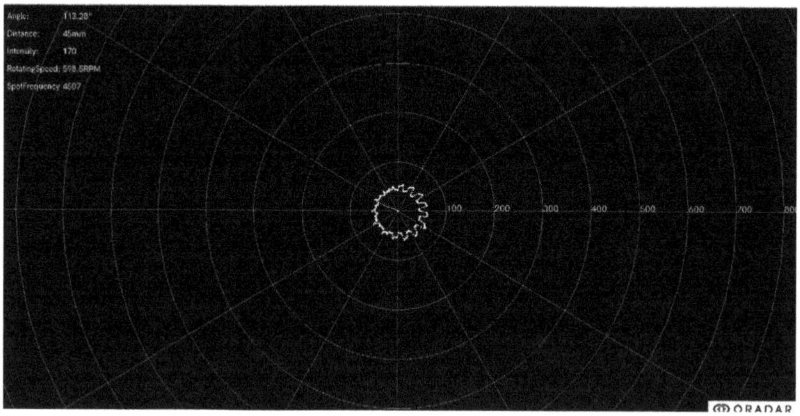

Figure 10. Data collection page window.

Table 1. Sample point cloud data of the pipeline interior.

Angle/°	Radius/mm	Angle/°	Radius/mm
70.016	40	109.151	46
74.490	40	115.377	44
78.091	40	117.761	39
80.182	47	121.319	39
92.877	40	128.384	41
102.468	40	131.445	46
103.923	41	136.382	46
104.151	43	139.850	43

Based on the data presented in the table, the point cloud data were first processed by transforming the X and Y coordinates. The Z coordinate was initialized from 0 and incremented by a value of 0.5 mm for each step, corresponding to the motor's step length. Finally, three-dimensional reconstruction was performed on the acquired point cloud data of the pipeline's internal bore, resulting in the point cloud model shown in Figure 11.

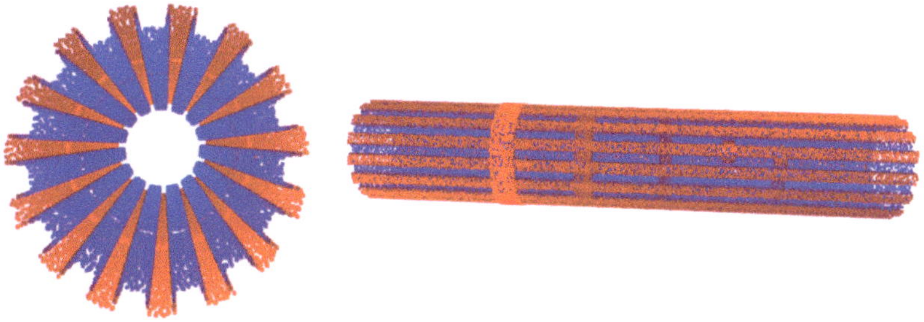

Figure 11. Internal pipeline point cloud model (The colors are mainly used to distinguish feature regions, with feature regions marked in red. The red fan-shaped area in the left figure mainly corresponds to the concave groove feature region, and the blue fan-shaped area corresponds to the convex groove feature region).

The model was then radially sectioned to obtain a cross-sectional view, as presented in Figure 12. This cross-sectional perspective allows for clearer observation of the various internal defects within the pipeline.

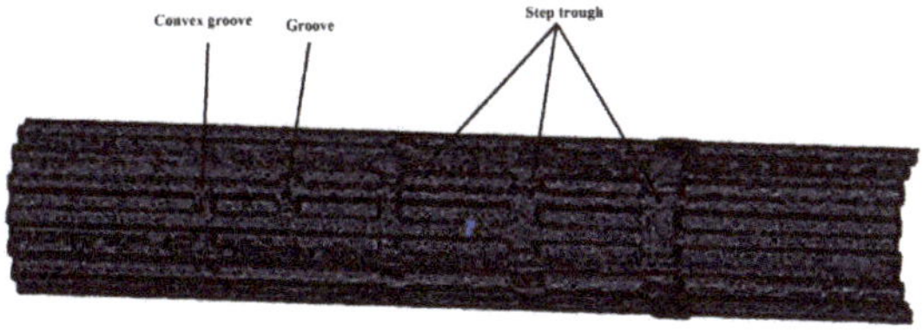

Figure 12. Cross-sectional view of the model.

A comparison between the actual pipeline interior and the point cloud model reveals that the model effectively captures the relevant defects present within the pipeline. However, further analysis of the detection device's measurement errors is required. Following the 3D reconstruction of the internal pipeline point cloud, cross-sectional data were selected to calculate the LiDAR's detection error. MATLAB 2020 was employed to fit the pipeline's internal cross-sections. The fitting results are shown in Figure 13, and a comparison of the data from distinct feature regions is presented in Table 2.

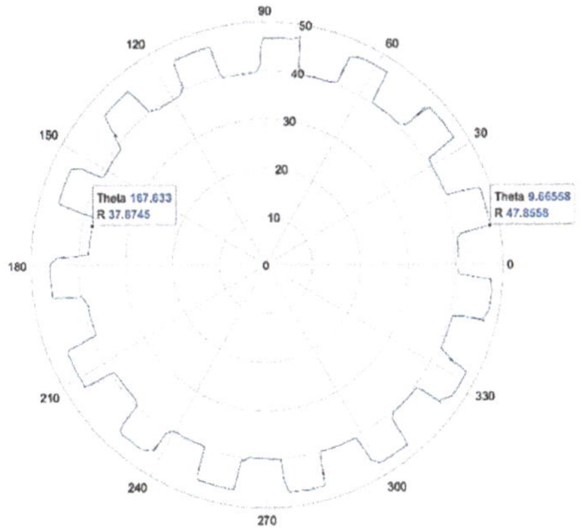

Figure 13. Fitting results for the pipeline's internal cross-section.

Table 2. Comparison of data from distinct feature regions.

Region	Defect Size/mm	Maximum Measured Dimension/mm
Groove	45	48
Convex groove	38	38
Step trough 1	46	48
Step trough 2	44	46
Step trough 3	42	44
Depression 1	45.5	47
Depression 2	41.8	43

By comparing the above fitting results, it can be seen that the detection data from the LiDAR exhibit a certain degree of error. In the aforementioned fitting graphs, the maximum radius is 47.8558 mm and the minimum radius is 37.8745 mm. Through the data comparison between the model dimensions and the maximum measured dimensions, it can be inferred that the maximum error of the internal pipeline inspection device is approximately 3 mm, with an average error of 1.8 mm. The main causes of the error are as follows: the internal pipeline inspection device may experience slight vibration during travel, thereby affecting the detection results obtained from the LiDAR; the data detected by the LiDAR contain some noise and missing data points; and the measurement accuracy of the LiDAR is limited.

Compared with the results obtained by Zang Chunhua et al. [30], whose average measurement error was 0.147 mm, the error measured in this work is large. This can be explained by the fact that the aforementioned study collected data on a pipeline's outer diameter in a different working scenario, where more sensor models were available. Moreover, they fused two types of data (LiDAR and camera), resulting in higher data accuracy. In the future, our team will consider conducting a feasibility analysis on a multi-source data fusion scheme to improve the accuracy and reduce the errors.

5. Conclusions

The subsea pipeline internal inspection device designed in this work can achieve relatively stable movement inside pipelines, synchronously collect high-density point cloud data on the pipeline's inner surface, and complete the 3D reconstruction of the internal structure of a pipeline with a diameter of 100 mm and above. This device provides an effective technical means for the visual inspection of internal pipeline conditions and has strong engineering applicability. However, to further improve the accuracy and reliability of the detection data, subsequent research is needed to verify its movement stability; it is also necessary to select LiDAR sensors with higher measurement accuracy and ntegrate a three-axis gyroscope to determine the impact of instability on the data measurement accuracy. Following these improvements, the system will facilitate the high-precision positioning, qualitative identification, and quantitative analysis of internal defects in subsea pipelines, providing strong data support for pipeline safety assessment, maintenance, and management.

Author Contributions: Conceptualization, W.L. and Q.M.; methodology, W.L. and Q.M.; validation, P.Z. and Q.W. (Qianshi Wang); formal analysis, P.Z. and Q.W. (Qianshi Wang); investigation, Q.W. (Qianshi Wang) and H.C.; resources, W.L.; data curation, H.C. and Q.W. (Qianshi Wang); writing—original draft preparation, Q.W. (Qianshi Wang) and Q.M.; writing—review and editing, W.L. and Q.W. (Qingshan Wang); supervision, W.L.; project administration, W.L.; funding acquisition, W.L. All authors have read and agreed to the published version of the manuscript.

Funding: The authors gratefully acknowledge the financial support from the Hubei Provincial Natural Science Foundation, 2023AFB900.

Data Availability Statement: The data provided in this study are available upon request from the corresponding author due to privacy restrictions.

Conflicts of Interest: The authors declare no conflicts of interest.

References

1. Du, F.; Li, C.; Wang, W. Development of Subsea Pipeline Buckling, Corrosion and Leakage Monitoring. *J. Mar. Sci. Eng.* **2023**, *11*, 188. [CrossRef]
2. Ho, M.; El-Borgi, S.; Patil, D.; Song, G. Inspection and Monitoring Systems Subsea Pipelines: A Review Paper. *Struct. Health Monit.* **2020**, *19*, 606–645. [CrossRef]
3. Davis, P.; Brockhurst, J. Subsea Pipeline Infrastructure Monitoring: A Framework for Technology Review and Selection. *Ocean Eng.* **2015**, *104*, 540–548. [CrossRef]
4. Zhang, H.; Zhang, J.; Lin, R.; Wang, Y. Advanced Inspection Techniques for Submarine Pipeline Integrity. In *Submarine Pipeline Integrity: Assessment of Failure Modes and Advanced Evaluation Techniques*; Zhang, H., Zhang, J., Lin, R., Wang, Y., Eds.; Springer Nature: Cham, Switzerland, 2025; pp. 1–14. ISBN 978-3-031-92092-9.
5. Ma, Q.; Liang, W.; Zhou, P. A Review on Pipeline In-Line Inspection Technologies. *Sensors* **2025**, *25*, 4873. [CrossRef]
6. Wang, C.; Wu, S.; Xin, J.; He, R.; Chen, J.; Wang, D. Numerical Simulation of Oil and Gas Pipeline Crack Detection Based on Pulsed Eddy Current Testing Technology. In Proceedings of the 2021 4th International Conference on Electron Device and Mechanical Engineering (ICEDME), Guangzhou, China, 19–21 March 2021; pp. 46–50.
7. Cao, W.Y. Research on Pipeline Acoustic Detection and Leak Localization Technology. Master's Thesis, North China Electric Power University (Beijing), Beijing, China, 2024.
8. Fahimipirehgalin, M.; Trunzer, E.; Odenweller, M.; Vogel-Heuser, B. Automatic Visual Leakage Detection and Localization from Pipelines in Chemical Process Plants Using Machine Vision Techniques. *Engineering* **2021**, *7*, 758–776. [CrossRef]
9. Liang, Y. Discussion on Subsea Pipeline Internal Inspection and Anti-Corrosion Strategies. *Petro. Chem. Equip.* **2024**, *27*, 160, 164–167.
10. Liu, Z. Series Design of Measuring Nodes for Subsea Pipeline Inspection Equipment. *Orient. Enterp. Cult.* **2013**, 241–242. Available online: https://kns.cnki.net/kcms2/article/abstract?v=DJp2nd4LPS1cscr9z09rI124I3CYnNQJe1xTnqfyzdM49v_Gkxj5LsjL_-Ohu98CnrTc8_EXRCCmOs-2qrGHgD24o0AjW4ijSoBqY1O4Vc273gnQ6qqJwCrszpPTaj50JWvBHFIo20kxe8FjgGFYHSjWZ2CKP3-elU5p7uLXNG98pD_vcVOK3RmA==&uniplatform=NZKPT&language=CHS (accessed on 17 October 2025).
11. Meniconi, S.; Brunone, B.; Tirello, L.; Rubin, A.; Cifrodelli, M.; Capponi, C. Transient Tests for Checking the Trieste Subsea Pipeline: Diving into Fault Detection. *J. Mar. Sci. Eng.* **2024**, *12*, 391. [CrossRef]
12. Meniconi, S.; Brunone, B.; Tirello, L.; Rubin, A.; Cifrodelli, M.; Capponi, C. Transient Tests for Checking the Trieste Subsea Pipeline: Toward Field Tests. *J. Mar. Sci. Eng.* **2024**, *12*, 374. [CrossRef]
13. Cawley, P. Practical Long Range Guided Wave Inspection–Applications to Pipes and Rail. *Mater. Eval.* **2003**, *61*, 66–74.
14. Davidson, R.; Pipeline, H. *An Introduction to Pipeline Pigging*; Gulf Publishing: Houston, TX, USA, 2002.
15. Bohorquez, J.; Lambert, M.F.; Alexander, B.; Simpson, A.R.; Abbott, D. Stochastic Resonance Enhancement for Leak Detection in Pipelines Using Fluid Transients and Convolutional Neural Networks. *J. Water Resour. Plan. Manag.* **2022**, *148*, 04022001. [CrossRef]
16. Covas, D.; Ramos, H. Case Studies of Leak Detection and Location in Water Pipe Systems by Inverse Transient Analysis. *J. Water Resour. Plan. Manag.* **2010**, *136*, 248–257. [CrossRef]
17. Mohapatra, P.K.; Chaudhry, M.H.; Kassem, A.A.; Moloo, J. Detection of Partial Blockage in Single Pipelines. *J. Hydraul. Eng.* **2006**, *132*, 200–206. [CrossRef]
18. Brockhaus, S.; Ginten, M.; Klein, S.; Teckert, M.; Stawicki, O.; Oevermann, D.; Meyer, S.; Storey, D. 10–in-Line Inspection (ILI) Methods for Detecting Corrosion in Underground Pipelines. In *Underground Pipeline Corrosion*; Orazem, M.E., Ed.; Woodhead Publishing: Cambridge, UK, 2014; pp. 255–285. ISBN 978-0-85709-509-1.
19. Mpesha, W.; Gassman, S.L.; Chaudhry, M.H. Leak Detection in Pipes by Frequency Response Method. *J. Hydraul. Eng.* **2001**, *127*, 134–147. [CrossRef]
20. Kazeminasab, S.; Sadeghi, N.; Janfaza, V.; Razavi, M.; Ziyadidegan, S.; Banks, M.K. Localization, Mapping, Navigation, and Inspection Methods in in-Pipe Robots: A Review. *IEEE Access* **2021**, *9*, 162035–162058. [CrossRef]
21. Liou, C.P. Pipeline Leak Detection by Impulse Response Extraction. *J. Fluids Eng.* **1998**, *120*, 833–838. [CrossRef]
22. Gazis, D.C. Three-dimensional Investigation of the Propagation of Waves in Hollow Circular Cylinders. I. Analytical Foundation. *J. Acoust. Soc. Am.* **1959**, *31*, 568–573. [CrossRef]

23. Jia, M.Y.; Shen, X.; Feng, X.X. Application of LiDAR Technology in Pipeline Inspection Robots. *Metrol. Meas. Technol.* **2025**, *51*, 64–67. [CrossRef]
24. Liu, X.X. Research on Defect Detection Method for Pipeline Inner Wall Based on Point Laser Sensor. Master's Thesis, Tianjin University of Science and Technology, Tianjin, China, 2025.
25. Li, Y.; Dong, S.Q.; Shi, Y.; Chen, Q. Quantitative Detection Technology for Deformation Defects in Drainage Pipes Based on Laser Point Cloud Data. *J. Eng. Geophys.* **2024**, *21*, 54–61.
26. Montoya, M.; Montelongo, Y.; Jiang, N.; Morris, S.M.; Parra-Michel, J. Free-Form Laser Profilometry for Pipeline Inspection and 3D Cylindrical Reconstructions. *IEEE Sens. J.* **2022**, *22*, 297–303. [CrossRef]
27. Tian, T.; Wang, L.; Yan, X.; Ruan, F.; Aadityaa, G.J.; Choset, H.; Li, L. Visual-Inertial-Laser-Lidar (VILL) SLAM: Real-Time Dense RGB-D Mapping for Pipe Environments. In Proceedings of the 2023 IEEE/RSJ International Conference on Intelligent Robots and Systems (IROS), Detroit, MI, USA, 1–5 October 2023; pp. 1525–1531.
28. Ambati, P.; Raj, K.M.S.; Joshuva, A. A Review on Pipeline Inspection Robot. *AIP Conf. Proc.* **2020**, *2311*, 060002. [CrossRef]
29. Abdelrahim, M.; Hassan, A.; Ojo, D.; Hosny, M.; Ammar, H.H.; El-Samanty, M. A Novel Design of a T-Model Three Mecanum Wheeled Mobile Robot. In Proceedings of the 2022 International Conference on Advanced Robotics and Mechatronics (ICARM), Guilin, China, 9–11 July 2022; pp. 509–513.
30. Zang, C.H.; Wang, H.J.; Gao, X.Y.; Ji, H.G.; Wang, K.; Zhao, L.X.; Qiu, N.C. Diameter and Roundness Measurement Method of Expanded Pipes Based on Laser Line Scanning 3D Reconstruction. *Dev. Innov. Mach. Electr. Prod.* **2021**, *34*, 79–81.

Journal of
*Marine Science
and Engineering*

MDPI

Article

Study on the Effect of Column Form on the Dynamic Response of Semi-Submersible Truss-Type Fish Culture Platforms

Kangyang Liang [1] and Dapeng Zhang [2,*]

[1] College of Ocean Engineering and Energy, Guangdong Ocean University, Zhanjiang 524088, China; liangky@gdou.edu.cn
[2] Ship and Maritime College, Guangdong Ocean University, Zhanjiang 524088, China
* Correspondence: zhangdapeng@gdou.edu.cn

Abstract

To investigate the effect of column form on the hydrodynamic performance of semi-submersible truss fishery aquaculture platforms, this study focused on an active semi-submersible aquaculture platform located in the South China Sea. Three platform models featuring distinct column structures were established. Employing three-dimensional potential flow theory and Morrison's equations, numerical simulation methods were utilised to analyse the dynamic response of the three types of column platforms in both the frequency and time domains under wind, wave, and current action. Consequently, relevant conclusions regarding the influence of column form on the hydrodynamic performance of semi-submersible platforms were derived. The results show that: The quasi-elliptical column platform exhibits superior frequency-domain response characteristics, with the circular column platform following, while the square column platform demonstrates the poorest performance. When subjected to the combined effects of waves and currents, the circular column platform shows the most favourable time-domain dynamic response, with the quasi-elliptical column platform next, and the square column platform lagging behind. In contrast, under the combined influence of wind, waves, and currents, the quasi-elliptical column platform excels in time-domain dynamic response, followed by the square column platform, with the circular column platform being the least effective. These variations in time-frequency dynamic response characteristics among the three column platforms are attributed to their distinct structural forms.

Keywords: semi-submersible truss fishery aquaculture platform; column form; quasi-elliptical; hydrodynamic; dynamic response

1. Introduction

In recent years, the gradual transition of fishery aquaculture to deeper and more remote marine environments has garnered considerable attention towards large-scale aquaculture platforms constructed from steel and concrete materials [1]. These platforms exhibit robust structural integrity and possess the capability to endure substantial environmental loads, making them well-suited to the complexities of oceanic conditions. A variety of large-scale offshore farms have been proposed and constructed, including semi-submersible aquaculture platforms such as Ocean Fishery 1, Deep Blue 1, and Jostein Albert, among others.

Semi-submersible aquaculture platforms primarily consist of frame structures, including columns, support rods, and floats, alongside net structures. The hydrodynamic loads acting on the columns and net structure represent the principal components of the overall loads on the aquaculture platform, thereby influencing the dynamic response characteristics of the entire structure [2].

In deep and remote marine environments, aquaculture platforms frequently encounter severe conditions [3], Strong winds, huge waves, and powerful currents significantly increase the risk of mooring failure and instability in movement. Periodic wave loads present a significant threat to the overall stability of floating aquaculture platforms when moored, compromising the integrity of the local structure [4]. To enhance the understanding of aquaculture platform structures, scholars have conducted extensive research on their hydrodynamic performance [1–18]. However, while many studies have concentrated on the sensitivity analysis of environmental loads on the platform, there remains a notable deficiency in in-depth investigations of the column structure. For example, Bai et al. [1] conducted numerical and experimental analyses of the hydrodynamic response of a ship-shaped semi-submersible aquaculture platform with square columns under wave action by using physical modeling experiments and numerical simulation, in which the effect of the net structure on the platform's motion was also investigated. However, this study failed to consider the column structure in relation to the platform motion; Wang et al. [2] investigated the hydrodynamic interactions between columns and nets for an aquaculture structure with circular columns and proposed a new method to evaluate the hydrodynamic interactions between the aquaculture structure and the environment. However, the study focused on the basic substructure containing only one cylinder and two symmetrically arranged nets, and the effects of neighboring cylinders and nets in the complete platform structure were not investigated, and the dynamic response characteristics of the column structure on the whole platform structure were not taken into account; Yue et al. [5] employed three-dimensional potential flow theory alongside the quasi-static finite element analysis (FEA) method to calculate wave-induced loads and dynamic responses of a semi-submersible aquaculture platform featuring square columns. Their analysis focused on parameters such as wave frequency, wave direction, and draft, but did not consider how column structure parameters might affect the platform's dynamic responses.

Existing studies on the hydrodynamic characteristics of semi-submersible aquaculture platforms primarily focus on the effects of wind and wave current conditions, mooring methods, and net coat design on the platforms' hydrodynamic performance. However, there is a notable lack of research examining the relationship between structural changes in the columns and the dynamic response of these platforms.

This study focuses on a semi-submersible truss-form fishery aquaculture platform operating in the southern waters of the South China Sea, specifically within a deep-sea net-pen aquaculture area in Zhanjiang City, China. This platform represents an innovative type of wind and wave-resistant green intelligent aquaculture equipment, characterised by its semi-submersible structural design. It comprises eight square columns, features four-corner mooring and positioning, and measures 86 m in length, 32 m in width, 16.50 m in height, and has an operating draft of 10.5 m. As a novel aquaculture solution, this semi-submersible platform is preferred for its substantial aquaculture capacity, high structural strength, robust resistance to wind and waves, and advanced level of intelligence [1].

The prevalent column types for fishery aquaculture platforms are square and circular structures [1,2,5]. Currently, there is no research documenting the use of quasi-elliptical column structures in fishery aquaculture platforms. Quasi-elliptical caissons have been employed in related engineering applications owing to their superior performance in reducing fluid resistance and suppressing vortex-induced vibrations [19]. This study presents a quasi-elliptical column structure, integrating it with both square and circular column designs. It thoroughly investigates how variations in column structure affect the dynamic response of a semi-submersible aquaculture platform. The aim is to precisely assess how column structure influences the platform's dynamic characteristics. This research is essential for supplying valuable data for the design and optimisation of aquaculture facilities, thereby mitigating the risk of structural instability or damage. These investigations hold significant importance.

This paper investigates a semi-submersible truss form fishery aquaculture platform, serving as the reference object for the study. Three platform models featuring square, circular, and quasi-elliptical column cross-sectional shapes are designed to examine their dynamic responses under the combined effects of wind, wave, and current. This analysis employs numerical simulation methods alongside three-dimensional potential flow theory and Morrison's equations. The study evaluates hydrodynamic characteristics, including additional mass, radiation damping, first-order wave excitation force, motion response amplitude operators (RAO), and motion responses under irregular waves. Relevant conclusions regarding the impact of column shape on the hydrodynamic performance of semi-submersible platforms are presented.

2. Research Objects

2.1. Platform Model Design

The semi-submersible fishery aquaculture platform model developed in this study comprises a floating body system, which includes columns, pontoons, cross beams, and spars, alongside a mooring system and a net suit. The platform frame adopts a truss design, with the lower float tank and upper deck rigidly connected by columns. These columns are further interconnected by cross beams, resulting in a cohesive structure, as illustrated in Figure 1. Notably, the platform is distinguished by the substantial size of individual columns and the considerable number of column groups.

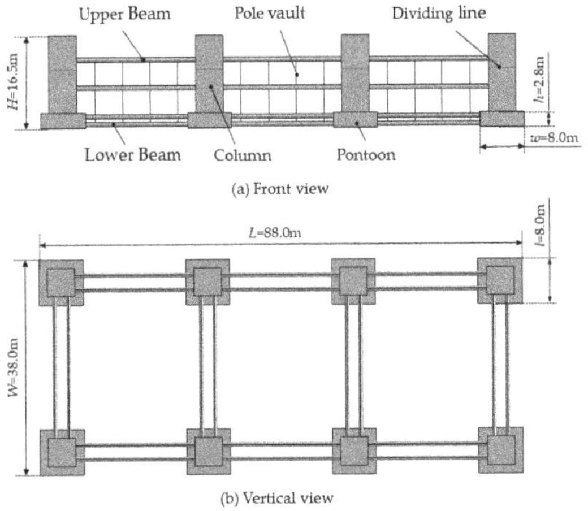

Figure 1. Schematic diagram of square column platform structure.

On the basis of the design of the square column platform, the columns are controlled to have the same or similar volume of the underwater submerged part, i.e., the buoyancy provided by the columns is equal, and the columns of circular section and quasi-elliptical section are modeled respectively. Figure 2 illustrates the cross-sectional views of three distinct column structures alongside the 3D model of the platform. Table 1 details the platform's pertinent parameters.

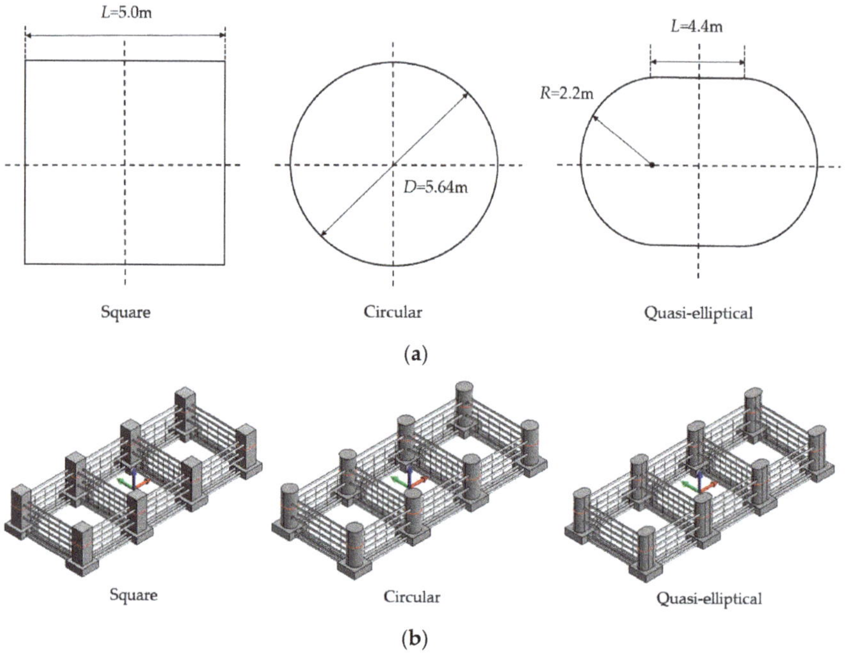

Figure 2. Three types of column structure cross-sectional diagrams and platform 3D models. (**a**) Column cross-sectional view; (**b**) 3D model of the platform.

Table 1. Main parameters of the platform.

Parameters	Square	Circular	Quasi-Elliptical
Platform scale (L × W × H)/m	88 × 38 × 16.5	88 × 38 × 16.5	88 × 38 × 16.5
Pontoon ($l \times w \times h$)/m	8 × 8 × 2.8	8 × 8 × 2.8	8 × 8 × 2.8
Column height/m	13.7	13.7	13.7
Roll Ixx/m	16.58	16.59	16.59
Pitch Iyy/m	30.36	30.38	30.31
Yaw Izz/m	32.93	32.93	32.88
Designed full load mass/t	3358	3360	3349
Design drainage volume/m^3	4575.72	4573.88	4563.15
Center of gravity (x,y,z)/m	0, 0, 4.59	0, 0, 4.59	0, 0, 4.59
Draught/m	10.5	10.5	10.5

2.2. Kinematic Modes and Coordinate Systems

The designed working depth is 100 m, and the environmental loads consider the effects of waves and currents. The semi-submersible aquaculture platform has 6 degrees of freedom under wave loads, which are translation along x-axis, y-axis and z-axis, and rotation around each axis, including Surge, Sway and Heave for translation, and Roll, Pitch and Yaw for rotation.

The right-angle coordinate system $Oxyz$ is established, with the Oxy plane parallel to the hydrostatic water surface. The origin O is positioned at the centre of gravity of the platform. The longitudinal direction of the structure is designated as the x-axis, oriented towards the bow, while the transverse direction is defined as the y-axis. The z-axis extends vertically upwards, passing through the centre of gravity of the platform. The definition of the coordinate system is illustrated in Figure 3.

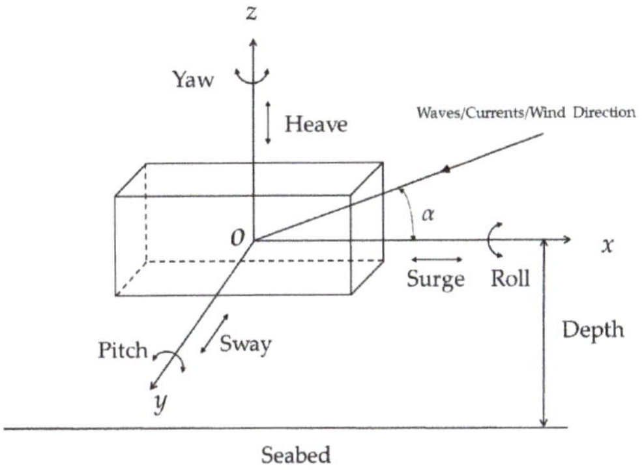

Figure 3. Coordinate system.

2.3. Mooring Systems

The longitudinal structure of the semi-submersible truss-type aquaculture platform is set up with the mooring fixed in the same direction as the 0° down-wave direction, which is along the x-axis positive direction. The platform is positioned by a multi-point mooring system, which consists of four mooring cables with suspended chain lines, arranged symmetrically. One end of the cable is connected to the platform foundation through a cable guide hole, and the other end is fixed to the seabed through an anchoring point.

Based on the consideration of control variables, the mooring system arrangement and cable parameters of the three platforms are kept consistent, Figure 4 shows the form of this mooring system arrangement, and the design parameters of the mooring cable are shown in Table 2.

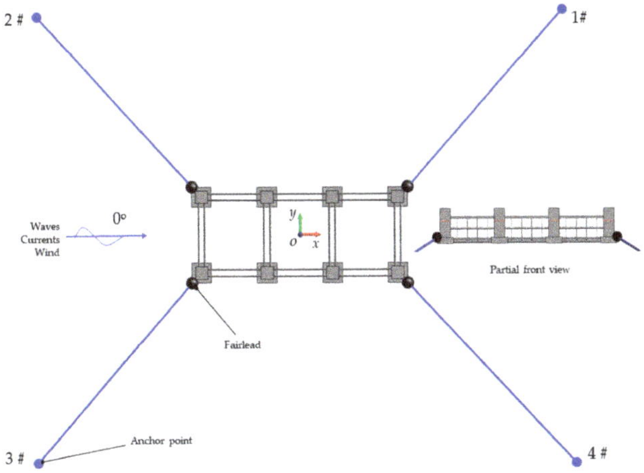

Figure 4. Mooring cable arrangement.

Table 2. Parameters of the mooring system.

Parameter	Length/m	Diameter/m	Wet Weight/(kg·m^{-1})	Axial Stiffness/kN	Additional Quality Factor	Drag Coefficient	Breaking Force/kN
Numerical value	508	0.076	20	23,300	1	1.2	3660

3. Research Methodology

3.1. Environmental Load Calculation

3.1.1. Ocean Random Waves

Wave spectra, as an effective means of describing complex waves, are commonly used, such as the Pierson–Moskowitz spectrum, the JONSWAP spectrum, the Wallops spectrum and the Bretshneider spectrum. Among them, JONSWAP spectrum is a wave spectrum developed on the basis of P-M spectrum, which takes into account the influence of wave development state, and is usually applicable to underdeveloped waves with limited wind distance. The southern part of the South China Sea is a semi-enclosed deep-sea area with a significant limited wind range.

The reference object of the study in the paper is located in the southern sea area of the South China Sea, and the JONSWAP spectrum used in this paper is used to generate random waves [5]. The wave energy spectral density function $S(\omega)$ is expressed as:

$$S(\omega) = \frac{A}{\omega^5} \exp\left[-\frac{5}{4}\left(\frac{\omega_0}{\omega}\right)^4\right] \gamma^{\exp\left[-\frac{(\omega-\omega_0)}{2\gamma\sigma^2\omega_0^2}\right]} \tag{1}$$

where A is the constant; ω_0 is the peak frequency; γ is the peak growth factor; and σ is the peak shape parameter.

3.1.2. Theory of Wave Load Calculation

Based on the ratio of the cross-section characteristic scale D to wavelength λ of semi-submersible fishery aquaculture platform components, the calculation method of platform wave loads is divided into two ways.

(1) For small-sized members with $D \leq 0.2\lambda$, the viscous effect and additional mass effect are mainly considered. For the wave force acting on the gusset, Morrison's equation is:

$$F = F_D + F_I = \frac{1}{2} C_d \rho D |u_r| u_r + C_m \rho \frac{\pi D^2}{4} \dot{u}_r + C_a \rho \frac{\pi D^2}{4} u_W \tag{2}$$

$$u_r = u_W - u_B \tag{3}$$

where ρ is the density of seawater; C_d is the coefficient of drag force; u_r is the velocity of relative motion in the horizontal direction; C_m is the coefficient of inertia; $\dot{u}_r$ is the acceleration of relative motion in the horizontal direction; u_W is the velocity of the water quality point; u_B is the velocity of the floating body motion; and C_a is the coefficient of additional mass, $C_a = C_m - 1$.

(2) For large-size members with $D > 0.2\lambda$, the scattering, bypassing and radiation effects of wave loads on the structure are mainly considered. Wave loads on columns, floats and beams are calculated using the potential flow theory method.

Using the Green function method, the boundary integral equation satisfying the fixed solution condition is solved to obtain the velocity potential function $\Phi(x,y,z,t)$. The velocity potential Φ is decomposed into the incident potential Φ^I, the number of bypassing potentials Φ^D and the radiation potential Φ^R. The wave dynamic pressure p_w distribution is obtained by Bernoulli's equation, and the wave force F_w as well as the wave moment M_w are obtained according to the linearized Bernoulli's equation:

$$p_w = -\rho(\partial\Phi/\partial t) \tag{4}$$

$$F_w = \iint_{S_B} -p_w \boldsymbol{n} \, ds \tag{5}$$

$$M_w = \iint_{S_B} -p_w(\boldsymbol{r} \times \boldsymbol{n})\mathrm{d}s \tag{6}$$

where ρ is the density of seawater; S_B is the wet surface of the platform; $\boldsymbol{n}$ is the outer normal vector of the wet surface of the platform; $\boldsymbol{r}$ is the tangent direction vector of the wet surface of the platform.

The first-order wave force of a semi-submersible fish culture platform consists of the first-order wave excitation force and the radiation force. A unified description of Equations (5) and (6) yields the first-order wave force F_k expression [20].

$$F_k = -\rho\mathrm{Re}\left[\mathrm{i}\omega \mathrm{e}^{-\mathrm{i}\omega t}\iint_{S_\mathrm{B}}\left(\Phi^\mathrm{I}+\Phi^\mathrm{D}\right)n_j\mathrm{d}s\right] - \rho\mathrm{Re}\left[\sum_{j=1}^{6}\mathrm{i}\omega\zeta_j \mathrm{e}^{-\mathrm{i}\omega t}\iint_{S_\mathrm{B}}\Phi_j^\mathrm{R} n_j\mathrm{d}s\right] \tag{7}$$

where i is the imaginary unit; ω is the wave circular frequency; ζ_j is the amplitude of the motion of the platform in the jth degree of freedom direction; $\boldsymbol{n}_j$ is the unit vector normal to the outside of the wet surface of the platform in the jth degree of freedom direction; and Φ_j^R is the spatial component of the radial wave velocity potential per unit amplitude in the jth degree of freedom direction.

The first term of Equation (7) is the first-order wave excitation force (F_EX), which characterizes the wave excitation forces and moments, and is composed of the Froude–Krylov force related to the incidence potential and the wrap-around force related to the wrap-around potential, which is the main wave loading.

The 2nd term of Equation (7) is the radiation force (F_R), expressed in terms of additional mass and radiation damping:

$$F_{\mathrm{R}jk} = \mu_{kj}\ddot{\zeta}_j - \lambda_{kj}\dot{\zeta}_j \tag{8}$$

where F_Rjk denotes the radiative force or moment in the kth degree of freedom due to the jth mode of motion under the action of unit wave amplitude; $\dot{\zeta}_j$ and $\ddot{\zeta}_j$ are the linear motion velocity vector and acceleration vector of the structure with forced motion, respectively; μ_{kj} and λ_{kj} are the additional mass of the platform and the radiative damping, respectively, which are usually frequency-dependent and satisfy the following relationship:

$$\mu_{kj} = \frac{\rho}{\omega}\iint_{S_B}\mathrm{Im}(\Phi_j^R)n_k\mathrm{d}s \tag{9}$$

$$\lambda_{kj} = \rho\iint_{S_B}\mathrm{Re}\left(\Phi_j^\mathrm{R}\right)n_k\mathrm{d}s \tag{10}$$

3.1.3. Current Loads

Currents are generated by a steady, continuous flow of seawater over a large area. The speed of movement of the water quality points in the current changes very slowly, and its force on the platform is simplified as the horizontal drag force F_current. The formula is as follows:

$$F_\mathrm{current} = \frac{1}{2}C_\mathrm{D}\rho A_\mathrm{c}V_\mathrm{c}^2 \tag{11}$$

where C_D is the drag force coefficient, V_c is the design current velocity, and A_c is the plane projection area of the member perpendicular to the flow direction.

3.1.4. Wind Loads

Wind loads predominantly affect the structural elements of the platform situated above the waterline. The calculation formula is as follows:

$$F_{\text{wind}} = \frac{1}{2}\rho_w C_s C_h A_w V_w^2 \tag{12}$$

where ρ_w is the air density; C_s is the shape coefficient of wind load; C_h is the height coefficient of wind pressure; A_w is the orthographic projection area of the wind-affected component; V_w is the wind speed.

3.1.5. Hydrodynamics of Net-Coat Systems

The main hydrodynamic models of netcoat are Morison model and Screen model [21]. In this paper, Morison units are used to simulate the equivalent netcoat. These Morison units are assumed to be small-scale cylinders and are rigidly connected to the platform frame structure so that the whole works as a rigid body [22]. The wave and current hydrodynamic forces acting on a fishing net can be categorized into inertial, drag and lift forces. In this paper, the inertial force [23,24] and the lift force [25,26] are ignored when calculating the hydrodynamic force on the net coat, and only the drag force on the net coat is considered. The force per unit length of netting can be expressed as [27].

$$f_n = \frac{1}{2} C_d \rho D u|u| \tag{13}$$

where C_d is the drag force coefficient in the Morrison model; D is the diameter of the structure; and u is the relative flow velocity perpendicular to the mesh rope, and where the drag force coefficient C_d is determined based on a screen model developed by Loran through mesh drag tests [28].

$$C_D = 0.04 + \left(-0.04 + 0.33S_n + 6.54S_n^2 - 4.88S_n^3\right)\cos\theta \tag{14}$$

where C_D is the drag coefficient in the screen model; θ is the angle between the normal direction of the net and the water flow; and S_n is the denseness of the net.

3.2. Equations of Motion

According to Newton's second law, the equation of motion in the frequency domain for a semi-submersible truss fishery aquaculture platform under the action of a unit amplitude regular wave is:

$$[\boldsymbol{M} + \boldsymbol{M}(\omega)]\ddot{\boldsymbol{x}}(\omega) + \boldsymbol{C}(\omega)\dot{\boldsymbol{x}}(\omega) + \boldsymbol{K}\boldsymbol{x}(\omega) = \boldsymbol{F}_{\text{EX}}(\omega) \tag{15}$$

where $\boldsymbol{M}$ is the platform structural mass matrix; $\boldsymbol{M}(\omega)$ is the additional mass matrix; $\boldsymbol{C}(\omega)$ is the platform structural damping matrix; $\boldsymbol{K}$ is the platform stiffness matrix; $F_{\text{EX}}(\omega)$ is the first-order wave excitation force acting on the platform; $\ddot{x}(\omega)$, $\ddot{x}(\omega)$ and $x(\omega)$ are the acceleration vector, velocity vector, and displacement vector of the platform, respectively.

In order to ensure the accuracy of the numerical simulation results, the indirect time-domain method is used to convert the frequency-domain calculation results to the time-domain for the coupling analysis. The frequency domain calculation results are converted to time domain by inverse Fourier transform, and the coupled equations of motion in time domain for the floating body system, the net-clothing system and the mooring system are:

$$[\boldsymbol{M} + \boldsymbol{M}(\infty)]\ddot{x}(t) + \boldsymbol{C}\dot{x}(t) + \boldsymbol{K}x(t) + \int_0^t \boldsymbol{R}(t-\tau)\dot{x}(\tau)\mathrm{d}\tau = \boldsymbol{F}(t) + \boldsymbol{F}_{\text{n}}(t) + \boldsymbol{F}_{\text{m}}(t) \tag{16}$$

where $\boldsymbol{M}(\infty)$ is the additional mass matrix at infinity frequency; $\boldsymbol{C}$ is the damping matrix; $\boldsymbol{K}$ is the stiffness matrix; $\boldsymbol{R}(t-\tau)$ is the hysteresis function matrix; $\ddot{x}(t)$, $\dot{x}(t)$, and $x(t)$ are the acceleration, velocity, and displacement vectors of the platform at the moment t, respectively; $F(t)$ is the environmental load of the platform's floatation system; $F_n(t)$ is the environmental load of the netsuit system; and $F_m(t)$ is the mooring cable force.

3.3. Hydrodynamic Modeling

The Ansys/Aqwa (ANSYS 2020 R2) software establishes the hydrodynamic calculation grid for the platform. In this grid, the columns, floats, and beams utilise surface cells, with grid sizes of 1.0 m for the upper part and 0.65 m for the lower part of the waterline. The spars employ Morrison cells, featuring a grid size of 0.2 m. Figure 5 illustrates the hydrodynamic model of the square column platform.

Figure 5. Hydrodynamic model of the square column platform.

4. Method Validation

4.1. Validation of Computational Mesh Accuracy

The Haskind relation is usually used to verify whether the mesh partitioning meets the hydrodynamic calculation accuracy requirements, or judged by comparing the consistency of the second-order drift force obtained from the near-field method and the far-field method [28]. In this study, the second-order mean drift force of the longitudinal oscillation of the platform model is solved by using the near-field method and the far-field method, and when the results of the two solution methods have the same trend and are close to the same order of magnitude, it can be assumed that the mesh delineation meets the requirements for the accuracy of hydrodynamic calculations. Taking the square column platform as an example, its calculation results are shown in Figure 6.

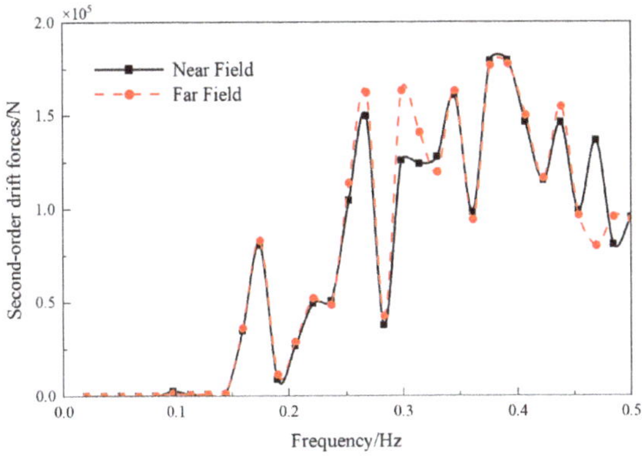

Figure 6. Comparison of second-order average drift forces in Surge.

Figure 6 illustrates that the near-field and far-field methods exhibit strong agreement in the solution results for the second-order mean drift force associated with longitudinal oscillations. The maximum discrepancy observed in the numerical difference is merely 1%. Therefore, it can be concluded that the mesh division sufficiently meets the accuracy requirements for hydrodynamic calculations.

4.2. Validation of Model Rationality

The hydrodynamic models of square, circular and quasi-elliptical column platforms are established, and the initial displacement in the longitudinal swing direction is 8 m, the initial displacement in the vertical swing direction is 2 m, and the initial inclination angle in the longitudinal swing direction is 8°. The free decay motions of three platform types were analysed, with simulations conducted over 600 s. The logarithmic decay rates for these platforms were subsequently calculated, and the findings are presented in Figure 7.

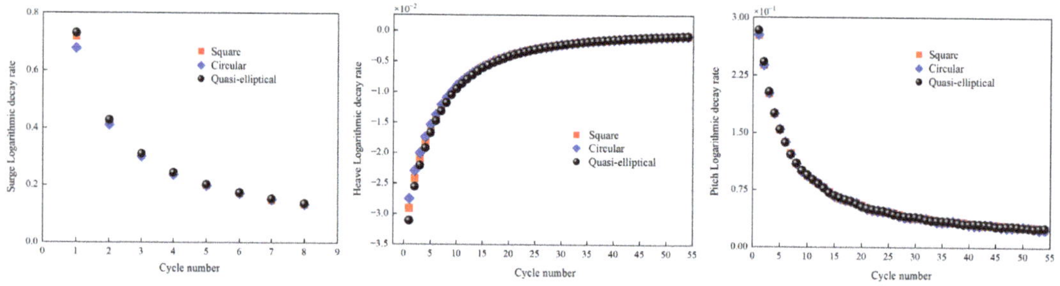

Figure 7. Comparison of logarithmic decay rates among three platforms.

From Figure 7, the free decay results of the three types of platforms in the longitudinal oscillation, pendulum oscillation and longitudinal rocking directions all show a gradual decrease in amplitude over time, all converge to the equilibrium position, and the trend of change is consistent. Based on the free decay results, the intrinsic frequencies of the square, circular and quasi-elliptical column platforms are 0.015 Hz in the longitudinal oscillation direction, 0.093 Hz in the pendulum direction and 0.093 Hz in the longitudinal rocking and pendulum direction. In summary, the three different column platforms all have stable convergence, highly consistent trends and the same intrinsic frequency, indicating the reasonable validity of the three platform hydrodynamic models.

4.3. Comparative Validation

To verify the reliability of the numerical model, the results of the sagging Response Amplitude Operator (RAO) calculation for the square-column semi-submersible aquaculture platform, without net suit mounting, were extracted and compared with data from the literature [5] pertaining to the square-column aquaculture platform. The main parameter comparisons are illustrated in Figure 8.

The comparison reveals that the resultant deviations of the mean and standard deviation between the two datasets are −8.11% and −7.47%, respectively, indicating a relatively small difference. Despite these discrepancies, both models exhibit similar structures, and the trends in the changes of the RAO curves are alike. Therefore, it can be concluded that the calculation method employed in this study demonstrates high accuracy and reliability.

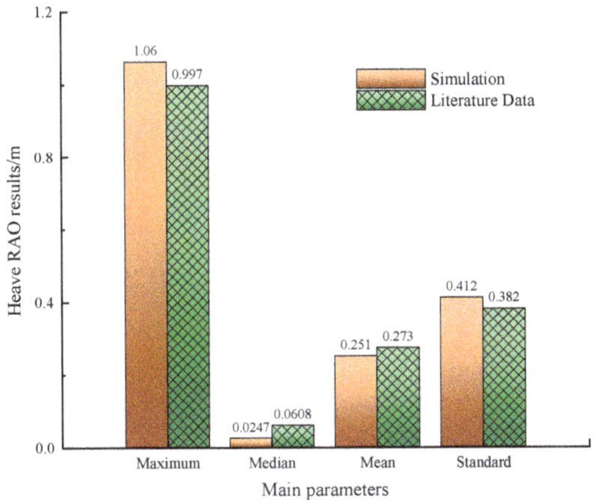

Figure 8. Comparison of simulation results with reference data.

5. Results

5.1. Frequency Domain Response Analysis

Thirty regular wave frequencies in the range 0.02–0.5 Hz were selected as the design waves, and wave directions from −180° to 180° at 15° intervals were used; frequency-domain analysis was performed with ANSYS/Aqwa. This paper presents only the dynamic responses in three degrees of freedom—surge, heave and pitch—under those waves.

5.1.1. Additional Mass and Radiation Damping

From Equation (8), it can be seen that the additional mass is directly proportional to the acceleration of the breeding platform motion, and the radiation damping is directly proportional to the speed of the breeding platform motion, and at the same time, the additional mass and the radiation damping are also a function of the shape of the breeding platform and the vibration frequency.

Figure 9 shows how added mass in surge, heave and pitch varies with wave frequency for square, circular and quasi-elliptical column platforms; the vertical axis denotes added mass per unit wave amplitude.

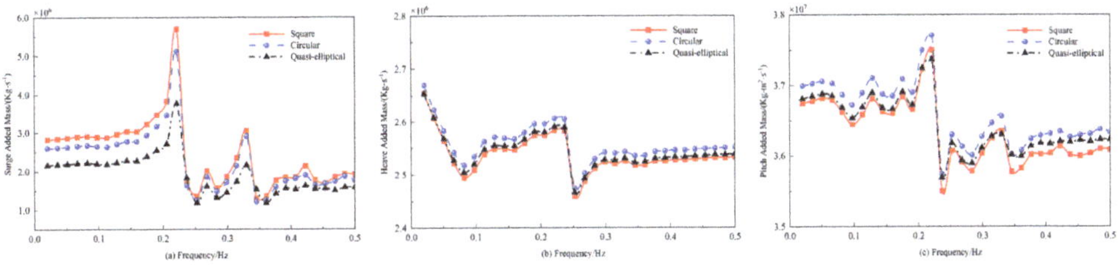

Figure 9. Comparison of added mass in frequency domain. (**a**) Surge added mass; (**b**) Heave added mass; (**c**) Pitch added mass.

Figure 9 shows that the added-mass trends for the three platforms are broadly consistent. In the surge direction, the added mass of all platforms is highly sensitive to frequency between 0.2 and 0.45 Hz, exhibiting marked oscillations and extreme values. Overall, the added-mass magnitudes follow the order: square-column platform > circular-column platform > quasi-elliptical-column platform. The maximum added-mass peaks for all three platforms occurred at 0.221 Hz, with substantial numerical differences: the square-column platform had the largest value, while the quasi-elliptical-column platform had the smallest, lower by approximately −40%.

In heave, the added mass of all three platforms varied relatively smoothly with frequency, and the overall ranking by magnitude was: circular-pillar platform > quasi-elliptical-pillar platform > square-pillar platform. The maximum added-mass peak for each platform occurred at the initial response frequency of 0.02 Hz, and the peak magnitudes differed only slightly among the platforms.

In pitch, the added-mass trends resembled those in surge, and the overall magnitude ranking matched that in heave, remaining circular-pillar platform > quasi-elliptical-pillar platform > square-pillar platform. The peak frequency coincided with that in surge (0.221 Hz), and the peak magnitudes again showed only small differences among the three platforms.

Figure 10 presents the response curves of radiation damping per unit wave amplitude for the square-column, circular-column and quasi-elliptical-column platforms under heave, surge and pitch motions, plotted against wave frequency. The vertical axis denotes the radiation damping per unit wave amplitude. Overall, the three platform types exhibit similar trends in radiation damping with frequency.

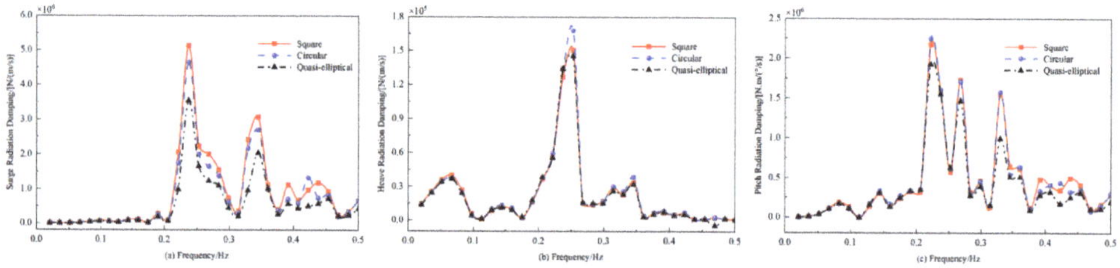

Figure 10. Comparison of radiation damping in frequency domain. (**a**) Surge radiation damping; (**b**) Heave radiation damping; (**c**) Pitch radiation damping.

For surge motion, when the frequency lies in the 0.2–0.469 Hz range, radiation damping is markedly sensitive to frequency, showing pronounced oscillations and multiple extrema. The primary peaks for all three platforms occur near 0.237 Hz, although their magnitudes differ substantially: the square-column platform attains the largest peak, while the quasi-elliptical-column platform has the smallest, with a difference of −30%. Secondary peaks appear near 0.345 Hz, where the values for the square-column and quasi-elliptical-column platforms differ by −33%.

In heave, the radiation damping of all three platform types is highly sensitive to frequency, with responses concentrated mainly in the 0.2 Hz to 0.3 Hz band. Each configuration exhibited its largest peak at 0.252 Hz, though the peak magnitudes differed markedly: the circular-column platform showed the highest value, while the square- and quasi-elliptical-column platforms showed the lowest, with a maximum difference of −13%.

In pitch, the radiation damping of each platform was strongly frequency-sensitive, with responses concentrated mainly in the 0.2–0.469 Hz band and accompanied by pronounced oscillations that produced multiple extrema. The magnitudes of the primary peak (0.221 Hz), the secondary peak (0.267 Hz) and the tertiary peak (0.329 Hz) followed the pattern: square-column platform highest and quasi-elliptical-column platform lowest, with inter-peak differences of −15%, −14% and −37%, respectively.

5.1.2. First-Order Wave Excitation Forces

Figure 11 illustrates the characteristics of first-order wave excitation forces on square, circular, and quasi-elliptical column platforms in surge, heave, and pitch directions as the wave frequency varies. The vertical axis represents the wave excitation force corresponding to a unit wave amplitude.

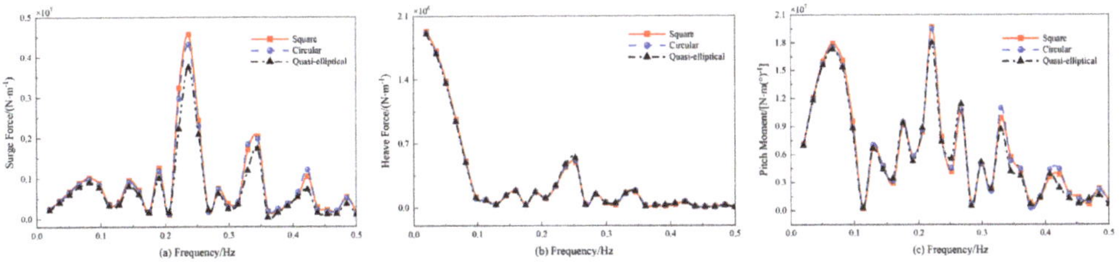

Figure 11. Comparison of first order wave excitation force. (**a**) Surge wave excitation force; (**b**) Heave wave excitation force; (**c**) Pitch wave excitation force.

Figure 11 illustrates that the wave-induced excitation forces in surge, heave, and pitch directions exhibit similar patterns across the three platforms. In the surge direction, all platforms demonstrate high sensitivity to frequency changes in the high-frequency range, with significant fluctuations and multiple extremum points during oscillation. The primary peak for all platforms occurs at a frequency of 0.237 Hz. Notably, the quasi-elliptical column platform has the smallest response amplitude, 18% lower than that of the largest, the square column platform. Other secondary peaks appear near frequencies of 0.19 Hz, 0.237 Hz, 0.345 Hz, and 0.422 Hz, where the response differences among the platforms are relatively minor.

In the heave direction, the wave-induced forces on the three platforms predominantly concentrated in the low-frequency range. The maximum peak occurred at the initial response frequency of 0.02Hz, with all three platforms exhibiting similar peak amplitudes. At a frequency of 0.252Hz, a significant oscillation peak was observed, where the square column platform displayed the smallest value, differing by −9.75% from the maximum value of the quasi-elliptical column platform. The study indicated the presence of minor local high-frequency responses in the heave direction.

In the pitch direction, the wave-induced forces on the three platforms demonstrated high sensitivity to high-frequency variations, exhibiting considerable fluctuations with multiple extrema. The main peak appeared at 0.221 Hz, where the quasi-elliptical column platform had the smallest value, differing by −8.5% from the maximum value of the square column platform. Other peaks were observed at frequencies of 0.066 Hz, 0.175 Hz, 0.267 Hz, and 0.329 Hz, with minimal differences among the three platforms at these peaks.

5.1.3. RAO Frequency Domain Motion Response

Figure 12 illustrates the variation in the Response Amplitude Operators (RAO) with wave frequency for square column platforms, circular column platforms, and quasi-elliptical column platforms in the surge, heave, and pitch directions. The vertical axis represents the RAO corresponding to a unit wave amplitude.

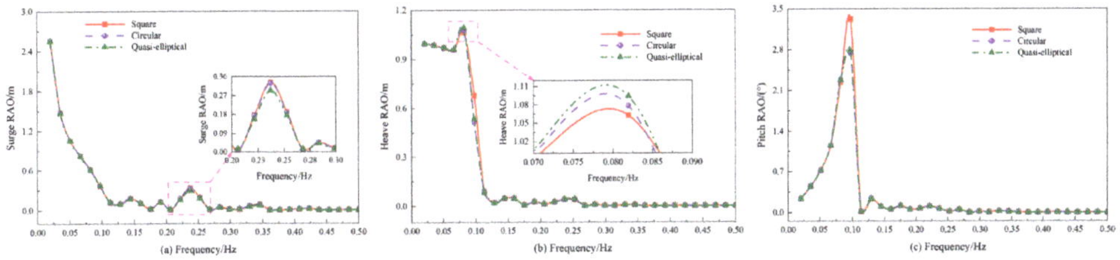

Figure 12. Comparison of RAO motion response. (**a**) Surge RAO; (**b**) Heave RAO; (**c**) Pitch RAO.

Figure 12 shows that, overall, the RAO amplitudes of the three platforms in all three degrees of freedom decrease gradually as wave frequency increases, and after small oscillations they stabilise; the patterns of variation are broadly consistent. In surge, the main responses concentrate in the low-frequency range, with the largest response peak occurring at the initial response frequency of 0.02 Hz for all platforms; the peak amplitudes are essentially identical (approximately 2.56 m). In the high-frequency region there is a small oscillation at 0.236 Hz. At that frequency the quasi-elliptical-column platform has the smallest peak (0.294 m), followed by the circular-column platform (0.329 m), and the largest peak is observed for the square-column platform (0.333 m). The peak values for the circular and square platforms are almost the same, but the difference between the smallest and largest values is −12%. In summary, the quasi-elliptical-column platform exhibits superior motion characteristics in surge.

In heave, the predominant response occurred below 0.128 Hz. All maximum response peaks were at 0.082 Hz, with the square-column platform showing the smallest peak (1.06 m), the circular-column platform 1.08 m, and the quasi-elliptical platform the largest at 1.09 m. The difference between the minimum and maximum peaks was −2.8%, indicating that the heave characteristics of the three platforms were essentially identical.

In pitch, the principal response occurred below 0.112 Hz. All maximum peaks occurred at 0.097 Hz, with the circular-column platform reaching 2.74°, the quasi-elliptical platform 2.78°, and the square-column platform 3.32°. The circular and quasi-elliptical platforms therefore had nearly identical, and the smallest, peaks; the difference between the minimum and maximum values was −17%. Thus, the circular and quasi-elliptical platforms exhibited the most favourable pitch motion characteristics.

5.2. *Time Domain Response Analysis*

The most notable dynamic response of the semi-submersible truss aquaculture platform is the change in spatial pose under environmental loading, specifically translational displacements and rotational angles. To evaluate the platform's dynamic response accurately, one must solve its global equations of motion in the time domain and account for the coupling between the mooring system and the platform structure, thereby revealing its motion characteristics.

5.2.1. Marine Environmental Parameters

According to the results of China's domestic literature, the maximum wave height of the sea state in the South China Sea in one year is 2.7 m. In order to better evaluate the kinematic performance of the semi-submersible truss fishery aquaculture platform, the influence of the one-year marine environment on the platform is selected to be studied in this paper. Irregular waves are generated based on the JONSWAP spectrum, with a meaningful wave height H_s of 2.7 m, a spectral peak period T_p of 8.3 s, and a spectral peak factor γ of 3.3. The current load velocity of 0.9 m/s. The frequency calculation range of the selected frequency is 0.07 Hz ~ 0.5 Hz, and the angle of incidence of the wave is 0°, and the length of the simulation is 3600s.

5.2.2. Time Domain Motion Response

Table 3 presents the time-domain statistics of the heave, sway and surge responses for the three platforms; the maximum value denotes the largest absolute response.

From Table 3 it can be seen that, in the surge direction, the circular-column platform exhibits the smallest maximum, mean and standard deviation. Compared with the square-column platform, these three metrics are reduced by 36.8%, 44.2% and 33.2%, respectively; compared with the quasi-circular column platform, they are reduced by 35.1%, 42.7% and

29.4%, respectively. These results indicate that the circular-column platform performs best in surge response, followed by the quasi-elliptical-column platform.

Table 3. Statistical results of time-domain motion response of three kinds of platforms.

Degree of Freedom	Column Form	Numerical Value		
		Maximum	Mean	Standard
Surge/m	square	6.36	4.67	0.45
	circular	4.02	2.61	0.30
	quasi-elliptic	6.20	4.56	0.43
Heave/m	square	6.32	−5.61	0.43
	circular	6.40	−5.55	0.51
	quasi-elliptic	6.48	−5.70	0.48
Pitch/(°)	square	1.75	−0.10	0.76
	circular	1.73	−0.07	0.79
	quasi-elliptic	1.86	−0.10	0.83

In the heave direction, the three platforms showed minimal differences in maximum and mean values, with all deviations within 4.5%. For standard deviation, the square-column platform recorded the smallest value, differing by −16.7% and 10.9% relative to the circular- and quasi-elliptical-column platforms, respectively. Overall, the circular-column platform produced the smallest heave motion, with the square-column platform ranking second.

In the pitch direction, maximum and mean values for all three platforms were very small and nearly identical. The square-column platform again had the smallest standard deviation, while the quasi-elliptical-column platform had the largest; the maximum difference between extremes was −8.2%. Thus, the circular-column platform exhibited the smallest pitch motion, followed by the square-column platform.

For clarity, Figure 13 shows the motion response curves only during the period when the square-column platform attains its maximum response. As shown in Figure 13, in the surge, heave and pitch directions all three platforms underwent aperiodic reciprocating motion, oscillating strongly about the median and exhibiting peaks in translational displacement and rotational deflection angle.

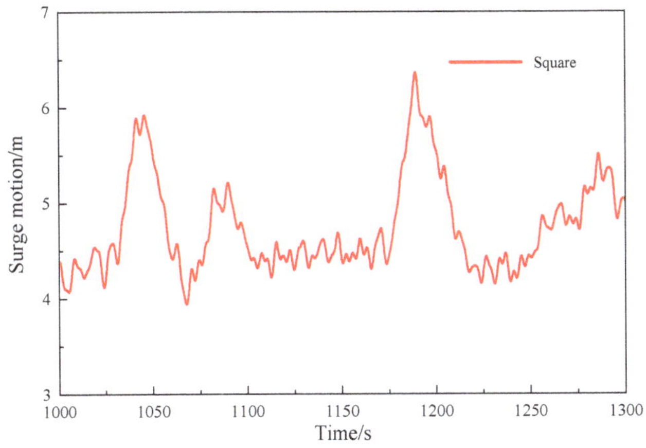

(a)

Figure 13. *Cont.*

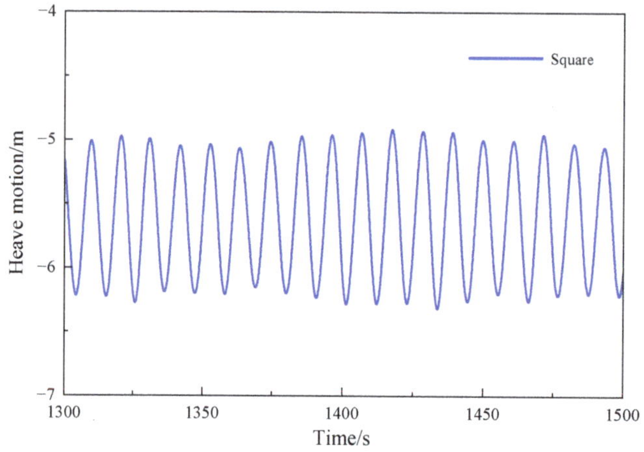

(**b**)

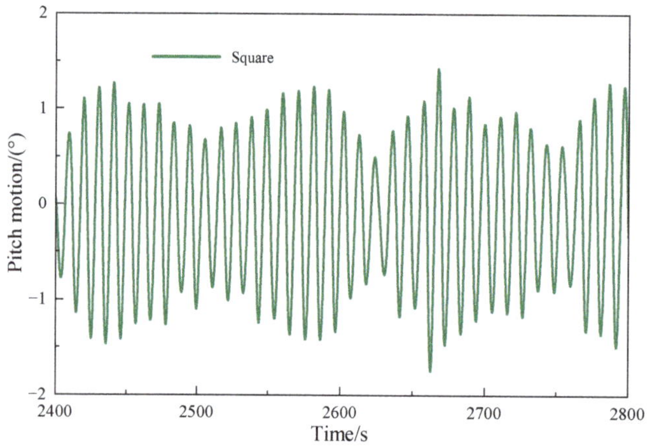

(**c**)

Figure 13. Comparison of three kinds of platform motion response time history curves. (**a**) Surge position; (**b**) Heave position; (**c**) Pitch position.

5.2.3. Power Spectrum Analysis

In order to be able to assess the difference in energy components of the time-domain course motion response at different frequencies, the time-domain course responses of the three platforms under the three degrees of freedom of longitudinal oscillation, vertical oscillation and longitudinal rocking were analyzed by spectrum analysis, and the characteristic curves of the spectrum analysis are shown in Figure 14, respectively, where the vertical coordinate is the value of the power spectral density (PSD).

Figure 14 illustrates that, in the surge direction, the energy distribution for the three platform types is primarily concentrated within the frequency range below 0.028 Hz. Within this range, several energy peaks are evident at approximately 0.0028 Hz, 0.0056 Hz, 0.011 Hz, 0.0167 Hz, 0.0194 Hz, 0.022 Hz, and 0.025 Hz. Notably, 0.011 Hz emerges as the dominant frequency of the motion spectrum. In terms of peak values, the square column platform exhibits the highest peak, followed by the circular column platform, with the quasi-elliptical column platform demonstrating the lowest peak. Conversely.

In the heave direction, the energy distribution for all three platforms is primarily centred around a frequency of 0.093 Hz. Regarding peak power spectral density values, the circular column platform records the highest, the quasi-elliptical column platform follows, and the square column platform registers the lowest.

In the pitch direction, the energy distribution of the three platforms aligns with that in the heave direction, with the dominant frequency of the motion spectrum at 0.093 Hz. A secondary peak emerged around 0.109 Hz, close to the spectral peak frequency of 0.12 Hz, demonstrating distinct wave frequency characteristics. Regarding peak values, the quasi-elliptical column platform exhibited the highest, followed by the circular column platform, with the square column platform displaying the lowest.

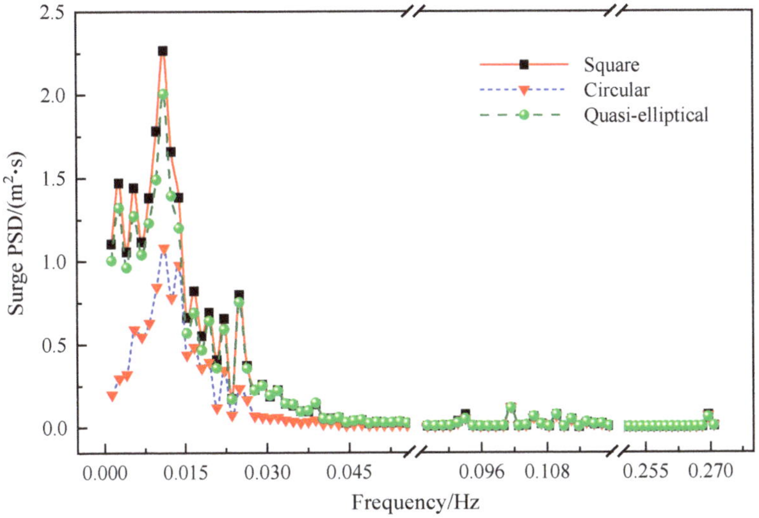

(a)

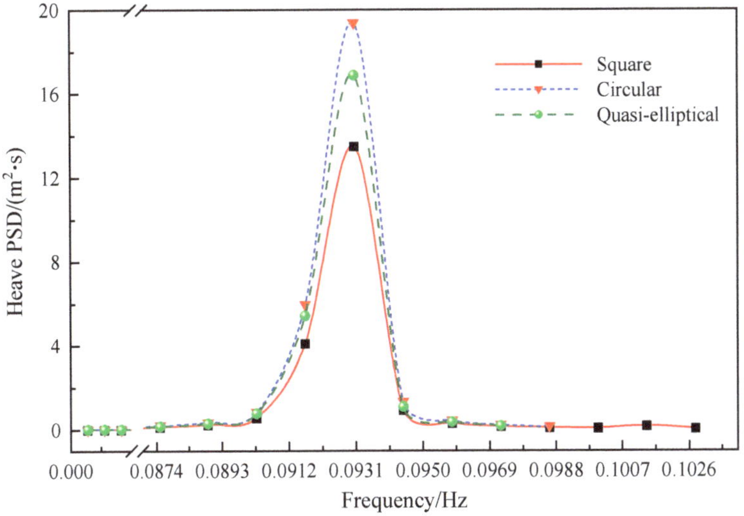

(b)

Figure 14. *Cont.*

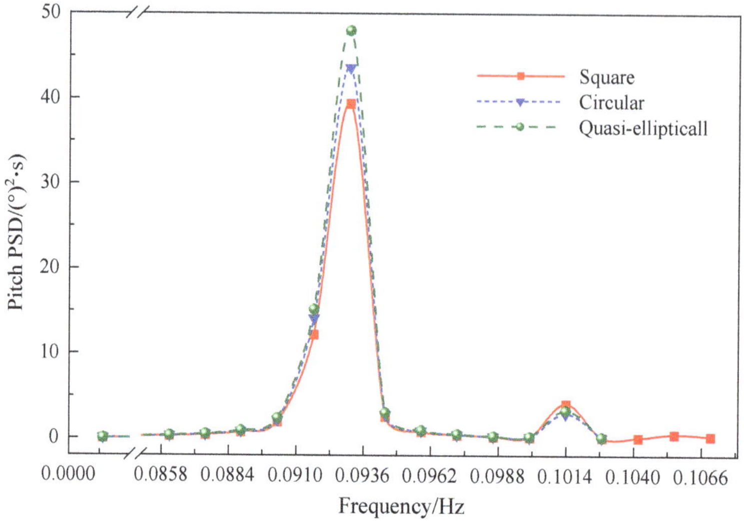

(c)

Figure 14. Comparison of Time-domain history spectrum analysis. (**a**) Surge PSD; (**b**) Heave PSD; (**c**) Pitch PSD.

5.2.4. Mooring Cable Tension Response

Mooring-line tension is also an important indicator for assessing the motion performance of the platform. Typically, the ratio of mooring-line breaking strength to mooring-line tension defines the safety factor; according to the API-RP-2SK design code, the platform must satisfy an allowable safety factor of 1.67. Table 4 presents the maximum mooring-line tensions and safety factors for a 0° wave heading.

Table 4. Maximum tension and safety factor of mooring cables.

Number	F_{max}/N			Safety Factor		
	Square	Circular	Quasi-Elliptical	Square	Circular	Quasi-Elliptical
Cable1	557,916.50	372,640.59	541,339.31	6.56	9.82	6.76
Cable2	199,291.77	219,349.08	196,547.42	18.37	16.69	18.62
Cable3	199,219.20	219,418.30	196,352.98	18.37	16.68	18.64
Cable4	558,366.00	373,843.19	537,501.69	6.55	9.79	6.81

From Table 5, the maximum tensions for all three platforms occur at Cable1 and Cable4; the square-column platform exhibits the highest values, followed by the quasi-elliptical-column platform, with the circular-column platform showing the lowest values. All three platforms meet the strength requirements; the circular-column platform has the highest safety factors at Cable1 and Cable4, the quasi-elliptical-column platform has intermediate values, and the square-column platform has the lowest.

5.2.5. Dynamic Response of the Platform Under the Combined Action of Wind, Waves, and Currents

Previous studies primarily focused on the dynamic response characteristics under wave and current influences, which may present significant limitations. Consequently, it is essential to investigate the platform's motion response under combined sea conditions of wind, waves, and currents. In light of the South China Sea's conditions, a rated wind speed of 11.4 m/s has been chosen. The wind, waves, and currents are aligned in the same direction, with the incident angle corresponding to the wave direction.

Table 5. Statistical Results of Three Platform Motion Responses Under Combined Wind, Wave, and Current Actions.

Degree of Freedom	Column Form	Numerical Value		
		Maximum	Mean	Standard
Surge/m	square	6.47	4.79	0.45
	circular	4.16	2.80	0.30
	quasi-elliptic	6.30	4.66	0.43
Heave/m	square	6.32	−5.61	0.42
	circular	6.40	−5.55	0.51
	quasi-elliptic	6.48	−5.70	0.48
Pitch/(°)	square	1.75	−0.10	0.76
	circular	1.73	−0.07	0.79
	quasi-elliptic	1.87	−0.10	0.83

Table 5 presents the statistical results of the time-domain motion responses for the three platforms in the surge, heave, and pitch directions. The maximum value indicates the highest absolute value of these responses.

Tables 3 and 5 reveal that, under the combined influence of wind, waves, and currents, the surge displacement of the three types of platform motion responses is slightly greater than when only waves and currents are considered. However, the differences in heave and pitch responses remain minimal. Specifically, the response of the square-column platform increased by approximately 1.73%, the circular-column platform by about 3.48%, and the quasi-elliptical-column platform by around 1.61%. These findings suggest that the dynamic response characteristics of the semi-submersible aquaculture platform do not significantly alter when wind is added to the combination of waves and currents. This may be attributed to the wind-exposed height above the waterline being just 6 m, with heave further reducing this height, thus resulting in a relatively small wind-exposed area.

Table 6 displays the maximum tensions experienced by the mooring lines of three different platform types when subjected to the combined effects of wind, waves, and currents.

Table 6. The maximum mooring tension of Cable.

Number	Square	Circular	Quasi-Elliptical
Cable1/N	569,087.75	381,168.91	550,840.31
Cable2/N	198,991.63	216,696.84	196,257.08
Cable3/N	198,913.08	216,768.31	196,068.02
Cable4/N	569,511.06	382,422.22	546,967.25

Tables 4 and 6 illustrate that, when subjected to the combined forces of wind, waves, and currents, the mooring tension in Cable1 and Cable4 increased, whereas Cable2 and Cable3 experienced a slight decrease. Among the platforms, the circular-column platform exhibited the most significant increase in tension at 2.29%, followed by the square-column platform at 2%, with the quasi-elliptical column platform showing the smallest increase at 1.76%. Conversely, the decrease in tension across the three platforms was negligible.

6. Discussion

6.1. Frequency-Domain Response Characteristics

(1) Added Mass and Radiation Damping

In terms of added mass, according to the calculation results in Section 5.1.1, the variation trends of the frequency affected by the added mass for the three platforms are similar, and the sensitivity is obvious in the motions in all directions.

In the surge direction, the added mass of the quasi-elliptical column platform is the smallest, which is related to the fact that the longitudinal projected area of this model is the smallest. The added mass of the square column platform is the largest, followed by the circular column platform. In the heave direction, although there are differences in the added masses of the three platforms, the deviations are very small, which is related to the fact that the vertical projected areas of the three platforms are almost the same. In the pitch direction, there are also differences among the three platforms, but the deviations are very small.

In terms of radiation damping, the variation trends of the three platforms with respect to frequency are the same as those of the added mass, also showing basically consistent trends, with an increasing degree of sensitivity and multiple peaks and troughs. In the surge direction, the radiation damping of the quasi-elliptical column platform is the smallest, which is also affected by the projected area. In the heave and pitch directions, the radiation damping of the circular column platform is the largest, that of the square column platform is moderate, and that of the quasi-elliptical column platform is the smallest.

The above results indicate that the added mass and radiation damping of the quasi-elliptical column platform are either the smallest or tend to be insufficient. For the operation of semi-submersible aquaculture platforms, heave and pitch are the primary motion modes of concern. The differences in the maximum peaks of the quasi-elliptical column platform in the heave and pitch directions compared with those of the square and circular column platforms are not very significant. These are all aspects worthy of attention.

In addition, the heave and pitch radiation damping values of the three platforms in the low-frequency region are extremely small, indicating that the platforms will not generate significant radiation waves under low-frequency response conditions. Moreover, the peak frequencies of the maximum added mass and maximum radiation damping of the three platforms are the same, and in the high-frequency region, the high-frequency response is obvious. These are also aspects worthy of attention.

(2) First-order wave excitation force

According to the calculation results in Section 5.1.2, the variation laws of wave excitation forces of the three platforms are similar. They are significantly sensitive to frequency and exhibit multiple wave peaks and troughs. The wave excitation forces of the quasi-elliptical column platform in the surge, heave, and pitch directions are all the smallest, while those of the square column platform are the largest. The wave excitation forces of the circular column platform are moderate, but the difference from those of the square column platform is not obvious.

The peak frequencies of the maximum wave excitation forces of the three platforms are all the same and are in the high-frequency region, showing obvious high-frequency characteristics. Additionally, the phenomenon of multiple wave crests and troughs in surge and pitch was observed, which may be attributed to the special internal structure of the semi-submersible aquaculture platform with multiple compartments and multiple columns. The cancellation effect of wave loads on the platform may lead to the occurrence of wave troughs.

(3) RAO frequency-domain motion response

The calculation results in Section 5.1.3 indicate that the motion variation patterns of the three types of platforms are similar. Among them, the quasi-elliptical column platform has the optimal motion characteristics, followed by the circular column platform, and the square column platform ranks last. It is observed that the peak frequencies of the maximum motion responses of the three platforms are all the same and occur in the low-frequency region, suggesting obvious low-frequency characteristics and the highest sensitivity to low-frequency waves. Meanwhile, the extreme frequencies of the aquaculture platform in the heave and pitch directions are around 0.082 Hz and 0.097 Hz, which are close to the natural frequencies, intensifying the platform's motion and even leading to resonance. Additionally, small-amplitude oscillations in the high-frequency region in the surge direction are noticed. Although the oscillation level is very small, the influence of wave frequencies in severe sea conditions needs attention.

6.2. Time-Domain Response Characteristics

Based on the calculation results in Section 5.2.2, in the directions of surge, heave, and pitch, all three types of platforms undergo non-periodic reciprocating motions around the median, with peaks in translational displacement and rotational deflection angle. The circular column platform has the smallest motion response values in all three directions. The square column platform has relatively small response values in the heave and pitch directions, while the quasi-elliptical column platform has relatively small response values in the surge direction. Focusing on the motion responses in the heave and pitch directions, the circular column platform shows the most significant advantages. There are differences between the square column platform and the quasi-elliptical column platform, but they are not very obvious.

In terms of the motion spectrum, the data results in Section 5.2.3 indicate that the energy peak frequency band in the surge direction is relatively wide, showing obvious low-frequency characteristics and being sensitive to low-frequency incident waves. The PSD peaks show that the motion of the square-column platform is the most intense, followed by that of the circular-column platform, and the quasi-elliptical column platform has the smallest motion. The dominant frequency (0.011 Hz) is close to the natural surge frequency (0.015 Hz), which is likely to cause resonance responses.

The prominent peaks in the heave direction indicate that the platform motion is extremely intense, which is most evident for the circular-column platform and least for the square-column platform. The dominant frequencies all coincide with the natural frequency in the heave direction (0.093 Hz), which can easily lead to severe resonance responses and intensify the platform's heave motion. The response in the pitch direction is similar to that in the heave direction, except that it is most evident for the quasi-elliptical-column platform and least for the square-column platform.

The research also found that the low-frequency components of the three types of platforms dominate the surge, heave, and pitch motions; there are local high-frequency motions, but they are not obvious. The dominant frequencies of the three types of platforms are close to or coincide with the natural frequencies, which are likely to cause resonance responses. In particular, they intensify the motions of the platform in the heave and pitch directions, posing a danger to the structural safety and operational efficiency of the platform.

In terms of mooring cable tension response, the results in Section 5.2.4 indicate that the circular column platform has the highest safety factor, while the square column platform has the lowest. The safety factor of the quasi-elliptical column platform is moderate, but the difference between it and the square column platform is small. It can be seen that the circular and quasi-elliptical column platforms have significant advantages in terms of mooring cables.

Section 5.2.5 reports the dynamic response of the platform under combined wind and wave currents. The quasi-elliptical column platform exhibited the most favourable dynamic response, followed by the square column platform. In particular, the quasi-elliptical riser platform performed best in mitigating longitudinal oscillations induced by wind and wave currents, thereby reducing the risk of platform motion instability and mooring failure. Comparing the frontal areas of the three column types in the incident direction, the quasi-elliptical column platform had the smallest area (161.4 m^2), the square column platform was larger (176 m^2), and the circular column platform was the largest (191.84 m^2). The superior dynamic responses of the quasi-elliptical and square column platforms may be attributed to their column geometries.

7. Conclusions

In this study, a semi-submersible truss-type fishery aquaculture platform was taken as the research object, and three platform models with square, circular, and quasi-elliptical column cross-sectional shapes were designed. Based on the potential flow theory and the Morrison equation, through numerical simulation methods, the response characteristics of the platforms with three column forms under the combined action of wind, waves, and currents were systematically studied from aspects such as theoretical methods and analysis means. These characteristics include added mass, radiation damping, first-order wave excitation force, motion RAO, and motion response under irregular waves. The differences in the responses of the semi-submersible platforms caused by different column forms were also compared and analyzed.

(1) The quasi-elliptical column platform exhibits the lowest first-order wave excitation forces in the surge, heave, and pitch directions when considering frequency-domain characteristics. The circular column platform follows, while the square column platform experiences the highest forces. In terms of frequency-domain motion characteristics, the quasi-elliptical column platform performs best, the circular column platform ranks second, and the square column platform shows relatively poor performance.

(2) Time-domain characteristics under the combined influence of waves and currents: The circular column platform exhibited the smallest motion response values across all three degrees of freedom. In contrast, the square column platform showed relatively low response values in the heave and pitch directions, while the quasi-elliptical column platform demonstrated a smaller response in the surge direction. Regarding mooring cable tension safety factors, the circular column platform achieved the highest, followed by the quasi-elliptical column platform with a moderate factor, and the square column platform with the lowest. Thus, the circular and quasi-elliptical column platforms offer notable advantages in cable mooring.

(3) Time-domain characteristics under the combined influence of wind, waves, and currents: The quasi-elliptical column platform exhibited the most favourable dynamic response, followed by the square column platform. In contrast, the circular column platform demonstrated a comparatively weaker response.

This paper presents a numerical analysis of the hydrodynamic responses of a semi-submersible aquaculture platform subjected to sea winds, waves, and ocean currents. However, certain limitations persist. For instance, the platform's responses under extreme wind, wave, and current conditions warrant further investigation. In particular, the impact of typhoons and the associated wave currents should be examined, given the frequent occurrence of typhoons affecting aquaculture platforms in the South China Sea. Moreover, future research could explore whether the mechanical properties of mooring lines in the direction of incident winds, waves, and currents differ from those of other mooring lines, especially when wind, wave, and current directions are non-collinear. This study aims to assess the influence of column forms on the dynamic responses of a semi-submersible truss-type fishery aquaculture platform, offering valuable insights for the design and optimisation of aquaculture facilities. Future research might also consider a 45° rotation of the square columns to examine the sensitivity of column angle changes to the incident direction of loads.

Although this study has identified some hydrodynamic performance advantages of the quasi-elliptical column structure compared to the commonly used circular and square column structures, experimental tests are lacking. Further preparation of test models should be carried out to provide feasibility guidance for the future engineering application of the quasi-elliptical column structure in semi-submersible aquaculture platforms.

Author Contributions: Conceptualization, K.L. and D.Z.; methodology, K.L.; software, K.L.; validation, K.L. and D.Z.; formal analysis, K.L.; investigation, D.Z.; resources, D.Z.; data curation, K.L.; writing—original draft preparation, K.L.; writing—review and editing, K.L.; visualization, D.Z.; supervision, D.Z.; project administration, D.Z.; funding acquisition, K.L. All authors have read and agreed to the published version of the manuscript.

Funding: This research was funded by the Zhanjiang Science and Technology Key Project of China (No. 2024B01083).

Data Availability Statement: The original contributions presented in this study are included in the article. Further inquiries can be directed to the corresponding author.

Conflicts of Interest: The authors declare no conflict of interest.

References

1. Bai, X.D.; Lu, Z.; Luo, H.B.; Guo, M.; Yang, C. Experimental and numerical investigations on hydrodynamic response of vessel-shaped semi-submersible aquacultural platform. *Aquac. Eng.* **2025**, *110*, 102540. [CrossRef]
2. Wang, H.B.; Zhang, X.Y.; Zhang, X.T.; Li, X.; Tian, X.; Shen, Y.; Song, W. Experimental investigations on hydrodynamic interactions between the cylinder and nets of a typical offshore aquacultural structure in steady current. *Mar. Struct.* **2023**, *88*, 103367. [CrossRef]
3. Liu, H.F.; Zhao, T.H.; Liu, Y. Investigation on dynamic performance of semi-submersible aquaculture platform in two mooring forms. *Ocean Eng.* **2024**, *297*, 117092. [CrossRef]
4. Liu, H.F.; Liu, Y. Numerical investigation on the dynamic response of the semi-submersible aquaculture platform in regular waves. *Ocean Eng.* **2024**, *294*, 116718. [CrossRef]
5. Yue, W.Z.; Wang, W.S.; Sheng, S.W.; Ye, Y.; Hong, T. Analysis of the wave load and dynamic response of a new semi-submersible wave-energy-powered aquaculture platform. *Ocean Eng.* **2022**, *248*, 110346. [CrossRef]
6. Wang, G.; Martin, T.; Huang, L.Y.; Bihs, H. Numerical investigation of the hydrodynamics of a submersible steel-frame offshore fish farm in regular waves using CFD. *Ocean Eng.* **2022**, *256*, 111528. [CrossRef]
7. Zhao, Y.P.; Liu, H.F.; Bi, C.W.; Cui, Y.; Guan, C.-T. Numerical study on the flow field inside and around a semi-submersible aquaculture platform. *Appl. Ocean Res.* **2021**, *115*, 102824. [CrossRef]
8. Zhao, Y.; Guan, C.; Bi, C.W.; Liu, H.; Cui, Y. Experimental investigations on hydrodynamic responses of a semi-submersible offshore fish farm in waves. *J. Mar. Sci. Eng.* **2019**, *7*, 238. [CrossRef]
9. Liu, H.; Bi, C.; Zhao, Y. Experimental and numerical study of the hydrodynamic characteristics of a semisubmersible aquaculture facility in waves. *Ocean Eng.* **2020**, *214*, 107714. [CrossRef]
10. Miao, Y.J.; Ding, J.; Tian, C.; Chen, X.-J.; Fan, Y.-L. Experimental and numerical study of a semi-submersible offshore fish farm under waves. *Ocean Eng.* **2021**, *225*, 108794. [CrossRef]
11. Ma, C.; Bi, C.W.; Xu, Z.; Zhao, Y.-P. Dynamic behaviors of a hinged multi-body floating aquaculture platform under regular waves. *Ocean Eng.* **2022**, *243*, 110278. [CrossRef]
12. Ding, W.W.; Jiang, J.Q.; Yue, W.Z.; Li, Y.-Z.; Wang, W.-S.; Sheng, S.-W.; Chen, M. Numerical study on hydrodynamic performance of a new semi-submersible aquaculture platform. *Appl. Sci.* **2022**, *13*, 12652. [CrossRef]
13. Liu, Q.; Chen, H.; Wang, B.L. A true double-body method based on porous media model for simulation and froude scaling verification of an aquaculture vessel resistance. *Ocean Eng.* **2024**, *310*, 118501. [CrossRef]
14. Wang, C.M.; Ma, M.; Chu, Y.; Jeng, D.-S.; Zhang, H. Developments in modeling techniques for reliability design of aquaculture cages: A review. *J. Mar. Sci. Eng.* **2024**, *12*, 103. [CrossRef]
15. Ma, C.; Zhao, Y.P.; Bi, C.W. Numerical study on hydrodynamic responses of a single-point moored vessel-shaped floating aquaculture platform in waves. *Aquac. Eng.* **2022**, *96*, 102216. [CrossRef]
16. Yu, J.; Cheng, X.M.; Fan, Y.L.; Ni, X.; Chen, Y.; Ye, Y. Mooring design of offshore aquaculture platform and its dynamic performance. *Ocean Eng.* **2023**, *275*, 114146. [CrossRef]
17. Bi, C.W.; Ma, C.; Zhao, Y.P.; Xin, L.-X. Physical model experimental study on the motion responses of a multi-module aquaculture platform. *Ocean Eng.* **2021**, *239*, 109862. [CrossRef]
18. Liu, J.B.; Guo, A.X. Wave force identification for a large-scale quasi-elliptical cylinder from monitored wave elevation. *Ocean Eng.* **2023**, *271*, 113769. [CrossRef]

19. Wang, S.; Cao, Y.; Fu, Q.; Li, H. Hydrodynamic performance of a novel semi-submersible platform with nonsymmetrical pontoons. *Ocean Eng.* **2015**, *110*, 106–115. [CrossRef]
20. Kristiansen, T.; Faltinsen, O.M. Modelling of current loadson aquaculture net cages. *J. Fluids Struct.* **2012**, *34*, 218–235. [CrossRef]
21. Lei, Y.; Zheng, X.Y.; Li, W.; Zheng, H.; Zhang, Q.; Zhao, S.; Cai, X.; Ci, X.; He, Q. Experimental study of the state-of-the-artoffshore system integrating a floating offshore wind turbine with a steel fish farmingcage. *Mar. Struct.* **2021**, *80*, 103076. [CrossRef]
22. Fan, Q.X.; Xu, Y.W.; Xie, Q.H.; Zhang, M.; Ren, H.; Sun, T. Investigation of the dynamic response of a floating wind-aquaculture platform under the combined actions of wind, waves and current. *J. Ocean Eng. Sci.* **2025**, *10*, 580–592. [CrossRef]
23. Cifuentes, C.; Kim, M.H. Hydrodynamic response of a cage system underwaves and currents using a Morison-force model. *Ocean Eng.* **2017**, *141*, 283–294. [CrossRef]
24. Cifuentes, C.; Kim, M.H. Numerical simulation of fish nets in currents using a Morison force model. *Ocean Syst. Eng.* **2017**, *7*, 143–155.
25. Cha, B.J.; Kim, H.Y.; Bae, J.H.; Yang, Y.S.; Kim, D.H. Analysis of the hydrodynamic characteristics of chain-link woven copper alloy nets for fish cages. *Aquac. Eng.* **2013**, *56*, 79–85. [CrossRef]
26. Lei, Y.; Zhao, S.X.; Zheng, X.Y.; Li, W. Effects of Fish Nets on the Nonlinear Dynamic Performance of a Floating Offshore Wind Turbine Integrated with a Steel Fish Farming Cage. *Int. J. Struct. Stab. Dyn.* **2020**, *20*, 2050042. [CrossRef]
27. Løland, G. Current Forces on and Flow Through Fish Farms. Ph.D. Thesis, Institutt for Marin Hydrodynamikk, Trondheim, Norway, 1991. [CrossRef]
28. Roddier, D.; Cermelli, C.; Aubault, A.; Weinstein, A. WindFloat: A floating foundation for offshore wind turbines. *J. Renew. Sustain. Energy* **2010**, *2*, 033104. [CrossRef]

Journal of
Marine Science and Engineering

MDPI

Article

Study on the Influence of 3D Printing Material Filling Patterns on Marine Photovoltaic Performance

Huiling Zhang [1], Shengqing Zeng [2], Yining Zhang [2], Sixing Guo [2], Huaxian Feng [2] and Dapeng Zhang [2,*]

[1] College of Ocean Engineering and Energy, Guangdong Ocean University, Zhanjiang 524088, China; zhanghl@gdou.edu.cn
[2] Ship and Maritime College, Guangdong Ocean University, Zhanjiang 524005, China
* Correspondence: zhangdapeng@gdou.edu.cn

Abstract

With the rapid development of offshore photovoltaic (PV) systems, PV support structures have become a critical component in offshore PV installations. The material properties of these structures significantly influence the safety and reliability of the entire system. 3D printing technology, leveraging its advantages such as rapid prototyping, complex structure manufacturing, and high material utilization, holds broad application prospects in the field of offshore PV. However, the infill pattern of 3D printing materials can significantly affect their mechanical properties. Marine PV systems require extremely high resistance to wave action, tensile strength, and torsional performance, while offshore PV support structures need sufficient compressive capacity. Therefore, this study aims to investigate how different infill patterns affect the compressive properties of 3D printed materials, thereby optimizing material selection and printing processes for offshore PV applications. Through experimental design, a variety of common infill patterns were selected. Universal testing machines and torsion testing machines were used to conduct systematic tests on compressive strength, elastic modulus, and compressive fracture strain. The results showed that different infill patterns have a significant impact on compressive properties, among which the honeycomb infill exhibited the best overall mechanical performance, effectively enhancing load-bearing capacity and stability. Based on the experimental results, appropriate infill configurations and material combinations for different components of offshore PV systems were proposed. The feasibility of optimizing 3D printing processes to improve the overall performance of offshore PV structures was further explored. The findings of this study not only provide a theoretical basis for material selection and process optimization in 3D printing for offshore PV systems but also offer important references for promoting the application of 3D printing technology in this field.

Keywords: 3D printing materials; infill patterns; compressive properties; marine photovoltaics

1. Introduction

The rapid advancement of technology, industrialization, and population growth has led to an ongoing increase in global energy demand, with projections suggesting a substantial rise by 2040 [1]. Non-renewable fossil fuels, however, encounter several challenges, including limited reserves [2], environmental pollution, and greenhouse gas emissions, particularly carbon dioxide. These issues contribute to adverse weather events and environmental concerns such as global warming and acid rain [3,4]. Consequently, there is an urgent need to investigate renewable and environmentally sustainable energy solutions. Various sources, including solar [5], hydro [6], geothermal [7], biomass [8], and wind [9] energy, have been proposed and extensively studied. Among these, solar energy emerges as a significant and promising alternative. Current floating photovoltaics (FPV) can be classified based on their deployment environment into three categories: floating PV, above-water PV, and submerged PV [10], as shown in Figure 1.

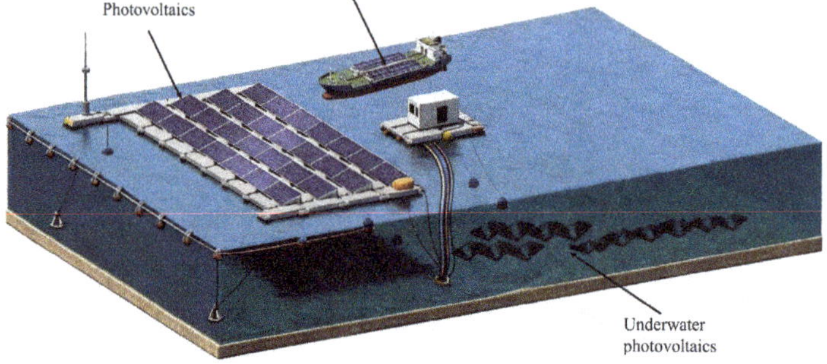

Figure 1. Different types of marine photovoltaics.

FPV denotes photovoltaic systems installed on water surfaces. These systems generally consist of floats, support structures, mooring systems, photovoltaic modules, electronic components, and tracking systems. Depending on the installation environments and system designs, above-water photovoltaic systems can be classified into fixed-pile PV systems and those specifically engineered for large ocean-going vessels. To mitigate environmental impacts on the panels, submerged underwater PV (UPV) systems have been developed. This approach stabilizes panel temperatures, streamlines the structure of photovoltaic inverters, and removes the necessity for additional cleaning equipment [10–12]. In marine environments, the support structure is essential for maintaining panels at a safe elevation above sea level [13]. Mooring systems secure the entire floating photovoltaic (FPV) array against environmental loads, limiting the equipment's movement to ensure stability and safety [14]. This configuration prevents the FPV system from capsizing or drifting away [15]. In freshwater FPV projects, synthetic fiber ropes, elastic rubber ropes, or combinations of both materials are commonly used [16]. Conversely, mooring lines for offshore floating platforms are generally made of steel cables or steel wire ropes [17]. Marine Photovoltaic (MPV) systems are exposed to severe environmental conditions, including storms and waves, which render them vulnerable to natural disasters that can result in equipment damage and interruptions in power generation [18]. As a result, MPV supports must possess the capability to endure such adverse conditions. Various materials exhibit unique properties, and the necessary characteristics of marine PV support structures differ according to the type of project and the surrounding environmental conditions [19–23]. Nonetheless, the high costs associated with physical testing of marine PV systems restrict research in this field. The emergence of 3D printing technology offers a cost-effective alternative. This approach enables the fabrication of diverse components using multiple materials, specifically designed to meet particular requirements. Consequently, it has facilitated experimental exploration in the field of marine photovoltaic research [24,25]. In the studies mentioned above and in most existing research, traditional steel materials have been employed for marine photovoltaics. The high cost of these materials significantly elevates both experimental and construction expenses in this domain. To address the financial challenges associated with research and practical implementation, this paper proposes the use of 3D printing technology with polylactic acid (PLA) as the material for fabricating components necessary for marine photovoltaics, thereby reducing overall costs. To assess the feasibility of PLA for marine photovoltaic components, this study printed standard compression test specimens from PLA and conducted strength testing. This material substitution has the potential to substantially lower both research and construction costs in this field. To enhance the mechanical properties of PLA for industrial applications, techniques such as plasticization, blending, and other modifications have been proposed to improve its degradability, mechanical strength, and physical and biological properties. Several researchers have documented methods for modifying PLA's hydrophilicity, degradability, and elongation at break. Blends of PLA with various synthetic and biopolymers have been developed to enhance its properties [26–28]. Reports suggest that the combination of PLA with substances like collagen improves its performance relative to the pure polymer, resulting in enhanced toughness, modulus, impact strength, and thermal stability [29,30]. These studies demonstrate that PLA, through ongoing research and optimization, has attained conditions suitable for complex engineering environments.

This paper begins by detailing our printing tools and specimen materials in Section 2, outlining the various parameters of the specimens, including printing angle, printing method, and infill density, as well as the procedures employed in conducting experiments on different specimens. Section 3 then analyzes the experimental data obtained and compares the conclusions of this study with existing research. Finally, the conclusion summarizes the research findings.

2. Experiment

2.1. Experimental Equipment

This experiment utilized the Bambu Lab's P1S 3D printer (Bambu Lab, Shenzhen, China) with polylactic acid (PLA) as the printing material to prepare samples. PLA has a melting point range of 190 to 220 °C and exhibits excellent printability and mechanical properties (as shown in Figure 2). Its eco-friendly and economical characteristics make it an ideal choice for this study. Default parameter settings were employed throughout the printing process: layer thickness of 0.2 mm, print speed of 60 mm/s, nozzle temperature of 210 °C, and heated bed temperature of 60 °C. Printing proceeded along the X-axis with material extrusion perpendicular to the XY plane. Material was kept dry during printing to ensure consistent print quality. The printer achieves positioning accuracy of ±0.02 mm along the XYZ axes. This means that under optimal settings, dimensional deviations between printed models and design specifications are typically controlled within 0.02 mm.

Compression tests were conducted using an ETM series electronic universal testing machine (WANCE, Shenzhen, Chian) with a maximum load capacity of 30 kN and an accuracy of ±0.5%. This testing machine is equipped with a compression fixture and a displacement transducer, enabling precise measurement of the load–displacement curve during the compression process. The data acquisition system records test data in real time and subsequently analyzes and processes it through the accompanying software.

2.2. Design of Experimental Parts

Referencing Jiang's research [31], PLA material was employed to print experimental specimens in this study. Different infill patterns, infill densities, and printing angles were also utilized for specimen printing.

The initial group of controlled experiments was structured to include three infill patterns—honeycomb, spiral, and linear infill—as depicted in Figure 3. Additionally, four infill ratios (25%, 35%, 50%, and 75%) were employed to systematically examine the effects of both infill patterns and infill ratios on the compression properties of 3D printed specimens. The specific design is as follows: the honeycomb infill utilizes a hexagonal cell structure, with filling ratios of 25%, 35%, 50%, and 75% achieved by adjusting the density of the hexagonal cells; the spiral infill employs a helical path design, with the same filling ratios controlled by varying the sparseness of the helical paths; and the linear infill consists of parallel linear paths, with filling ratios of 25%, 35%, 50%, and 75% realized by modifying the spacing between the linear paths. Multiple sets of standard compression specimens were printed for each combination of infill pattern and infill ratio. Based on international standards [32], we developed a model for compressed specimens. The dimensions and shape of the specimens were rigorously designed to meet the requirements of standard compression testing, as shown in Figure 4.

Figure 2. Printer used to print the specimen.

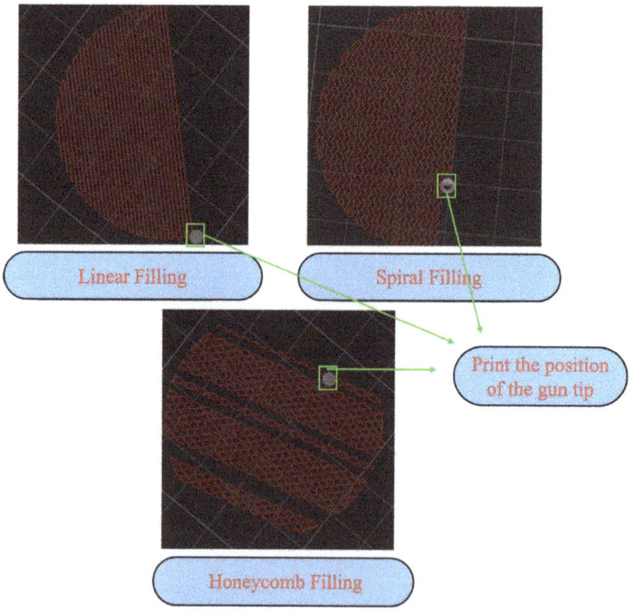

Figure 3. Three different fill styles.

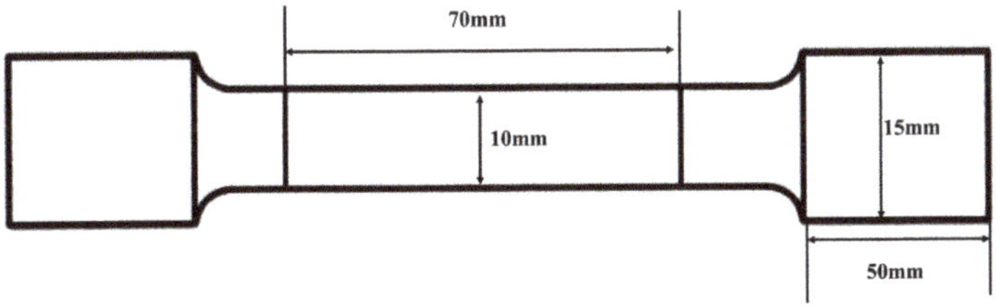

Figure 4. Component Dimensions and Styles.

This approach ensured the accuracy and comparability of the experimental results. The experimental design allowed for a comprehensive evaluation of the combined effects of various filling patterns and filling ratios on the compressive performance of the specimens, thereby providing reliable data to support subsequent analyses and optimizations of mechanical properties.

The second set of control tests was derived from the initial experiments. Each filling form—honeycomb, spiral, and linear—was printed according to the printing directions of 0°, 15°, 30°, 45°, 60°, 75°, and 90°. One specimen was printed for each group while maintaining consistent conditions across all tests. The objective of these experiments was to examine the effects of printing direction on the compressive properties of the materials, specifically focusing on the maximum compressive stress, modulus of elasticity, and damage modes of the specimens under varying printing orientations. It is anticipated that a significant difference in compressive properties will exist between specimens printed at 0° and those printed at 90°. Specimens printed at 0° are expected to demonstrate higher compressive strength and stiffness, whereas those printed at 90° may exhibit more extensive compressive damage. Honeycomb infill and spiral infill exhibit smaller variations in performance across different print orientations, whereas linear infill is more significantly influenced by print orientation.

This tilt angle refers to the angle between the sample axis and the print bed, as shown in Figure 5 below, representing a straight line forming an angle with the surface. This angle is called the print tilt angle or the sample tilt angle.

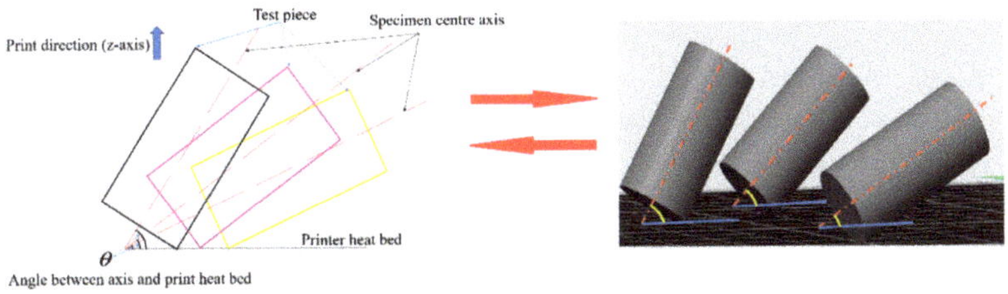

Figure 5. Print the diagram of the tilt angle. The red arrows indicate the transition between the 3D and 2D schematics, and the blue arrows indicate the positive direction of the Z-axis.

This experiment comprehensively evaluated the impact of printing direction on the compressive properties of 3D-printed materials, providing reliable data for optimizing marine photovoltaic designs. In addition, it lays the foundation for further optimization of 3D printing process parameters, thereby enhancing the performance and reliability of marine photovoltaics. Future research should explore the interactions between printing direction and other process parameters to further advance the application and development of 3D printing technology in the field of marine photovoltaics.

2.3. *Experimental Process and Data Processing*

The test was conducted at room temperature (25 °C) under normal environmental conditions with the specimen kept dry is shown Figure 6. Prior to starting the test, the lower platform was leveled to ensure horizontal alignment. After confirming platform levelness, the specimen was placed into the compression fixture of the universal testing machine. The compression device was then activated, applying a compressive load at a rate of 2 mm per minute until specimen failure. The machine was stopped upon specimen failure. Load–displacement data is recorded in real time throughout the test, while the specimen's deformation and damage patterns are observed.

Figure 6. Experimental apparatus.

The test data were collected via the data acquisition system of the universal material testing machine, encompassing key parameters such as maximum compressive strength, modulus of elasticity, and damage mode. Statistical analysis of the data was performed using Origin software 2024 to generate load–displacement curves and to calculate the average mechanical properties for various filling forms and filling ratios.

This study employs the previously mentioned experimental approaches to investigate three key influencing factors of the components. The first factor analyzed is the filling structure, which includes a total of three variables. The second factor is the filling rate, comprising four variables. The third factor relates to the specimen inclination angle, totaling seven variables, as shown in Table 1.

Table 1. Study variables.

Research Factors							
Filling structure		Honeycomb		Spiral		Linear	
Filling rate	25%		35%		50%		75%
Specimen tilt angle	0°	15°	30°	45	60°	75°	90°

2.4. Theoretical Equation

Hooke's Law: Within the elastic range, the relationship between stress (σ) and strain (ε) can be expressed by the following Equation (1), where $\boldsymbol{E}$ is the material's modulus of elasticity, F is the applied force, and $\boldsymbol{A}$ is the cross-sectional area in Equation (2).

$$\sigma = E\varepsilon \tag{1}$$

$$\sigma = \frac{F}{A} \tag{2}$$

$$\varepsilon = \frac{\Delta L}{L_0}, \quad v = -\frac{\varepsilon_{lat}}{\varepsilon_{lang}}, W = \frac{1}{2}\sigma\varepsilon \tag{3}$$

In Equation (3), ΔL denotes the change in length of the specimen when loaded, and L_0 is the original length of the specimen. v denotes Poisson's ratio, which describes the ratio of the strain in one direction to the strain perpendicular to that direction, and ε_{lat} denotes transverse strain denotes longitudinal strain. ε_{long} The stored energy density is denoted as W.

Consider the microelementary hexahedron in one of the specimens, and let it have three main directions of stress components in xyz coordinate system: positive stress $\sigma_x, \sigma_y, \sigma_z$ and shear stress component $\tau_{xy}, \tau_{xz}, \tau_{yz}$; therefore, the stress state of the microelementary hexahedron can be expressed by the following stress tensor as the following Equation (4):

$$\sigma = \begin{pmatrix} \sigma_x & \tau_{xy} & \tau_{xz} \\ \tau_{xy} & \sigma_y & \tau_{yz} \\ \tau_{xz} & \tau_{yz} & \sigma_z \end{pmatrix} \tag{4}$$

The maximum stress can be obtained from the characteristic Equation (5), where λ is the principal stress and I is the identity matrix.

$$d(\sigma - \lambda I) = 0d = \begin{pmatrix} \sigma_x - \lambda & \tau_{xy} & \tau_{xz} \\ \tau_{xy} & \sigma_y - \lambda & \tau_{yz} \\ \tau_{xz} & \tau_{yz} & \sigma_z - \lambda \end{pmatrix} \tag{5}$$

Yao et al. pointed out that 3D printing is stacked layer by layer, which can be considered transversely isotropic in the plane of any layer, and provided the elastic intrinsic relationship of the material [33], as shown in Equation (6) below.

$$\begin{bmatrix} \varepsilon_1 \\ \varepsilon_2 \\ \varepsilon_3 \\ \varepsilon_4 \\ \varepsilon_5 \\ \varepsilon_6 \end{bmatrix} = \begin{bmatrix} S_{11} & S_{12} & S_{12} & 0 & 0 & 0 \\ S_{12} & S_{22} & S_{23} & 0 & 0 & 0 \\ S_{12} & S_{23} & S_{22} & 0 & 0 & 0 \\ 0 & 0 & 0 & 2(S_{22} - S_{23}) & 0 & 0 \\ 0 & 0 & 0 & 0 & S_{56} & 0 \\ 0 & 0 & 0 & 0 & 0 & S_{65} \end{bmatrix} \begin{bmatrix} \sigma_1 \\ \sigma_2 \\ \sigma_3 \\ \sigma_4 \\ \sigma_5 \\ \sigma_6 \end{bmatrix} \tag{6}$$

For the plane stress state, then we have $\sigma_3 = 0$, $\tau_{23} = \sigma_4 = 0$, $\tau_{31} = \sigma_5 = 0$; the eigenstructure matrix can be written as Equation (7).

$$\begin{bmatrix} \varepsilon_1 \\ \varepsilon_2 \\ \gamma_{12} \end{bmatrix} = \begin{bmatrix} S_{11} & S_{12} & 0 \\ S_{12} & S_{22} & 0 \\ 0 & 0 & S_{55} \end{bmatrix} \begin{bmatrix} \sigma_1 \\ \sigma_2 \\ \tau_{12} \end{bmatrix} \gamma_{\text{al}} = \gamma_{\text{al}} = 0, \varepsilon_3 = S_{13}\sigma_1 + S_{23}\sigma_2 \tag{7}$$

where the E_1, E_2 engineering constants are Young's modulus; Poisson's ratio, ν_{12}, $\nu_{21} G_{12}$; and shear modulus. The matrix of four elastic constants can be expressed as Equation (8):

$$\begin{aligned} S_{11} &= \tfrac{1}{E_1} \\ S_{22} &= \tfrac{1}{E_2} \\ S_{56} &= \tfrac{1}{G_{12}} \\ S_{12} &= -\tfrac{\nu_{21}}{E_1} = -\tfrac{\nu_{12}}{E_2} \end{aligned} \tag{8}$$

The volume of the model we created in SolidWorks 2024 is referred to as the model volume, denoted by V. The fill volume represents the portion of the model volume that has been hollowed out, with the remaining volume classified as the fill volume, which directly indicates the amount of material used. The filling rate is defined as the ratio of the total fill volume to the model volume, multiplied by 100%. The specimen is constructed layer by layer, where V_i represents the fill volume of the i layer, M_i denotes the model volume of the i layer, S_i refers to the filling cross-sectional area of the i layer, and A_i signifies the model cross-sectional area of the i layer. The filling rate is represented by ρ, the height of the layer is indicated by h, and the extruded material of the 3D printer for the i layer is expressed as L_{i}. The total length of the filament is also denoted as L_{i}, while α represents the width of the extruded filament, determined by the print nozzle. In summary, we can derive the definition in Equation (9) for the fill rate.

$$\rho = \left(\frac{\sum V_i}{V}\right) \times 100\% \tag{9}$$

$$V_i = S_i \cdot h,\ M_i = A_i \cdot h \tag{10}$$

$$\sum V_i = \sum S_i \cdot h, \sum M_i = \sum A_i \cdot h \tag{11}$$

$$S_i = L_{\text{i}} \cdot \alpha \tag{12}$$

$$\rho = \frac{\sum L_i \cdot \alpha}{\sum A_i} \times 100\% \tag{13}$$

As indicated in the combined formula, Formula (14), let a, m and s represent the number of variables for the four research factors: filling structure, filling rate, specimen inclination angle, and experimental load, respectively. The total number of specimens, denoted as P in Equation (15), is established with $b = n = f = 1$ using the control variable method.

$$C_x^y = \frac{x!}{y!(x-y)!} \tag{14}$$

$$P = C_a^b C_m^n C_s^f \tag{15}$$

Substituting the data $b = n = f = 1$, $a = 3$, $m = 4$, $s = 7$ to find the total number of specimens, $P = 84$, we get 84 independent data points. The distribution of data points and the number of experiments conducted for each data point are shown in Table 2.

Table 2. Distribution of Data Points and Number of Experiments.

Filling Structure	**Filling Rate**	**Specimen Tilt Angle**	**Number of Samples**	**Number of Experiments**
Honeycomb	25%	0°/15°/30°/45° /60°/75°/90°	7	12
Honeycomb	35%	0°/15°/30°/45° /60°/75°/90°	7	12

Table 2. *Cont.*

Filling Structure	Filling Rate	Specimen Tilt Angle	Number of Samples	Number of Experiments
Honeycomb	50%	0°/15°/30°/45° /60°/75°/90°	7	12
Honeycomb	75%	0°/15°/30°/45° /60°/75°/90°	7	12
Spiral	25–75% (Consistent with the honeycomb structure	0°/15°/30°/45° /60°/75°/90°	28	12
Linear	25–75% (Consistent with the honeycomb structure	0°/15°/30°/45° /60°/75°/90°	28	12

3. Results and Discussion

3.1. Data Analysis

The compression testing machine we use is the ETM series electronic universal testing machine. Before commencing the experiment, the lower platform must be leveled to ensure horizontal alignment; the upper platform compresses the test specimen downward at a rate of 2 mm per minute. The force–compression displacement curve exhibits only one form, as shown in Figure 7; the initial stage involves force increasing the compression within the specimen, enabling it to withstand a rapid rise in pressure after reaching a certain pressure value during the slow increase phase. At both ends of the platform where force is applied, the specimen's constraints prevent deformation; the cross-section in the specimen's center, subjected to increasing pressure, continues to thicken, and its load-bearing capacity increases accordingly. When deformation approaches 5 mm, the test specimen fails, and the experiment is terminated. The damaged specimen is shown in Figure 8.

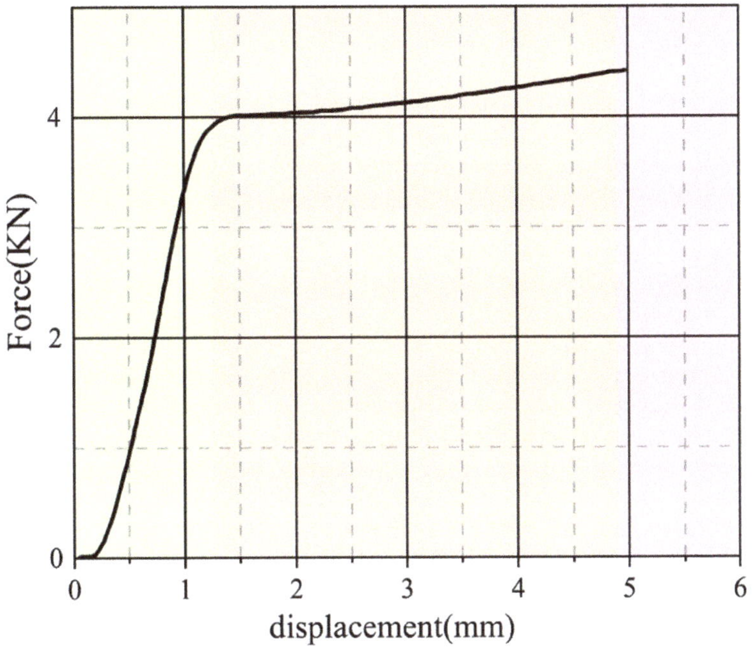

Figure 7. Stress–strain curve of the compressed specimen. The green area represents the linear stage, the orange area represents the yielding stage, and the pink area represents the fracture stage.

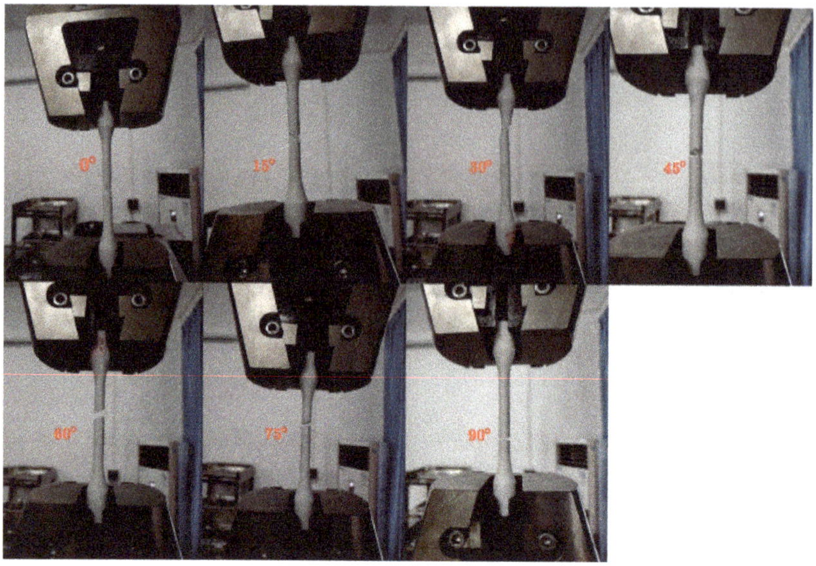

Figure 8. Picture of the test specimen damaged due to compression.

Analysis of experimental data indicates a significant correlation between the mechanical properties of various filler structures and their topological design. Under consistent filling ratios, the honeycomb filling structure exhibits optimal compressive performance due to its biomimetic hexagonal cell design, as illustrated in Figure 9. When the filling rate increases from 25% to 50%, the maximum compressive stress rises from 3.304 kN to 4.818 kN, representing a 45.8% increase. This enhancement reflects the ability of hexagonal cells to effectively delay material damage mechanisms by uniformly distributing stress. A stress relaxation of 2.3% (from 4.818 kN to 4.708 kN) occurs once the filling rate reaches 50%. According to Formula (4), hexagonal pressure is determined jointly by three directions. As filling increases, differential units in each direction acquire more material, thereby enhancing their compressive performance and effectively mitigating structural damage. A similar stress shrinkage of 2.3% (from 4.818 kN to 4.708 kN) is observed after surpassing the 50% filling rate. The honeycomb filling employs a biomimetic hexagonal cell structure whose topological design inherently possesses "trisubstructural stress balance" characteristics. This aligns with the synergistic interaction of the three stress components as defined by stress tensor Equation (4). When the filling rate increases from 25% to 50%, the wall thickness of hexagonal cells grows and cell density rises. This allows stress to be uniformly distributed throughout the structure via the hexagonal edges under compression, preventing localized stress concentration. Simultaneously, according to Hooke's law (Equation (1)), the material's elastic modulus is positively correlated with stress transfer efficiency. The increased packing ratio means more "differential elements" bear the load and a higher proportion of material is engaged. Each element can more effectively share the load, delaying the damage progression of "elastic deformation $\longrightarrow$ plastic deformation $\longrightarrow$ fracture". This ultimately manifests as a significant increase in compressive stress. The printing angle affects the degree of fiber interlacing between layers. When printed at 75°, the cross-angle of fibers in the honeycomb structure approaches 45°, enhancing the structure's shear resistance and thus achieving the highest maximum compressive stress.

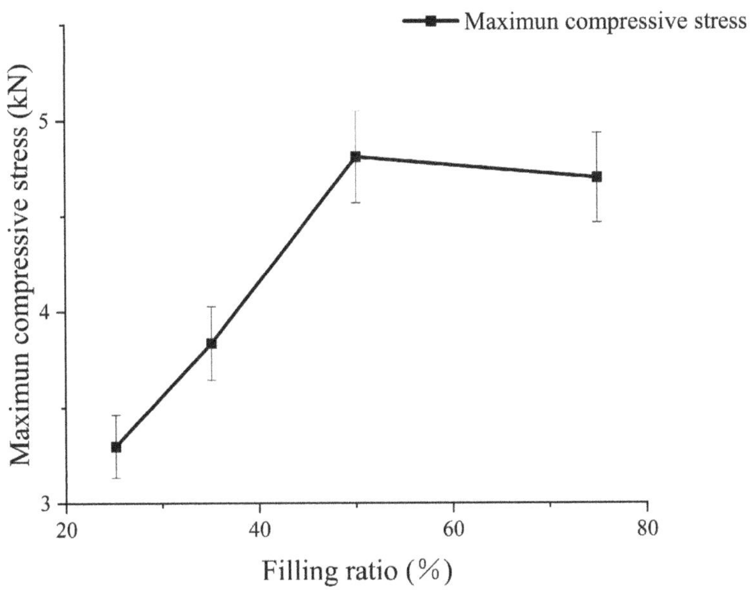

Figure 9. Trend of maximum compressive stress of honeycomb-filled specimen with increasing filling ratio.

As shown in Figure 10, the compressive curve (3.078 ⟶ 3.966 kN) of the helical-filled structure exhibits a honeycomb-like decay trend. The observed 32.8% gain in axial strength, relative to straight-line filling, at 50% filling supports the geometrical advantage of the helical structure in resisting bending-torsional coupling loads. However, the 5.4% stress decay at 75% filling indicates a threshold effect of helical spacing on plastic flow. The helical path design with spiral filling endows it with "bending–torsion coupled load resistance capability"—under compression, the spiral's intertwined structure partially converts axial pressure into tangential stress, dispersing concentrated loads from a single direction. When the filling rate increases from 25% to 50%, the pitch of the spiral path decreases and the linear density increases, enhancing the structure's "spatial entanglement stiffness". This creates a more continuous load transfer path, resulting in a sustained increase in axial compressive stress. Compared to linear filling, the spiral structure exhibits tighter interlayer bonding (with no distinct parallel interfaces), resulting in weaker stress concentration effects. Consequently, at 50% filling rate, compressive stress is 32.8% higher than in linear filling.

As shown in Figure 11, the low mechanical efficiency of the linearly filled body, which is 27.3% lower than that of the honeycomb structure at 50% filling, highlights an inherent flaw in the anisotropic design. This flaw arises from the stress-concentration effect at the interlayer interfaces, which complicates the ability to surpass the 50% filling threshold of the honeycomb structure, even under the strong constraint of 75% filling (3.39 kN). The quantitative results of these structure-function relationships offer a critical parameter optimization space for the additive manufacturing-based reverse design of porous materials. Linear filling employs a parallel straight-line path design, forming "parallel interlayer interfaces" after 3D printing. This structure exhibits unidirectional load transfer pathways, effectively transmitting stress only along the linear path direction, while interlayer bonding perpendicular to the path direction is extremely weak. According to the "stress transfer principle for anisotropic materials" in materials mechanics, the transverse (perpendicular to the straight path) elastic modulus of linear structures is significantly lower than the longitudinal modulus. Under compression, stress tends to accumulate at the interlayer interfaces, creating a "potential shear failure risk". Even with increased fill density, interlayer defects persist, resulting in inherently inadequate mechanical efficiency.

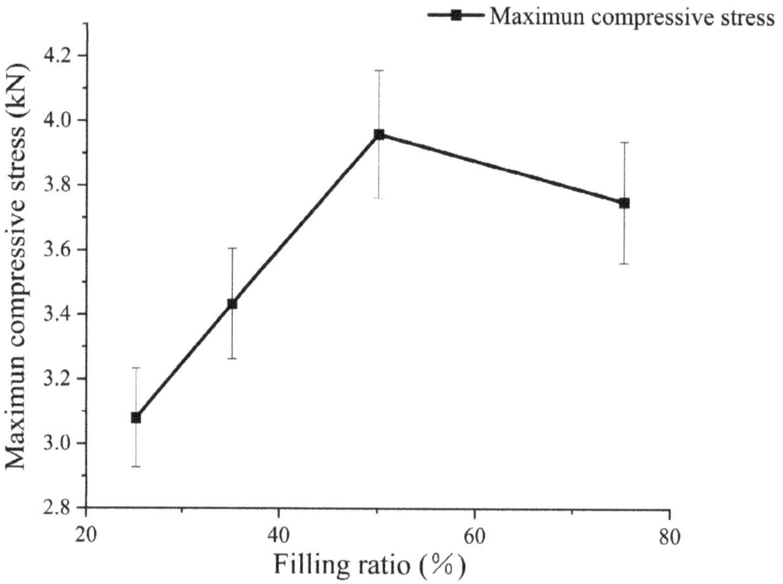

Figure 10. Trend of maximum compressive stress of spiral-filled specimen with increasing filling ratio.

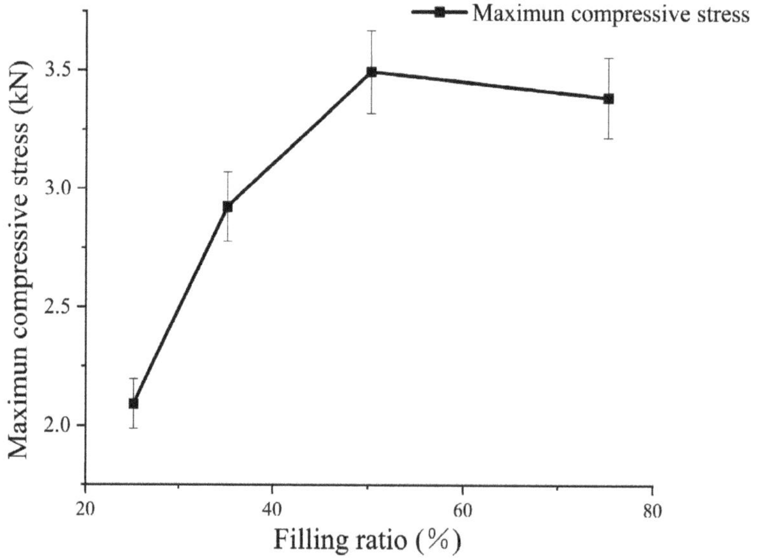

Figure 11. Trend of maximum compressive stress of linearly filled specimens with increasing filling ratio.

As shown in Figures 12–14. By analyzing the maximum compressive stress data of different filling forms (honeycomb, spiral and linear filling) at different printing angles (0°, 15°, 30°, 45°, 60°, 75°, 90°), it can be found that the influence of printing angle on the compression performance of the specimen shows a certain regularity. In general, the maximum compressive stress with the change in printing angle usually shows a decreasing and then increasing trend; 15° or 30° is usually the lowest point of performance, while 0° or 90° is the highest point of performance. Specifically, the honeycomb infill specimen has the best performance at 75° or 90° printing direction, for example, the maximum compressive stress of 75% honeycomb infill specimen reaches 5.095 kN at 75°, which is the highest value among all the specimens, indicating that the honeycomb infill has excellent load-bearing capacity and impact resistance at high angle printing direction; while the spiral infill and linear infill specimens have better performance at 0° or 90° printing direction. For example, the maximum compressive stress of the 75% spiral-filled specimen at 90° was 4.38 kN, and that of the 75% linear-filled specimen at 90° was 4.061 kN, indicating that the spiral-filled and linear-filled specimens have high compressive performance under these printing directions. In addition, from the above three graphs, it can be seen once again that the compressive properties of the specimens generally show an upward trend with the increase in the filling ratio, e.g., the maximum compressive stress of the specimens with a 75% filling ratio is significantly higher than that of the specimens with a 25% filling ratio in most of the printing orientations, which indicates that a high filling ratio can effectively increase the load-carrying capacity of the material. However, in some printing directions (e.g., 15°and 30°), the increase in filling ratio has limited effect on the compression performance, e.g., the maximum compressive stress of 25% cellular filled specimen at 15°is 1.989 kN, while that of 75% cellular filled specimen at 15°is 4.988 kN, which is improved but still lower than that of the performance in other printing directions. Taken together, the honeycomb-filled specimen performs best under the high-angle printing direction (75° or 90°), which is suitable for application scenarios that require high load-bearing capacity and impact resistance; the spiral-filled and linear-filled specimens perform better under the 0° or 90° printing direction, which is suitable for parts requiring torsional resistance and lightweight design.

Due to the "multi-directional stress adaptability" of the hexagonal structure in honeycombs, during high-angle printing, the edges of hexagonal cells form obtuse angles with the load direction. This allows stress dispersion through elastic deformation of the edges. Furthermore, when printing at 90°, the stacking direction aligns parallel to the vertical edges of the hexagons. This maximizes the "effective stress-bearing length" of the load-bearing cells, resulting in the highest load-carrying capacity. At the same time, when the specimen is subjected to compression, the deposition direction of PLA aligns with the printing tilt angle of the specimen. At 0° printing, the layers bond along the X-axis, and at 90°, they bond along the Y-axis. Since the longitudinal strength of PLA is 20–30% higher than the transverse strength, the performance of the specimens printed at 0° and 90° is better than at intermediate angles.

The winding direction of the spiral structure is directly related to the printing angle. At 0° printing, the spiral winds axially, directly resisting axial pressure. At 90° printing, the spiral winds radially, forming a "ring-shaped load-bearing structure" that enhances the synergistic effect of torsion and compression resistance. When printing at an oblique angle, the winding direction of the spiral becomes misaligned with the load direction, disrupting the stress transfer path and resulting in reduced performance.

When the linear path aligns with the print angle, the path direction is either parallel (0°) or perpendicular (90°) to the load direction. When parallel, stress transfers directly along the path; when perpendicular, the path forms a "transverse support" that prevents layer delamination. During angled printing, the path forms an acute angle with the load direction, concentrating stress at path intersections and resulting in the lowest performance.

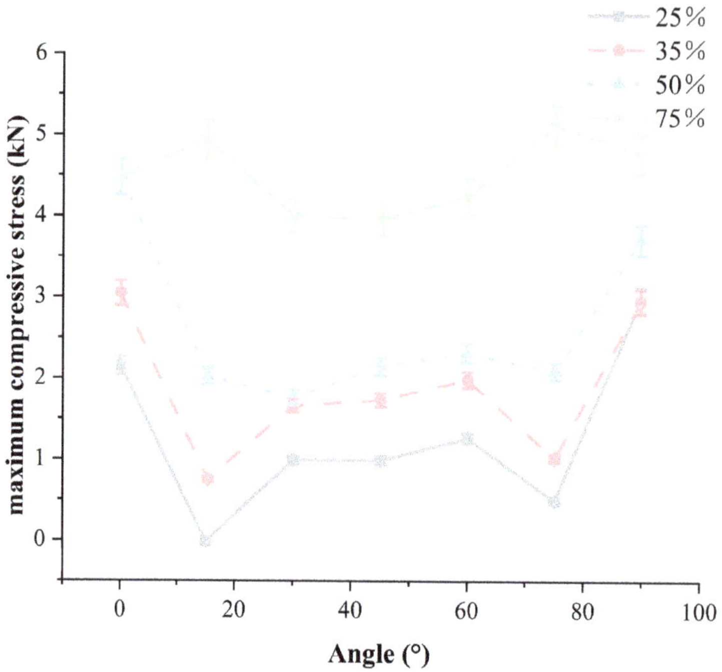

Figure 12. Variation in maximum compressive stress with printing angle for honeycomb-filled specimens with different filling ratios.

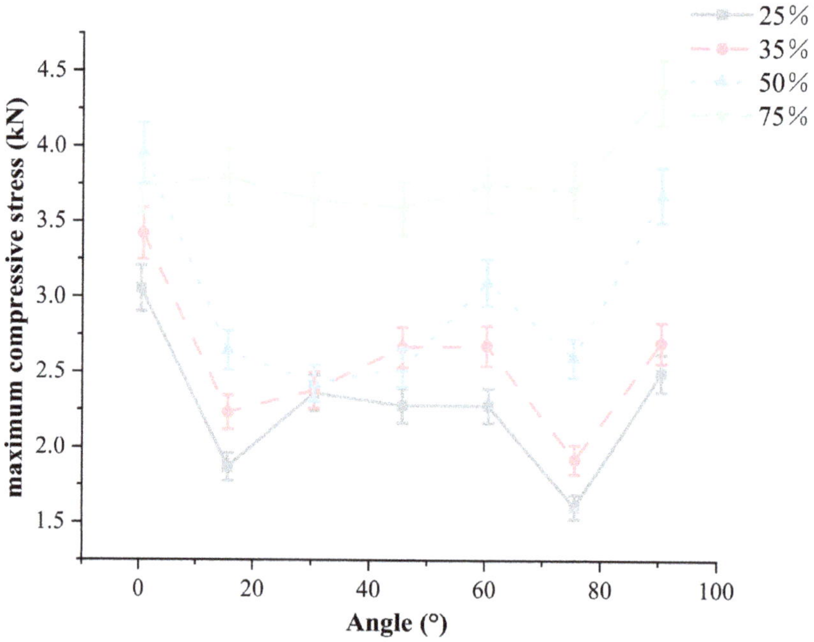

Figure 13. Variation in maximum compressive stress with printing angle for different filling ratios of spiral-filled specimens.

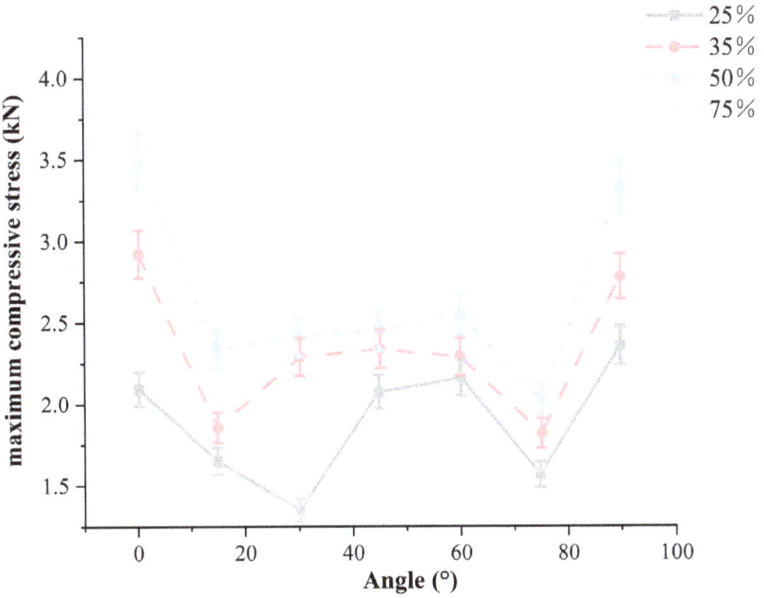

Figure 14. Variation in maximum compressive stress with printing angle for linearly filled specimens with different filling ratios.

As shown in Figure 15, the maximum compressive stress at angles ranging from 0° to 90° for honeycomb, spiral, and linear filling structures, all exhibiting a 75% filling rate. The honeycomb filling structure demonstrates significantly higher compressive stress than both the spiral and linear structures, indicating superior compressive performance among the three. At 0°, the stress reaches approximately 4.5 kN, peaking near 5.0 kN at 15° and 75°. Although stress declines at 45° and 60°, it remains greater than that of the other two structures, with corresponding error bars reflecting notable stress fluctuations. The helical filling structure displays intermediate compressive stress levels with minimal overall variation, typically maintaining values between 3.5 and 3.8 kN, and showing only a slight increase beyond 75°. In contrast, the linear filling structure exhibits the lowest overall compressive stress, indicating the weakest performance among the three. At 0°, the stress is approximately 3.4 kN, decreasing to about 3.1 kN at 15° before gradually recovering. Between 75° and 90°, it increases to around 4.0 kN, with error bars suggesting relatively stable performance. Overall, the honeycomb structure shows the highest compressive strength but with more pronounced angular fluctuations, while the linear structure presents the lowest compressive strength with more consistent performance across varying angles.

The hexagonal topology of honeycomb structures represents "nature's optimal load-bearing configuration," characterized by "uniform stress distribution across three adjacent unit edges under forces applied in any direction" (fulfilling the triaxial stress balance in the stress tensor equation). A 75% fill rate provides ample material foundation for this topology. Even when printing angles vary, hexagonal cells adapt to stress directions through geometric deformation, ensuring compressive stress remains maximized. However, at high angles (75°/90°), the force direction within hexagonal cells aligns more closely with the load-bearing edges. At low angles (15°/30°), stress must traverse layer interfaces for transfer, resulting in the greatest fluctuation amplitude.

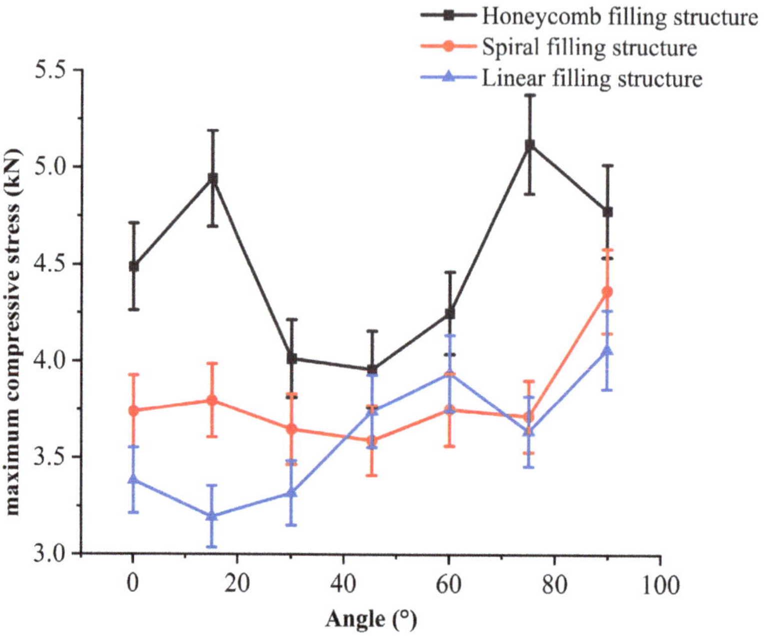

Figure 15. Compression performance curves of three different specimens as a function of printing angle at a fill rate of 75%.

The "enveloping characteristic" of helical structures confers a degree of 'isotropy'—regardless of printing angle, the helical lines form continuous load-transfer paths. This avoids the "unidirectional weakness" of linear structures while circumventing the "high angle dependency" of honeycomb structures. However, due to "tangential losses" in spiral stress transfer, the stress transmission efficiency of spiral structures is lower than that of hexagonal structures. This results in consistently lower compressive strength compared to honeycomb structures. Additionally, owing to the stability of spiral structures, their stress fluctuations exhibit smaller amplitudes than those of honeycomb structures.

Interlayer stress concentration in linear structures is a universal phenomenon across all printing angles. Regardless of angle variation, the interlayer interfaces along parallel paths remain stress weak points, resulting in the lowest overall compressive stress. Simultaneously, the issue of weak interlayer bonding persists irrespective of angle, leading to minimal performance fluctuation in linear structures and manifesting as the flattest curve.

Table 3 shows the variance of three different infill patterns when the infill method is the same, with the printing tilt angle as the only variable. Table 4 shows the variance for specimens with the infill pattern as the variable, across seven different printing tilt angles. By comparing the data, it can be intuitively seen that the variance at tilt angles of 30°, 45°, and 90° is similar to the data in Table 3. This indicates that under the conditions of 30°, 45°, and 90°, the influence of the infill pattern on the compressive performance of the specimens is similar in significance to the influence of the tilt angle on the compressive performance. However, when the printing tilt angles are 0°, 15°, and 75°, the variance in Table 4 is significantly larger than that in Table 3. This indicates that at tilt angles of 0°, 15°, and 75°, the effect of the infill pattern on the compressive performance of the specimens is significantly more pronounced than the effect of the printing tilt angle on the compressive performance.

Table 3. Maximum compressive stress variance at different angles for identical filling structures.

Filling Structure	Variance
Honeycomb	0.181
Spiral	0.067
Linear	0.128

Table 4. Maximum compressive stress variance for different filling structures at the same angle.

Angle (°)	Variance
0	0.341
15	0.791
30	0.123
45	0.063
75	0.678
90	0.143

3.2. Discussion on Existing Research

In the study conducted by Liu et al. [34], experiments were performed on five distinct specimens. The authors defined the relative density of the specimens as F_V, where a higher F_V value corresponds to increased wall thickness and, consequently, greater packing density. Their experiments produced stress–strain curves for the five specimens with differing relative densities, as shown in Figure 16. The figure clearly indicates that the specimen with a relative density of 60% exhibits the most favorable compressive performance among the five, followed by those with relative densities of 50%, 40%, 30%, and 20%, as depicted in Figure 16. Liu's findings demonstrate that specimens with higher relative densities, indicative of greater packing density, exhibit enhanced mechanical properties.

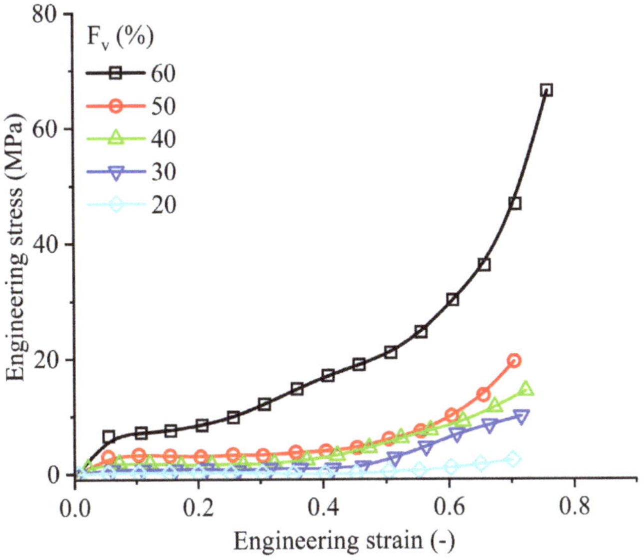

Figure 16. Stress–strain curves of specimens with different relative densities in Liu's study.

In the study by Wang et al. [35], compression tests were performed on three specimens: PDMS/PLA-3, PDMS/PLA-5, and PDMS/PLA-7. The fundamental data for these specimens are summarized in Table 5. Utilizing the same filler material, the researchers assessed the compressive properties of specimens featuring three distinct pillar diameters.

The results indicated that increasing the pillar diameter enhanced the filling rate of the specimens. The experimental data obtained by Wang et al. are illustrated in Figure 17. Their findings demonstrated that the mechanical properties of the specimens were directly affected by the pillar diameter. The specimen exhibiting the most favorable mechanical properties was the one with a 7 mm diameter, followed by the 5 mm diameter specimen. Conversely, the specimen with the least favorable mechanical properties was the 3 mm diameter specimen, as depicted in Figure 17. The study concluded that larger pillar diameters, which corresponded to higher filling rates, led to improved mechanical properties of the specimens.

Table 5. Sample labeling, filler, pillar diameter, mass, and density of PDMS/PLA lattice composites in Wang et al.'s study [35].

Specimen Label	m/g	ρ/g·cm^{-3}
PDMS/PLA-3	49.743 ± 0.142	1.066 ± 0.003
PDMS/PLA-5	50.902 ± 0.176	1.091 ± 0.004
PDMS/PLA-7	53.346 ± 0.285	1.143 ± 0.006

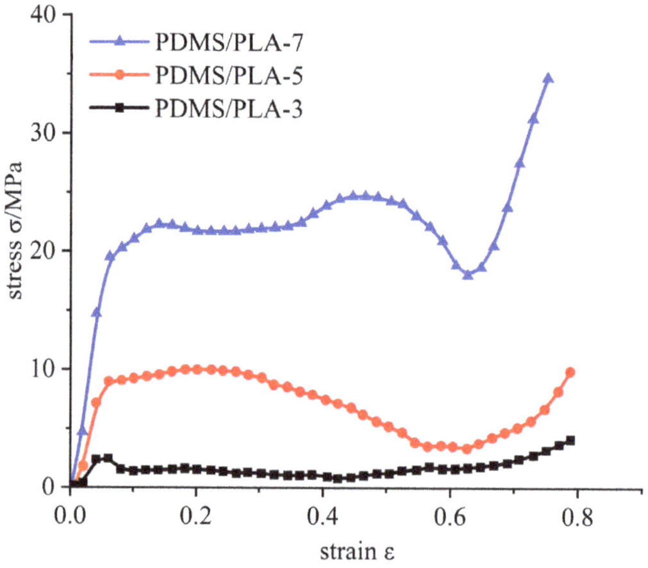

Figure 17. Stress–strain curves of different specimens in Wang's research.

In the studies conducted by Liu and Wang, the researchers modified the filling rate of components by adjusting their relative density and pillar diameter, respectively. Both Liu and Wang performed compression tests on specimens with varying filling rates. An analysis of these experiments reveals consistent conclusions: the mechanical properties of the studied samples improved as the filling rate increased. This finding is consistent with the results of the present study, which utilized different printing angles, infill patterns, and infill ratios for specimen fabrication. As shown in Figures 9–11, when both the printing angle and infill pattern were held constant, specimens with higher infill density demonstrated enhanced compressive strength.

Marine photovoltaic structures must consider not only compressive properties but also tensile and flexural characteristics. The packing ratio serves as a crucial factor that influences the compressive performance of specimens and significantly impacts their tensile and flexural properties.

Kilinc's research [36] indicates that a reduction in layer thickness and hatch spacing leads to an increase in the deposition of reinforcing fibers per unit area. The study involved the fabrication of nine specimens; for instance, "1-08" denotes a specimen with a layer thickness of 1 mm and a hatch spacing of 0.8 mm, whereas "08-06" signifies a specimen with a layer thickness of 0.8 mm and a hatch spacing of 0.6 mm. The fiber content for various specimen types is presented in Table 6.

Table 6. Fiber volume fraction of composites.

Sample Code	Fiber Content (vol%)
1-08	11.297
1-06	18.150
1-04	21.833
08-08	16.395
08-06	19.266
08-04	23.371
06-08	17.884
06-06	20.165
06-04	26.851

Kilinc et al. conducted tensile and flexural tests on nine specimens, which were categorized into three groups according to thickness (1 mm, 0.8 mm, and 0.6 mm). Each group underwent both tensile and flexural testing. As illustrated in Figures 18–23, the specimens demonstrating the highest performance in both tests across the three thickness categories were 1-04, 08-04, and 06-04, followed by 1-06, 08-06, and 06-06. The lowest-performing specimens were 1-08, 08-08, and 06-08. Their research indicated that increased fiber content, specifically higher filler loading, led to enhanced tensile and flexural strengths in the specimens.

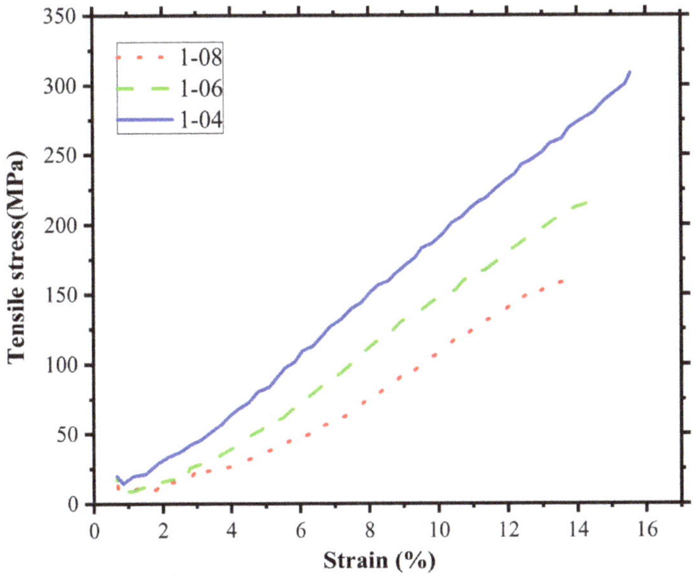

Figure 18. The stress–strain curves of specimens printed with a thickness of 1 mm and different hatch spacing values.

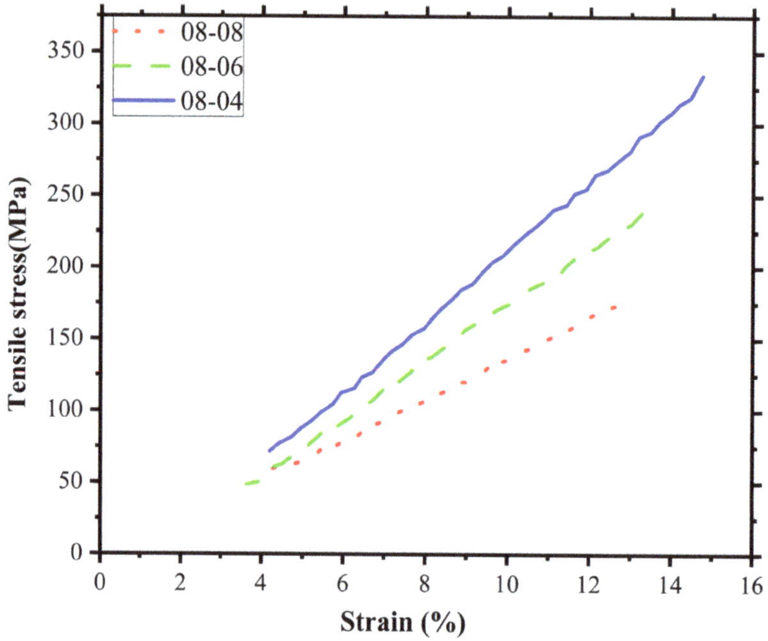

Figure 19. The stress–strain curves of specimens printed with a thickness of 0.8 mm and different hatch spacing values.

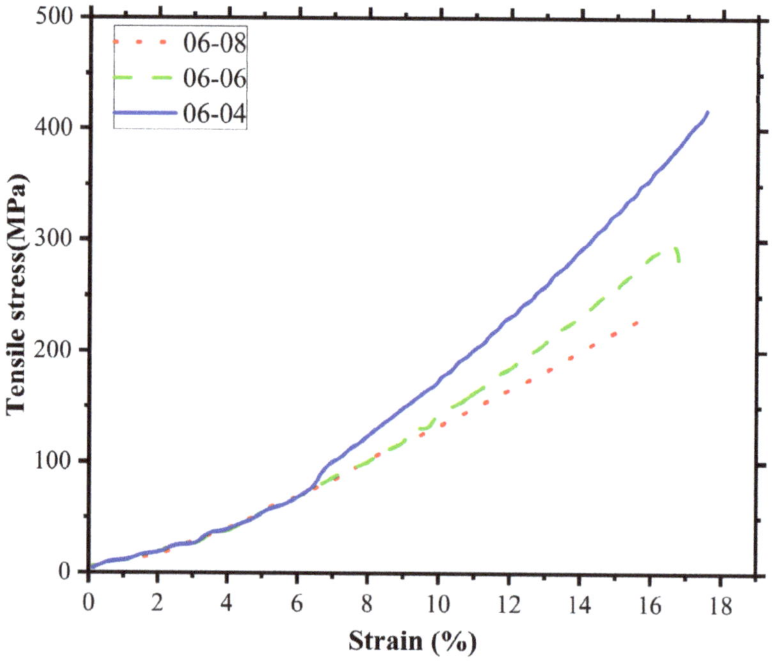

Figure 20. The stress–strain curves of specimens printed with a thickness of 0.6 mm and different hatch spacing values.

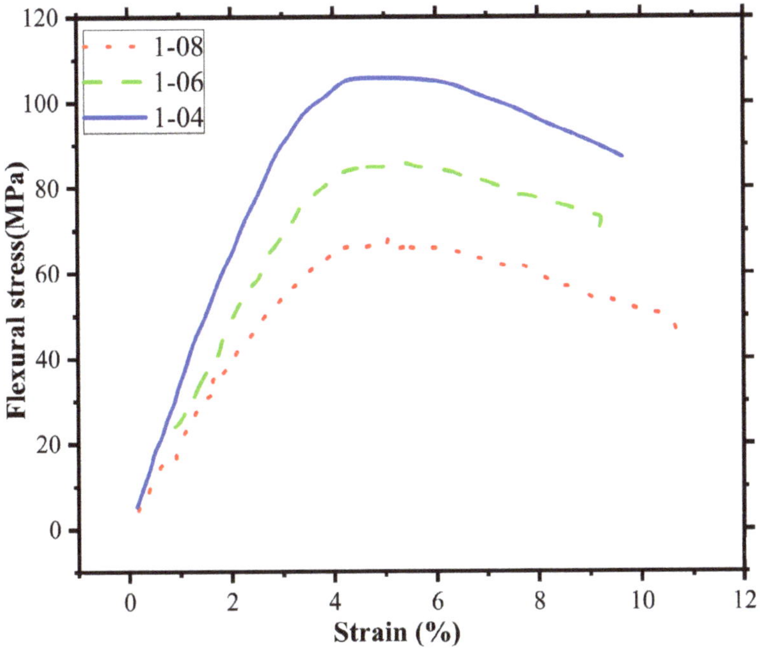

Figure 21. The flexural stress–strain curves of specimens printed with a thickness of 1 mm and different hatch spacing values.

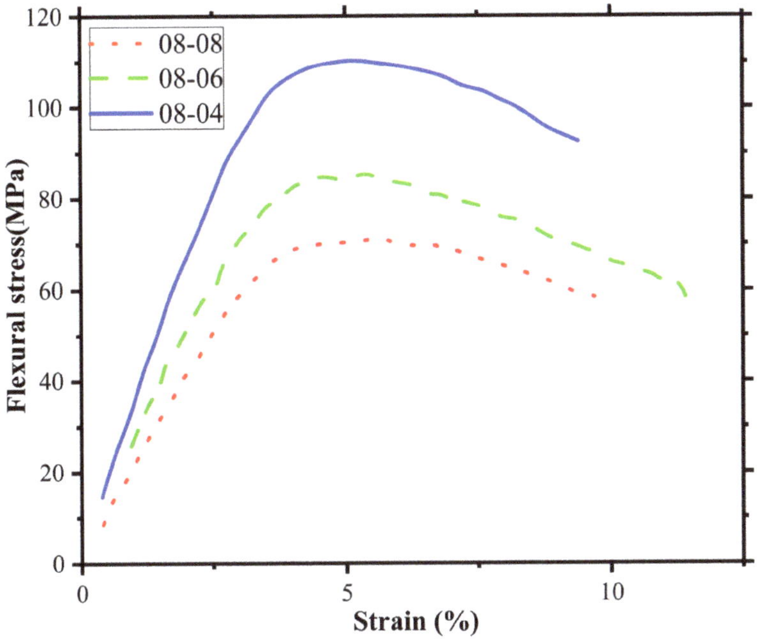

Figure 22. The flexural stress–strain curves of specimens printed with a thickness of 0.8 mm and different hatch spacing values.

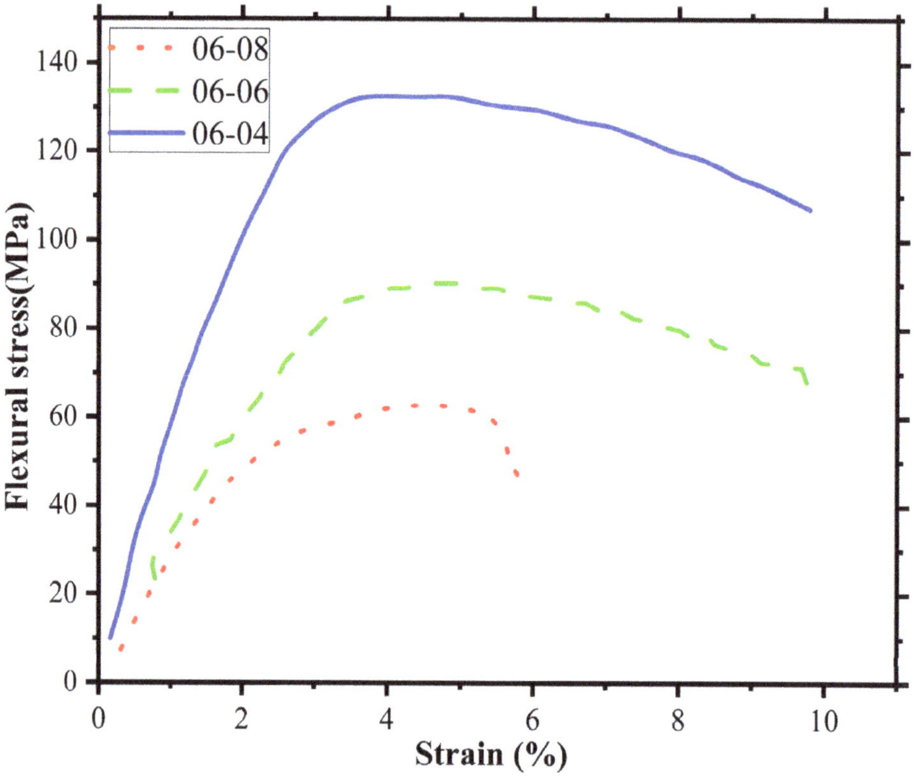

Figure 23. The flexural stress–strain curves of specimens printed with a thickness of 0.6 mm and different hatch spacing values.

The study conducted by Abdulsalam et al. [37] summarized the impact of various printing parameters on the mechanical properties of specimens, as detailed in Table 7. As the parameters increase, they influence the mechanical properties of the components, with "+" denoting an increase in response values and "-" indicating a decrease in response values.

Table 7. The Influence of Various Printing Factors on the Mechanical Properties of Test Specimens in Abdulsalam's Research.

	Tensile		Flexural		Compressive	
Printing Parameter	$F_{m,t}$, **N**	E_t, **GPa**	$F_{m,f}$, **N**	E_f, **GPa**	$F_{m,c}$, **N**	E_c, **GPa**
Infill density	+	+	+	+	+	+

The table clearly indicates that packing density is positively correlated with all three mechanical properties: compression, tensile strength, and flexural strength. Specifically, as packing density increases, the mechanical properties of the specimens also improve.

A summary of the above studies reveals that increasing the specimen filling rate significantly enhances the mechanical properties of the specimens. Integrating the experimental data yields Table 8. As mentioned earlier, this study encompasses a total of 84 data points. Twelve experiments were conducted for each of the 84 data points in this study. The average values for all data points are presented in Table 8. The table directly demonstrates that, under identical conditions, the specimen with a 75% filling rate exhibits the best compressive performance among the four filling rates. Comparing the remaining three filling rates, it is also observed that the compressive performance of the specimens improves to some extent as the filling rate increases. Through the integration of data points, along with the error analysis and variance analysis of the data discussed earlier, it is not difficult to arrive at the same conclusions as those found in existing literature. This demonstrates that the research presented in this paper is reproducible and has repeatability.

Table 8. Maximum compressive stress of different specimens in this study.

		Fill Ratio			
Filling Structure	**Tilt Angle (°)**	**75%**	**50%**	**35%**	**25%**
Honeycomb	0	4.48734	4.48722	3.04724	2.15478
	15	4.94227	2.02658	0.75117	0.01473
	30	4.01366	1.77183	1.66335	0.98785
	45	3.95942	2.15478	1.73567	0.98785
	60	4.25033	2.31914	1.99042	1.26232
	75	5.12471	2.10054	1.02401	0.49642
	90	4.77956	3.7211	2.95684	2.97492
Spiral	0	3.7397	3.95683	3.42044	3.05518
	15	3.79674	2.64484	2.23357	1.86831
	30	3.64861	2.4277	2.3817	2.35962
	45	3.59157	2.53075	2.66784	2.27957
	60	3.75166	3.10118	2.67888	2.27957
	75	3.71669	2.59883	1.92536	1.6061
	90	4.36717	3.68265	2.70188	2.49579
Linear	0	3.38441	3.49737	2.92323	2.09429
	15	3.19671	2.33974	1.85903	1.65179
	30	3.31902	2.41448	2.29218	1.35028
	45	3.74283	2.4527	2.33974	2.0756
	60	3.94072	2.54698	2.29218	2.16054
	75	3.63921	2.02889	1.82166	1.56686
	90	4.06302	3.32836	2.78224	2.35843

Liu et al. verified the positive correlation between increased infill rate and improved mechanical properties by adjusting relative density, while Wang et al. explored the impact of changing pillar diameters on material mechanical performance; both findings align with the core conclusions of this study. However, these studies generally lack an in-depth exploration of the synergistic effects of multiple parameters and their application to specific scenarios. Kilinc's research focused on the effect of fiber content on tensile and bending properties, without addressing the relationship among compressive performance, infill patterns, and print angles. Although Abdulsalam et al. summarized the trends in the influence of printing parameters, they did not analyze the mechanical performance differentiation mechanism caused by differences in the topology of various infill patterns, nor did they conduct targeted analysis for the harsh environmental requirements of marine photovoltaic support structures. In contrast, this study not only systematically investigates the synergistic effects of three key parameters—fill density, infill pattern, and print angle—but also, by comparing the stress transmission paths of honeycomb, spiral, and linear infill patterns, reveals the iso-stress distribution advantage of hexagonal topologies. It clarifies the suitability of honeycomb infill at high fill density (75%) with specific print angles (75° or 90°), filling a gap in existing research regarding multi-parameter optimization and scenarios specific to marine photovoltaics. Additionally, this study verifies the significance of differences in parameter effects through variance analysis, further deepening the understanding of the relationship between 3D printing processes and material performance. For instance, Liu et al. (2024) [34] found that specimens with a relative density (equivalent to the filling rate) of 60% exhibited optimal compressive properties. In contrast, this study demonstrates that the compressive stress of honeycomb-filled specimens continues to increase even at a filling rate of 75% (with only slight stress relaxation occurring after 50%). This discrepancy likely stems from differences in the matrix material (PLA/PHA-wood composite in Liu et al.'s study [34] versus pure PLA in this study) and the filler pattern (lattice structure in Liu et al [34]. versus honeycomb/spiral/linear fill in this study), indicating that the optimal filling rate is highly dependent on material type and topological structure. Regarding optimal printing angles, Jiang et al. (2025) [31] observed that honeycomb-filled specimens at 75% filling exhibited stable superior performance across all printing angles, aligning with the general trend in this study. However, this research further clarifies that honeycomb filling achieves peak compressive stress at 75° or 90°, whereas Jiang et al [31]. did not specify the peak angle. This refinement may stem from the finer print angle gradient employed in this study (7 angles versus fewer in Jiang et al.'s work [31]). Regarding performance gains, the maximum compressive stress (5.12 kN) achieved by honeycomb filling at 75% packing density and 75° printing angle in this study represents an approximately 28% improvement over the peak value (approximately 3.99 kN) of the lattice structure in Liu et al.'s research [34] and a roughly 12% enhancement over the performance (approximately 4.57 kN) of honeycomb filling in Jiang et al.'s study [31]. This disparity highlights the synergistic optimization effect of filling patterns, filling ratios, and printing angles—an aspect insufficiently explored in prior studies. Placing this work within a broader literature context reveals that multi-parameter synergistic optimization is a key direction for enhancing the mechanical properties of 3D-printed marine photovoltaic components. Subsequent studies on the mechanical properties of PLA-based components can reference this multi-parameter synergistic approach to optimize component performance.

Offshore photovoltaic structures are typically large in scale and must endure harsh environmental conditions, which necessitate exceptional mechanical properties in their construction. As previously noted, a higher infill density in 3D-printed specimens is associated with enhanced mechanical properties. However, beyond a certain threshold, further increases in infill density result in only marginal improvements in mechanical performance. At this juncture, alternative methods are required to improve mechanical properties, ensuring compliance with performance standards while also aiming to reduce structural mass and conserve materials. This study examines the effects of fill rate on the compressive properties of specimens, as well as the influence of printing method and angle. As shown in Figure 12, at a 75% fill rate, the honeycomb printing method consistently exhibits superior compressive performance compared to the other two methods, irrespective of the specimen's printing angle. Jiang et al. [38] utilized the same three printing methods as this study for specimen design and analysis. Compression tests were performed on specimens with a 75% fill rate. As shown, specimens featuring a honeycomb structure consistently demonstrated superior compressive performance compared to the other two structures, regardless of the printing angle, when the fill rate was maintained at 75%, as depicted in Figure 24. Jiang et al [31]. provided an explanation for this phenomenon in their research. The hexagonal topology of the honeycomb structure constitutes an 'equistress structure,' enabling uniform distribution of external forces throughout the entire structure and preventing localized stress concentration (refer to stress distribution simulation results in Jiang et al., 2025 [31]). Its interlayer bonding area is 35% larger than that of linear structures, enhancing resistance to interlaminar delamination.

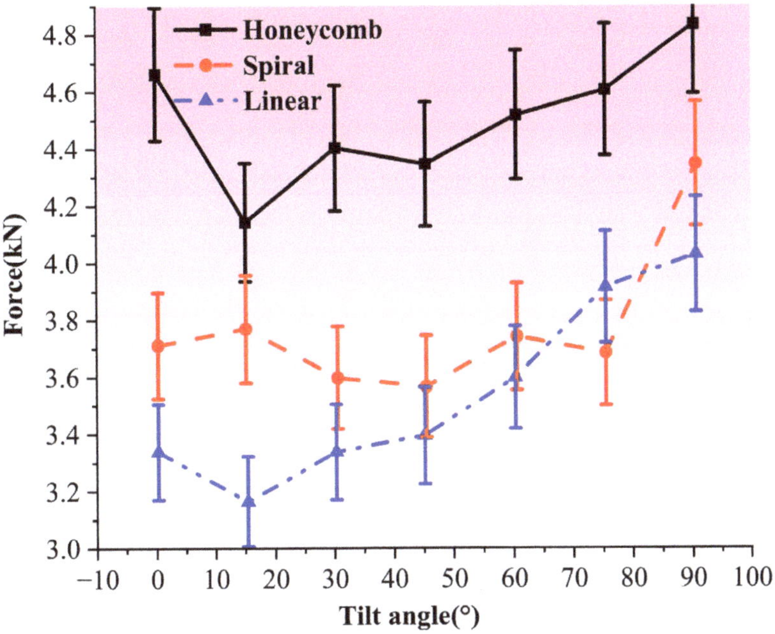

Figure 24. The compressive performance curves of three different specimens in Jiang et al.'s study [38], with the printing angle at a fill rate of 75% as the independent variable.

Although this study chose PLA as the 3D printing material and it demonstrated certain feasibility in laboratory compression performance tests, PLA, as a typical biodegradable polymer, has characteristics that make it susceptible to hydrolysis, which poses significant limitations in harsh marine environments. Under conditions of long-term immersion in seawater, high humidity, and salt spray erosion, PLA molecular chains are prone to breakage and degradation, leading to a significant decline in key mechanical properties such as compressive strength and tensile strength over service time. This makes it difficult to meet the engineering requirements of marine photovoltaic (MPV) systems, which typically need stable operation for 15–20 years. This is a core issue that this research must address at the material selection level. To address this limitation, polymers more suitable for marine environments, such as PETG, Nylon, and fiber-reinforced composites, should be included in subsequent studies and applications [38–41]. PETG has excellent hydrolysis resistance and UV aging resistance, maintaining structural stability even under fluctuations in daily marine temperatures and long-term seawater immersion, and its melt flow properties are compatible with 3D printing processes. Nylon combines high mechanical strength with seawater corrosion resistance, and its impact and fatigue resistance better respond to dynamic loads such as waves and wind. Fiber-reinforced composites (such as carbon fiber-reinforced PP and glass fiber-reinforced epoxy resin), with their high strength-to-weight ratio and excellent environmental aging resistance, have been widely used in marine engineering, with mechanical performance and environmental adaptability far exceeding that of PLA. It should be clarified that this study uses PLA as the target to explore the effects of infill pattern, density, and printing angle on compression performance. The core conclusions (such as the stress uniformity advantage of honeycomb infill and the threshold effect of high infill density) can, in principle, be transferred to optimize the 3D printing processes of the aforementioned marine-grade polymers. However, follow-up studies need to conduct targeted experiments based on the printing parameter sensitivities of different materials (such as the printing temperature window of PETG and the pre-drying requirements of Nylon). At the same time, accelerated tests such as seawater immersion and salt spray aging should be carried out to evaluate long-term performance, so that the research conclusions better align with the actual engineering needs of MPV systems, addressing the current study's shortcomings in material environmental adaptability analysis. Furthermore, a key limitation of this study must be clarified: marine photovoltaic support structures encounter complex dynamic loads from waves, wind forces, and corrosion during actual service, rather than purely static compressive loads. Therefore, conclusions drawn solely from compression tests are insufficiently comprehensive to fully guide engineering applications. This research should be regarded as one component within the overall design framework for marine photovoltaic support structures. Regarding the potential influence of filling patterns on other critical mechanical properties: Honeycomb filling, with its stress-distribution characteristics, may enhance tensile strength and fatigue performance under cyclic loading by reducing local stress concentrations—consistent with the stress transfer mechanism observed in compression tests. In contrast, linear filling, characterized by its anisotropic structure and weak interlayer bonding, may exhibit lower tensile strength and poorer fatigue performance under cyclic tensile or bending loads due to crack propagation at interfacial interfaces. Helical filling, with its ability to resist combined bending and torsional loads, may exhibit tensile and fatigue properties intermediate between honeycomb and linear filling. However, these inferences require validation through subsequent tensile and fatigue testing to provide comprehensive mechanical performance data supporting the design of marine photovoltaic structures.

4. Conclusions

The main conclusions of this paper are as follows:

The mechanical properties of test specimens fabricated from PLA are influenced by three primary factors: infill density, printing method, and printing angle. Notably, infill density exerts the most substantial effect on the specimens. As infill density increases, the specimens demonstrate enhanced compressive properties and other mechanical characteristics.

When the infill density of a specimen attains a specific threshold, further enhancement of its mechanical properties through infill density becomes markedly constrained. At this juncture, modifying the printing method is essential to achieve improved compressive strength and other mechanical characteristics. This study unequivocally illustrates that at 75% infill density, honeycomb infill exhibits significantly superior compressive performance compared to grid or linear infill patterns.

To satisfy the mechanical performance requirements of marine photovoltaics, components are generally printed with a high infill density. In specimens exhibiting high infill density, the printing orientation has a minimal impact on improving their mechanical properties.

This study demonstrates that under laboratory testing conditions, components printed using PLA material can only meet the basic compressive strength requirements for static loading scenarios. However, the marine environment is complex and variable, subjecting structures not only to dynamic loads such as waves and wind forces but also to corrosive factors like salt spray and seawater immersion. Therefore, the specimens in this research cannot be directly applied in marine engineering. During actual service, PLA's mechanical properties (such as tensile strength, flexural strength, and fatigue performance) and chemical stability (such as hydrolysis resistance and UV aging resistance) have not been verified under real marine loading conditions. The safety margin for its long-term operation (marine photovoltaic systems require 15–20 years) remains unclear. Consequently, PLA should only be considered a potential candidate material for marine photovoltaic support structures under specific conditions. Its practical application necessitates supplementary testing, including dynamic load testing, long-term aging tests, and safety margin calculations. Additionally, material modification—such as blending with hydrolysis-resistant polymers or incorporating reinforcing fibers—is required to further enhance its suitability for marine environments. The application of 3D printing technology and PLA materials in marine engineering necessitates a series of studies, including salinity resistance testing, high-temperature resistance testing, fatigue testing of PLA materials, and bending, torsion, and tensile tests. This represents a complex and time-consuming research process. This study tested only the compressive strength of PLA specimens under laboratory conditions. It represents a small part of the broader research exploring PLA materials for marine structures.

Author Contributions: Conceptualization, H.Z. and D.Z.; methodology, H.Z. and D.Z.; software, S.Z.; validation, S.Z. and Y.Z.; formal analysis, S.Z.; investigation, S.G.; resources, D.Z.; data curation, H.F.; writing—original draft preparation, S.Z.; writing—review and editing, D.Z.; visualization, D.Z.; supervision, D.Z.; project administration, D.Z.; funding acquisition, D.Z. All authors have read and agreed to the published version of the manuscript.

Funding: This study received no external funding.

Data Availability Statement: The original contributions presented in this study are included in the article. Further inquiries can be directed to the corresponding author.

Conflicts of Interest: The authors declare no conflicts of interest.

References

1. Pasten, C.; Santamarina, J.C. Energy and quality of life. *Energy Policy* **2012**, *49*, 468–476. [CrossRef]
2. Perez, M.; Perez, R. Update 2022—A fundamental look at supply side energy reserves for the planet. *Sol. Energy Adv.* **2022**, *2*, 100014. [CrossRef]
3. Saha, S.; Ruslan, A.R.M.; Morshed, A.K.M.M.; Hasanuzzaman, M. Global prospects and challenges of latent heat thermal energy storage: A review. *Clean Technol. Environ. Policy* **2021**, *23*, 531–559. [CrossRef]
4. Raza, M.Q.; Nadarajah, M.; Ekanayake, C. On recent advances in PV output power forecast. *Sol. Energy* **2016**, *136*, 125–144. [CrossRef]
5. Al-Ali, M.; Dincer, I. Energetic and exergetic studies of a multigenerational solar–geothermal system. *Appl. Therm. Eng.* **2014**, *71*, 16–23. [CrossRef]
6. Li, X.-Z.; Fan, X.-C.; Cheng, Z.-J. Hydropower development situation and prospects in China. *Renew. Sustain. Energy Rev.* **2018**, *82*, 232–239. [CrossRef]
7. Anderson, A.; Rezaie, B. Geothermal technology: Trends and potential role in a sustainable future. *Appl. Energy* **2019**, *248*, 18–34. [CrossRef]
8. Bilgili, F.; Koçak, E.; Bulut, Ü.; Kuşkaya, S. Can biomass energy be an efficient policy tool for sustainable development? *Renew. Sustain. Energy Rev.* **2017**, *71*, 830–845. [CrossRef]
9. Islam, M.; Mekhilef, S.; Saidur, R. Progress and recent trends of wind energy technology. *Renew. Sustain. Energy Rev.* **2013**, *21*, 456–468. [CrossRef]
10. Lv, S.; Lu, M.; Liu, W.; Li, X.; Lv, W.; Liu, Z.; Dong, X.; Lu, T.; Yang, B. Recent advances in longitudinal spatial area marine photovoltaics. *Renew. Sustain. Energy Rev.* **2025**, *208*, 115036. [CrossRef]
11. Claus, R.; López, M. Key issues in the design of floating photovoltaic structures for the marine environment. *Renew. Sustain. Energy Rev.* **2022**, *164*, 112502. [CrossRef]
12. Rosa-Clot, M.; Tina, G.M. Chapter 4—Submerged PV systems. In *Submerged and Floating Photovoltaic Systems*; Academic Press: Cambridge, MA, USA, 2018; pp. 65–87.
13. López, M.; Claus, R.; Soto, F.; Cebada, A.; Hernández-Garrastacho, Z.; Simancas, O. Structural reliability of a novel offshore floating photovoltaic system to supply energy demands of ports. In *Developments in the Analysis and Design of Marine Structures*; CRC Press: Boca Raton, FL, USA, 2021; pp. 505–512.
14. Jubayer, C.M.; Hangan, H. A numerical approach to the investigation of wind loading on an array of ground mounted solar photovoltaic (PV) panels. *J. Wind. Eng. Ind. Aerodyn.* **2016**, *153*, 60–70. [CrossRef]

15. Alrashidi, H.; Ghosh, A.; Issa, W.; Sellami, N.; Mallick, T.K.; Sundaram, S. Evaluation of solar factor using spectral analysis for CdTe photovoltaic glazing. *Mater. Lett.* **2019**, *237*, 332–335. [CrossRef]
16. *Where Sun Meets Water: Floating Solar Handbook for Practitioners*; World Bank Group: Sydney, NSW, Australia, 2019; ESMAP, SERIS.
17. Villa-Caro, R.; Carral, J.C.; Fraguela, J.Á.; López, M.; Carral, L. A review of ship mooring systems. *Brodogr. Int. J. Nav. Archit. Ocean. Eng. Res. Dev.* **2018**, *69*, 123–149. [CrossRef]
18. Huang, G.; Tang, Y.; Chen, X.; Chen, M.; Jiang, Y. A comprehensive review of floating solar plants and potentials for offshore applications. *J. Mar. Sci. Eng.* **2023**, *11*, 2064. [CrossRef]
19. Nogami, S.; Watanabe, S.; Reiser, J.; Rieth, M.; Sickinger, S.; Hasegawa, A. A review of impact properties of tungsten materials. *Fusion Eng. Des.* **2018**, *135*, 196–203. [CrossRef]
20. Xu, F.; Lin, X.; Zhou, A. Performance of internal curing materials in high-performance concrete: A review. *Constr. Build. Mater.* **2021**, *311*, 125250. [CrossRef]
21. Chen, Z.; Qiao, J.; Yang, X.; Sun, Y.; Sun, D. A review of grouting materials for pouring semi-flexible pavement: Materials, design and performance. *Constr. Build. Mater.* **2023**, *379*, 131235. [CrossRef]
22. Sun, X.; Zhong, J.; Gao, P.; Li, G.; Xiao, D.; Zhang, Z.; Gan, T. Properties of High-Performance Materials for the Crack Repair of Segment Structures. *Sustainability* **2023**, *15*, 15933. [CrossRef]
23. Bentur, A.; Mitchell, D. Material performance lessons. *Cem. Concr. Res.* **2008**, *38*, 259–272. [CrossRef]
24. Ahn, S.-J.; Lee, H.; Cho, K.-J. 3D printing with a 3D printed digital material filament for programming functional gradients. *Nat. Commun.* **2024**, *15*, 3605. [CrossRef]
25. Sun, L.; Hua, G.; Cheng, T.; Wang, Y. How to price 3D-printed products? Pricing strategy for 3D printing platforms. *Int. J. Prod. Econ.* **2020**, *226*, 107600. [CrossRef]
26. Lee, S.G.; young An, E.; Lee, J.B.; Park, J.C.; Shin, J.W.; Kim, J.K. Enhanced cell affinity of poly (d, l-lactic-co-glycolic acid)(50/50) by plasma treatment with β-(1→ 3)(1→ 6)-glucan. *Surf. Coat. Technol.* **2007**, *201*, 5128–5131. [CrossRef]
27. Ogaki, R.; Green, F.; Li, S.; Vert, M.; Alexander, M.; Gilmore, I.; Davies, M. G-SIMS of biodegradable homo-polyesters. *Appl. Surf. Sci.* **2006**, *252*, 6797–6800. [CrossRef]
28. Sasatsu, M.; Onishi, H.; Machida, Y. In vitro and in vivo characterization of nanoparticles made of MeO-PEG amine/PLA block copolymer and PLA. *Int. J. Pharm.* **2006**, *317*, 167–174. [CrossRef] [PubMed]
29. Zhao, H.; Cui, Z.; Sun, X.; Turng, L.-S.; Peng, X. Morphology and properties of injection molded solid and microcellular polylactic acid/polyhydroxybutyrate-valerate (PLA/PHBV) blends. *Ind. Eng. Chem. Res.* **2013**, *52*, 2569–2581. [CrossRef]
30. Yu, L.; Dean, K.; Li, L. Polymer blends and composites from renewable resources. *Prog. Polym. Sci.* **2006**, *31*, 576–602. [CrossRef]
31. Jiang, H.; Xie, Y.; Zeng, S.; Guo, S.; Chen, Z.; Liang, Z.; Zhang, D. Research on Lightweight Design Performance of Offshore Structures Based on 3D Printing Technology. *J. Mar. Sci. Eng.* **2025**, *13*, 2007. [CrossRef]
32. *ASTM D1621*; Standard Test Method for Compressive Properties of Rigid Cellular Plastics. American Society for Testing and Materials: New York, NY, USA, 2010.
33. Yao, T.; Deng, Z.; Zhang, K.; Li, S. A method to predict the ultimate tensile strength of 3D printing polylactic acid (PLA) materials. *Compos. Part B Eng.* **2019**, *163*, 393–402. [CrossRef]
34. Liu, T.; Zhu, J.-H.; Zhang, W.; Belhabib, S.; Guessasma, S. Microstructure and compressive behaviour of PLA/PHA-wood lattice structures processed using additive manufacturing. *Polym. Test.* **2024**, *141*, 108612. [CrossRef]
35. Wang, H.; Wang, K.; Lei, J.; Fan, X. Compressive Properties and Energy Absorption Characteristics of Co-Continuous Interlocking PDMS/PLA Lattice Composites. *Materials* **2024**, *17*, 3894. [CrossRef] [PubMed]
36. Kilinc, F.B.; Bozaci, E.; Kilinc, A.C.; Turkoglu, T. Effect of Atmospheric Plasma Treatment on Mechanical Properties of 3D-Printed Continuous Aramid Fiber/PLA Composites. *Polymers* **2025**, *17*, 397. [CrossRef]
37. Al-Tamimi, A.A.; Pandžić, A.; Kadrić, E. Investigation and prediction of tensile, flexural, and compressive properties of tough PLA material using definitive screening design. *Polymers* **2023**, *15*, 4169. [CrossRef] [PubMed]
38. El Mehtedi, M.; Buonadonna, P.; El Mohtadi, R.; Loi, G.; Aymerich, F.; Carta, M. Optimizing milling parameters for enhanced machinability of 3D-printed materials: An analysis of PLA, PETG, and carbon-fiber-reinforced PETG. *J. Manuf. Mater. Process.* **2024**, *8*, 131. [CrossRef]
39. Techawinyutham, L.; Tengsuthiwat, J.; Srisuk, R.; Techawinyutham, W.; Rangappa, S.M.; Siengchin, S. Recycled LDPE/PETG blends and HDPE/PETG blends: Mechanical, thermal, and rheological properties. *J. Mater. Res. Technol.* **2021**, *15*, 2445–2458. [CrossRef]

40. Liu, F.F.; McAuley, K.B. Improved kinetic rate constants for nylon 6, nylon 6, 6, and nylon 6/6, 6 copolymer. *Macromol. React. Eng.* **2020**, *14*, 1900037. [CrossRef]
41. Rubino, F.; Nisticò, A.; Tucci, F.; Carlone, P. Marine application of fiber reinforced composites: A review. *J. Mar. Sci. Eng.* **2020**, *8*, 26. [CrossRef]

Journal of
Marine Science and Engineering

MDPI

Review

Pipeline Systems in Floating Offshore Production Systems: Hydrodynamics, Corrosion, Design and Maintenance

Jin Yan [1,2], Yining Zhang [1], Zehan Chen [1], Pengji Li [1], Yuting Li [1], Zeyu Cao [1], Jiaming Wu [1], Kefan Yang [1] and Dapeng Zhang [1,*]

1 Ship and Maritime College, Guangdong Ocean University, Zhanjiang 524005, China; yanj@gdou.edu.cn (J.Y.); zhangyining_1029@163.com (Y.Z.); 18319556625@163.com (Z.C.); pengjigdou@163.com (P.L.); lyt77978@outlook.com (Y.L.); 19002039756@163.com (Z.C.); wu1328688209@163.com (J.W.); 18102811435@163.com (K.Y.)

2 Guangdong Provincial Key Laboratory of Intelligent Equipment for South China Sea Marine Ranching, Guangdong Ocean University, Zhanjiang 524088, China

* Correspondence: zhangdapeng@gdou.edu.cn

Abstract

Floating offshore production systems play a critical role in offshore resource development, where the structural integrity and operational safety of risers, umbilical cables, and mooring cables are of paramount importance. Focusing on the failure risks of these key components under harsh marine environments, this paper systematically reviews the coupled mechanisms of wave-induced loading, electrochemical corrosion, and material fatigue. Unlike traditional reviews on offshore pipelines and cables, this study not only examines the mechanical performance of deepwater pipelines and cables along with representative research cases but also discusses corrosion mechanisms in marine environments and corresponding repair and mitigation strategies. In addition, recent advances in machine learning-based digital twin frameworks and real-time monitoring technologies are reviewed, with an analysis of representative application cases. The findings indicate that interdisciplinary material innovations combined with data-driven predictive models are essential for addressing maintenance challenges under extreme ocean conditions. Furthermore, this review identifies existing research gaps in data fusion for monitoring technologies and outlines clear directions for the intelligent operation and maintenance of future deep-sea infrastructure.

Keywords: floating offshore production systems; risers; umbilical cables; mooring cables; deep-sea infrastructure; corrosion mechanisms; digital twin

1. Introduction

Fossil fuels are essential to human production and daily life [1,2]. While several renewable energy technologies have begun to emerge in recent years [3–5], they cannot fully replace fossil fuels in the short term [6]. The oceans and seas, as vast reservoirs of resources, are particularly rich in fossil fuel reserves [7–9]. Floating Offshore Production Systems play a critical role in the extraction of marine resources [10,11]. Marine pipelines and cables within these floating production systems are essential for ensuring their reliable and efficient operation [12–14]. There are three main types of marine pipelines and cables that play a critical role in floating offshore production systems: mooring cables, umbilical cables, and risers. Mooring cables are used to anchor the floating platform to the seabed, ensuring its stability and maintaining its position [15,16]. Umbilical cables connect the platform to subsea facilities, transmitting power, communications, and other essential services [17,18].

Risers are responsible for transporting oil, gas, or other resources from the seabed to the platform for processing or production operations [19,20]. The complex environment in which floating offshore production platforms operate presents several threats to the health of marine pipelines and cables within these systems. These threats include the impact of waves and currents [21,22], collisions with other offshore structures [23,24], environmental corrosion [25,26], variations in temperature and pressure [27,28], transporting pipeline and cables corrosion [29,30], and biological erosion [31,32]. If not adequately addressed, these factors can lead to environmental pollution [33,34], significant economic losses [35,36], and even pose risks to the lives of operators [37,38]. Figure 1 shows an offshore floating production system and common health and safety issues related to pipelines and cables in the system. Consequently, it is crucial to study the negative impacts of these threats on offshore pipelines and explore effective strategies for their placement and protection.

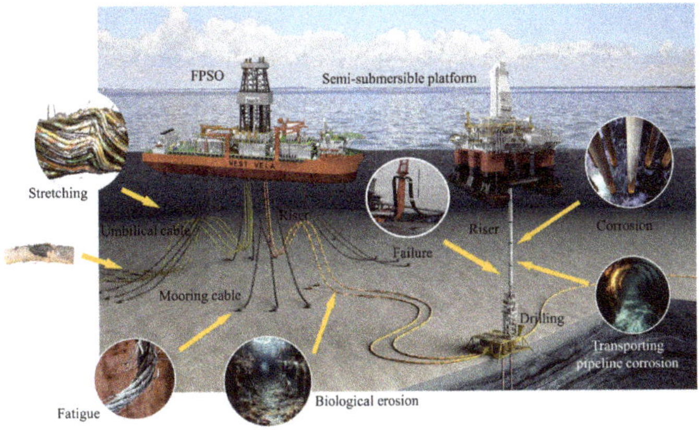

Figure 1. Offshore Floating Production Systems and common health and safety issues related to pipelines in such systems.

An initial analysis of health and safety issues related to marine pipelines highlights the need for these concerns to be addressed through interdisciplinary research, incorporating insights from multiple fields. The causes of damage to submarine pipelines can be divided into the following broad categories.

(1) Deformation or breakage of pipelines due to sea currents during laying.
(2) In the process of transportation due to the influence of third-party external forces leading to damage.
(3) For submarine pipelines, the occurrence of pipe spans is inevitable during operation. Submarine pipelines can develop hanging spans due to current scouring. The hanging pipe sections are subjected to alternating loads from water flow, leading to vortex-induced vibrations and fatigue fractures, ultimately resulting in pipeline fatigue damage [39–42].
(4) Submarine pipelines are susceptible to corrosion due to the acidic composition of the ocean, inherent pipeline defects, and the presence of sand in the transmission substances, leading to perforation and fracture.

Marine pipelines and cables serve as the critical lifelines for offshore energy transmission, yet they are increasingly exposed to multi-hazard environments, including extreme hydrodynamic loads, complex deep-sea corrosion, and mechanical failures. Despite extensive localized research, the field currently lacks a holistic, interdisciplinary framework that integrates fluid mechanics, material science, and structural integrity to guide the entire

https://doi.org/10.3390/jmse14020176

lifecycle of these assets. This paper mainly reviews three major types of marine pipelines and cables used in offshore floating production systems. This review aims to address this critical gap by developing a comprehensive research manual that fosters cross-disciplinary synergy. Moving beyond isolated studies, this work systematically maps the research trajectory starting from wave-induced effects and extending to advanced material design and anti-corrosion strategies. The scope encompasses an in-depth analysis of environmental impacts, corrosion mechanisms, maintenance protocols, and innovative design methodologies. By identifying current research bottlenecks and emerging trends, this review provides a strategic roadmap for the sustainable development and intelligent management of marine infrastructure.

2. Literature Selection Methods

This study selected the Web of Science Core Collection (WoSCC), indexed in the Science Citation Index Expanded (SCIE), and the China National Knowledge Infrastructure (CNKI) as data sources. The relevant literature was retrieved on 28 January 2025, through the online library of Guangdong Ocean University, China. The publication period was limited to studies published between 1950 and 2025.

To systematically explore multidisciplinary research on the safety of pipelines and cables in floating offshore production systems, a carefully designed search strategy was developed to integrate studies from engineering as well as environmental science and technology. Key structural components, including risers, mooring cables, and umbilical cables, were selected as primary keywords to ensure comprehensive coverage of the core elements of floating production systems. Safety-related terms such as corrosion, failure, and fatigue were incorporated to capture major threats to structural integrity. In addition, environmental factors including waves, currents, and vortex-induced vibration were included to highlight the influence of complex marine conditions on system safety.

The final search query was refined as follows:

(KP = (Riser OR Mooring Cable OR Umbilical Cable) OR TS = (Floating Offshore Production)) AND (TS = (Health OR Safety) OR KP = (Corrosion OR Failure OR Fatigue OR Biological Erosion OR Collisions OR Temperature OR Pressure OR Pipeline Contamination OR Wave OR Current OR Vortex-Induced Vibration)).

The literature selected for this review consists of reviews and research articles, with documents published in both Chinese and English. This is because the authors' research background is primarily based in Chinese higher education institutions, making them more familiar with studies in both languages. Based on the above search strategy, a total of 657 papers were retrieved from WoSCC and 109 papers from CNKI. First, duplicate records were initially removed using the bibliometric analysis software CiteSpace (Version 6.3R1). Second, through manual screening, papers whose titles involved risers, umbilical cables, and mooring lines were selected. Subsequently, the literature was further filtered based on three criteria: research relevance, citation count, and journal reputation, with journal reputation assessed considering database indexing, impact factor, JCR quartile, and Chinese Academy of Sciences (CAS) ranking. Finally, a total of 260 papers were included. In addition, during the review process and based on recommendations from experts in the field, five additional DNV standards were incorporated. Meanwhile, due to the limited existing research on the performance and protection of pipelines and cables specifically within FPSO systems, this review incorporates high-quality case studies involving risers, umbilical cables, and mooring cables to provide a comprehensive analysis.

https://doi.org/10.3390/jmse14020176

3. Wave and Current Impact

The impact of waves and currents are critical considerations in ocean engineering, particularly when studying marine pipeline and cable issues [43–45]. During the key operations of pipe-laying [46,47] and pipe-lifting [48,49], waves can compromise operational accuracy and even cause damage. These environmental factors not only directly impact the structural stability and mechanical properties of offshore pipelines and cables but also influence the reliability and safety of FPSO system long-term operations. Periodic loads from waves [50,51], vortex-induced vibrations [52,53], and the pressure and drag forces [54,55] from ocean currents create complex dynamic effects on pipeline and cables systems. These can lead to fatigue damage [56,57] and structural failure [58,59].

3.1. Morison Equation

The Morison Equation [60] is a crucial tool in marine engineering for calculating wave-induced forces on submerged structures, as expressed in Equation (1).

$$dF = C_m \rho dV \dot{u}_n + \frac{1}{2} C_d \rho dA |u_n| u_n \tag{1}$$

Here, dF represents the total wave force acting on a micro-section of the member with a projected area dA. The variables $\dot{u}_n$ and u_n denote the instantaneous wave-induced fluid acceleration and velocity perpendicular to the object's axis. The coefficients C_d and C_m correspond to the drag force coefficient and inertia force coefficient, respectively.

The equation incorporates both inertial and drag forces in hydrodynamics to predict fluid forces acting on small-diameter structures. It enables quick estimations of wave and fluid forces on elongated structures with a straightforward calculation process that avoids complex numerical solutions [61,62]. However, its applicability is typically limited to elongated structures whose diameters are much smaller than the wavelength. Recent intensive research has revealed potential extensions of this equation's use to larger marine structures.

Zhou et al. [63] applied Morrison's equations to calculate wave loads on large-scale round-ended sinkhole foundations for sea-crossing bridges subjected to regular waves. They compared the results obtained from Morrison's equations with those from other methods. The findings indicated that Morrison's equations might overestimate the contribution of wave drag forces, particularly for large-scale structures. The team also analyzed the sensitivity of Morrison's equations to the relative scale D/L of the structure by varying the wave period and comparing the results with those of alternative methods.

Luo (2023) [64] used Morrison's equations to calculate the wave forces in both horizontal and vertical directions for a suspension tunnel tube in a wave environment. The study revealed that Morrison's equation might introduce significant errors when calculating wave loads on large-scale structures. To address this, the hydrodynamic coefficients, including additional mass and inertia coefficients, were optimized and corrected using the least-squares method. This adjustment made the wave forces predicted by Morrison's equation more consistent with those calculated by the theory of wrap-around potential flow. The accuracy and validity of the modified Morrison's equation in wave force prediction were verified by comparing its results with those obtained from the bypassing potential flow theory.

For calculating wave loads on large semi-submersible platforms, Liu and Xiao (2003) [65] combined Morrison's formula with potential flow theory, particularly for the elongated portion of the platform. In this region, the forces are determined by the wave motion around the structure, influenced by the large-volume section of the platform.

https://doi.org/10.3390/jmse14020176

Although the Morison equation is widely applied due to its simplicity, it exhibits significant limitations when used for large-scale marine structures. By neglecting wave diffraction and radiation effects, the conventional formulation fails to accurately capture wave–structure interactions when the relative diameter-to-wavelength ratio (D/L) becomes large, leading to distorted force estimations. At present, although coefficient calibration or hybrid approaches incorporating potential flow theory can improve prediction accuracy to some extent, a universally applicable unified criterion is still lacking. Therefore, in engineering design, it is necessary to balance computational efficiency and accuracy according to structural scale and wave conditions, and the establishment of standardized guidelines for model selection and parameter calibration remains a key focus for future research.

3.2. Current Scour

Current-induced scour can result in the removal of sediment around marine pipelines, exposing their surfaces and increasing abrasion and erosion. This can lead to pipeline overhangs, stress concentrations, buckling, or fractures. Additionally, scour can alter seabed morphology, cause uneven stresses and vibrations on the pipeline, reduce stability, and accelerate the corrosion process. Meanwhile, current scour poses a severe threat to the structural integrity and operational safety of FPSO systems by weakening the anchoring capacity of mooring systems, inducing fatigue failure in free-spanning pipelines and risers through vortex-induced vibrations (VIV), and causing instability in subsea infrastructure foundations. Figure 2 illustrates a typical free-spanning offshore pipeline sketch.

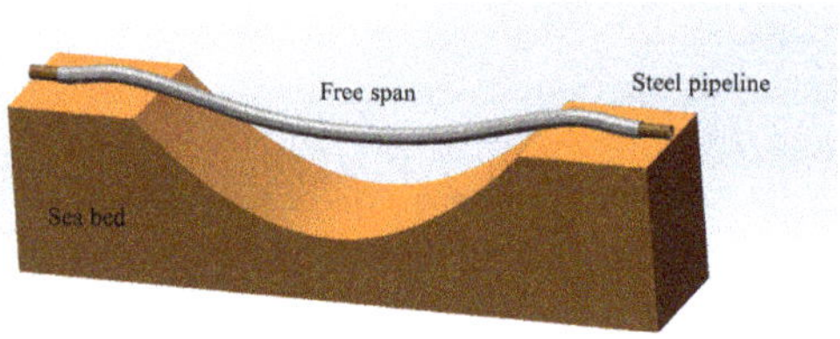

Figure 2. A typical free spanning offshore pipeline sketch.

Zhang from Shanghai Jiao Tong University has summarized recent research findings on the local scour problem of fixed pipelines in marine environments [66].

Lucassen and Kjeldsen et al. [67,68] conducted experimental research on the current scour problem of submarine pipelines. They analyzed the effects of flow velocity, flume size, and pipeline burial depth on the equilibrium depth of scouring of submarine pipelines. They discussed the equilibrium depth of scouring as a function of flow velocity, flume size, and sand size and proposed a formula to predict the relationship between the equilibrium depth of scouring of pipelines and the diameter of the pipe and the flow velocity.

Chao and Hennesy [69] derived an analytical method for determining the equilibrium depth of local scour in pipelines under uniform flow based on two-dimensional potential flow theory. They proposed an analytical method specifically for clear water scour, which involves solving the following equations:

$$\left[1+\frac{R}{S}-\frac{R^2}{2S^2+RS}\right]^2 u_c^2\rho_w f_c=\tau_{cr} \tag{2}$$

where τ_{cr} is critical shear stress, u_c^2 is critical velocity and f_c is friction factor.

Li and others (2012) [70] improved upon the Chao-Hennesy resolution method by incorporating the factor of sediment inflow into the pit. Based on two-dimensional potential

https://doi.org/10.3390/jmse14020176

flow theory and the principle of energy balance, they combined this with the formula for flow-type and power-type sediment transport rates. This led to the development of a semi-empirical and semi-theoretical prognostic method applicable to determining the equilibrium depth of localized scour around pipelines under homogeneous flow conditions in a moving bed scour situation.

Subsequently, Mao (1987) [71] investigated the interaction mechanism between submarine pipelines and the seabed through indoor experiments. He proposed a mathematical model for the scour depth of pipelines based on the improved potential flow theory, discussed the effect of vortex shedding on scour downstream of the pipeline, and conducted a preliminary analysis of the impact of pipeline vibration on scour.

Sumer et al. (1988) [72] investigated the scour in the wake behind the pipeline using both experimental and numerical methods. They revealed the mechanism of wake scour and discussed the effect of bed shear on wake scour.

Van Beek (1990) [73] investigated the problem of localized scour around a submarine pipeline using numerical methods. The numerical model employed the Navier–Stokes (N-S) equation and the sediment movement equation to calculate fluid movement and sediment scour.

Chiew et al. [74] discussed the mechanism of localized scouring of pipelines in cohesionless seabed's using experimental methods. They concluded that the pressure difference between the stagnation pressure upstream of the pipeline and the wake pressure downstream is the primary cause of scouring. They suggested that placing impermeable membranes in the seabed below the pipeline could prevent scouring. Additionally, Chiew [75] examined the effect of spoilers on pipeline scour, finding that spoilers installed at the top of the pipeline increase the blocking effect on the water flow, causing more water to be deflected downward and thus increasing the scour depth. The presence of spoilers also enhances wake scour, which is significant for the self-burial phenomenon of pipelines. Furthermore, the local scour of pipelines is strongly influenced by the parameters of both the fluid and the pipeline [76–80].

Hu et al. [81] investigated the effect of the incidence angle of water flow on the scouring of submarine pipelines, based on two-dimensional studies predicting scour depth. They proposed a machine learning approach to predict the equilibrium scour depth of submarine pipelines, considering the incidence angle of water flow. In their study, three models were proposed: GA-BP neural network, RBF neural network, and SVM. They found that the Froude number is the most influential factor affecting the scour depth of submarine pipelines.

Zhao et al. [82] discussed the effects of isolated waves and currents on localized scour. They proposed a coupled numerical model incorporating current and sediment scour modules. The hydrodynamics and local scour around various submarine pipelines, under the combined influence of isolated waves and surface back currents, were investigated using the immersed boundary method.

Dogan et al. [83] conducted an experimental study on the development of scour holes along the spreading direction of the pipe and the velocity of scour propagation. They found that the development along the spreading direction of the pipe was inversely proportional to the Keulegan–Carpenter number and directly proportional to the Shields parameter. Additionally, a new expression based on the Keulegan–Carpenter number was provided to evaluate the clear water scour condition data for this study.

$$V_h^* = 4.5KC^{-1.67} \tag{3}$$

The effective range is $0.01 < \theta < \theta_{cr}$.

https://doi.org/10.3390/jmse14020176

They also proposed that the propagation velocity of scour increases with the increase in pipe diameter and decreases with the increase in particle size.

Chen et al. [84] proposed a distributed model to evaluate the damage of submarine pipeline networks caused by regional mudflows. Figure 3 illustrates the framework of the proposed distributed model. The model comprises three components: a spatial discretization module, a regional submarine mudflow dynamics module, and a pipeline network impact force and damage assessment module. This model is designed to assess the response of pipeline networks to regional submarine mudflows, aiding in risk-informed decision-making, planning, and optimization of pipeline routes.

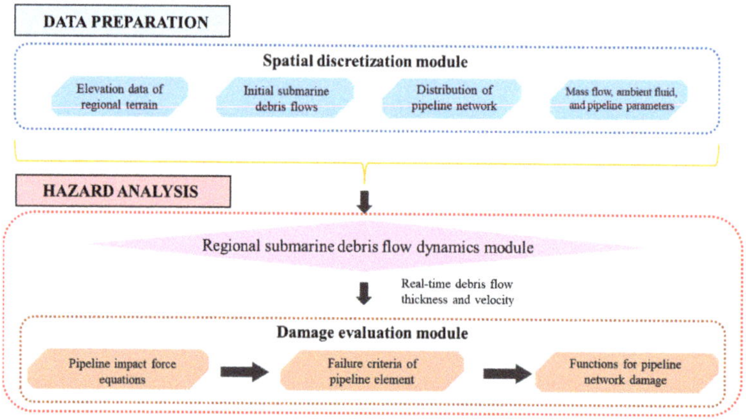

Figure 3. Framework of the proposed distributed model.

Although significant progress has been made in the study of local scour around submarine pipelines, most existing research is based on simplified assumptions such as two-dimensional steady flow, rigid pipelines, and idealized seabed conditions. These assumptions limit the ability to fully represent the complex marine environment, including wave–current coupling, oblique flow incidence, and seabed heterogeneity, and introduce pronounced scale effects and restrictions on model applicability. Local scour–induced free spans and flow-induced vibrations pose serious threats to pipeline structural integrity; however, the accuracy and general applicability of existing prediction models and mitigation measures remain insufficient under multi-field coupled conditions. Future research should integrate large-scale experimental studies, high-fidelity numerical simulations, and field observations to develop a unified evaluation framework that incorporates structural dynamic response and reliability-based design, thereby establishing more robust engineering design guidelines and mitigation strategies.

3.3. *The Vortex-Induced Vibration*

Nonlinear hydrodynamic effects induced by wave action can lead to complex kinematic responses, among which vortex-induced vibration (VIV) is a representative phenomenon. VIV significantly compromises FPSO systems by driving high-frequency cyclic oscillations in risers, umbilical cables, and mooring lines, thereby accelerating fatigue damage and increasing the risk of structural failure or loss of containment.

An experimental study on the vortex-induced vibration (VIV) of submarine pipelines with long free spans was conducted. When submarine pipelines and cables are laid on cohesionless soil, uneven seabed's or scouring can lead to the development of local scour around the pipeline. This local scour results in the periodic shedding of vortices across the pipeline, subjecting it to transverse and downstream periodic fluid forces. These

 https://doi.org/10.3390/jmse14020176

periodic fluid forces cause the pipeline to vibrate periodically, a phenomenon known as "Vortex-Induced Vibration" (VIV) [85–87].

Numerous researchers conducted an experimental study on the vortex-excited vibration of low mass ratio pipelines. The experimental results revealed three branches in the variation in amplitude with the approach velocity in these pipelines. The vortex-excited cylindrical amplitude and vibration frequency were found to be related to the approach velocity. As the approach velocity changed, the performance of the three branches—initial, lower, and upper—also changed. Notably, there was a significant hysteresis effect observed before the transitions between these branches [88–92].

Zhang [66] summarized the equations and proposed several forms of vortex shedding for cylindrical vortex-excited vibrations, such as 2S, 2P, P + S, 2C, 2T, etc., as shown in Figure 4.

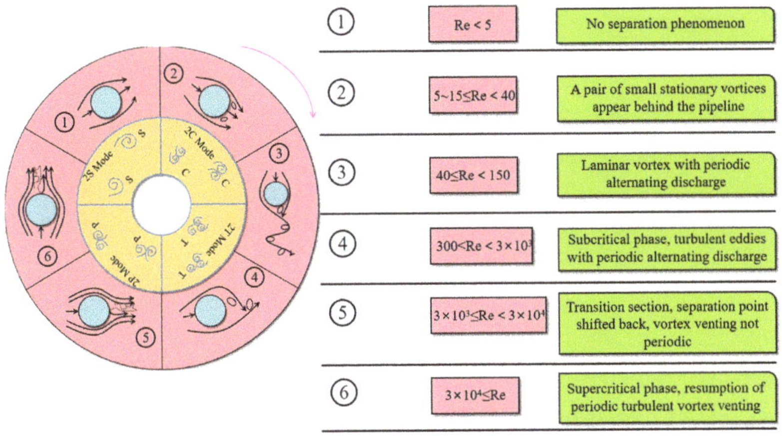

Figure 4. The vortices shedding for various Reynolds number and vortex shedding mode.

Fan D et al. [93] conducted an experimental study on the steady-state hydrodynamic vibrations of small flexible cylinders (aspect ratio L/d = 61, mass ratio m0 = 1.38) within a backed structure on the semi-buried seafloor of a main pipeline. The study investigated the effect of the gap-to-diameter ratio for five different cases, including the open-flow case (G/d = ∞) at various folding velocities. The following conclusions were drawn:

(1) Compared to the narrow-band single-frequency response in the CF (first harmonic) and IL (second harmonic) directions for the open-flow case, the presence of a larger semi-buried pipe increases the mean drag coefficient of the smaller flexible cylinders in the backed structure. Additionally, it exhibits a wider-band response when the gap-to-diameter ratio is small.

(2) For small gap-to-diameter ratios, this leads to larger average IL displacements at the same rate of reduction.

(3) At G/d = 1.0, a stronger third harmonic motion is observed in the CF direction. In the IL direction, the frequency response is much richer, exhibiting stronger first and fourth harmonic motions, detectable third harmonic motions, and larger second harmonic motions.

Xu [94] and colleagues investigated the multi-span classification method and interaction mechanism of vortex-excited vibration in submarine pipelines. They studied the interaction mechanism of multi-span pipelines from both static and dynamic perspectives and proposed a multi-span pipeline classification method based on modal frequency and

 https://doi.org/10.3390/jmse14020176

energy transfer. Figures 5 and 6 illustrate the processes of the modal frequency-based classification method and the energy transfer-based classification method, respectively.

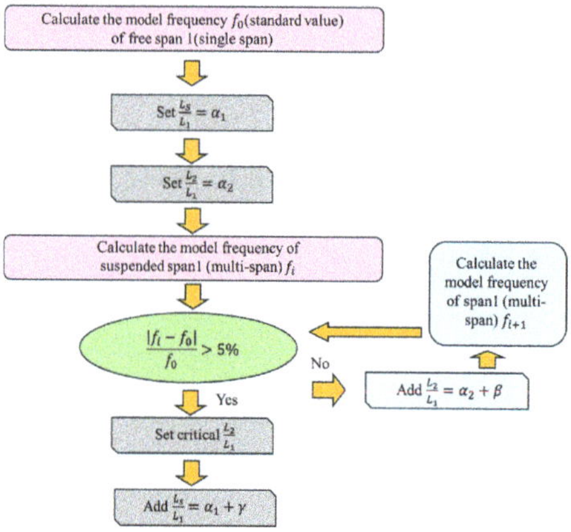

Figure 5. Process of the classification method based on modal frequencies.

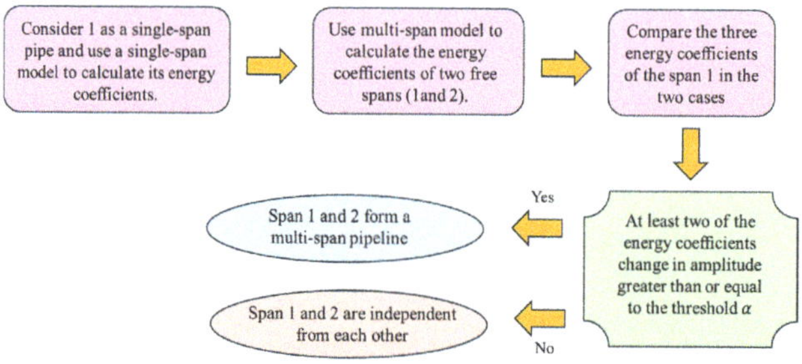

Figure 6. Process of the classification method based on energy transfer.

Zhang et al. [95] numerically investigated the inward flow effect (IFE) and outward flow effect (ECE) of vortex-excited vibrations (VIVs) on the axial inward flow of free-spanning submarine pipelines with mis-flow vortex-excited vibrations. They explored the competing mechanisms involved. The partial differential equations for the pipeline system were developed using Euler-Bernoulli beam theory and a wake oscillator model.

Although significant progress has been made in the study of vortex-induced vibration (VIV) of submarine pipelines, most research is limited to idealized conditions such as uniform flow, two-dimensional assumptions, and isolated free spans, which cannot fully capture wave–current coupling, seabed irregularities, and complex pipe–soil interactions. Existing predictive models are constrained by parameter sensitivity and scale effects, and often neglect the nonlinear coupling between scour evolution and vibration response, limiting their applicability to full-scale engineering scenarios. VIV-induced fatigue damage and resonance pose serious threats to pipeline integrity, while the long-term effectiveness of current mitigation measures under dynamic seabed conditions remains uncertain. Future research should employ large-scale experiments and high-fidelity fluid–structure–soil simulations to develop an integrated predictive framework encompassing hydrodynamics, structural dynamics, and seabed evolution, thereby establishing more robust engineering design guidelines.

https://doi.org/10.3390/jmse14020176

4. Corrosion

The corrosion behavior of marine pipelines is primarily influenced by seawater [96,97] and seabed sediments [98,99]. Seabed sediments exert a significant influence on the integrity of FPSO systems by creating complex electrochemical environments that induce differential aeration cells and promote microbiologically influenced corrosion (MIC). These mechanisms adversely affect the long-term structural reliability of key subsea components, including risers, umbilical cables, and mooring lines, particularly in sediment-covered or partially buried conditions.

The complexity of the seawater environment complicates the corrosion process, which involves electrochemical reactions affected by factors such as the composition of the rust layer and the nature of the solution [100–103]. To better understand corrosion behavior, it is essential to use relevant corrosion models that can explain these patterns and interactions.

Compared to the seawater environment, seabed sediments present a non-uniform system consisting of solid and liquid phases [104–106]. In submarine sediments, steel primarily undergoes electrochemical and microbiological corrosion. The limited dissolved oxygen in seafloor sediments and the poor exchange between seafloor water and ocean water usually result in more oxygen-poor corrosive environments [107,108]. Different types of seafloor sediments affect the rate of oxygen diffusion, leading to uneven oxygen distribution [109–111], large potential differences, corrosion currents, and galvanic coupling corrosion. Additionally, variations in oxygen supply can initiate autocatalytic processes of localized corrosion, such as pitting [112,113] and crevice corrosion [114,115]. Figure 7 illustrates the analysis of corrosion in submarine pipelines.

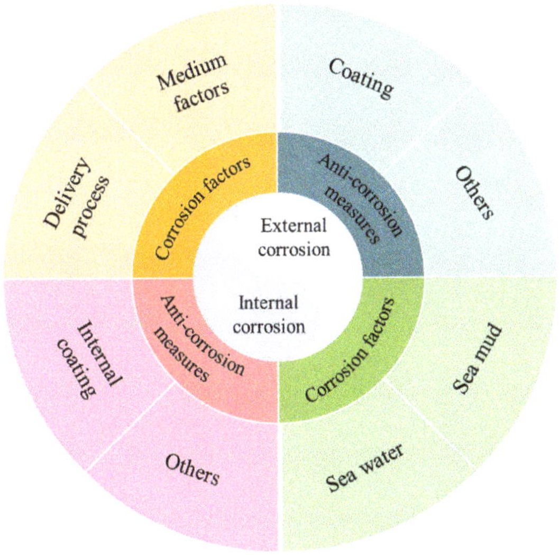

Figure 7. Corrosion scenarios analysis of submarine pipeline.

4.1. Corrosion Behavior in Seawater

Figure 8 illustrates the environmental distribution of metals in the oceans. Depending on the environment in which metals are located in the ocean, they can be divided into the marine atmospheric zone, splash zone, tidal zone, submerged zone, and mud zone. The corrosion behavior of metals in the marine environment varies depending on the zone, and thus, marine pipelines are subject to different degrees of corrosion [116–118].

https://doi.org/10.3390/jmse14020176

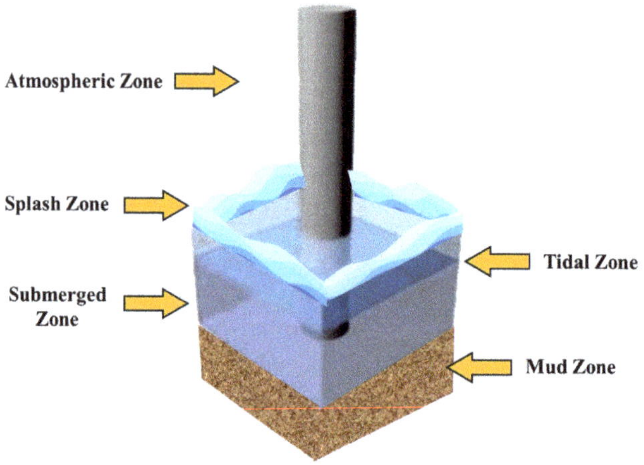

Figure 8. Distribution of metals in the marine environment.

In the marine atmospheric and splash zones, frequent wet-dry cycles and high oxygen levels lead to faster corrosion rates, with uniform corrosion and pitting corrosion being common [119–121]. The tidal zone, influenced by tidal changes, presents a complex environment where differential aeration corrosion and crevice corrosion are more likely to occur [122,123]. The submerged zone, where metals are constantly submerged in seawater, experiences slower corrosion rates due to limited oxygen supply, but uniform corrosion and biofouling remain significant issues [124,125]. In the mud zone, metals buried in sediment face minimal oxygen exposure, making microbial corrosion, such as sulfide corrosion caused by sulfate-reducing bacteria, the primary concern [126,127].

Melchers R E et al. [128] established a corrosion model for mild steel in seawater by conducting long-term experiments on its early corrosion behavior. Figure 9 illustrates the time model of corrosion loss for multiphase phenomena. They concluded that the long-term immersion of carbon steel in seawater can be divided into five stages:

(1) The activation control stage:

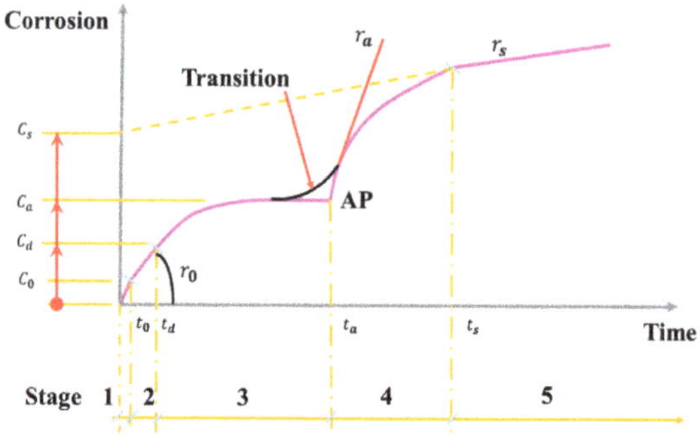

Figure 9. The multi-phase phenomenological corrosion loss–time model.

In the initial stage, during a relatively short period of time when carbon steel is immersed in seawater, metal activation is driven by the ample supply of oxygen.

(2) The concentration control stage:

https://doi.org/10.3390/jmse14020176

Dissolved oxygen diffuses and generates thin and loose corrosion products. Due to the thin rust layer, there is no obstruction to dissolved oxygen diffusion, and the corrosion process can be described by a linear function.

(3) Oxygen diffusion control:

As the rust layer becomes thicker, the diffusion of dissolved oxygen through the rust layer becomes the controlling step of the corrosion reaction. This stage is characterized by oxygen diffusion control.

(4) The growth control stage of SRB:

Due to the gradual thickening of the rust layer, it becomes difficult for oxygen molecules to reach the surface of the metal. Consequently, sulfate-reducing bacteria (SRB) start to grow in this stage.

(5) The stabilization control stage of SRB:

As bacterial growth stabilizes, the corrosion rate of carbon steel also gradually stabilizes. This stage is characterized by a prolonged period of stable corrosion rate.

As early as 1925, Evans et al. [129–131] used vertical steel and zinc plates half-immersed in NaCl solution to study the effect of waterline action on metal corrosion. They found that the metal at the gas–liquid junction exhibited slight corrosion, while the corrosion at the waterline was particularly pronounced. This was attributed to different oxygenation states in the waterline region. The narrow, well-oxygenated waterline area became a cathodic zone, while the under-oxygenated area beneath the waterline became an anodic zone. This disparity created a macroscopic oxygen concentration cell, leading to galvanic corrosion of the metal.

Tomashov et al. [132] believe that the cathodic area of the macro-galvanic cell, which is the waterline region, contains multiple micro-galvanic cells. These two types of cells together influence the corrosion behavior of the metal. As the corrosion behavior of metals in the waterline region receives increasing attention, researchers worldwide have conducted extensive studies on the corrosion of metals in seawater and atmospheric environments [133–138].

Jeffrey et al. [139] conducted a study on waterline corrosion by immersing a 1-m-long vertical steel plate in seawater. Based on the different lengths of the steel plate immersed in seawater, they obtained a classic mass loss profile. The results showed that the mass loss near the waterline was minimal, indicating relatively mild corrosion. The most severe corrosion occurred in the splash zone and the fully immersed zone below the waterline. The splash zone experienced the greatest mass loss due to the influence of sea wind and waves, while the fully immersed zone below the waterline also showed significant mass loss. The study described the distribution areas of anodes and cathodes, and the electron transfer paths under different aeration conditions.

4.2. Corrosion Rate Prediction

Researchers have developed various corrosion prediction models to better understand the corrosion mechanisms in marine pipelines and enhance their protection [140]. Commonly used prediction methods include probabilistic statistical methods [141,142], reliability function analysis [143,144], regression analysis methods [145,146], time series prediction methods [147,148], gray model prediction methods [149,150], and artificial intelligence techniques [151–153].

Traditional prediction methods, such as probabilistic statistics and regression analysis, often require high-quality data and may not accurately capture the relationships between variables when fitting regression functions. Reliability function analysis, frequently employing the Monte Carlo method [154], involves substantial computational

https://doi.org/10.3390/jmse14020176

effort. Time series prediction methods, including autoregressive models [155], sliding average models [156], ARMA models [157], and ARIMA models [158], aim to predict future outcomes by constructing models based on the corrosion development law. However, time series-based methods face limitations in predicting multiple influencing factors that are not time-dependent. Additionally, artificial intelligence algorithms like the Least Squares Support Vector Machine (LSSVM) model [159] and gray models, such as the GM (1, N) model [160], offer alternative approaches to prediction.

Deng et al. [161] proposed a novel prediction method for the corrosion rate of submarine pipelines, combining gray correlation analysis and fuzzy neural networks (FNN). The FNN integrates fuzzy theory with artificial neural networks, addressing some of the limitations of traditional neural networks [162]. The theoretical foundation of FNN is rooted in fuzzy mathematics and the learning mechanisms of neural networks. In this context, the degree of affiliation refers to how closely an element u belongs to a fuzzy subset f. The fuzzy affiliation function is used to quantitatively calculate this degree of affiliation [163]. The T-S fuzzy system can continuously modify the affiliation function of the fuzzy subset. The experimental steps are shown in Figure 10. The experiment demonstrated that this hybrid prediction method maintains high prediction accuracy even when the number of environmental factors is reduced. Future work can focus on extensive testing and training with large datasets on the corrosion rate of submarine pipelines to enhance the generalization capability of this hybrid prediction method.

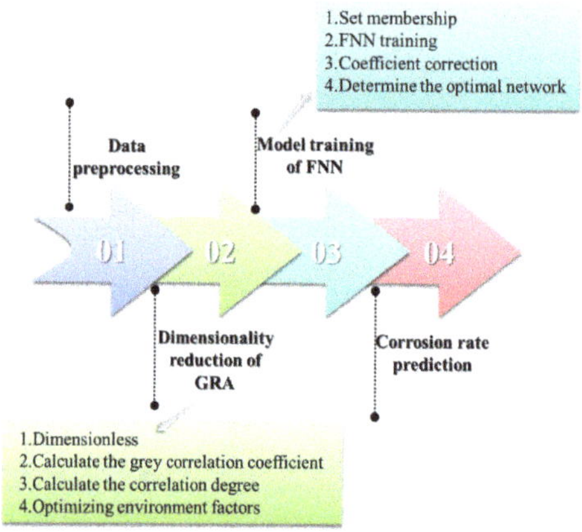

Figure 10. Experimental procedure.

Wang et al. [164] conducted a research experiment utilizing a hybrid Bayesian network to predict the corrosion risk of submarine pipelines. The results indicated that the model effectively captures the dynamic nature of corrosion risk in submarine pipelines and offers actionable recommendations for mitigating this risk.

5. Maintenance and Repair of Marine Pipelines

5.1. Buckling Protection

Buckling protection constitutes a critical design consideration for marine pipelines and cables within FPSO systems. Risers and umbilical cables may experience thermally induced compressive loads under high-pressure, high-temperature (HPHT) operating conditions, while mooring lines are susceptible to instability under combined environmental and operational loading. Effective buckling protection and stability control measures are therefore essential to prevent excessive deformation, ensure structural integrity, and

 https://doi.org/10.3390/jmse14020176

maintain long-term service safety. As marine development progresses into deeper waters, submarine pipelines must withstand the complex pressures, temperature fluctuations, and currents of the deep-sea environment. Effective buckling protection ensures the stability and safety of these pipelines through careful design and construction, which includes selecting suitable materials and structures, predicting and assessing buckling phenomena, and designing reinforcement measures such as additional supports and anti-buckling devices. Regular monitoring and maintenance are also critical. The significance of these efforts lies in extending the service life of pipelines, reducing maintenance costs, minimizing the risk of environmental pollution, and ensuring the efficient and safe exploitation of marine resources [165–169]. Figure 11 illustrates five typical buckling shapes for pipelines.

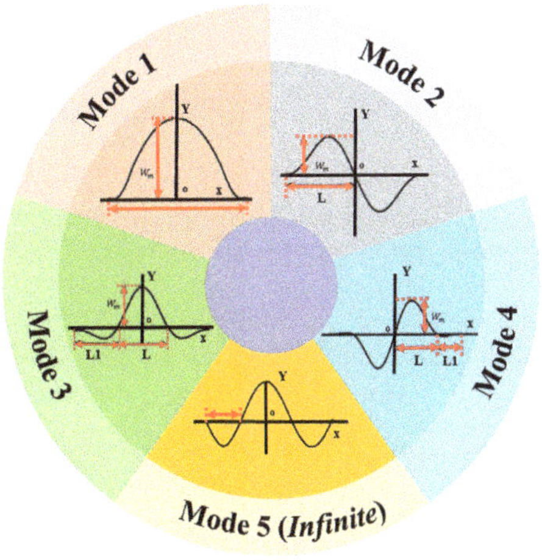

Figure 11. Five typical buckling shapes in pipelines.

In 1977, Fabian [170] investigated the nonlinear pre-buckling state, bifurcation, and initial post-buckling behavior of an infinitely long cylindrical elastic tube subjected to bending, pressure, and axial loads. He analyzed the collapse behavior by determining the ultimate load and exploring the possibility and significance of axial wrinkling in the compression region of the shell before reaching the ultimate load. However, his study only considered linear buckling mode forms.

In 1984, Kyriakides et al. [171] discussed the activation of propagation buckles in submarine pipelines. If the external pressure is sufficiently high, a propagation buckle can be initiated through a localized depression in the pipe. This buckle will propagate at any pressure higher than the propagation pressure of the pipe. The pressure at which a localized geometric defect transforms into a propagating buckle is called the initiating pressure, which depends on the geometric characteristics of the damage. Kyriakides et al. [171] calculated the buckling initiation pressure and the buckling propagation pressure of a submarine pipeline with a localized depression under external pressure based on the "double bell" mode of a point connection.

In 1990, Huang et al. [172] investigated the post-buckling behavior of offshore pipelines with vortex-shaped geometric defects under the combined effects of axial force and external pressure. The authors employed a combination of dual-scale uptake techniques and Fourier expansion to derive the initial post-buckling coefficients of circular shells with general vortex-shaped defects. These defects exhibited an exponential decay in the circumferential direction and took the form of an arbitrary function in the longitudinal direction. Their study provided insights into the relationship between extreme loads and defect

https://doi.org/10.3390/jmse14020176

parameters under the combined action of axial force and external pressure. Additionally, the sensitive zone of shell response to vortex defects was identified based on extensive example calculations.

In 2022, Ning et al. [173] proposed an efficient pipe cell model to simulate the bulging and buckling behavior of pipelines based on Euler-Bernoulli beam theory. They accurately considered the continuously distributed pipe-soil interaction (PSI) using the Gauss-Legendre integration method. Nonlinear PSI, thermal expansion forces due to temperature increases, and axial loads due to internal pressure differences were integrated into the pipe element. Continuously distributed zero-length Winkler-type springs were employed to simulate the interaction between the pipe and the surrounding soil, including both vertical and axial PSI. The researchers derived the stiffness matrices for expansion forces caused by temperature rise and internal pressure differences. They designed three sets of validation cases to verify the correctness and effectiveness of the pipeline model, leading to the following conclusions:

(1) Compared to traditional solid and shell elements, the pipeline unit requires fewer elements to simulate a long pipeline and eliminates the need for distributed springs or solid elements to represent the surrounding soil in traditional finite element analysis. The pipeline unit can describe and synthesize nonlinear pipe-soil interaction (PSI), thermal expansion, temperature effects, and internal pressure effects. Consequently, using pipe cells is more effective for simulating the bulging and flexure of pipelines.
(2) For pipes with small out-of-straightness (OOS), bulge-flexure behavior may occur when the temperature reaches a critical value. Therefore, when using the Newton-Raphson iterative method, the load increment needs to be reduced to accurately simulate the pipe's bulge-flexure behavior.

Figure 12 illustrates the turbulent buckling induced by cumulative expansion due to high temperature and high pressure (HTHP) and geometric defects.

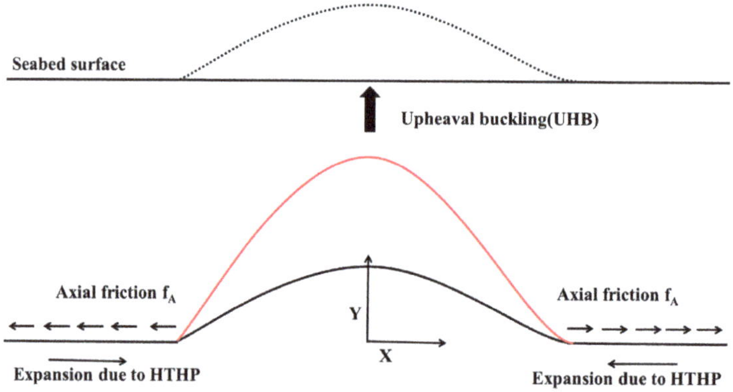

Figure 12. An upheaval buckling triggered by the accumulative expansion due to high-temperature high-pressure (HTHP) and geometric imperfections.

5.2. Overhaul and Monitoring of Submarine Pipelines

Overhaul and monitoring strategies for marine pipelines and cables play a vital role in enhancing the operational resilience of FPSO systems. Key components, including risers, umbilical cables, and mooring lines, benefit from a transition from reactive to predictive maintenance enabled by condition monitoring and integrity management, thereby minimizing the risks of production downtime and environmental disasters associated with structural failures.

https://doi.org/10.3390/jmse14020176

Wang [174] studied the current status of corrosion monitoring technology research in submarine pipelines. The monitoring methods for marine pipelines are mainly divided into full coverage monitoring and local corrosion monitoring.

The full coverage monitoring methods for submarine pipelines mainly include ultrasonic internal detection technology [175,176], vortex current inner inspection technology [177,178], magnetic leakage inner detection technology [179,180], and Magnetic Tomography Method (MTM) [181,182]. Although the implementation costs of these methods are high, they provide comprehensive, intuitive, and reliable data that cover the entire submarine pipeline, offering significant reference value.

Localized corrosion monitoring has become a crucial aspect of assessing the condition of sea pipelines. Key methods include field signature method (FSM) [183,184], fixed-point thickness monitoring [185,186], the inductive probe method [187,188], the hanging piece method [189,190], the linear polarization method [191,192], the electrochemical noise method [193,194], the electrochemical impedance method [195,196], and the resistance probe method [197,198]. These techniques enable more frequent monitoring while maintaining relatively low costs, with some allowing for continuous data collection.

Employing a combination of these localized monitoring methods enhances the comprehensiveness of the data gathered, providing insights such as the uniform corrosion rate, pitting corrosion rate, types of corrosion, and information on corrosion products. This multi-method approach ensures a richer and more detailed understanding of corrosion processes affecting sea pipelines.

Li et al. [199] conducted research on third-party damage monitoring technology for submarine pipelines based on ship dynamic tracking. By dynamically monitoring and identifying ship behaviors, this technology can effectively determine whether submarine pipelines are subjected to damage from ships, thereby enabling risk monitoring and early warning of third-party damage. However, due to various technical challenges, the monitoring and early warning of submarine pipeline damage caused by external forces from ships remain in the research and experimental stages.

At present, ship dynamic monitoring technology is primarily implemented using AIS (Automatic Identification System), satellite, radar, and other related technologies [200,201]. Developing early warning technology for third-party damage to submarine pipelines based on ship dynamic tracking will create an anti-damage barrier for these pipelines. This advancement will shift the safety management of submarine pipelines from experience-based methods to digital management, providing a reliable guarantee for their safe operation.

Zhang et al. [202] investigated the dynamic characteristics of pipeline inspection gauges (PIGs) under the excitation of submarine pipeline girth welds. Internal inspection technology for submarine pipelines forms the foundation for comprehensive inspection and evaluation of pipeline integrity. Figure 13 illustrates the PIG model. PIGs perform online detection of corrosion, deformation, and cracks within pipelines. The researchers proposed a dynamic model of the PIG sealing disk based on Kelvin Spring damping and a nonlinear dynamic model of the PIG system. Additionally, they used MSC/Adams software (version 2021.0.1) to simulate the dynamic response of the PIG when colliding with a circumferential weld at different velocities and center of mass positions.

 https://doi.org/10.3390/jmse14020176

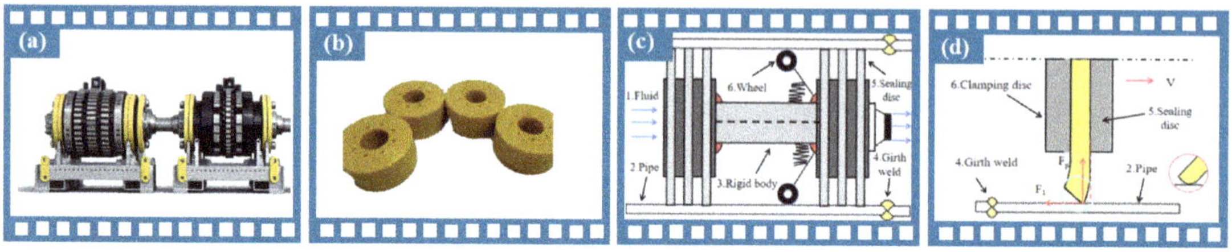

Figure 13. (**a**) A typical pipeline inspection gauges. (**b**) Polyurethane sealing disks. (**c**) The basic structure of a PIG, as well as the pipe exhibiting a girth weld. (**d**) Deformation of the sealing disk by forcing it to pass through the girth weld.

Feng et al. [203] proposed an automatic cable tracking method (ACTM) based on side-scan sonar (SSS) to equip autonomous underwater vehicles (AUVs) with autonomous decision-making capabilities for the rapid localization and stable tracking of submarine cables under significant uncertainties. They designed a behavior-based planning method to enable AUVs to make online decisions such as cable search, tracking, and recapture after tracking loss, facilitating autonomous surveys of flexible cables in uncertain seafloor environments.

When a cable is detected, the tracking task is innovatively approached as a path tracking problem in the horizontal plane. The virtual mass extracted from the cable is modeled using a uniform motion model based on target tracking theory, and the state of the cable is estimated using a Kalman filter. The control objective is established based on the relative geometrical relationship between the underwater robot and the cable to ensure optimal imaging. Multiple cable states are incorporated into the guidance law to enhance tracking performance. The method's effectiveness was validated through numerical simulations and field experiments.

Zhang et al. [204] proposed using forward-looking sonar to obtain the precise position of submarine pipelines and a tracking algorithm to maintain the tracking of these pipelines. This approach addresses the issue of positional deviation caused by the accumulation of errors in GPS signals within the vehicle's combined navigation system during extended operations, which can result in significant inaccuracies when real-time measurements are made by an autonomous underwater vehicle (AUV) using a multi-beam sonar system.

Conventional tracking algorithms often fail to detect curved pipes, losing the tracking target at certain curved sections. To address this, they propose an algorithm to identify and fit curved pipes separately. The algorithm employs an optimized Rosenfeld algorithm to efficiently extract the center line of the pipeline in a binary image and uses straight-line fitting and power function fitting to obtain more accurate paths for straight and curved submarine pipelines, respectively.

A realistic tracking scenario of the submarine pipeline in Qingdao Bay was simulated. Numerous experiments concluded that the target point position of this method is easy to calculate and can be continuously tracked, demonstrating the effectiveness of the submarine pipeline tracking technology. Figure 14 shows the pipeline tracking system and Figure 15 shows the flow chart of the subsea pipeline tracking algorithm.

 https://doi.org/10.3390/jmse14020176

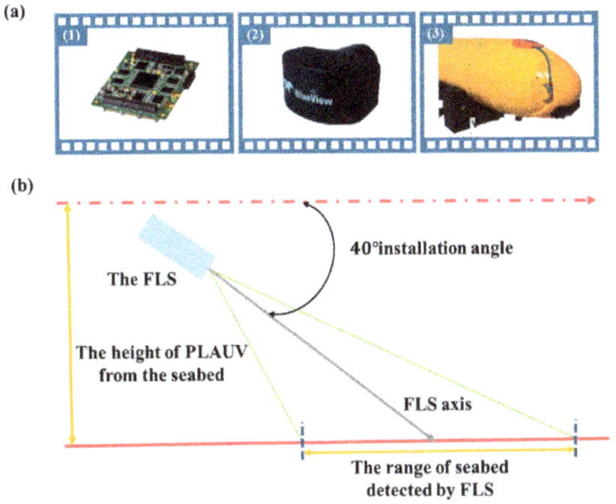

Figure 14. (**a1**) Image processing computers. (**a2**) The M450-130 FLS. (**a3**) The PLAUV developed by Tianjin University. (**b**) The range of seabed detected by FLS.

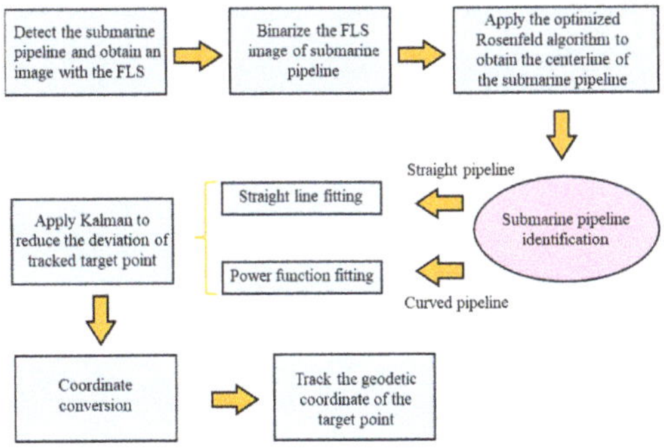

Figure 15. The flowchart of submarine pipeline tracking algorithm.

6. Marine Pipeline Design

6.1. Risk Analysis of Marine Pipelines and Cables

Submarine pipelines are essential systems in the process industry, designed for long-term use in marine environments. However, these harsh environments often pose significant threats to the condition of submarine pipelines [205]. For instance, these pipelines are vulnerable to scouring by currents and waves, third-party damage, earthquakes, and design flaws, all contributing to their high likelihood of failure [206]. Therefore, risk assessment of submarine pipelines is crucial due to the potentially devastating impacts of leakage failures on human safety, the environment, and the economy [207]. Within FPSO systems, marine pipelines and cables—comprising risers, umbilical cables, and mooring lines—play distinct yet interdependent roles in hydrocarbon production and station keeping. Risk analysis provides critical decision support by identifying and evaluating multi-dimensional threats such as environmental loads, structural failures, and accidental events across these components. While the subsequent analysis focuses primarily on leakage failure in submarine pipelines due to its high likelihood and severe consequences, the adopted risk assessment framework is applicable to other FPSO components with appropriate adaptation.

Yu et al. [208] aimed to obtain reliable brittleness values through expert assessment in the absence of historical failure data and to perform probabilistic risk assessment of submarine pipeline leakage failure. They proposed a probabilistic Petri net (PPN) method based on intuitive fuzzy evidential reasoning, which quantitatively evaluates the probability of

 https://doi.org/10.3390/jmse14020176

submarine pipeline leakage failure by determining the failure probability of the basic event through the experience of domain experts. In the same year, Yu et al. [205] proposed an improved Failure Modes and Effects Analysis (FMEA) method based on cloud modeling and the extended risk assessment methodology (VIKOR) to enhance the effectiveness of submarine pipeline failure risk assessment. This FMEA method addresses the uncertainty and ambiguity of risk assessment information while leveraging the advantages of the extended VIKOR method in solving complex risk analysis problems. Additionally, an extended two-level submarine pipeline failure risk factor hierarchy was established to improve the comprehensiveness of risk assessment. The improved integrated dynamic weighting algorithm considers the weights of different FMEA team members, making the results more reliable and valid.

Failure Modes and Effects Analysis (FMEA) is a proactive risk management tool designed to identify and evaluate the causes and effects of potential failure modes and implement necessary measures to prevent critical failures. Despite being a widely recognized method for analyzing and preventing failure modes in engineering, FMEA has not been extensively studied in the context of submarine pipeline failure risk assessment [209,210].

Among the existing studies, Li et al. [211] used the Decision-Making Trial and Evaluation Laboratory (DEMATEL) method to analyze the interdependence between failure factors of submarine pipelines and combined it with a fuzzy inference algorithm to calculate the risk levels of these factors. To improve the reliability of submarine pipelines.

Current FMEA methods applied to submarine pipeline risk assessment are essentially based on Multi-Criteria Decision Making (MCDM) techniques, which are also prevalent in other fields [212–215]. MCDM is a well-known branch of operations research and is considered a valuable tool for improving FMEA performance and addressing the limitations of the traditional Risk Priority Number (RPN) approach [216]. MCDM has been widely used to solve selection problems involving multiple conflicting evaluation criteria [217].

Yu et al. [218] extended the TODIM method to submarine pipeline risk analysis, highlighting the need for improvements to ensure its accuracy and applicability. Additionally, for multilevel index systems, the fuzzy comprehensive evaluation method has been utilized for analysis.

6.2. Simulation and Analysis Methods

Simulation and analysis methods provide essential support for FPSO systems by modeling the structural responses of marine pipelines and cables, including risers, umbilical cables, and mooring lines, under complex operational conditions such as scour, vortex-induced vibration (VIV), and thermal buckling. Although specific mechanisms may dominate different components, these methods enable comprehensive prediction of structural behavior, providing essential guidance for safe design and reliable operation of FPSO systems.

Hanging submarine pipelines are a common configuration in the transport of paint and other materials [219]. To predict the mechanical properties and ensure the structural integrity of these hanging pipelines, it is essential to model and analyze them accurately. Wang et al. [220] reviewed various methods for modeling and analyzing hanging submarine pipelines. Continuous modeling is typically employed for static morphology analysis of hanging pipelines. In contrast, discrete models can be used for both static and dynamic analysis of complex deep-water hanging pipelines, addressing geometric non-linearity, material non-linearity, boundary non-linearity, and other intricate issues. Figure 16 shows the classification of trailer line modeling methods.

https://doi.org/10.3390/jmse14020176

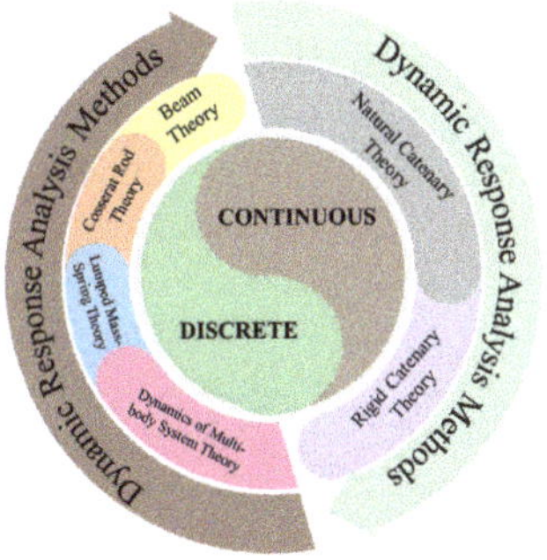

Figure 16. Trailer line modeling method classification.

For the static analysis of hanging submarine pipelines, the natural catenary theory method is often chosen due to its theoretical simplicity and ease of computation. However, when bending stiffness or higher computational accuracy is required, finite element methods such as the rigid catenary method and the nonlinear beam method can be utilized. These methods offer more precise analysis by accounting for factors like bending stiffness and other complex behaviors.

The natural catenary theory method has a straightforward theoretical structure, making it an excellent starting point for more refined solution methods. However, its calculation results in shallow water are not sufficiently accurate due to several factors: it neglects the bending stiffness of the pipeline, it assumes a discontinuity of the bending moment at the point of contact, and it fails to account for dynamic loads on the pipeline. Additionally, the model cannot solve for hydrodynamic forces and cable characteristic parameters (e.g., mass, elasticity) that vary along the cable. The natural catenary chain line morphology equation [204] for horizontal force H is:

$$z = \frac{H}{\omega}\left[\cos\left(\frac{\omega x}{H}\right)h - 1\right] \tag{4}$$

In this catenary equation, ω denotes the unit weight (load per meter), while H signifies the constant horizontal tension throughout the span.

In 1967, Plunkett [221] enhanced the existing natural catenary theory by considering the bending stiffness of the pipeline and proposed the rigid catenary method. This approach addressed some of the shortcomings of the natural catenary theory method, yet it still could not analyze the dynamic response of the pipeline. The nonlinear beam theory, which discretizes the pipeline into beam unit micro-segments, neglects the influence of pipeline cross-section deformation and only considers pipeline bending deformation and axial tensile pressure. Compared to the catenary chain line concept, the nonlinear beam theory excels in handling complex boundary conditions and environmental loads. Furthermore, this modeling method is well-suited for numerical calculations using computer language, making pipeline correlation analysis more convenient.

The dynamic response analysis of hanging line pipelines often involves nonlinear elements. Currently, only various finite element methods combined with numerical solutions can effectively analyze the dynamic response of hanging submarine pipelines. Among the finite element methods, the nonlinear beam theory model and the Cosserat rod theory model exhibit higher accuracy. In contrast, the centralized mass method and the multi-rigid body pipeline model offer higher computational efficiency, with the advantage becoming more pronounced with increasing water depth. Different numerical algorithms are selected based on the specific simplification results of the model equations of motion.

https://doi.org/10.3390/jmse14020176

de Lucena et al. [222] conducted a study on the optimized design of submarine pipeline routes using genetic algorithms, focusing on different constraint handling techniques. The study specialized in incorporating more relevant design criteria, categorized as "soft" and "hard" based on the consequences of their violation. It evaluated more efficient techniques to address constraints related to the "hard" criteria. The researchers found that classical static penalty methods were insufficient, especially in more complex cases. Therefore, they emphasized the importance of associating these criteria with more advanced constraint handling techniques, such as Adaptive Penalty Method (APM), e-constraints, and the Ho-Shimizu method. These advanced techniques demonstrated better performance in selecting the best feasible routes and required fewer evaluations.

7. Marine Pipeline Materials

The selection of marine pipeline materials fundamentally dictates the corrosion resistance, mechanical strength, and thermal stability of the system, serving as the primary defense against structural failure and environmental leakage while ensuring the durability and fatigue life of FPSO assets in extreme deepwater environments. Pipeline steel for submarine applications can be divided into three categories: cold [223,224], high sulfur areas [225,226], and submarine laying [227,228]. A comprehensive evaluation of pipeline laying conditions and the main forms and causes of failure reveals that pipeline steel must possess excellent mechanical properties. It should also have a large caliber, weldability, resistance to cold and low temperatures, corrosion resistance, and resistance to seawater and hydrogen-induced cracking (HIC), among other properties. The working environment for submarine pipelines is harsh, with long lines that are difficult to maintain, making stringent quality requirements for the steel essential.

The composition design of submarine pipeline steel aims to minimize the content of impurity elements such as phosphorus (P), sulfur (S), oxygen (O), nitrogen (N), and hydrogen (H). During the steelmaking process, precise control of the composition is required to meet design specifications. Additionally, efforts must be made to remove non-metallic inclusions from the steel, thereby improving its purity.

7.1. Materials Development

The growing energy demand is driving the development of marine oil and gas resources, with submarine pipelines playing an increasingly crucial role as the arteries for the collection and transportation of sea-based oil and gas. Submarine pipelines operate long-term in low-temperature, high-pressure, and highly corrosive marine environments. They must withstand internal and external pressure, axial force, bending moments, and other static loads, as well as the combined effects of temperature loads. Additionally, they endure alternating external pressures, waves, currents, and other dynamic loads, which subject the pipeline to various forms of stress and potential damage.

Given the extremely high safety requirements of submarine pipeline projects, the specifications for the size, shape, surface quality, and physical and chemical properties of thick-walled submarine pipeline steel are very stringent. These demanding conditions necessitate meticulous design and manufacturing processes to ensure the pipeline's reliability and integrity in its challenging operating environment [229].

Therefore, the development of thick-walled submarine pipeline steel plates and steel pipes is particularly challenging, mainly reflected in the following points:

(1) The substantial thickness of steel plates and pipes makes it challenging to achieve adequate low-temperature fracture toughness. Meeting the technical conditions [230] poses high demands, marking a significant milestone globally. Enhancing the low-

 https://doi.org/10.3390/jmse14020176

temperature fracture toughness of thick-walled pipeline steel remains a key technical challenge in the development of pipeline steel worldwide.

(2) Achieving high strength and high toughness in steel plates and pipes, while also ensuring excellent deformation resistance, presents a significant challenge. The precise matching of these properties is particularly difficult. The stability of performance requirements is extremely high, permitting only a small range of fluctuations. Maintaining consistent product quality on a large scale further complicates the process.

(3) The wall thickness of the steel pipe is large, and the diameter-to-thickness ratio (D/t) is small, so it is difficult to make the pipe.

(4) For safety reasons, the size, shape and surface quality of steel pipe are strictly required, and it is difficult to control the mass production. In industrial production, it is very difficult to meet the strict requirements of thick-walled submarine pipeline steel pipe and realize stable, mass production and control [231–235].

In contrast to land-based pipelines, submarine pipelines must meet more stringent requirements due to the harsh marine environment. These include higher demands on the strength, toughness, compressive performance, and dimensional accuracy of both the steel and the steel pipes used. According to the DNV-ST-F101 standard for submarine pipelines, there are rigorous requirements not only for the transverse and longitudinal strength of the steel pipe but also for its resistance to low-temperature dynamic tearing. As offshore oil exploitation advances from shallow to deep waters, deep-sea pipelines impose even stricter demands on pipeline materials, welding, construction, and maintenance. Enhancing the strength, toughness, compressive properties, corrosion resistance, and dimensional accuracy of steel pipes is crucial to improving the safety of pipeline installation and operation [236,237].

Zhang et al. [238] have developed and studied X70 and X65 thick-walled submarine pipelines for the Li wan project in the South China Sea. These pipelines must meet the yield strength requirements specified by DNV offshore pipeline technical standards [239]. The standard allows for a yield strength fluctuation range of 120 MPa, which is more than 20% smaller than the permissible fluctuation range for conventional land pipelines (150 MPa). Additionally, the transverse and longitudinal strength fluctuations of the same steel pipe must not exceed 40 MPa.

The main challenges in developing pipeline steel for Li wan, South China Sea, included the use of "a small amount of ferrite + bainite" in the needle-like ferrite microstructure design. To ensure good welding performance, low carbon and low carbon equivalent alloys were employed. Clean steel smelting and continuous casting technologies were used to meet high toughness requirements, while two-stage controlled rolling technology was applied to ensure the desired properties of the pipeline steel.

Lang et al. [240] investigated the HE sensitivity of a new generation of X70 grade acid-resistant submarine pipeline steel treated with varying Mg contents, based on titanium and aluminum deoxidation rolling. The study used constant load and slow strain rate tensile (SSRT) methods to assess the hydrogen embrittlement (HE) sensitivity. Table 1 illustrates the chemical composition of the tested steels.

Table 1. Chemical compositions of tested steels %.

Steel	C	Si	P	S	Mn + Cr + Ni + Mo	Nb + V	Ti + Ai	Mg
Mg0	0.044	0.26	≤0.05	≤0.003	≤1.80	≤0.10	≤0.03	0
Mg3	0.041	0.25	≤0.05	≤0.003	≤1.80	≤0.10	≤0.03	0.003
Mg5	0.043	0.25	≤0.05	≤0.003	≤1.80	≤0.10	≤0.03	0.005

https://doi.org/10.3390/jmse14020176

With a 0.003% Mg treatment, the morphology of Al_2O_3 and MnS inclusions improved, resulting in finer, more diffuse inclusions with a "core–shell" structure. However, a 0.005% Mg treatment increased the number of larger inclusions and led to the formation of chain-like inclusions. The "core–shell" structure from the Mg treatment exhibited higher dehydrogenation activation energy (greater than 100 kJ/mol), making it more difficult for hydrogen to desorb from the inclusion-substrate interface. This increased hydrogen trapping efficiency in the steel. Conversely, further increasing the Mg content from 0.003% to 0.005% reduced the hydrogen trapping efficiency in the tested steels.

7.2. Material Applications

The exploration and development of marine oil and gas resources have led to a growing demand for submarine pipeline construction. The industry for submarine pipeline materials has promising development prospects, particularly in the need for new materials and technologies. With advancements in marine pipeline laying technology, the depth of submarine pipeline installations has progressed from shallow waters of a few dozen meters to more than 3000 m in the deep sea [241–244].

The development of marine pipeline laying technology poses significant challenges for the industry. Notably, the Gulf of Mexico features a famous pipeline, 2412 m deep and 222 km long, using X65 grade steel, with an outer diameter of 610 mm and a wall thickness ranging from 24.1 to 34.3 mm, completed in 2014. Another record-setting pipeline in the Mediterranean Sea extends from Algeria to Sardinia, reaching a water depth of 2824 m. Currently, the highest steel grade used in foreign submarine pipeline projects in non-acidic environments is X70, while in acidic environments, the highest grade is X65 [244].

In recent years, the development of deep-sea pipelines has been rapid, with increasing depths. The design water depth of newly constructed submarine pipelines abroad has reached 3500 m, with most using X70 steel grade, large diameter, and thick-walled steel pipes. The maximum pipe diameter has reached 1219 mm, and the maximum wall thickness has reached 44.0 mm [245]. The application of typical foreign submarine pipelines is shown in Table 2.

Table 2. Typical application of seabed pipeline abroad.

Line Name	Laying Depth/m	Grade of Steel	Pipe Diameter/mm	Wall Thickness/mm
The Gulf of Mexico ITP Pipeline	2412	X65	610	24.1~34.3
The Mediterranean Medgaz Pipeline	2160	X65	640	-
Ichthys Gas Pipeline	2160	X65	1067	29.6~40.1
The North Sea Lang Grad Pipeline	1000	X70	1016	34

The laying of submarine pipelines in China commenced relatively late due to the slow development of marine engineering and the relatively backward state of equipment and technology. Consequently, there remains a significant gap between the overall level of China's offshore pipeline design and construction and the international advanced standards [246].

In recent years, China has rapidly advanced its offshore pipeline construction efforts. By the end of the "Twelfth Five-Year Plan," China had laid hundreds of submarine pipelines, with a total length exceeding 6000 km. Submarine pipeline steel grades X56 and X60, suitable for H2S acidic service environments, have been applied in several Chinese sub-

 https://doi.org/10.3390/jmse14020176

marine pipeline engineering projects, accompanied by the development of related welding process research.

By summarizing foreign experience and conducting extensive experimental research, several domestic pipeline steel and steel pipe manufacturers in China have mastered the production technology for high-strength, high-toughness pipeline steel. Consequently, the steel grade for long-distance submarine pipelines has been elevated to the X70 level, alleviating China's long-term reliance on imported high-grade oil and gas pipeline steel [247,248].

Ma et al. [249] conducted research on the production and application of 10MnVNbNiMo steel for marine pipeline materials. This steel is produced using high-quality scrap and first-grade molten iron, with phosphorus content controlled within 0.010% during electric furnace tapping. The refining process is enhanced with deoxidation to create white slag, keeping the sulfur content of the finished product controlled within 0.002%. Refining involves precise composition adjustments to ensure the carbon equivalent is controlled between 0.17% and 0.19%. During the later stages of VD (vacuum degassing), calcium treatment and soft blowing are employed to promote the adequate flotation of inclusions, resulting in inclusion ratings of less than or equal to grade 1.0 in the pipe body samples. Reasonable process parameters are set according to the characteristics of the continuous casting of this steel type, ensuring that the surface and internal quality of the cast billets fully meet the rolling requirements.

8. Digital Twin and Real-Time Monitoring Technologies

As the wave of AI continues to surge, artificial intelligence is increasingly being applied across various scientific and engineering fields. Similarly, AI holds immense potential for the protection of offshore pipelines and subsea cables. The marine environment is complex and volatile; traditional periodic manual inspection models are not only prohibitively expensive but also suffer from significant response lags. Consequently, Digital Twin technology and real-time monitoring powered by AI are transforming protection strategies from reactive "after-the-fact" repairs to predictive maintenance. By significantly reducing costs, this shift is becoming the future trend and a key potential solution for the industry [250–252].

Chen et al. (2025) [253] developed a digital twin–based predictive diagnosis method for subsea suspended pipelines to overcome the limitations of traditional inspection approaches, which typically lack long-term and multi-directional diagnostic capabilities. For these pipelines, a unit-level digital twin framework was proposed, enabling real-time monitoring, fault prediction, and operational optimization through digital twin functionalities. In addition, a PFD-TCN (Polynomial Fitting Denoising–Temporal Convolutional Network) model was introduced. By integrating polynomial fitting functions with a conventional TCN, the proposed model effectively reduces noise and improves pipeline strain prediction accuracy by more than 50% compared with the traditional TCN model. However, acquiring more comprehensive data generally requires the installation of additional subsea sensors, which can be costly. Moreover, the actual stress states of subsea suspended pipelines may deviate from idealized models, indicating the need for further investigation into pipeline force characteristics. Furthermore, subsequent validation of modeling approaches under real offshore conditions can also be economically expensive.

Similarly, Chen et al. (2024) [254] integrated physical monitoring with virtual simulation to develop an integrated platform management system. This system periodically collects monitoring data on environmental conditions, offshore platforms, mooring lines, and risers from field sensors, and employs long short-term memory (LSTM) neural networks to reconstruct virtual simulation models of the dynamic responses of offshore platforms under environmental loads.

https://doi.org/10.3390/jmse14020176

Repalle et al. (2020) [255] proposed a novel approach that combines refined finite element analysis (FEA) with artificial neural networks (ANNs) to construct a digital twin model for risers. This model serves as a decision-support tool for integrity management and life-extension operations. By training advanced neural networks using partially available metocean and vessel motion data, the digital twin can be used to predict fatigue damage under various metocean conditions and internal pressure scenarios. Compared with conventional finite element analysis, this approach significantly reduces the time required for fatigue damage assessment while improving prediction accuracy.

Bhowmik (2021) [256] proposed a novel digital twin concept based on computer vision, in which convolutional neural network (CNN) algorithms are employed to perform real-time corrosion detection and classification using remotely operated vehicle (ROV) images and online inspection data. This approach significantly improves the performance of existing corrosion identification methods, reduces inspection time, and supports predictive and prescriptive maintenance strategies, achieving an accuracy of approximately 81% in corrosion detection and classification. However, certain limitations remain, as CNN-based corrosion detection methods do not inherently capture the uncertainty and risk associated with the detected corrosion.

Imran et al. (2023) [257] investigated recent artificial intelligence (AI) techniques for marine-related corrosion prediction and detection, and highlighted that both real-time and historical data are crucial for predicting corrosion failures within data-driven predictive maintenance (PdM) models. In some approaches, real-time data are used to frequently update prediction models, while historical data are leveraged to support the characterization of corrosion defect distributions.

Meniconi et al. (2024) [258,259] conducted a two-part companion study on fault detection in the Trieste subsea pipeline in Italy. The first paper focuses on the off-site preparatory work, including the selection of a small Side Discharge Valve (SDV) through hydraulic characterization to generate small but sharp pressure waves, and the formulation of field test procedures to ensure the system reaches a stable steady state. In the second paper, the authors analyze actual field test data using 1D numerical modeling and Inverse Transient Analysis (ITA) to identify pipe wall deterioration in certain sections, a finding subsequently confirmed by professional divers' field investigations.

Current research on Digital Twin and real-time monitoring technologies for FPSO systems focuses primarily on the hull, topside structures, and production planning. In contrast, research concerning the real-time monitoring of subsea pipelines and cables within FPSOs remains in its infancy, representing a significant research gap [260–262].

9. Conclusions and Prospects

A wide survey of the literature for risers, umbilical cables, and mooring cables for a FPS in offshore floating production would indicate that, although current results provide a robust engineering platform, there are still considerable knowledge gaps in theories, models, and their applicability. Based on the review presented in this paper, the hazards, underlying mechanisms, typical models, and monitoring/mitigation measures related to marine pipelines and cables—particularly those in FPSO systems—are summarized in Table 3.

Morison's equation is very popular for basic hydrodynamic calculation work but fails to account for either wave diffraction or radiation effects. This is particularly a concern when dealing with very large structures or with situations where the diameter is a large fraction of the wavelength, leading to inaccurate predictions of hydrodynamic force. Semi-empirical models combining potential flow theory with adjustments provide a

 https://doi.org/10.3390/jmse14020176

means to improve accuracy, though models and a method to establish calibration constants appropriate to specific structures and environment conditions are not established.

Table 3. Synthesis of key failure hazards, mechanisms, typical models, and normative standards for pipelines and cables in marine and FPSO systems.

Hazard	Key Mechanism	Typical Model	Normative Standard	Monitoring and Mitigation Strategy
Vortex-Induced Vibration (VIV)	Periodic vortex shedding causing resonance and fatigue damage in free spans or risers.	Wake oscillator models; Morison Equation	DNV-RP-F105 [263]	Accelerometers, strain gauges, Digital Twin prediction, Helical strakes, fairings, span rectification
Local Scour and Free Spans	Sediment transport due to wave interaction removing seabed support.	Potential flow theory; Empirical scour depth formulas	DNV-RP-F105 [263], DNV-RP-F109 [264]	Side-scan sonar, multi-beam echo sounder, Rock dumping, concrete mattresses, sediment stabilization.
Corrosion	Electrochemical oxidation; Microbiologically Influenced Corrosion in anaerobic mud.	Melchers' Multi-phase Model; Bayesian Network	DNV-RP-B401 [265]	Linear Polarization Resistance (LPR), Electrical Resistance (ER), Cathodic protection (sacrificial anodes), anti-corrosion coatings
Buckling	Thermal expansion under High-Pressure High-Temperature (HPHT) conditions constrained by pipe-soil friction.	Hobbs' analytical method; Euler-Bernoulli beam theory.	DNV-RP-F110 [266]	Distributed Temperature/Strain Sensing, Snake-lay installation, intermittent rock dumping
Fatigue Failure	Cyclic loading from waves/currents or accidental loads	S-N Curve analysis; Palmgren-Miner rule.	DNV-RP-C203 [267]	Intelligent PIGs for crack detection, Stress relief joints, fatigue-resistant welding procedures.

Secondly, for seabed scour and free spans of seabed pipes, the current models involve simplistic theories such as Two-Dimensional Steady Flow, rigid pipes, and empty seabed. Additionally, they fail to involve oceanic phenomena such as Interaction between Currents and Waves, Non-Uniform Current Flow, and different seabed materials. Essentially, this implies a less accurate method of predicting the scour and span changes with respect to safety. Additionally, research on Vortex-Induced Vibration (VIV) assumes a uniform flow condition and a span setup without putting adequate emphasis on the combination of scour, soil interactions, and the phenomenon of vibration.

Third, the factors in buried or partially buried pipelines involving the influence of sediments on electrochemical and microbiologically influenced corrosion processes are not quantified. The influence of different sediments on the diffusion of oxygen, bioactivity, and the rate of corrosion is not modeled.

Finally, while AI and Digital Twin technologies offer new ways to sense condition, warn early about risks, and manage operations, using them safely in offshore engineering has challenges. These include a lack of high-quality training data, trouble collecting real-time data, unclear model validation paths, and the "black-box" nature of many neural networks.

Looking ahead, cross-disciplinary teamwork is essential. Future work should combine large-scale physical tests, high-fidelity multi-physics simulations, and long-term field data. It is important to promote Explainable AI and Physics-Informed Neural Networks (PINN) within reliability-based design. These advances will strengthen safety, durability, and sustainable growth of floating offshore energy systems. At the same time, this study has certain limitations. Due to the relative scarcity of research specifically focusing on subsea pipelines and cables compared to FPSO hulls or topside systems, this paper incorporates a broad range of studies related to the three primary types of lines and cables to ensure

 https://doi.org/10.3390/jmse14020176

comprehensiveness and credibility. Future efforts should focus on further expanding the database of research cases in this specific field.

Author Contributions: Conceptualization, D.Z. and J.Y.; methodology, J.Y.; software, J.Y. and Y.Z.; validation, J.Y., Z.C. (Zehan Chen) and P.L.; formal analysis, J.Y.; investigation, Y.L., Z.C. (Zehan Chen) and J.W.; resources, D.Z.; data curation, Y.Z. and K.Y.; writing—original draft preparation, J.Y.; writing—review and editing, D.Z., Y.Z. and K.Y.; visualization, Z.C. (Zeyu Cao); supervision, D.Z.; project administration, D.Z.; funding acquisition, D.Z. All authors have read and agreed to the published version of the manuscript.

Funding: This study was financially supported by Research on the Structural Optimization and Fish-Friendly Adaptability of Gravity-Based Cage Nets in the South China Sea Based on Multi-physics Coupling, Program for Scientific Research Start-up Funds of Guangdong Ocean University (060302072101), Comparative Study and Optimization of Horizontal Lifting of Subsea Pipeline (2021E05011). The authors gratefully acknowledge the support from the Natural Science Foundation of Guangdong Province (2022A1515011562), and the Guangdong Provincial Special Fund for promoting high-quality economic development (GDNRC [2021]56, Yuerong Office Letter [2020]161).

Data Availability Statement: No new data were created or analyzed in this study. Data sharing is not applicable to this article.

Conflicts of Interest: The authors declare no conflicts of interest.

References

1. Hubbert, M.K. Energy from fossil fuels. *Science* **1949**, *109*, 103–109. [CrossRef]
2. Ediger, V.Ş.; Hoşgör, E.; Sürmeli, A.N.; Tatlıdil, H. Fossil fuel sustainability index: An application of resource management. *Energy Policy* **2007**, *35*, 2969–2977. [CrossRef]
3. Zhang, Y.; Zhang, D.; Jiang, H. A Review of Offshore Wind and Wave Installations in Some Areas with an Eye towards Generating Economic Benefits and Offering Commercial Inspiration. *Sustainability* **2023**, *15*, 8429. [CrossRef]
4. Zhang, D.; Zhao, B.; Jiang, H. Wave energy converter with floating-point absorber and catenary mooring: Dynamic coupling analysis. *Front. Mar. Sci.* **2024**, *11*, 1338330. [CrossRef]
5. Mitchell, C. Momentum is increasing towards a flexible electricity system based on renewables. *Nat. Energy* **2016**, *1*, 15030. [CrossRef]
6. Oliveira-Pinto, S.; Rosa-Santos, P.; Taveira-Pinto, F. Electricity supply to offshore oil and gas platforms from renewable ocean wave energy: Overview and case study analysis. *Energy Convers. Manag.* **2019**, *186*, 556–569. [CrossRef]
7. Zhang, D.; Ma, Y.; Zhang, H.; Zhang, Y. Marine Equipment Siting Using Machine-Learning-Based Ocean Remote Sensing Data: Current Status and Future Prospects. *Sustainability* **2024**, *16*, 8889. [CrossRef]
8. Rona, P.A. Resources of the Sea Floor. *Science* **2003**, *299*, 673–674. [CrossRef]
9. Zhang, Y.; Zhang, D.; Xie, Y.; Zhang, Y.; Liang, Z.; Zhu, K.; Zhang, S. Dynamic Configuration Simulation of Multi-Branches Towed Array System. *Mar. Struct.* **2025**, *103*, 103819. [CrossRef]
10. Zhang, D.; Zhao, B.; Zhu, K.; Jiang, H. Dynamic analysis of the subsea production system with lazy-wave risers attached to FPSO. *PLoS ONE* **2023**, *18*, e0291603. [CrossRef] [PubMed]
11. Meng, H.; Kloul, L.; Rauzy, A. Production availability analysis of Floating Production Storage and Offloading (FPSO) systems. *Appl. Ocean Res.* **2018**, *74*, 117–126. [CrossRef]
12. Zhang, D.; Zhang, Y.; Zhao, B.; Ma, Y.; Si, K. Exploring subsea dynamics: A comprehensive review of underwater pipelines and cables. *Phys. Fluids* **2024**, *36*, 101304. [CrossRef]
13. Zhang, D.; Zhao, B.; Sun, J.; Zhang, Y.; Zhu, K.; Jiang, H. Influence of Different Static Equilibrium Calculation Methods on the Dynamic Response of Marine Cables during the Releasing Process: Review and a Case Study. *J. Mar. Sci. Eng.* **2023**, *11*, 764. [CrossRef]
14. Shah, K.A.; Meng, F.; Li, Y.; Nagamune, R.; Zhou, Y.; Ren, Z.; Jiang, Z. A synthesis of feasible control methods for floating offshore wind turbine system dynamics. *Renew. Sustain. Energy Rev.* **2021**, *151*, 111525. [CrossRef]
15. Chen, J.; Zhou, L.; Ding, S.; Li, F. Numerical simulation of moored ships in level ice considering dynamic behavior of mooring cable. *Mar. Struct.* **2025**, *99*, 103716. [CrossRef]
16. Tang, Y.; Chen, X.; Huang, G.; Chen, M.; Zhang, K.; Jiang, Y.; Li, T.; Tao, T.; Yang, Y. Dynamic analysis of multi-module floating photovoltaic platforms with composite mooring system by considering tidal variation and platform configuration. *Ocean Eng.* **2024**, *312*, 119243. [CrossRef]

 https://doi.org/10.3390/jmse14020176

17. Mao, Y.; Yang, Z.; Wang, G.; Chen, J.; Yan, J.; Wang, Z. A study on the mechanical behavior of umbilical cables under impact loads using experimental and numerical methods. *Mar. Struct.* **2025**, *99*, 103700. [CrossRef]
18. Deng, Y.; Chen, S.; Wang, T.; Hao, N.; Sun, S.; Liang, X. Lateral buckling of umbilical cable armour under compression and bending. *Ships Offshore Struct.* **2024**, *20*, 1757–1772. [CrossRef]
19. Zhang, D.; Zhao, B.; Zhu, K. Dynamic analysis of the umbilical cable pull-in operation through J-tube under different wave directions. *Ocean Eng.* **2023**, *280*, 114838. [CrossRef]
20. Amaechi, C.V.; Reda, A.; Shahin, M.A.; Sultan, I.A.; Beddu, S.B.; Ja'E, I.A. State-of-the-art review of composite marine risers for floating and fixed platforms in deep seas. *Appl. Ocean Res.* **2023**, *138*, 103624. [CrossRef]
21. Hong, K.-S.; Shah, U.H. Vortex-induced vibrations and control of marine risers: A review. *Ocean Eng.* **2018**, *152*, 300–315. [CrossRef]
22. Chen, H.C. Time-domain simulation of nonlinear wave impact loads on fixed offshore platform and decks. *Int. J. Offshore Polar Eng.* **2010**, *20*, 275–283.
23. Zhang, Y.; Zhang, D.; Jiang, H. A Review of Artificial Intelligence-Based Optimization Applications in Traditional Active Maritime Collision Avoidance. *Sustainability* **2023**, *15*, 13384. [CrossRef]
24. Zhang, Y.; Zhang, D.; Zhang, Y.; Xie, Y.; Xie, B.; Jiang, H. A Comprehensive Review of Simulation Software and Experimental Modeling on Exploring Marine Collision Analysis. *ENG Trans.* **2023**, *4*, 2869. [CrossRef]
25. Bhandari, J.; Khan, F.; Abbassi, R.; Garaniya, V.; Ojeda, R. Modelling of pitting corrosion in marine and offshore steel structures—A technical review. *J. Loss Prev. Process. Ind.* **2015**, *37*, 39–62. [CrossRef]
26. Wang, Z.; Zhou, Z.; Xu, W.; Yang, D.; Xu, Y.; Yang, L.; Ren, J.; Li, Y.; Huang, Y. Research status and development trends in the field of marine environment corrosion: A new perspective. *Environ. Sci. Pollut. Res.* **2021**, *28*, 54403–54428. [CrossRef]
27. Cai, J.; Le Grognec, P. Lateral buckling of submarine pipelines under high temperature and high pressure—A literature review. *Ocean Eng.* **2022**, *244*, 110254. [CrossRef]
28. Kershenbaum, N.; Mebarkia, S.; Choi, H. Behavior of marine pipelines under seismic faults. *Ocean Eng.* **2000**, *27*, 473–487. [CrossRef]
29. Khan, A.; Ali, S.S.A.; Anwer, A.; Adil, S.H.; Meriaudeau, F. Subsea pipeline corrosion estimation by restoring and enhancing degraded underwater images. *IEEE Access* **2018**, *6*, 40585–40601. [CrossRef]
30. Varela, F.; Tan, M.Y.; Forsyth, M. An overview of major methods for inspecting and monitoring external corrosion of on-shore transportation pipelines. *Corros. Eng. Sci. Technol.* **2015**, *50*, 226–235. [CrossRef]
31. Valdez, B.; Ramirez, J.; Eliezer, A.; Schorr, M.; Ramos, R.; Salinas, R. Corrosion assessment of infrastructure assets in coastal seas. *J. Mar. Eng. Technol.* **2016**, *15*, 124–134. [CrossRef]
32. Drumond, G.P.; Pasqualino, I.P.; Pinheiro, B.C.; Estefen, S.F. Pipelines, risers and umbilicals failures: A literature review. *Ocean Eng.* **2018**, *148*, 412–425. [CrossRef]
33. Pham, D.-C.; Sridhar, N.; Qian, X.; Sobey, A.J.; Achintha, M.; Shenoi, A. A review on design, manufacture and mechanics of composite risers. *Ocean Eng.* **2016**, *112*, 82–96. [CrossRef]
34. Ho, M.; El-Borgi, S.; Patil, D.; Song, G. Inspection and monitoring systems subsea pipelines: A review paper. *Struct. Health Monit.* **2020**, *19*, 606–645. [CrossRef]
35. Yang, K.; Yang, K.; Zeng, S.; Zhang, Y.; Zhang, D. Research on Intelligent Monitoring of Offshore Structure Damage Through the Integration of Multimodal Sensing and Edge Computing. *Eng. Proc.* **2025**, *118*, 65. [CrossRef]
36. Li, X.; Liu, Y.; Chen, G.; Abbassi, R. Dynamic risk-based methodology for economic life assessment of aging subsea pipelines. *Ocean Eng.* **2024**, *294*, 116687. [CrossRef]
37. Aljaroudi, A.; Khan, F.; Akinturk, A.; Haddara, M.; Thodi, P. Risk assessment of offshore crude oil pipeline failure. *J. Loss Prev. Process. Ind.* **2015**, *37*, 101–109. [CrossRef]
38. Figueredo, A.K.M.; Coelho, D.G.; Miranda, P.P.; de Souza Junior, M.B.; e Melo, P.F.F.F.; Junior, C.A.V. Subsea pipelines incidents prevention: A case study in Brazil. *J. Loss Prev. Process. Ind.* **2023**, *83*, 105007. [CrossRef]
39. Yu, J.; Yu, Y.; Li, H.; Wu, H. Fatigue reliability study of submarine pipe across vortex-excited vibration. *Ship Mech.* **2005**, *2*, 109–114. (In Chinese)
40. Shi, M.; Chen, G.; Sun, Y. Analysis of vortex-excited vibration across pipe spans of submarine pipelines based on pipe-soil coupling model. *Pet. Min. Mach.* **2007**, *10*, 5–8. (In Chinese)
41. Zhu, H.; Chen, G. Comprehensive assessment method for fatigue reliability of subsea pipeline systems with pipe span vortex-excited vibration. *China Offshore Oil Gas* **2009**, *21*, 133–137. (In Chinese)
42. Furnes, G.; Berntsen, J. On the response of a free span pipeline subjected to ocean currents. *Ocean Eng.* **2003**, *30*, 1553–1577. [CrossRef]
43. Xu, P.; Du, Z.; Gong, S. Numerical Investigation into Freak Wave Effects on Deepwater Pipeline Installation. *J. Mar. Sci. Eng.* **2020**, *8*, 119. [CrossRef]

https://doi.org/10.3390/jmse14020176

44. Yang, K.; Zeng, S.; Yang, K.; Zhang, D.; Zhang, Y. From Multi-Field Coupling Behaviors to Self-Powered Monitoring: Triboelectric Nanogenerator Arrays for Deep-Sea Large-Scale Cages. *J. Mar. Sci. Eng.* **2025**, *13*, 2042. [CrossRef]
45. Zhang, D.; Bai, Y.; Zhu, K.; Liu, J. Dynamic Characteristics Analysis of the Rigid Flexible Coupling Multi-body System of Wharf-cable-dynamic Positioning Vessel. *IOP Conf. Ser. Mater. Sci. Eng.* **2019**, *612*, 32068. [CrossRef]
46. Zhang, D.; Bai, Y.; Zhao, W.; Zhu, K. Dynamic analysis of submarine cable during the process of laying back to the seabed. *Ships Offshore Struct.* **2020**, *15*, 153–161. [CrossRef]
47. Szczotka, M. Pipe laying simulation with an active reel drive. *Ocean Eng.* **2010**, *37*, 539–548. [CrossRef]
48. Zhao, B.; Zhang, D.; Zhu, K. Lifting of horizontal subsea pipeline with diverse lifting points in waves: Hydrodynamic analysis. *Phys. Fluids* **2024**, *36*, 87140. [CrossRef]
49. Zhang, D.; Zhao, B.; Zhu, K. Mechanical characteristics analysis of horizontal lifting of subsea pipeline with different burial depths. *Front. Earth Sci.* **2022**, *10*, 1011291. [CrossRef]
50. Ding, L.; Zhu, H.-P.; Wu, L. Effects of axial load and structural damping on wave propagation in periodic Timoshenko beams on elastic foundations under moving loads. *Phys. Lett. A* **2016**, *380*, 2335–2341. [CrossRef]
51. Ruzzene, M.; Baz, A. Dynamic stability of periodic shells with moving loads. *J. Sound Vib.* **2006**, *296*, 830–844. [CrossRef]
52. Williamson, C.H.; Govardhan, R. Vortex-induced vibrations. *Annu. Rev. Fluid Mech.* **2004**, *36*, 413–455. [CrossRef]
53. Zhang, D.; Zhao, B.; Zhu, K. Dynamic Analysis of Pipeline Lifting Operations for Different Current Velocities and Wave Heights. *Fluid Dyn. Mater. Process.* **2023**, *19*, 603–617. [CrossRef]
54. Zakeri, A. Review of state-of-the-art: Drag forces on submarine pipelines and piles caused by landslide or debris flow impact. *J. Offshore Mech. Arct. Eng.* **2009**, *131*, 014001. [CrossRef]
55. Zhou, C.; Li, G.; Dong, P.; Shi, J.; Xu, J. An experimental study of seabed responses around a marine pipeline under wave and current conditions. *Ocean Eng.* **2011**, *38*, 226–234. [CrossRef]
56. Yeganeh-Bakhtiary, A.; Shabakhty, N.; Valipour, A.; Gol-Zaroudi, H. Assessment of Wave-induced Fatigue Damage to Marine Pipelines in Intermediate Seas. *Int. J. Offshore Polar Eng.* **2016**, *26*, 385–391. [CrossRef]
57. Li, Z.; Zhang, H.; Litak, G.; Zhou, S. Periodic Solutions and Frequency Lock-in of Vortex-Induced Vibration Energy Harvesters with Nonlinear Stiffness. *J. Sound Vib.* **2024**, *568*, 117952. [CrossRef]
58. Ibrion, M.; Paltrinieri, N.; Nejad, A.R. Learning from failures: Accidents of marine structures on Norwegian continental shelf over 40 years time period. *Eng. Fail. Anal.* **2020**, *111*, 104487. [CrossRef]
59. Vukelić, G.; Vizentin, G. *Common Case Studies of Marine Structural Failures. Failure Analysis and Prevention*; InTech: Tokyo, Japan, 2017; pp. 135–151.
60. Chung, J.S. Morison Equation in Practice and Hydrodynamic Validity. *Int. J. Offshore Polar Eng.* **2018**, *28*, 11–18. [CrossRef]
61. Zhang, D.; Luo, Y.; Zhang, Y.; Ma, Y.; Zhu, K.; Zeng, S. A Comprehensive Review of an Underwater Towing Cable Array: A Discussion on the Dynamic Characteristics of the Towing Cable Array During the Outspread Process. *J. Mar. Sci. Eng.* **2024**, *12*, 1880. [CrossRef]
62. Bai, Y.; Zhang, D.; Zhu, K.; Zhang, T. Dynamic analysis of umbilical cable under interference with riser. *Ships Offshore Struct.* **2018**, *13*, 809–821. [CrossRef]
63. Zhou, Y.; Zhiaoping, Z.; Zhang, M.; Wei, K. Comparative study of wave load calculation methods for large-scale foundations of cross-sea bridges. *J. Disaster Prev. Mitig. Eng.* **2011**, *43*, 387–394. (In Chinese)
64. Luo, M.L. Dynamic Characteristics and Theory of Suspension Tunnel under Wave Environment. Master's Thesis, Chongqing Jiaotong University, Chongqing, China, 2023. (In Chinese)
65. Liu, H.X.; Xiao, X. Wave load calculation in structural strength analysis of semi-submersible platforms. *China Offshore Platf.* **2003**, *2*, 3–6. (In Chinese)
66. Zhang, Q. Research on Local Scour Problem of Seabed Around Marine Pile Foundation and Pipeline. Ph.D. Thesis, Shanghai Jiao Tong University, Shanghai, China, 2017. (In Chinese)
67. Kjeldsen, S.P.; Guøsvik, O.; Bringaker, K.G.; Jacobsen, J. Local scour near offshore pipelines. In Proceedings of the Second International Conference on Port and Ocean Engineering Under Arctic Conditions (POAC), Trondheim, Norway, 27–30 August 1973.
68. Lucassen, R.J. Scour Underneath Submarine Pipelines. Master's Thesis, Delft University of Technology, Delft, The Netherlands, 1984.
69. Chao, J.L.; Hennessy, P.V. Local scour under ocean outfall pipelines. *J. (Water Pollut. Control. Fed.)* **1972**, *44*, 1443–1447.
70. Li, N.; Gu, F. Improvement of Chao's analytical method for local scour equilibrium depth of submarine pipelines. *J. Dalian Marit. Univ.* **2012**, *38*, 77–80. (In Chinese)
71. Mao, Y. The interaction between a pipeline and an erodible bed. *Series Pap. Tech. Univ. Den.* **1987**, *39*, 169.
72. Sumer, B.M.; Jensen, H.R.; Mao, Y.; Fredsøe, J. Effect of lee-wake on scour below pipelines in current. *J. Waterw. Port Coastal, Ocean Eng.* **1988**, *114*, 599–614. [CrossRef]

73. Van Beek, F.; Wind, H. Numerical modelling of erosion and sedimentation around offshore pipelines. *Coast. Eng.* **1990**, *14*, 107–128. [CrossRef]
74. Chiew, Y. Mechanics of local scour around submarine pipelines. *J. Hydraul. Eng.* **1990**, *116*, 515–529. [CrossRef]
75. Chiew, Y. Effect of spoilers on scour at submarine pipelines. *J. Hydraul. Eng.* **1992**, *118*, 1311–1317. [CrossRef]
76. Ajdehak, E.; Zhao, M.; Cheng, L.; Draper, S. Numerical investigation of local scour beneath a sagging subsea pipeline in steady currents. *Coast. Eng.* **2018**, *136*, 106–118. [CrossRef]
77. Liu, M.; Lu, L.; Teng, B.; Zhao, M.; Tang, G. Numerical modeling of local scour and forces for submarine pipeline under surface waves. *Coast. Eng.* **2016**, *116*, 275–288. [CrossRef]
78. Li, F.; Cheng, L. Numerical model for local scour under offshore pipelines. *J. Hydraul. Eng.* **1999**, *125*, 400–406. [CrossRef]
79. Yang, J.; Low, Y.M.; Lee, C.-H.; Chiew, Y.-M. Numerical simulation of scour around a submarine pipeline using computational fluid dynamics and discrete element method. *Appl. Math. Model.* **2018**, *55*, 400–416. [CrossRef]
80. Mirmohammadi, A.; Ketabdari, M. Numerical simulation of wave scouring beneath marine pipeline using smoothed particle hydrodynamics. *Int. J. Sediment Res.* **2011**, *26*, 331–342. [CrossRef]
81. Hu, K.; Bai, X.; Zhang, Z.; Vaz, M.A. Prediction of submarine pipeline equilibrium scour depth based on machine learning applications considering the flow incident angle. *Appl. Ocean Res.* **2021**, *112*, 102717. [CrossRef]
82. Zhao, E.; Dong, Y.; Tang, Y.; Sun, J. Numerical investigation of hydrodynamic characteristics and local scour mechanism around submarine pipelines under joint effect of solitary waves and currents. *Ocean Eng.* **2021**, *222*, 108553. [CrossRef]
83. Dogan, M.; Arisoy, Y. The propagation of wave scour along the spanwise direction of submarine pipelines in case of clear-water regime. *Coast. Eng.* **2021**, *168*, 103958. [CrossRef]
84. Chen, Y.; Zhang, L.; Wei, X.; Xu, J.; Fu, S.; Liao, C. Debris-flow-induced damage assessment for a submarine pipeline network in regional-scale natural terrain. *Eng. Geol.* **2022**, *311*, 106917. [CrossRef]
85. Cui, W.; Zhao, L.; Ge, Y.; Xu, K. A generalized van der Pol nonlinear model of vortex-induced vibrations of bridge decks with multistability. *Nonlinear Dyn.* **2023**, *112*, 259–272. [CrossRef]
86. Li, X.; Wang, Y.; Li, G.; Jiang, M.; He, X. Experimental investigation of vortex-induced vibrations of long free spans near seabed. *Sci. China Technol. Sci.* **2011**, *54*, 698–704. [CrossRef]
87. Zhao, M.; Cheng, L.; An, H. Numerical investigation of vortex-induced vibration of a circular cylinder in transverse direction in oscillatory flow. *Ocean Eng.* **2012**, *41*, 39–52. [CrossRef]
88. Govardhan, R.; Williamson, C.H.K. Modes of vortex formation and frequency response of a freely vibrating cylinder. *J. Fluid Mech.* **2000**, *420*, 85–130. [CrossRef]
89. Williamson, C.; Govardhan, R. A brief review of recent results in vortex-induced vibrations. *J. Wind. Eng. Ind. Aerodyn.* **2008**, *96*, 713–735. [CrossRef]
90. Khalak, A.; Williamson, C. Motions, forces and mode transitions in vortex-induced vibrations at low mass-damping. *J. Fluids Struct.* **1999**, *13*, 813–851. [CrossRef]
91. Khalak, A.; Williamson, C. Dynamics of a hydroelastic cylinder with very low mass and damping. *J. Fluids Struct.* **1996**, *10*, 455–472. [CrossRef]
92. Feng, C.C. The Measurement of Vortex Induced Effects in Flow Past Stationary and Oscillating Circular and D-Section Cylinders. Ph.D. Thesis, University of British Columbia, Vancouver, BC, Canada, 1968.
93. Fan, D.; Wu, B.; Bachina, D.; Triantafyllou, M.S. Vortex-induced vibration of a piggyback pipeline half buried in the seabed. *J. Sound Vib.* **2019**, *449*, 182–195. [CrossRef]
94. Xu, W.; Jia, K.; Ma, Y.; Wang, Y.; Song, Z. Multispan classification methods and interaction mechanism of submarine pipelines undergoing vortex-induced vibration. *Appl. Ocean Res.* **2022**, *120*, 103027. [CrossRef]
95. Zhang, T.; Zhang, S.; Yang, D.; Huang, G. Numerical investigation on competitive mechanism between internal and external effects of submarine pipeline undergoing vortex-induced vibration. *Ocean Eng.* **2022**, *266*, 112744. [CrossRef]
96. Huang, Y.; Ji, D. Experimental study on seawater-pipeline internal corrosion monitoring system. *Sens. Actuators B Chem.* **2008**, *135*, 375–380. [CrossRef]
97. Mathew, A.; Kumar, A.; Velu, V. Predictive Modeling for Corrosion Rate on Pipeline Corrosion in Seawater. In Proceedings of the Abu Dhabi International Petroleum Exhibition and Conference, Abu Dhabi, United Arab Emirates, 3–6 November 2025; SPE: Richardson, TX, USA, 2023.
98. Lawal, S.L.; Afolalu, S.A.; Jen, T.C.; Akinlabi, E.T. Corrosion Control and Its Application in Marine Environment—A Review. *Solid State Phenom.* **2024**, *355*, 61–73. [CrossRef]
99. Caines, S.; Khan, F.; Shirokoff, J. Analysis of Pitting Corrosion on Steel under Insulation in Marine Environments. *J. Loss Prev. Process Ind.* **2013**, *26*, 1466–1483. [CrossRef]
100. Liu, H.; Ma, Y.; Du, Y.; Li, M.; Shen, Y. A numerical analysis of the mechano-electrochemical effects on localised corrosion in pipelines using a multi-physical field coupling approach. *Ocean Eng.* **2024**, *310*, 118522. [CrossRef]

 https://doi.org/10.3390/jmse14020176

101. Zhao, M.; Wang, J.; Ma, Y.; Xu, Y.; Zhang, X.; Li, Z.; Xu, D.; Wang, F. Electroactive Shewanella algae accelerates the crevice corrosion of X70 pipeline steel in marine environment. *Corros. Sci.* **2024**, *235*, 112226. [CrossRef]
102. Xia, D.-H.; Deng, C.-M.; Macdonald, D.; Jamali, S.; Mills, D.; Luo, J.-L.; Strebl, M.G.; Amiri, M.; Jin, W.; Song, S.; et al. Electrochemical measurements used for assessment of corrosion and protection of metallic materials in the field: A critical review. *J. Mater. Sci. Technol.* **2022**, *112*, 151–183. [CrossRef]
103. Abdelbar, M.; El-Shamy, A.M. Understanding Soil Factors in Corrosion and Conservation of Buried Bronze Statuettes: Insights for Preservation Strategies. *Sci. Rep.* **2024**, *14*, 19230. [CrossRef] [PubMed]
104. Bale, A.J.; Kenny, A.J. Sediment analysis and seabed characterisation. In *Methods for the Study of Marine Benthos*, 3rd ed.; Blackwell Science Ltd.: Oxford, UK, 2005; pp. 43–86.
105. Jiang, H.; Ma, Q.; Zhou, F.; Shu, J.; Yang, Y. Horizontal and vertical motion at the surface of seawater-gas hydrate bearing sediment-seabed under obliquely incident P waves. *Appl. Ocean Res.* **2023**, *142*, 103844. [CrossRef]
106. Zhu, C.; Zhu, C.; Li, Q.; Li, Z.; Duan, M.; Li, S.; Zhou, Q.; Geng, M.; Chen, J.; Jia, Y. Seabed fluid flow in the China Seas. *Front. Mar. Sci.* **2023**, *10*, 1158685. [CrossRef]
107. Yue, Z.; Zhao, G.; Xiao, L.; Liu, M. Comparative study on collection performance of three nodule collection methods in seawater and sediment-seawater mixture. *Appl. Ocean Res.* **2021**, *110*, 102606. [CrossRef]
108. Ekdale, A.A.; Mason, T.R. Characteristic trace-fossil associations in oxygen-poor sedimentary environments. *Geology* **1988**, *16*, 720–723. [CrossRef]
109. Glud, R.N. Oxygen dynamics of marine sediments. *Mar. Biol. Res.* **2008**, *4*, 243–289. [CrossRef]
110. Gundersen, J.K.; Jorgensen, B.B. Microstructure of diffusive boundary layers and the oxygen uptake of the sea floor. *Nature* **1990**, *345*, 604–607. [CrossRef]
111. Hulthe, G.; Hulth, S.; Hall, P.O. Effect of oxygen on degradation rate of refractory and labile organic matter in continental margin sediments. *Geochim. et Cosmochim. Acta* **1998**, *62*, 1319–1328. [CrossRef]
112. Ake, L.M.Q.; Liu, H.B.; Alamilla, J.; Sosa, E.; Contreras, A. Electrochemical study of external corrosion of three API steels exposed to seawater and sediment from seabed of the gulf of México. *Surf. Eng. Appl. Electrochem.* **2020**, *56*, 365–380. [CrossRef]
113. Shen, Y.; Ma, R.; Wei, J.; Wang, C.; Dong, J.; Xu, D. Influence of sea sand and Shewanella algae on the corrosion behavior of 316 stainless steel in marine environment. *Corros. Sci.* **2024**, *235*, 112200. [CrossRef]
114. Zhang, T.; Xu, Z.; Wan, H.; Chen, X.; Liu, H. Dual corrosion promotion of pipeline steel in sea mud induced by sulfate reducing bacteria: Bacteria concentration cell and electronic conduction of the biofilms covered sand grains. *Corros. Sci.* **2024**, *232*, 112005. [CrossRef]
115. Melchers, R.E.; Tan, M.Y. Long-term corrosion of abandoned offshore steel infrastructure. *Corros. Eng. Sci. Technol.* **2023**, *58*, 712–722. [CrossRef]
116. Khodabux, W.; Causon, P.; Brennan, F. Profiling corrosion rates for offshore wind turbines with depth in the north sea. *Energies* **2020**, *13*, 2518. [CrossRef]
117. Adedipe, O.; Brennan, F.; Kolios, A. Review of corrosion fatigue in offshore structures: Present status and challenges in the offshore wind sector. *Renew. Sustain. Energy Rev.* **2016**, *61*, 141–154. [CrossRef]
118. Farro, N.W.; Veleva, L.; Aguilar, P. Copper Marine Corrosion: I. Corrosion rates in atmospheric and seawater environments of peruvian port. *Open Corros. J.* **2009**, *2*, 130–138. [CrossRef]
119. Tian, H.; Cui, Z.; Ma, H.; Zhao, P.; Yan, M.; Wang, X.; Cui, H. Corrosion evolution and stress corrosion cracking behavior of a low carbon bainite steel in the marine environments: Effect of the marine zones. *Corros. Sci.* **2022**, *206*, 110490. [CrossRef]
120. Creamer, E.V. Splash zone protection of marine structures. In Proceedings of the Offshore Technology Conference, Houston, TX, USA, 21–23 April 1970.
121. Xia, R.; Jia, C.; Garbatov, Y. Deterioration of marine offshore structures and subsea installations subjected to severely corrosive environment: A review. *Mater. Corros.* **2024**, *76*, 758–775. [CrossRef]
122. Zhou, X.; Wang, X.; Wang, Q.; Wu, T.; Li, C.; Luo, J.; Yin, F. Study on corrosion behavior of Q235 steel in a simulated marine tidal environment. *J. Mater. Eng. Perform.* **2022**, *31*, 4459–4471. [CrossRef]
123. Zhou, X.; Zhou, X.; Wu, T.; Tan, L.; Deng, J.; Li, C.; Qiu, W. A study on corrosion of X80 steel in a simulated tidal zone. *J. Mater. Res. Technol.* **2021**, *12*, 2224–2237. [CrossRef]
124. Kurth, J.C.; Krauss, P.D.; Foster, S.W. Corrosion management of maritime infrastructure. *Transp. Res. Rec. J. Transp. Res. Board* **2019**, *2673*, 2–14. [CrossRef]
125. Budiea, A.M.; Yahaya, N.; Noor, N.M.; Abdullah, A.H.; Ibrahim, M.H.; Suratkon, A.; Muthusamy, K. Corrosion study of pipeline material for seabed sediment in tropical climate. *IOP Conf. Ser. Mater. Sci. Eng.* **2020**, *849*, 012023. [CrossRef]
126. Wu, M.; Gong, K.; Xie, F. Effect of sulfate reducing bacteria and stress on corrosion behavior of X100 steel in sea mud environment. *J. Electroanal. Chem.* **2021**, *886*, 115129. [CrossRef]
127. Liu, P.; Liu, C.; Zhang, S.; Wang, Y.; Wang, Q. Depth-varying corrosion characteristics and stability bearing capacity of steel pipe piles under aggressive marine environment. *Ocean Eng.* **2022**, *266*, 112649. [CrossRef]

128. Melchers, R.E.; Jeffrey, R. Early corrosion of mild steel in seawater. *Corros. Sci.* **2005**, *47*, 1678–1693. [CrossRef]
129. Evans, U.R. Water-line corrosion of iron and steel with special reference to the addition of the so-called 'inhibitors' of corrosion. *J. Soc. Chem. Ind.* **1925**, *44*, 163.
130. Goldstein, E.M. The corrosion and oxidation of metals: Scientific principles and practical applications (Evans, Ulick R.). *J. Chem. Educ.* **1960**, *37*, 662. [CrossRef]
131. Evans, U.R. Report on corrosion research work at Cambridge University interrupted by the outbreak of war. *J. Iron Steel Inst.* **1940**, *141*, 219–234.
132. Tomashov, N.D. *Theory of Corrosion and Protection of Metals: The Science of Corrosion*; Macmillan Publishers Ltd.: London, UK, 1966.
133. Gubner, R.J. Biofilms and Accelerated Low-Water Corrosion of Carbon Steel Piling in Tidal Waters. Ph.D. Thesis, University of Portsmouth, Portsmouth, UK, 1998.
134. Tan, Y.-J.; Bailey, S.; Kinsella, B. Mapping non-uniform corrosion using the wire beam electrode method. I. Multi-phase carbon dioxide corrosion. *Corros. Sci.* **2001**, *43*, 1905–1918. [CrossRef]
135. Liu, Z.; Wang, W.; Wang, J.; Peng, X.; Wang, Y.; Zhang, P.; Wang, H.; Gao, C. Study of corrosion behavior of carbon steel under seawater film using the wire beam electrode method. *Corros. Sci.* **2014**, *80*, 523–527. [CrossRef]
136. Chen, Y.; Zhang, W.; Wang, Q.; Wang, J. WBE technique to study the stripping mechanism of broken coatings in the waterline zone-II. *Chin. J. Corros. Prot.* **2017**, *37*, 322–328. (In Chinese)
137. Hu, J.Z.; Li, X.G.; Deng, P.C.; Zhang, J.B.; Wang, G.; Su, L.H. Corrosion behaviour of carbon steel in the seawater-atmosphere interfacial zone studied by array electrodes (WBE) combined with linear polarization (LP) technique. *Corros. Prot.* **2015**, *36*, 1014–1020. (In Chinese)
138. Chen, Y.; Zhang, W.; Wang, W.; Wang, J.; Wang, Q.; Cai, G.X. WBE technique to study corrosion of Q235 carbon steel in waterline area. *Chin. J. Corros. Prot.* **2014**, *34*, 451–458. (In Chinese)
139. Jeffrey, R.; Melchers, R.E. Corrosion of vertical mild steel strips in seawater. *Corros. Sci.* **2009**, *51*, 2291–2297. [CrossRef]
140. Qin, Y. Corrosion Failure Prediction Study of Offshore Pipelines Based on LSSVM-GM(1, N). Master's Thesis, Xi'an University of Archi-tecture and Technology, Xi'an, China, 2021. (In Chinese)
141. Joel, A.N.; Spencer, H. *The Probabilistic Method*; John Wiley & Sons: Hoboken, NJ, USA, 2015.
142. Melchers, R.E. New insights from probabilistic modelling of corrosion in structural reliability analysis. *Struct. Saf.* **2021**, *88*, 102034. [CrossRef]
143. Kakaie, A.; Soares, C.G.; Ariffin, A.K.; Punurai, W. Fatigue reliability analysis of submarine pipelines using the bayesian approach. *J. Mar. Sci. Eng.* **2023**, *11*, 580. [CrossRef]
144. Motta, R.d.S.; Leal, C.F.; Ferreira, A.D.; Afonso, S.M.; Lyra, P.R.; Willmersdorf, R.B. Reliability analysis of ovalized deep-water pipelines with corrosion defects. *Mar. Struct.* **2021**, *77*, 102969. [CrossRef]
145. El-Abbasy, M.S.; Senouci, A.; Zayed, T.; Mirahadi, F.; Parvizsedghy, L. Condition prediction models for oil and gas pipelines using regression analysis. *J. Constr. Eng. Manag.* **2014**, *140*, 838. [CrossRef]
146. Chatterjee, S.; Hadi, A.S.; Chatterjee, S.; Hadi, A.S. *Regression Analysis by Example*, 5th ed.; John Wiley & Sons, Inc.: Hoboken, NJ, USA, 2006.
147. Bhaskaran, P.E.; Chennippan, M.; Subramaniam, T. Future prediction & estimation of faults occurrences in oil pipelines by using data clustering with time series forecasting. *J. Loss Prev. Process. Ind.* **2020**, *66*, 104203. [CrossRef]
148. Han, Z.; Zhao, J.; Leung, H.; Ma, K.F.; Wang, W. A review of deep learning models for time series prediction. *IEEE Sens. J.* **2019**, *21*, 7833–7848. [CrossRef]
149. Jin, W.; Xiong, X.; Fan, C.; Kong, X. Corrosion depth prediction of pipeline based on improved gray model. *Pet. Sci. Technol.* **2023**, *41*, 802–817. [CrossRef]
150. Zhang, D.-B.; Li, X.; Zhang, Y.; Zhang, H. Prediction method of asphalt pavement performance and corrosion based on grey system theory. *Int. J. Corros.* **2019**, *2019*, 1–9. [CrossRef]
151. De Masi, G.; Gentile, M.; Vichi, R.; Bruschi, R.; Gabetta, G. Machine Learning Approach to Corrosion Assessment in Subsea Pipelines. In Proceedings of the OCEANS 2015-Genova, Genova, Italy, 18–21 May 2015.
152. Du, J.; Zheng, J.; Liang, Y.; Xu, N.; Liao, Q.; Wang, B.; Zhang, H. Deeppipe: Theory-guided prediction method based automatic machine learning for maximum pitting corrosion depth of oil and gas pipeline. *Chem. Eng. Sci.* **2023**, *278*, 118927. [CrossRef]
153. Ferreira, A.D.M.; Afonso, S.M.; Willmersdorf, R.B.; Lyra, P.R. Multiresolution analysis and deep learning for corroded pipeline failure assessment. *Adv. Eng. Softw.* **2021**, *162–163*, 103066. [CrossRef]
154. Kroese, D.P.; Brereton, T.; Taimre, T.; Botev, Z.I. Why the Monte Carlo method is so important today. *WIREs Comput. Stat.* **2014**, *6*, 386–392. [CrossRef]
155. Helmut, L. Vector autoregressive models. In *Handbook of Research Methods and Applications in Empirical Macroe-Conomics*; Edward Elgar Publishing: Cheltenham, UK, 2013; pp. 139–164.
156. Boldin, M.V. On Testing Hypotheses in the Sliding Average Scheme by the Kolmogorov–Smirnov and ω^2 Tests. *Theory Probab. Its Appl.* **1990**, *34*, 699–704. [CrossRef]

 https://doi.org/10.3390/jmse14020176

157. ByoungSeon, C. *ARMA Model Identification*; Springer Science & Business Media: Berlin, Germany, 2012.
158. Shumway, R.H.; Stoffer, D.S. ARIMA models. In *Time Series Analysis and Its Applications: With R Examples*; Springer: Berlin/Heidelberg, Germany, 2017; pp. 75–163.
159. Suykens, J.A.K.; Vandewalle, J. Least Squares Support Vector Machine Classifiers. *Neural Process. Lett.* **1999**, *9*, 293–300. [CrossRef]
160. Tien, T.-L. A research on the grey prediction model GM(1,n). *Appl. Math. Comput.* **2012**, *218*, 4903–4916. [CrossRef]
161. Deng, C.A.; Li, Z.Y.; Li, X.S.; Wang, S.; Wang, X.J. A new method for corrosion rate prediction of marine pipelines based on fuzzy neural networks. *Chin. J. Corros. Prot.* **2015**, *35*, 571–576. (In Chinese)
162. Li, H. Research status review of fuzzy neural networks. *J. Liaoning Inst. Sci. Technol.* **2010**, *12*, 15–17.
163. Li, J.; Kong, L.; Liang, H.; Li, W. *MATLAB Neural Network Analysis of 30 Cases*; Beijing University of Aeronautics and Astronautics Press: Beijing, China, 2010.
164. Wang, W.; He, X.; Li, Y.; Shuai, J. Risk analysis on corrosion of submarine oil and gas pipelines based on hybrid Bayesian network. *Ocean Eng.* **2022**, *260*, 111957. [CrossRef]
165. Stelios, K.; Corona, E. *Mechanics of Offshore Pipelines: Volume I: Buckling and Collapse*; Gulf Professional Publishing: Houston, TX, USA, 2023.
166. Du, F.; Li, C.; Wang, W. Development of Subsea Pipeline Buckling, Corrosion and Leakage Monitoring. *J. Mar. Sci. Eng.* **2023**, *11*, 188. [CrossRef]
167. Triantafyllaki, A.; Loukidis, D.; Papanastasiou, P. Design of offshore gas pipelines against active tectonic fault movement. *Soil Dyn. Earthq. Eng.* **2022**, *166*, 107702. [CrossRef]
168. Li, Z.; Yu, Y.; Liu, X.; Liu, X.; Wang, X.; Xu, L.; Yu, J. Propagation and arrest of collapse failures in a buried offshore pipeline crossing reverse fault areas. *Mar. Struct.* **2024**, *93*, 103522. [CrossRef]
169. Wang, X.; Yuan, L.; Gong, S.; Liu, C. On the arresting efficiency of CFRP-winding buckle arrestors for subsea pipelines. *Thin-Walled Struct.* **2023**, *182*, 110154. [CrossRef]
170. Fabian, O. Collapse of cylindrical, elastic tubes under combined bending, pressure and axial loads. *Int. J. Solids Struct.* **1977**, *13*, 1257–1270. [CrossRef]
171. Kyriakides, S.; Babcock, C.D.; Elyada, D. Initiation of propagating buckles from local pipeline damages. *J. Energy Resour. Technol.* **1984**, *106*, 79–87. [CrossRef]
172. Huang, Y.Y.; Le, D.Y. Sensitivity of vortex defects in marine pipelines under combined axial force and external pressure. *J. Oceanogr.* **1990**, *6*, 783–793. (In Chinese)
173. Ning, J.; Liu, S.; Lin, T.; Huang, W. Pipeline element for upheaval buckling analysis of submarine pipelines with geometric-imperfections under high temperature high pressure. *Ocean Eng.* **2022**, *264*, 112456. [CrossRef]
174. Wang, W.J. Research status and development of corrosion monitoring technology in submarine pipelines. *Coat. Prot.* **2021**, *42*, 37–42. (In Chinese)
175. Zang, X.; Xu, Z.-D.; Lu, H.; Zhu, C.; Zhang, Z. Ultrasonic guided wave techniques and applications in pipeline defect detection: A review. *Int. J. Press. Vessel. Pip.* **2023**, *206*, 105033. [CrossRef]
176. Zhu, H.; Chen, J.; Lin, Y.; Guo, J.; Gao, X.; Chen, Y.; Ge, Y.; Wang, W. In-Line Inspection (ILI) Techniques for Subsea Pipelines: State-of-the-Art. *J. Mar. Sci. Eng.* **2024**, *12*, 417. [CrossRef]
177. Jiao, D.; Gao, B.; Ru, G.; Ma, Q.; Lei, J.; Zhang, Y.; Lu, J.F.; Woo, W.L. Instrumentation of Cross Electromagnetic Eddy Current Sensing for Internal Pipeline Inspection. *IEEE Sens. J.* **2024**, *24*, 22745–22757. [CrossRef]
178. Zhou, Z.; Li, B.; Sun, Z.; Liu, H.; Wang, J. Differential eddy current method for full circumferential defect detection of small diameter steel pipe: Numerical analysis and experimental study. *J. Magn. Magn. Mater.* **2023**, *579*, 170867. [CrossRef]
179. Lukonge, A.B.; Cao, X. Leak detection system for long-distance onshore and offshore gas pipeline using acoustic emission Technology. A review. *Trans. Indian Inst. Met.* **2020**, *73*, 1715–1727. [CrossRef]
180. Ma, Y.; Dong, S.; Zhang, H.; Xu, Q.; Wei, H.; Qian, W. Study on small leaks in submarine liquid pipelines based on passive acoustic internal detection method. *Energy Sources Part A Recover. Util. Environ. Eff.* **2023**, *45*, 9642–9667. [CrossRef]
181. Liu, L.; Yang, L.; Gao, S. Spatial propagation law of magnetic memory signals detected by using magnetic tomography method. *IEEE Access* **2022**, *10*, 64106–64113. [CrossRef]
182. Onuoha, C.; Nnamdi, D.; Eslami, A.; Kolesnikov, I.; Ilmi, M.I. Enhancement of pipeline integrity assessment of buried unpiggable pipelines with non-contact magnetic gradient tomography method (MTM-G). *J. Mater. Manuf.* **2024**, *3*, 25–37.
183. Gan, F.; Tian, G.; Wan, Z.; Liao, J.; Li, W. Investigation of pitting corrosion monitoring using field signature method. *Measurement* **2016**, *82*, 46–54. [CrossRef]
184. Gan, F.; Wan, Z.; Liao, J.; Li, W. High-accuracy evaluation of pitting corrosion in field signature method. *Insight-Non-Destr. Test. Cond. Monit.* **2014**, *56*, 613–616. [CrossRef]
185. Chen, L.; Yang, J.-F.; Lu, X.-Y. Development and Application of Integrated Corrosion Monitoring System in Refining Unit. In Proceedings of the International Petroleum and Petrochemical Technology Conference 2019 3rd, Beijing, China, 27–29 March 2019; Springer: Singapore, 2020.

186. Gao, B.; Li, Z.-H.; Gao, S.-H. Research and Application of Online Corrosion Monitoring System for Gas Pipelines. In Proceedings of the International Field Exploration and Development Conference, Wuhan, China, 19–21 September 2023; Springer Nature: Singapore, 2023.
187. Hu, B.; Yu, R.; Liu, J. Experimental study on the corrosion testing of a buried metal pipeline by transient electromagnetic method. *Anti-Corrosion Methods Mater.* **2016**, *63*, 262–268. [CrossRef]
188. Islam, M.; Pojtanabuntoeng, T.; Gubner, R. Study of the top-of-the-line corrosion using a novel electrochemical probe. *Corrosion* **2017**, *74*, 588–598. [CrossRef]
189. Yang, Y.; Li, W.; Liang, B. External Corrosion Analysis of Gas Pipeline Based on Gray Prediction Models. *J. Fail. Anal. Prev.* **2024**, *24*, 1277–1294. [CrossRef]
190. Chen, B.; Kang, Q.; Ju, R. Calculation of Electrochemical Corrosion Rate at Elbow of Supercritical CO_2 Pipeline. *IOP Conf. Ser. Mater. Sci. Eng.* **2020**, *772*, 012042. [CrossRef]
191. Soleymani, F. Internal Corrosion Monitoring of Pipelines Using Electrochemical Techniques: A Comprehensive Review. *Prot. Met. Phys. Chem. Surfaces* **2023**, *59*, 1045–1061. [CrossRef]
192. Liu, K.; Belarbi, Z.; Dogan, Ö.N.; Nakatsuka, M.A.; Arumugam, G.K. Rethinking the Future of Hydrogen Transport in Pipelines Using an Advanced Liner Material. In Proceedings of the Offshore Technology Conference 2024, Houston, TX, USA, 6–9 May 2024.
193. Permeh, S.; Lau, K. Identification of steel corrosion associated with sulfate-reducing bacteria by electrochemical noise technique. *Mater. Corros.* **2022**, *74*, 20–32. [CrossRef]
194. Caines, S.; Khan, F.; Zhang, Y.; Shirokoff, J. Simplified electrochemical potential noise method to predict corrosion and corrosion rate. *J. Loss Prev. Process. Ind.* **2017**, *47*, 72–84. [CrossRef]
195. Hoseinieh, S.; Homborg, A.; Shahrabi, T.; Mol, J.; Ramezanzadeh, B. A novel approach for the evaluation of under deposit corrosion in marine environments using combined analysis by electrochemical impedance spectroscopy and electrochemical noise. *Electrochim. Acta* **2016**, *217*, 226–241. [CrossRef]
196. Sun, F.; Han, P.; He, B. An analysis of electrochemical corrosion on pipeline steel in silty soil under salt-temperature coupling environments. *Chem. Eng. Sci.* **2023**, *274*, 118704. [CrossRef]
197. EC, B.G.; Moore, C.G.; Williams, S. Development and application of ductile iron pipe electrical resistance probes for monitoring underground external pipeline corrosion. In Proceedings of the NACE International Corrosion Conference 2007, Houston, TX, USA, 11–15 March 2007.
198. Laleh, M.; Xu, Y.; Tan, M.Y. A three-dimensional electrode array probe designed for visualising complex and dynamically changing internal pipeline corrosion. *Corros. Sci.* **2023**, *211*, 110924. [CrossRef]
199. Li, X.; Liu, H.; Li, S.; Guo, T.; Xiao, Y. Research on accident tree analysis and monitoring and warning method for third party damage of submarine pipelines. *Yunnan Chem.* **2021**, *48*, 101–103. (In Chinese)
200. Huang, Z.; Qi, X. Synthetic aperture radar image ship detection and analysis system. *Microcomput. Form.* **2009**, *25*, 298–300. (In Chinese)
201. Huang, H.W.; Yu, L.; Pan, Z.; Sheng, C.; Shuang, Y. Design and realisation of ship dynamic monitoring system based on BeiDou satellite. *Surv. Mapp. Spat. Geogr. Inf.* **2015**, *38*, 7–9. (In Chinese)
202. Zhang, H.; Gao, M.-Q.; Tang, B.; Cui, C.; Xu, X.-F. Dynamic characteristics of the pipeline inspection gauge under girth weld excitation in submarine pipeline. *Pet. Sci.* **2022**, *19*, 774–788. [CrossRef]
203. Feng, H.; Yu, J.; Huang, Y.; Cui, J.; Qiao, J.; Wang, Z.; Xie, Z.; Ren, K. Automatic tracking method for submarine cables and pipelines of AUV based on side scan sonar. *Ocean Eng.* **2023**, *280*, 114689. [CrossRef]
204. Zhang, Y.; Zhang, H.; Liu, J.; Zhang, S.; Liu, Z.; Lyu, E.; Chen, W. Submarine pipeline tracking technology based on AUVs with forward looking sonar. *Appl. Ocean Res.* **2022**, *122*, 103128. [CrossRef]
205. Yu, J.; Wu, S.; Chen, H.; Yu, Y.; Fan, H.; Liu, J. Risk assessment of submarine pipelines using modified FMEA approach based on cloud model and extended VIKOR method. *Process. Saf. Environ. Prot.* **2021**, *155*, 555–574. [CrossRef]
206. Li, X.; Chen, G.; Zhu, H. Quantitative risk analysis on leakage failure of submarine oil and gas pipelines using Bayesian network. *Process. Saf. Environ. Prot.* **2016**, *103*, 163–173. [CrossRef]
207. Aulia, R.; Tan, H.; Sriramula, S. Dynamic reliability model for subsea pipeline risk assessment due to third-party interference. *J. Pipeline Sci. Eng.* **2021**, *1*, 277–289. [CrossRef]
208. Yu, J.; Zeng, Q.; Yu, Y.; Wu, S.; Ding, H.; Gao, H.; Yang, J. An intuitionistic fuzzy probabilistic Petri net method for risk assessment on submarine pipeline leakage failure. *Ocean Eng.* **2022**, *266*, 112788. [CrossRef]
209. Shi, H.; Liu, Z.; Liu, H.-C. A new linguistic preference relation-based approach for failure mode and effect analysis with dynamic consensus reaching process. *Inf. Sci.* **2022**, *610*, 977–993. [CrossRef]
210. Resende, B.A.D.; Dedini, F.G.; Eckert, J.J.; Sigahi, T.F.; Pinto, J.D.S.; Anholon, R. Proposal of a facilitating methodology for fuzzy FMEA implementation with application in process risk analysis in the aeronautical sector. *Int. J. Qual. Reliab. Manag.* **2024**, *41*, 1063–1088. [CrossRef]

211. Li, X.; Yang, M.; Chen, G. An integrated framework for subsea pipelines safety analysis considering causation dependencies. *Ocean Eng.* **2019**, *183*, 175–186. [CrossRef]
212. Bhattacharjee, P.; Dey, V.; Mandal, U.K. Failure Mode and Effects Analysis (FMEA) using interval number based BWM—MCDM approach: Risk Expected Value (REV) method. *Soft Comput.* **2022**, *26*, 12667–12688. [CrossRef]
213. Lo, H.-W.; Shiue, W.; Liou, J.J.H.; Tzeng, G.-H. A hybrid MCDM-based FMEA model for identification of critical failure modes in manufacturing. *Soft Comput.* **2020**, *24*, 15733–15745. [CrossRef]
214. Boral, S.; Chakraborty, S. Failure analysis of CNC machines due to human errors: An integrated IT2F-MCDM-based FMEA approach. *Eng. Fail. Anal.* **2021**, *130*, 105768. [CrossRef]
215. Shakibaei, H.; Seifi, S.; Zhuang, J. A data-driven and cost-oriented FMEA–MCDM approach to risk assessment and ranking in a fuzzy environment: A hydraulic pump factory case study. *Risk Anal.* **2024**, *44*, 2629–2648. [CrossRef]
216. Liu, H.-C.; Chen, X.-Q.; Duan, C.-Y.; Wang, Y.-M. Failure mode and effect analysis using multi-criteria decision making methods: A systematic literature review. *Comput. Ind. Eng.* **2019**, *135*, 881–897. [CrossRef]
217. Yu, Y.; Wu, S.; Yu, J.; Xu, Y.; Song, L.; Xu, W. A hybrid multi-criteria decision-making framework for offshore wind turbine selection: A case study in China. *Appl. Energy* **2022**, *328*, 120173. [CrossRef]
218. Yu, Y.; Yang, J.; Wu, S. A novel FMEA approach for submarine pipeline risk analysis based on IVIFRN and ExpTODIM-PROMETHEE-II. *Appl. Soft Comput.* **2023**, *136*, 110065. [CrossRef]
219. Lenci, S.; Callegari, M. Simple analytical models for the J-lay problem. *Acta Mech.* **2005**, *178*, 23–39. [CrossRef]
220. Wang, W.W.; Zhang, M.; Liang, S. A review of modelling and analysis methods for overhanging marine pipelines. *China Ocean. Platf.* **2017**, *32*, 1–6. (In Chinese)
221. Plunkett, R. Static Bending Stresses in Catenaries and Drill Strings. *J. Eng. Ind.* **1967**, *89*, 31–36. [CrossRef]
222. de Lucena, R.R.; Baioco, J.S.; de Lima, B.S.L.P.; Albrecht, C.H.; Jacob, B.P. Optimal design of submarine pipeline routes by genetic algorithm with different constraint handling techniques. *Adv. Eng. Softw.* **2014**, *76*, 110–124. [CrossRef]
223. Ohaeri, E.G.; Szpunar, J.A. An overview on pipeline steel development for cold climate applications. *J. Pipeline Sci. Eng.* **2022**, *2*, 1–17. [CrossRef]
224. Li, X.; Zhao, J.; Jia, S.; Lu, G.; Misra, R.; Liu, Q.; Li, B. Ultrafine microstructure design of high strength pipeline steel for low temperature service: The significant impact on toughness. *Mater. Lett.* **2021**, *303*, 130429. [CrossRef]
225. Fan, Z.; Hu, X.; Liu, J.; Li, H.; Fu, J. Stress corrosion cracking of L360NS pipeline steel in sulfur environment. *Petroleum* **2017**, *3*, 377–383. [CrossRef]
226. Zhao, M.C.; Zhao, M.C.; Shan, Y.Y.; Li, Y.H.; Yang, K. Effect of microstructure on sulfide stress corrosion cracking of pipeline steels. *Acta Metall. Sin.* **2001**, *37*, 1087–1092.
227. Zhang, L.; Zhao, X.; Yan, X.; Yang, X. A semi-analytical method of stress-strain analysis of buried steel pipelines under submarine landslides. *Appl. Ocean Res.* **2016**, *59*, 38–52. [CrossRef]
228. Wang, F.-C.; Han, L.-H. Analytical behavior of carbon steel-concrete-stainless steel double-skin tube (DST) used in submarine pipeline structure. *Mar. Struct.* **2019**, *63*, 99–116. [CrossRef]
229. Li, S.; Liu, M.; Zuo, X. Prospects for the development and application of steel for deep-sea pipelines. *Therm. Process. Technol.* **2013**, *42*, 23–26. (In Chinese)
230. Xiao, Q.; Xie, Y.; Hu, F.; Hu, C. Current Status and Trends of Low-Temperature Steel Used in Polar Regions. *Materials* **2024**, *17*, 3117. [CrossRef]
231. Zheng, L.; Gao, S.; Lu, M. Development and application of steel for submarine pipelines. *Welded Pipe* **2006**, *29*, 36–39. (In Chinese)
232. Li, S.; Jiang, Z.; Li, Y.; Xu, X.; Bai, X. Development of corrosion-resistant large-wall X70 pipeline steel for marine applications. *World Met. Lett.* **2011**, *11*, 1–6. (In Chinese)
233. Yong, B.; Knauf, G.; Hillenbrand, H.-G. Materials and design of high strength pipelines. In Proceedings of the ISOPE International Ocean and Polar Engineering Conference, Seattle, DC, USA, 27 May–2 June 2000; ISOPE: Mountain View, CA, USA, 2000.
234. Gao, H. *Pipeline Steel and Pipeline Steel Pipes*; Sinopec Press: Beijing, China, 2012; pp. 10–12. (In Chinese)
235. Ma, K.; Sun, G.; Lei, Z.; Li, X.; Liu, X. Discussion on pipe selection for submarine pipelines. *Mater. Dev. Appl.* **2012**, *27*, 106–110. (In Chinese)
236. Li, S.; Jiang, Z.; Li, Y.; Xu, X.; Bai, X. Development of Corrosion-Resistant Thick-Walled X70 Pipeline Steel for Marine Applications. In Proceedings of the 8th (2011) Steel Conference, Perth, Australia, 5–6 October 2011; Metallurgical Industry Press: Beijing, China, 2011.
237. Zhao, D.; Yu, J.; Wang, Z.; Li, X. Research on risk-based safety assessment method for submarine pipelines. *Mar. Technol.* **2010**, *29*, 56–59. (In Chinese)
238. Zhang, B.; Yang, J.; Wang, B. Research on the development and application of X70 and X65 large-wall-thickness offshore pipelines in Liwan, South China Sea. *Welded Pipe* **2013**, *36*, 30–37. (In Chinese)
239. Det Norske Veritas. DNV-ST-F101 Submarine Pipeline Systems. Available online: https://www.dnv.com/energy/standards-guidelines/dnv-st-f101-submarine-pipeline-systems/ (accessed on 8 January 2026).

240. Lang, F.; Huang, F.; Yue, J.; Li, L.; Xu, J.; Liu, J. Hydrogen trapping and hydrogen embrittlement (HE) susceptibility of X70 grade high-strength, acid-resistant, submarine pipeline steel with Mg treatment. *J. Mater. Res. Technol.* **2023**, *24*, 623–638. [CrossRef]
241. Fan, N.; Jiang, J.; Nian, T.; Dong, Y.; Guo, L.; Fu, C.; Tian, Z.; Guo, X. Impact action of submarine slides on pipelines: A review of the state-of-the-art since 2008. *Ocean Eng.* **2023**, *286*, 115532. [CrossRef]
242. Kyew, A.Y. Problems and Solutions in Improving Subsea Pipeline Competency and Quality: A Comprehensive Literature Review. *Collab. Eng. Dly. Book Ser.* **2023**, *1*, 93–96. [CrossRef]
243. Vincent Connor, R. Sensor Analytics for Subsea Pipeline and Cable Inspection: A Review. Master's Thesis, Louisiana State University Louisiana State University, Baton Rouge, LA, USA, 2024.
244. Wang, H.; Chi, Q.; Li, H.; Du, W.; Huang, C.; Zhang, J. Current status of submarine oil and gas transmission pipeline material development and application. *Welded Pipe* **2014**, *37*, 25–29. (In Chinese)
245. Zheng, L.; Fu, J. Current status of the development of high-grade pipeline steel. *Steel* **2006**, *10*, 1–10. (In Chinese)
246. Qu, J.; Liu, W.; Cai, L. China's submarine pipeline laying and the development of pipe-laying vessels. *China Water Transp.* **2013**, *13*, 311–313. (In Chinese)
247. Lu, X.; Zhao, L.; Wang, J.; Wang, W.; Ma, H. Welding process development for submarine pipelines in H_2S acidic service environment. *Weld. Technol.* **2012**, *41*, 27–29. (In Chinese)
248. Wen, Z.; Lu, X.; Ren, S. Experimental study on butt welding process of X56 and F70 grade steel for submarine pipeline. *China Offshore Oil Gas* **2007**, *2*, 131–133+138. (In Chinese)
249. Ma, Q. Production practice of 10MnVNbNiMos steel for offshore pipelines. *Ind. Heat.* **2022**, *51*, 21–23+27. (In Chinese)
250. Chen, B.-Q.; Videiro, P.M.; Soares, C.G. Opportunities and Challenges to Develop Digital Twins for Subsea Pipelines. *J. Mar. Sci. Eng.* **2022**, *10*, 739. [CrossRef]
251. Chen, B.Q.; Soares, C.G.; Videiro, P.M. Review of digital twin of ships and offshore structures. *Marit. Technol. Eng.* **2021**, *1*, 445–451.
252. Chen, B.-Q.; Liu, K.; Yu, T.; Li, R. Enhancing Reliability in Floating Offshore Wind Turbines through Digital Twin Technology: A Comprehensive Review. *Energies* **2024**, *17*, 1964. [CrossRef]
253. Chen, F.; Wei, H.; Tang, J.; Sun, W.; Zhao, X.; Li, Y.; Dong, S.; Zhang, H.; Liu, G. Digital twin based predictive diagnosis approach for submarine suspended pipelines. *Int. J. Press. Vessel. Pip.* **2025**, *214*, 105451. [CrossRef]
254. Chen, K.; Yang, J.; Yin, Q.; Mayank, T.; Gao, B.; Song, Z.; Jia, S.; Liu, S.; Xie, B.; Chen, H. Digital Twin: Hybrid Virtual Simulation and Physical Monitoring Strategy for Long-Term Safe Operation of Floating Production Platform-Hybrid Polyester Mooring Lines-Steel Catenary Risers Multibody System in the South China Sea. In Proceedings of the Offshore Technology Conference, 2024, Houston, TX, USA, 6–9 May 2024; 2024.
255. Repalle, N.; Thethi, R.; Viana, P.; Tellier, E. Application of machine learning for fatigue prediction of flexible risers-digital twin approach. In Proceedings of the SPE Asia Pacific Oil and Gas Conference and Exhibition, Online, 17–19 November 2020; SPE: Richardson, TX, USA, 2020.
256. Bhowmik, S. Digital twin for offshore pipeline corrosion monitoring: A deep learning approach. In Proceedings of the Offshore Technology Conference, 2021, Houston, TX, USA, 16–19 August 2021.
257. Imran, M.H.; Jamaludin, S.; Ayob, A.F.M.; Ali, A.A.I.M.; Ahmad, S.Z.A.S.; Akhbar, M.F.A.; Suhrab, M.I.R.; Zainal, N.; Norzeli, S.M.; Mohamed, S.B. Application of artificial intelligence in marine corrosion prediction and detection. *J. Mar. Sci. Eng.* **2023**, *11*, 256. [CrossRef]
258. Meniconi, S.; Brunone, B.; Tirello, L.; Rubin, A.; Cifrodelli, M.; Capponi, C. Transient Tests for Checking the Trieste Subsea Pipeline: Toward Field Tests. *J. Mar. Sci. Eng.* **2024**, *12*, 374. [CrossRef]
259. Meniconi, S.; Brunone, B.; Tirello, L.; Rubin, A.; Cifrodelli, M.; Capponi, C. Transient Tests for Checking the Trieste Subsea Pipeline: Diving into Fault Detection. *J. Mar. Sci. Eng.* **2024**, *12*, 391. [CrossRef]
260. Shankar, B.; Nadathur, V.; Knezevic, D.; Aalberts, P.; Kolsters, H.; Amuda, D.; Atebe, O.; Pasala, D.; Hoang, T.; Luong, T.; et al. Structural digital twin of FPSO for monitoring the hull and topsides based on inspection data and load measurement. In Proceedings of the Offshore Technology Conference, 2021, Houston, TX, USA, 16–19 August 2021.
261. Ofoedu, A.T.; Ozor, J.E.; Sofoluwe, O.; Jambol, D.D. A SCADA-Integrated Framework for Real-Time Production Monitoring and Operational Intelligence in FPSO Units. *Int. J. Adv. Multidiscip. Res. Stud.* **2024**, *4*, 2453–2472. [CrossRef]
262. Neves, K.L.S.; Dotta, R.; Malta, E.B.; Neto, A.G.; Franzini, G.R.; Bitencourt, L.A.G. A digital twin to predict failure probability of an FPSO hull based on corrosion models. *J. Mar. Sci. Technol.* **2023**, *28*, 862–875. [CrossRef]
263. Det Norske Veritas. DNV-RP-F105 Free Spanning Pipelines. Available online: https://www.dnv.com/energy/standards-guidelines/dnv-rp-f105-free-spanning-pipelines/ (accessed on 8 January 2026).
264. Det Norske Veritas. DNV-RP-F109 On-Bottom Stability Design of Submarine Pipelines. Available online: https://www.dnv.com/energy/standards-guidelines/dnv-rp-f109-on-bottom-stability-design-of-submarine-pipelines/ (accessed on 8 January 2026).

 https://doi.org/10.3390/jmse14020176

265. Det Norske Veritas. DNV-RP-B401 Cathodic Protection Design. Available online: https://www.dnv.com/energy/standards-guidelines/dnv-rp-b401-cathodic-protection-design/ (accessed on 8 January 2026).
266. Det Norske Veritas. DNV-RP-F110 Global Buckling of Submarine Pipelines. Available online: https://www.dnv.com/energy/standards-guidelines/dnv-rp-f110-global-buckling-of-submarine-pipelines/ (accessed on 8 January 2026).
267. Det Norske Veritas. DNV-RP-C203 Fatigue Design of Offshore Steel Structures. Available online: https://www.dnv.com/energy/standards-guidelines/dnv-rp-c203-fatigue-design-of-offshore-steel-structures/ (accessed on 8 January 2026).

https://doi.org/10.3390/jmse14020176

Journal of
Marine Science and Engineering

MDPI

Article

A Comparative CFD Study on the Wave-Making Characteristics and Resistance Performance of Two Representative Naval Vessel Designs

Yutao Tian, Hai Shou, Sixing Guo, Zehan Chen, Zhengxun Zhou, Yuxing Zheng, Kunpeng Shi and Dapeng Zhang *

Ship and Maritime College, Guangdong Ocean University, Zhanjiang 524005, China; 15814748185@163.com (Y.T.); m15578093767@163.com (H.S.); 18027398113@139.com (S.G.); 18319556625@163.com (Z.C.); z3289630388@163.com (Z.Z.); zyx487814934@163.com (Y.Z.); chongqinggaofen@163.com (K.S.)

* Correspondence: zhangdapeng@gdou.edu.cn

Abstract

The wave-making characteristics and resistance performance of a naval vessel are fundamental to its hydrodynamic design, directly impacting its speed, stealth, and energy efficiency. To reveal the performance trade-offs inherent in different design philosophies, a systematic comparative study on the hydrodynamic performance of two representative mainstream naval destroyers from China and the United States was conducted using Computational Fluid Dynamics (CFD). Full-scale three-dimensional models of both vessels were established based on publicly available data. Their flow fields in calm water were numerically simulated at both economical (18 knots) and maximum (30 knots) speeds using an unsteady Reynolds-Averaged Navier–Stokes (RANS) solver, the Volume of Fluid (VOF) method for free-surface capturing, and the SST k-ω turbulence model. The performance differences were meticulously compared through qualitative observation of wave patterns, quantitative measurements (such as the transverse width of the wave-making region), and analysis of resistance data. Numerical results indicated that the wave-making generated by the vessel of the United States was more pronounced during steady navigation. To validate the reliability of the CFD results, supplementary towing tank tests were performed using a small-scale model (1.1 m in length) of the vessel from China. The test speed (1.5 m/s) was scaled to correspond to the full-scale ship speed through dimensional analysis. The experimental data showed good agreement with the simulation results, jointly confirming the aforementioned performance trade-off. This study clearly demonstrates that, at the economic speed, the design of the mainstream vessel from China tends to prioritize superior wave stealth performance at the expense of higher resistance, whereas the mainstream vessel from the U.S. exhibits the characteristics of lower resistance coupled with more significant wave-making features. These findings provide an important theoretical basis and data support for the future multi-objective optimization design of surface vessels concerning stealth, speed, and comprehensive energy efficiency.

Keywords: computational fluid dynamics (CFD); naval vessels; hydrodynamic performance

1. Introduction

The design of modern surface vessels represents a classic complex multi-objective optimization problem, requiring a delicate balance among sailing performance, combat effectiveness, and survivability [1]. Hydrodynamic performance serves as the foundation, directly determining a vessel's speed, seakeeping, and maneuverability, while profoundly influencing the physical field signatures generated during its operation, such as its wave wake [2]. Wave-making is not only a primary component of ship resistance but also forms distinctive spatial wave patterns that can serve as critical clues for the remote detection of vessels via optical or radar sensing methods [3,4]. Therefore, the accurate prediction and effective control of a vessel's wave-making characteristics and resistance performance are of paramount importance for enhancing its stealth, survivability, and overall operational effectiveness.

Numerical simulation is a widely adopted approach in ship hydrodynamic analysis. Lutfi et al. focused on optimizing energy efficiency in transport by investigating the influence of internal design parameters on monohull resistance. Utilizing Maxsurf Resistance software and the Savitsky method, they systematically analyzed the contributions of frontal area, appendage area, and appendage coefficients to the total resistance of a hull equipped with an additional rudder. Their quantitative evaluation aimed to identify the most critical factors affecting hull resistance, providing a theoretical foundation for improving fuel efficiency [5]. Similarly, Putranto et al. developed a solar-powered catamaran fishing vessel integrated with batteries, photovoltaic panels, and electric propulsion systems designed for a service speed of 5 knots. Their study employed numerical methods to assess stability according to International Maritime Organization (IMO) regulations, which specify minimum righting lever (GZ) values at defined angles. Comparative stability analyses were conducted under both full-load and lightship conditions [6].

With the continuous advancement of high-performance computing technology, Computational Fluid Dynamics (CFD) has evolved into a core tool for research on ship hydrodynamic performance. By numerically solving the governing equations for viscous flow, CFD enables high-fidelity simulation of the flow field around a vessel, allowing for precise prediction of its hydrodynamic characteristics [7]. Hakim et al. employed an analytical Reynolds-Averaged Navier–Stokes (RANS) computational fluid dynamics (CFD) approach to evaluate the effects of recently cleaned and painted hull surface roughness on ship resistance [8]. In particular, Oyuela S.'s team has demonstrated the excellent capability of CFD in predicting ship wave-making, resistance, and viscous flow fields by combining the VOF free-surface capturing technique with advanced turbulence models such as SST k-ω [9]. Researchers have utilized CFD technology to conduct extensive studies, including Cheng X.'s research on fully parametric hull–propeller–rudder cooperative optimization [10], Chen et al.'s advanced numerical simulation of ship hydroelastic responses in short-crested irregular waves [11], and Song S. et al.'s analysis of the impact of hull roughness and operational data on resistance and energy consumption [12]. These studies reflect the cutting-edge applications of CFD technology in the field of naval architecture and ocean engineering.

Despite the increasing maturity of CFD technology, traditional model testing remains an indispensable approach for validating numerical methods and acquiring reliable data [13]. In recent years, significant progress has been made in experimental platform design and model fabrication techniques within related research. For instance, Shi Kunpeng et al. developed a professional ship model resistance testing platform [14]; Shou Hai et al. showcased modern high-precision model-making technology through a 093-type nuclear submarine model [15]; Zhang Shutian et al. [16,17], Zhao Guanyixuan et al. [18], and Xie Yifan et al. [19] have accumulated extensive experience in the construction and application of models for aircraft carriers, large commercial vessels, and comprehensive supply ships. Furthermore, research on biomimetic underwater robots by Li Pengji et al. [20] offers new insights for the hydrodynamic design of marine equipment. Regarding hydrodynamic characteristic studies, Zhang et al. conducted fine simulations of the wake characteristics of Unmanned Underwater Vehicles (UUVs) operating near the free surface [21].

https://doi.org/10.3390/jmse14020212

However, a review of existing research reveals that most studies still focus on performance analysis of standardized ship models, parametric optimization of single hull forms, or exploration of mechanisms behind specific hydrodynamic phenomena. Systematic full-scale CFD comparative studies on main combat vessels that are already in large-scale service and represent the advanced design philosophies of different nations, coupled with validation via reliable model tests, remain a research gap. Such research cannot only quantitatively reveal performance differences between designs but also deeply interpret the underlying design philosophies and performance trade-offs, holding unique academic value and engineering significance for advancing ship hydrodynamic design.

Based on this, this study selects a representative mainstream destroyer of the Chinese Navy and a representative destroyer of the same class from the U.S. Navy as comparative subjects. The aim is to systematically compare and analyze the wave development process, steady-state wave pattern characteristics, and resistance variation patterns of the two vessel types in calm water through full-scale CFD numerical simulation. The study introduces quantitative indicators such as the transverse width of the wave-making region to objectively assess their wave stealth performance. Combined with resistance data, it delves into the trade-off relationship in hull form design between "low resistance" and "low detectability." Of particular importance, to verify the reliability of the numerical method, this study additionally designed and conducted towing tests using a small-scale mainstream vessel model (1.1 m in length) from China. The test speed (1.5 m/s) was scaled to correspond to the full-scale ship speed via dimensional analysis. Section 4 will report in detail the resistance data measured from these model tests, the full-scale resistance prediction based on dimensional analysis, and the comparative verification analysis with full-scale CFD simulation results, thereby forming a complete research loop where "numerical simulation—model testing—theoretical analysis" mutually corroborate.

This study performs a comparative CFD analysis based on high-fidelity, full-scale 3D [22] models of mainstream Chinese and U.S. destroyers, reconstructed from public technical data. Unlike previous research often limited to a single hull type, this work offers a direct cross-platform comparison. To ensure the reliability of the numerical simulations, a small-scale towing tank experiment was designed and conducted for model calibration and validation, thereby enhancing the credibility of the full-scale hydrodynamic results. The outcomes of this work are expected to provide a solid theoretical basis, valuable design references, and reliable experimental data support for the research and development of future high-performance, low-signature surface vessels.

2. Theoretical Derivation

This study employs Computational Fluid Dynamics (CFD) to investigate the hydrodynamic performance of two representative naval hull forms. Numerical simulations are based on the three-dimensional unsteady Reynolds-Averaged Navier–Stokes (RANS) equations to solve the complex viscous flow around the hull. The free surface is modeled using the Volume of Fluid (VOF) method, which tracks the evolution of the air–water interface by solving the transport equations for phase volume fractions. The VOF method effectively describes nonlinear wave-making phenomena and complex deformations of the free surface, making it suitable for simulating the intense waves and localized breaking common in high-speed vessel navigation.

2.1. Calculation of y+

In Computational Fluid Dynamics (CFD), particularly in turbulence simulation and numerical modeling of near-wall regions, y^+ is a very important non-dimensional parameter. It reflects the relationship between mesh generation and the boundary layer, especially

https://doi.org/10.3390/jmse14020212

regarding mesh refinement near the wall. y^+ is a dimensionless distance used to describe the distance from a mesh point to the wall in fluid flow, specifically in the near-wall region. It is defined as

$$y^+ = \frac{u_r \Delta y_1}{v} \tag{1}$$

where Δy_1 is the distance from the first layer grid node to the wall. u_r is the wall shear velocity (unit: m/s). v is the kinematic viscosity of the fluid.

The wall shear velocity can be calculated using the following formula:

$$u_r = \sqrt{\frac{\tau_w}{\rho}} \tag{2}$$

where τ_w is the magnitude of the wall shear stress. ρ is the density of the fluid. The calculation of wall shear stress is related to the skin friction coefficient, which is defined as

$$C_f = \frac{2\tau_w}{\rho u^2} \tag{3}$$

Using the skin friction coefficient and the fundamental formula for wall shear stress, the wall shear stress can be derived:

$$\tau_w = \left(\frac{\partial u}{\partial y}\right)_{y=0} = \frac{1}{2} C_f \rho u^2 \tag{4}$$

In the above equation, τ_w is the magnitude of the wall shear stress, u is the dynamic viscosity of the fluid, and y is the distance perpendicular to the wall.

In turbulence simulation, the y^+ value is used to determine the applicability of turbulence models for different regions in the flow field and the suitability of the mesh, especially the refinement level near the wall. However, in practical finite element analysis software for fluid dynamics simulation, this is achieved through wall functions or near-wall resolution.

Enhanced wall treatment combines a two-layer model with an enhanced wall function, ensuring minimal error for both coarse wall meshes ($y^+ > 15$, fully turbulent region) and fine meshes ($y^+ \approx 1$, viscous sublayer).

The two-layer model divides the boundary layer into a viscous sublayer and a fully turbulent layer, with the demarcation line distinguished by the Reynolds number:

Where Re is the dimensionless Reynolds number:

$$Re_y = \frac{\rho y \sqrt{k}}{\mu} \tag{5}$$

$$Re = \frac{u L_{pp}}{v} \tag{6}$$

where y is the distance from the cell center to the wall. u is the ship's speed, and L_{pp} is the length between perpendiculars. Crucially, k denotes the turbulent kinetic energy, and the term $\sqrt{k}$ serves as the characteristic velocity scale of turbulence. The value of Re_y indicates whether the near-wall cell lies within the viscous sublayer or the turbulent region, guiding the application of wall functions. v is the kinematic viscosity of water. If $Re_y < 200$ the fluid is in the viscous sublayer region, and the Wolfshtein equation is used for solution; if $Re_y > 200$, the fluid is in the fully turbulent region, and the k-ω model or Reynolds Stress Model is used for solution. To accurately resolve the viscous sublayer, the Wolfshtein one-equation model is employed as the near-wall treatment. In this approach, the turbulent kinetic energy transport equation is solved throughout the entire computational domain,

https://doi.org/10.3390/jmse14020212

while the turbulent length scale in the near-wall region is prescribed algebraically rather than derived from a dedicated dissipation-rate equation.

Specifically, a seamless coupling between the Wolfshtein model and the SST k–ω model is achieved. When the grid resolution is sufficient to resolve the near-wall region, the Wolfshtein model provides a prescribed near-wall dissipation scale in the low-Reynolds-number limit, effectively replacing the solution of the ω transport equation. This strategy avoids numerical stiffness and instability associated with the ω equation near the wall, while improving the accuracy of near-wall turbulence representation.

2.2. Derivation of Ship Resistance Coefficient

All meaningful performance comparisons must be based on dimensionless coefficients to eliminate the effects of differences in scale, speed, fluid density, etc. This is the mathematical foundation for fair comparison. To deeply explore the physical origins of the performance differences between the two vessel types, the following mathematical decomposition is performed based on ship resistance theory. This paper directly adopts the most straightforward performance indicator—the total resistance coefficient.

The CFD solver employed in this study is based on separately calculating the pressure resistance and the shear resistance, which are then summed to obtain the total resistance result; therefore, the decomposition of total resistance is introduced here.

According to ship resistance theory, the total resistance R_T can be expressed as

$$R_T = R_F + R_R \tag{7}$$

where R_T is the frictional resistance, which is primarily determined by the wetted surface area of the hull and the Reynolds number of the ship model. R_R is the residual resistance, which, in the calm water simulations of this study, mainly consisted of the pressure resistance. Residual resistance includes pressure-related components, specifically form drag—resulting from flow separation and wake formation—and wave-making resistance. In this study, form drag plays a significant role, as evidenced by the distinct wake structures observed within the flow field.

Frictional resistance, R_F, constitutes the primary component of viscous resistance. It is derived from the integration of the shear stress (tangential force) generated by fluid viscosity over the wetted surface area of the hull. The total frictional resistance is obtained by integrating over the entire wetted surface area, A, of the hull:

$$R_F = \iint_A \tau_w dA = \frac{1}{2}\rho u^2 A C_F \tag{8}$$

The application of this formula is based on the principle of equivalent simplification. Although the local flow velocity on the hull surface varies spatially, in engineering practice, we define u as the overall speed of the vessel and incorporate the complex variations in the local velocity gradient into the friction coefficient C_F. For the precise calculation of the frictional resistance coefficient, C_F, the ITTC-1957 resistance correlation formula was employed:

$$C_F = \frac{0.075}{(log_{10}\text{Re} - 2)^2} \tag{9}$$

In the CFD simulations of this study, the total resistance on the hull at specified speeds was directly calculated by solving the unsteady, incompressible Reynolds-Averaged Navier–Stokes (RANS) equations, combined with the Volume of Fluid (VOF) method for accurate free-surface capturing. The CFD methodology inherently accounts for both viscous effects and wave-making effects simultaneously, eliminating the need for artificial decomposition

https://doi.org/10.3390/jmse14020212

of the resistance components. Consequently, it provides high-fidelity resistance predictions that are in close agreement with physical reality. Through post-processing, the various resistance components can be further isolated and quantified. This enables a deeper analysis of the performance differences and their physical origins between mainstream vessel of the U.S. and the mainstream vessel of China at different speeds (economic and maximum).

3. Model Preprocessing Design

3.1. Ship Model Design

Based on various publicly available materials, including but not limited to images and reports, the basic parameters required for modeling the mainstream surface vessel of China and its U.S. counterpart are listed in Table 1.

Table 1. Basic parameters of the mainstream surface vessel of China and the U.S.

Parameter	China	U.S.
Overall Length (m)	155	154
Draft (m)	6	6
Length Between Perpendiculars (m)	154.6	151.7
Molded Breadth (m)	17.6	19.1

Based on the fundamental parameters outlined above and actual imagery from public sources [23], the mainstream surface vessel and its U.S. counterpart of the same class were modeled in accordance with core principles of engineering drawing—specifically, the consistent alignment of length, height, and width across orthogonal views. The resulting models are presented in Figure 1a,b. The renderings of the two ship models shown in Figure 1a,b were generated directly within the 3D modeling software using its built-in rendering mode. No post-processing image manipulation techniques (such as edge detection or contour extraction) were applied to these images. The visualizations presented are unaltered screenshots of the rendered 3D models, accurately reflecting the geometric features as constructed. First, the general framework of the model was determined based on the basic parameters to ensure that the geometric shape of the vessel basically met the parameter requirements. Second, based on various public materials, three-view drawings were compiled, and the main curvature of the hull lines for various parts of the model was kept as consistent as possible with the actual situation. Furthermore, considerations were made based on the vessel's actual functional design. For example, the stern area of the mainstream surface vessel of China is designed for helicopter takeoff and landing. Compared to the large curvature design at the bow intended to prevent wave-breaking and green water, this platform should be very smooth and flat to meet the requirements for aircraft operations. Another example is the mainstream vessel of the U.S., where the superstructure houses numerous heavy weapons for combat missions. Attention was paid to the coupling between the horizontal weapon platforms in the superstructure and the forecastle deck, which involves certain curvature changes.

Based on the numerical data provided in Table 1 and the two operating conditions at different speeds described in the text, the y^+ values for the model were calculated using the aforementioned Equations (1)–(6). The results are summarized in Table 2.

https://doi.org/10.3390/jmse14020212

Table 2. Summary of the calculated results.

Vessel	Speed (Knot)	Re	y^+
China	18	1.364×10^9	2393
China	30	2.272×10^9	3868
U.S.	18	1.338×10^9	2397
U.S.	30	2.229×10^9	3873

(a) (b)

Figure 1. Schematic diagrams of the two ship models are provided: (**a**) mainstream vessel model of the U.S.; (**b**) mainstream vessel model of China.

In ship CFD, particularly in full-scale simulations, the first-layer grid height is often insufficient to resolve the viscous sublayer ($y^+ \approx 1$) due to constraints on computational resources, resulting in y^+ values typically on the order of hundreds or even thousands. This falls within the applicable regime of "wall functions". In most commercial CFD solvers, when $y^+ > 30$, wall functions based on the logarithmic law are automatically activated to model the inner boundary layer (including the viscous sublayer and the buffer layer), without requiring direct resolution. For external flow simulations involving large-scale vessels, aerospace vehicles, and other applications with extremely high Reynolds numbers (Re > 10^8), the adoption of wall functions is considered a routine and acceptable practice to balance accuracy and computational cost.

3.2. Computational Domain and Mesh Generation

After completing the ship model construction in 3D modeling software and importing it into the simulation software, preprocessing is required to ensure the geometric structure of the model conforms to finite element computational theory. Simultaneously, minor details on the ship model's bow, such as anchor windlasses, deck railings, etc., which do not interfere with CFD calculations of ship-generated waves or resistance monitoring, can be removed. Removing these components can avoid overall penetrations of the ship model by small parts and effectively reduce the number of mesh elements and computational complexity during meshing, making the ship model easier to converge during simulation.

After clearing penetrations and minor details of the superstructure, the computational domain and mesh for the ship model need to be defined. To ensure calculation accuracy, the computational domain was planned as follows: Taking the midpoint of the transom stern as the origin, a three-dimensional Cartesian coordinate system was established, with the X-axis pointing towards the bow, the Y-axis pointing towards the port side, and the Z-axis perpendicular to the XOY plane pointing vertically upward. The computational domain consists of a region extending twice the ship's overall length from the origin towards the

https://doi.org/10.3390/jmse14020212

bow and stern, twice the ship's overall length towards port and starboard, once the ship's overall length vertically upward, and twice the ship's overall length downward.

Mesh generation for the computational domain primarily employed the cut-cell method and prism-layer mesh functions. The unique geometric shape of the hull significantly influences the entire flow field. For special parts of the ship model that generate noteworthy phenomena in the flow field, such as the bulbous bow at the front and the propeller at the stern, finer mesh scales were set in these areas and their surrounding flow fields to capture the surface changes or other details caused by disturbances during motion. However, applying the same fine mesh parameters to all objects in the entire computational domain would drastically increase computation time and hardware requirements. To balance computational efficiency while reflecting the influence of these special parts on the flow field and ensuring result accuracy, more refined mesh treatment was applied to these special parts and their nearby flow fields.

According to fundamental fluid dynamics knowledge, the flow field around the hull can be divided into laminar and turbulent boundary layers with increasing distance from the hull. Further subdivision along the Z-axis direction was performed for these two layers. The purpose of this operation is to better highlight the influence of hull motion on the free surface, more clearly demonstrate the velocity gradient change from the hull bottom to the free surface and improve the calculation accuracy of the hull frictional resistance. Simultaneously, during the computation process, the real-time changes in the velocity field and resistance field can be observed to judge whether the mesh parameters meet the requirements for result accuracy. When the computation has theoretically reached a steady state but exhibits significant residual instability or chaotic wave patterns, the computation can be stopped to recheck mesh parameter settings, enabling timely intervention and improving computational efficiency.

In Figure 2, the specific details of mesh refinement are presented for the bulbous bow, stern propeller, hull turbulent boundary layer, free surface, and the Kelvin wake system generated by the ship's motion.

Figure 2. Schematic of the computational mesh for the ship model.

An implicit unsteady solver was employed for the simulations conducted in this study. To efficiently obtain the flow field and resistance under steady navigation, the initial conditions were prescribed as follows: The fluids (water and air) within the entire computational domain were assigned an initial velocity identical to the target cruising speed (18 knots or 30 knots), corresponding to the vessel's motion at that speed. This approach eliminated the need for a prolonged physical acceleration process, allowing the simulation to rapidly reach a steady-state sailing condition. The inlet, top, and side boundaries of the computational domain were set as velocity inlets with values corresponding to the target speed, while the outlet boundary was defined as a pressure outlet. The free surface was captured using the VOF method, which was sufficient for the full development and attainment of a quasi-steady state of the free surface around the hull.

3.3. Mesh Independence Verification

To ensure the accuracy and reliability of the numerical results, a grid independence study was conducted following the Verification and Validation (V&V) framework outlined in the ITTC procedures [24].

A representative case of the mainstream vessel of the U.S. at an economic speed of 18 knots ($Fr \approx 0.24$) was selected. Three sets of systematically refined grids (coarse, medium, and fine) were generated while maintaining identical boundary layer settings (first-layer thickness targeting $y^+ \approx 1$) and local refinement zones. All simulations were performed on a computational platform equipped with a 12th Gen Intel Core i7-12700F processor. The grid parameters and the calculated total resistance values are summarized in Table 3.

Table 3. Grid independence study: configurations and results.

Grid Scheme	Total Cells ($\times 10^4$)	Total Resistance (N)	Meshing Time (s)	Relative Charge
Coarse (Gc)	$N_c = 696.7$	$R_{Tc} = 637,926$	171.5	-
Medium (Gm)	$N_m = 923.6$	$R_{Tm} = 598,306$	310.2	−6.21%
Fine (Gf)	$N_f = 1522.8$	$R_{Tf} = 583,678$	335.6	−2.44%

Based on the analysis method of grid convergence, calculate the convergence factor R_G and the observed order of accuracy p:

$$R_G = \frac{\varepsilon_{21}}{\varepsilon_{32}} = \frac{R_{Tm} - R_{Tf}}{R_{Tc} - R_{Tm}} \approx 0.369 \tag{10}$$

where R_{Tc}, R_{Tm}, and R_{Tf} represent the total resistance values obtained from the coarse, medium, and fine grids, respectively. The condition $0 < R_G < 1$ indicates monotonic convergence. The average refinement ratio r, a key parameter in grid convergence analysis, is defined as the ratio of the characteristic cell sizes between successive grid schemes. In this study, as the three grid sets were generated by non-uniformly scaling the global base size while maintaining consistent boundary layer and local refinement settings, a strictly constant theoretical size ratio was not achieved. Therefore, the relative grid size for each set was estimated from the inverse cube root of the total cell count (N). The refinement ratios between adjacent grids were calculated as:

$$r_{21} = \left(\frac{N_f}{N_m}\right)^{\frac{1}{3}} \quad r_{32} = \left(\frac{N_m}{N_c}\right)^{\frac{1}{3}} \tag{11}$$

The average refinement ratio, used for the subsequent Richardson extrapolation analysis, was then taken as their geometric mean:

$$r = \sqrt{r_{21} r_{32}} \approx 1.14 \tag{12}$$

https://doi.org/10.3390/jmse14020212

Subsequently, the observed order of accuracy was computed:

$$p = \frac{\ln\left(\frac{1}{R_G}\right)}{\ln r} \approx 7.64 \tag{13}$$

Subsequently, the grid discretization error δ_{RE} and the error correction factor C_G were determined based on ITTC-recommended practices:

$$\delta_{RE} = \frac{\varepsilon_{21}}{r^p - 1} \approx 8555\text{N} \tag{14}$$

$$C_G = \frac{r^p - 1}{r^2 - 1} \approx 5.74 \tag{15}$$

In the equation for C_G, r^2 is the theoretical order of accuracy, as suggested by the ITTC. Substituting the values yields $C_G = 5.74$. The numerical uncertainty associated with the grid U_G was then estimated:

$$U_G = [|C_G| + |1 - C_G|]\delta_{RE} \approx 89650\text{N} \tag{16}$$

Although the formal numerical uncertainty appears relatively large—primarily attributable to the high observed order p resulting from non-uniform refinement factors—the relative difference in total resistance between the medium and fine grids is only 2.44%. This value is below the common engineering threshold of 3%, suggesting that further grid refinement yields diminishing returns on result improvement.

While the fine grid (15.23 million cells) provides a solution closer to the grid-independent limit, it requires approximately 64.9% more cells than the medium grid, leading to a substantial increase in computational cost. Considering that the key variable (total resistance) changes by less than 3% between the medium and fine grids, a balance between computational accuracy and efficiency was sought. Consequently, the medium grid scheme (approximately 9.24 million cells) was selected for all subsequent simulations in this study. This choice ensures the reliability of results for comparative engineering analysis while maintaining computational feasibility for the multi-case study.

4. Results and Analysis

Two operating conditions were designed for each of the two vessels at two different speeds. The first condition is the economic speed of 18 knots, which is typically adopted for routine patrols. This speed is commonly used by vessels for maritime patrols in specific areas, as it aligns with the daily operational efficiency of the ship's power plant and ensures a good balance between fuel consumption and travel efficiency. The second condition is the vessel's maximum speed. This condition was primarily designed to reflect the state of the free surface and the resistance experienced when the vessel operates at full power. Both conditions correspond to different operational scenarios of the vessel, demonstrating its motion state under specific environments. For all free-surface deformation contour plots mentioned below, the images were generated directly by the solver based on the computational results, without the application of any third-party tools for complex edge detection or image segmentation. A globally consistent color bar range was applied to all free-surface contour plots for both the economic and maximum speeds. This range ([−1.85 m, +4.55 m] for the economic speed and [−2.39 m, +8.15 m] for the maximum speed) was determined based on the maximum positive and negative fluctuations of the free-surface elevation across all simulation data, ensuring clear visualization of wave height variations under all operating conditions.

https://doi.org/10.3390/jmse14020212

4.1. Free Surface Analysis at Economic Cruising Speed

This simulation condition was designed to simulate both ship models advancing at the economical cruising speed of 18 knots (9.252 m/s) in calm water, with a draft of 6 m for both vessels, representing the scenario of actual vessels patrolling a specific maritime area under normal circumstances. The motion trajectory of the ships from initiation to steady sailing can be obtained by analyzing representative wave pattern images generated during the operation of the different ship models.

Two operating conditions were designed for each of the two vessels at two different speeds. The first condition is the economic speed of 18 knots, which is typically adopted for routine patrols. This speed is commonly used by vessels for maritime patrols in specific areas, as it aligns with the daily operational efficiency of the ship's power plant and ensures a good balance between fuel consumption and travel efficiency. The second condition is the vessel's maximum speed. This condition was primarily designed to reflect the state of the free surface and the resistance experienced when the vessel operates at full power. Both conditions correspond to different operational scenarios of the vessel, demonstrating its motion state under specific environments.

The wave patterns generated after both ship models had advanced for 10 s and 20 s are shown in Figures 3 and 4. At this point, both vessels had sailed a distance exceeding one times their overall length. The significant seawater disturbance phenomenon induced at the bulbous bow of the mainstream vessel model of China was still ongoing. Areas with larger wave amplitude fluctuations were concentrated at the bow and stern regions, and distinct traces of the Kelvin wake system had already formed. Considerable vertical fluctuations were observed at both the bow and stern regions of the mainstream vessel of the U.S., while a fully developed Kelvin wave pattern was stretched out amidships. Pronounced wake undulations generated at the stern were also evident.

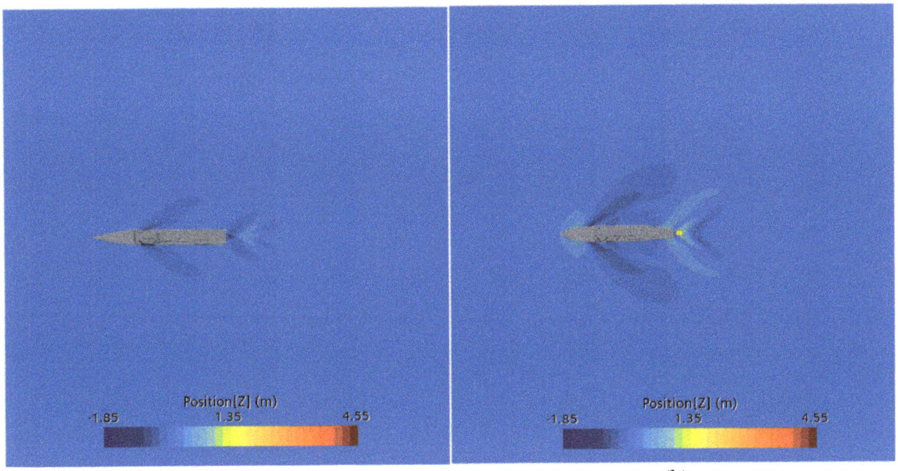

(a) (b)

Figure 3. Wave patterns and free surface deformation generated by the two ships advancing at economic speed (t = 10 s): (**a**) mainstream vessel model of China; (**b**) mainstream vessel model of the U.S.

As shown in Figure 5, the wave patterns generated after the two ship models had advanced for 50 s are presented. At this stage, the sailing distance had reached twice the overall length of both ship models. The disturbance generated by the bulbous bow of the mainstream vessel model of the U.S. for drag reduction was observed to be relatively

https://doi.org/10.3390/jmse14020212

mild, while the seawater disturbance ahead of the bulbous bow of the mainstream surface vessel model of China was also seen to stabilize, with the high-water-level disturbance concentrated at the stem area. Furthermore, wave disturbances amidships were noted to stabilize, and a marked symmetry in the wake of the Kelvin wake system was evident. Compared to the wave patterns at 30 s, the wave-making area generated on the port and starboard sides of the mainstream vessel model of the U.S. was slightly reduced, with more pronounced disturbances concentrated towards the stern. Similar to the mainstream vessel model of China, a clear symmetry in the generated waves was observed.

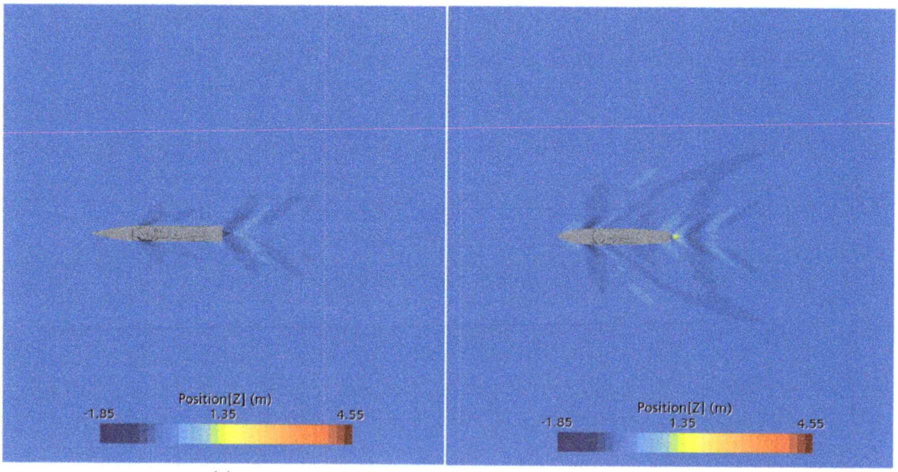

(a) (b)

Figure 4. Wave patterns and free surface deformation generated by the two ships advancing at economic speed (t = 20 s): (**a**) mainstream vessel model of China; (**b**) mainstream vessel model of the U.S.

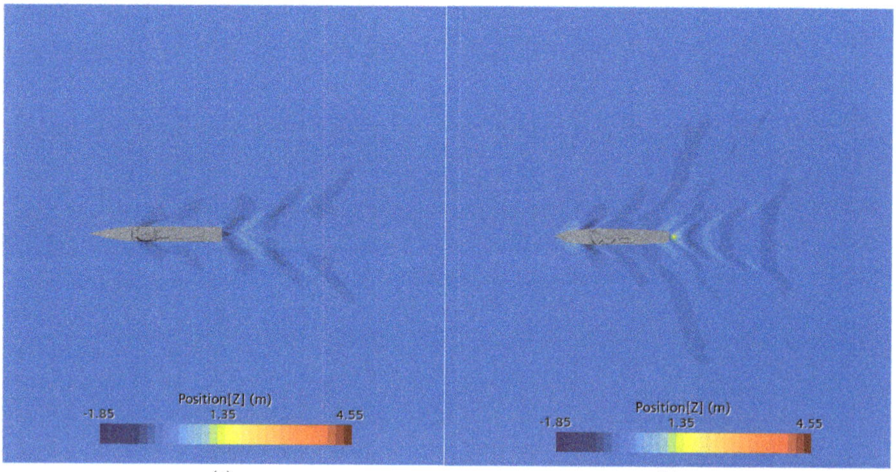

(a) (b)

Figure 5. Wave patterns and free surface deformation generated by the two ships advancing at economic speed (t = 50 s): (**a**) mainstream vessel model of China; (**b**) mainstream vessel model of the U.S.

 https://doi.org/10.3390/jmse14020212

The wave patterns generated after both ship models had advanced for 100 s are shown in Figure 6. The sailing distance at this point had reached six times the overall length of both ship models. The motion of both vessels had stabilized, with the symmetry of motion being particularly pronounced. Compared to the wave patterns at 50 s, a smaller wave-making region was produced by the motion of the mainstream vessel model of China relative to that generated by the mainstream vessel model of the U.S. The Kelvin wave patterns of both were observed to originate from the bow and gradually develop towards the stern, but more intense disturbances in the wave-making were generated at the stern by the mainstream vessel model of the U.S.

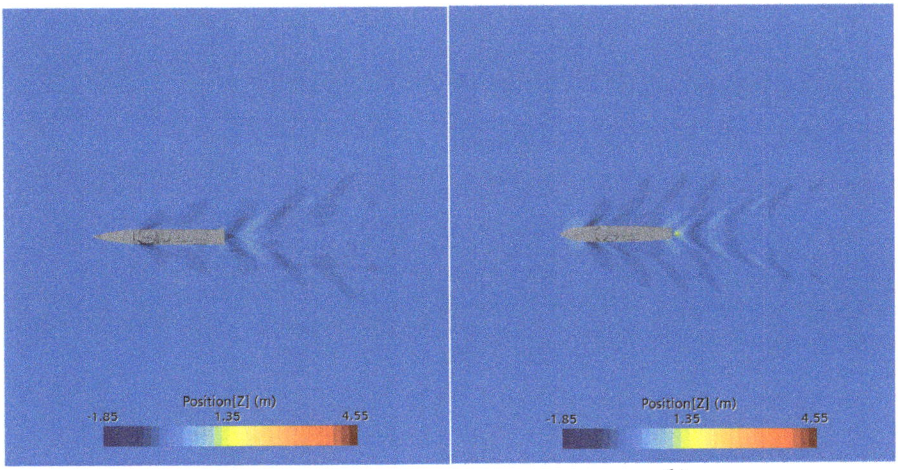

(a) (b)

Figure 6. Wave patterns and free surface deformation generated by the two ships advancing at economic speed (t = 100 s): (**a**) mainstream vessel model of China; (**b**) mainstream vessel model of the U.S.

After prolonged motion, a steady-state representation of the free surface was gradually reached by both ship models. The free surface at 180 s and 360 s for the two vessels is shown in Figure 7. Theoretically, a steady state is considered to have been entered for the motion of both vessels at this point. It can be observed that the wave-making range and wave density generated by the motion of the mainstream vessel of China show almost no change compared to those in Figure 6. This indicates that the mainstream vessel of China reaches a steady state more easily at economic speed compared to the mainstream vessel of the U.S., forming a stable wave influence range. A longer time is required for the mainstream vessel of the U.S. to stabilize the wave influence range generated by its motion. However, concurrently, the free surface area influenced on the port and starboard sides of the mainstream vessel of the U.S. at 360 s shows some convergence compared to the area at 180 s.

4.2. Analysis of Free Surface and Resistance at Economic Cruising Speed

In Figure 7, the free surface disturbance ranges for the two ship models after 360 s of motion at economic speed are outlined. The measured data indicate that the maximum transverse disturbance distance for the mainstream vessel of China corresponds to 10.5 times its molded breadth, while that for the mainstream vessel of the U.S. corresponds to 11.6 times its molded breadth. A very pronounced difference in the transverse wave-making disturbance is not observed between the two. The hull resistance variation

 https://doi.org/10.3390/jmse14020212

experienced by the two ship models at economic speed is clearly shown in the plot presented in Figure 8. From the initiation of motion until 360 s, the total resistance experienced by the hull of the mainstream vessel of the U.S. is approximately 610 kN, whereas that for the mainstream vessel of China is approximately 850 kN, resulting in a difference of about 240 kN between the total hull resistances of the two vessels. We define the transverse disturbance width as the full width at half maximum of the transverse displacement profile measured perpendicular to the propagation direction. Physically, it characterizes the lateral spatial extent of the perturbation introduced into the wavefield.

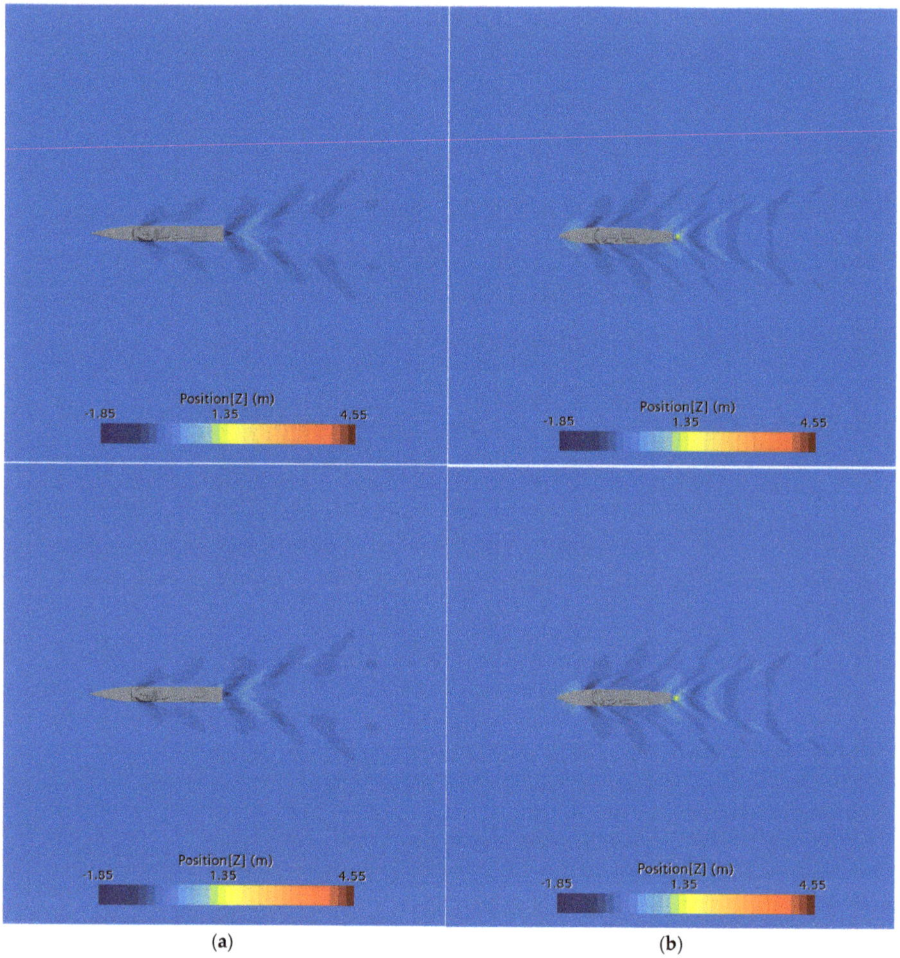

Figure 7. Wave patterns and free surface deformation generated by the two ships advancing at economic speed (t = 180 s and t = 360 s): (**a**) mainstream vessel model of China; (**b**) mainstream vessel model of the U.S.

Minor oscillations in the resistance values are commonly observed in transient CFD simulations. These fluctuations are primarily attributed to the convergence residuals of numerical iterations, temporal discretization in the time-stepping scheme, and the inherent high-frequency unsteadiness of turbulence itself. Such variations are regarded as numerical artifacts rather than physical phenomena. Moreover, the amplitude of these oscillations is significantly smaller than the steady-state mean resistance value, and thus does not

https://doi.org/10.3390/jmse14020212

affect the comparative analysis of the steady-state resistance performance between the two vessels.

4.3. Free Surface Analysis at Maximum Cruising Speed

This simulation condition was designed for the two ship models sailing at the maximum cruising speed of 30 knots (15.433 m/s) in calm water. Identical to the condition in Section 3.1, the draft of both vessels was maintained at 6 m, simulating a scenario where vessels are performing special missions at sea at full speed. The motion trajectory of the ships from initiation to normal sailing at maximum speed can be obtained by analyzing representative wave pattern images generated during the operation of the different ship models.

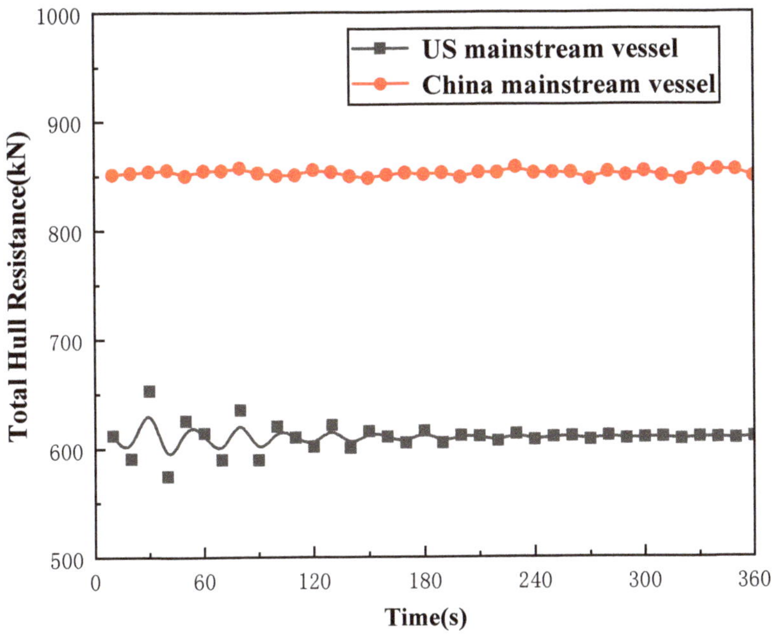

Figure 8. Variation in simulated total hull resistance with time for both vessels at economic speed.

The wave patterns generated after the two ship models had advanced for 10 s are shown in Figure 9. The sailing distance at this point is approximately equal to the overall length of both ship models. A slight but noticeable seawater disturbance phenomenon is observed ahead of the bulbous bow of the Arleigh Burke-class vessel model. The wave-making generated at the bulbous bow of the mainstream vessel of China is considerably calmer compared to that of the mainstream vessel of the U.S. However, based on the free-surface elevation difference, although a wider range of wave disturbance is generated by the mainstream vessel of the U.S., the free-surface elevation difference of the mainstream vessel of China is greater than that of the mainstream vessel of the U.S. This indicates that at this stage, changes in the free surface for the mainstream vessel of China are concentrated near the hull.

The wave patterns generated after both ship models had advanced for 20 s are shown in Figure 10. The sailing distance for both vessels at this point is approximately twice their overall length. Compared to the state at 10 s, the significant seawater disturbance phenomenon induced at the bulbous bow of the mainstream vessel model of China was still ongoing. Areas with larger wave amplitude fluctuations were concentrated at the bow and

https://doi.org/10.3390/jmse14020212

stern, and distinct traces of the Kelvin wake system were already present. Considerable vertical fluctuations were observed at both the bow and stern regions of the mainstream vessel model of the U.S., while a fully developed Kelvin wave pattern was stretched out amidships. Pronounced wake undulations generated at the stern were also evident.

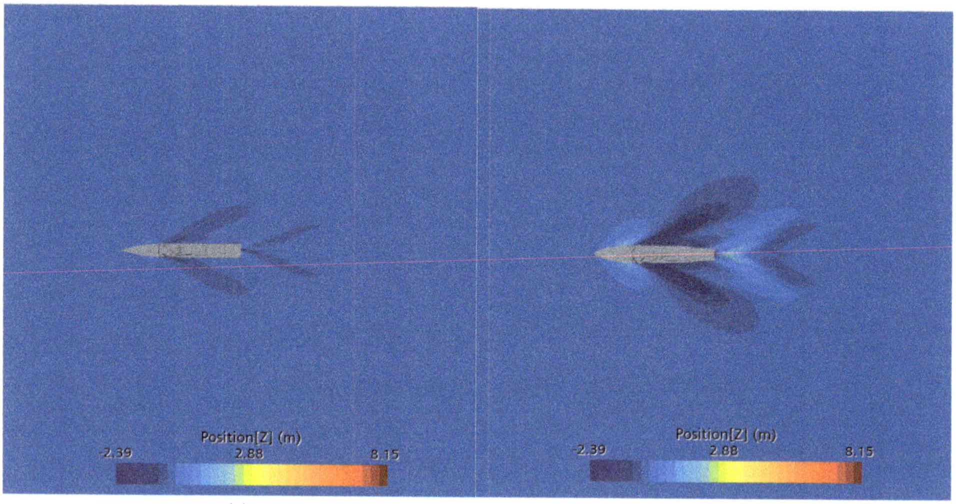

(a) (b)

Figure 9. Wave patterns and free surface deformation generated by the two ships advancing at maximum speed (t = 10 s): (**a**) mainstream vessel model of China; (**b**) mainstream vessel model of the U.S.

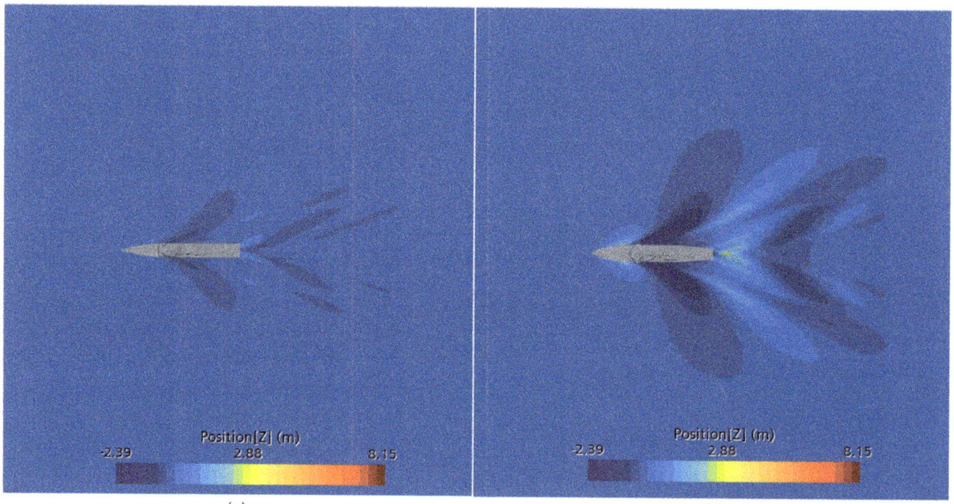

(a) (b)

Figure 10. Wave patterns and free surface deformation generated by the two ships advancing at maximum speed (t = 20 s): (**a**) mainstream vessel model of China; (**b**) mainstream vessel model of the U.S.

The free-surface condition after the two ship models had advanced for 50 s is shown in Figure 11. The sailing distance at this point had reached five times the overall length of both ship models. The disturbance generated by the bulbous bow of the mainstream

https://doi.org/10.3390/jmse14020212

vessel model of the U.S. for drag reduction was observed to be relatively mild, while the seawater disturbance ahead of the bulbous bow of the mainstream vessel model of China was also seen to stabilize. The high-water-level disturbance for both hulls was concentrated on the port and starboard sides. Furthermore, the wave disturbance amidships of the mainstream vessel model of the U.S. was noted to stabilize, and a marked symmetry in the wake of the Kelvin wake system was evident. Compared to the wave patterns at 30 s, the disturbance generated by the mainstream vessel of the U.S. was extremely intense, with larger amplitude disturbances being more concentrated towards the stern. In contrast, the mainstream vessel model of China exhibited relatively calm conditions, and a clear symmetry in the motion-generated waves was observed. Similar to the situation in Figure 10, the range over which the free surface was influenced by the mainstream vessel of the U.S. was far greater than that of the mainstream vessel of China. However, the free-surface elevation difference of the mainstream vessel of China at this stage remained greater than that of the mainstream vessel of the U.S.

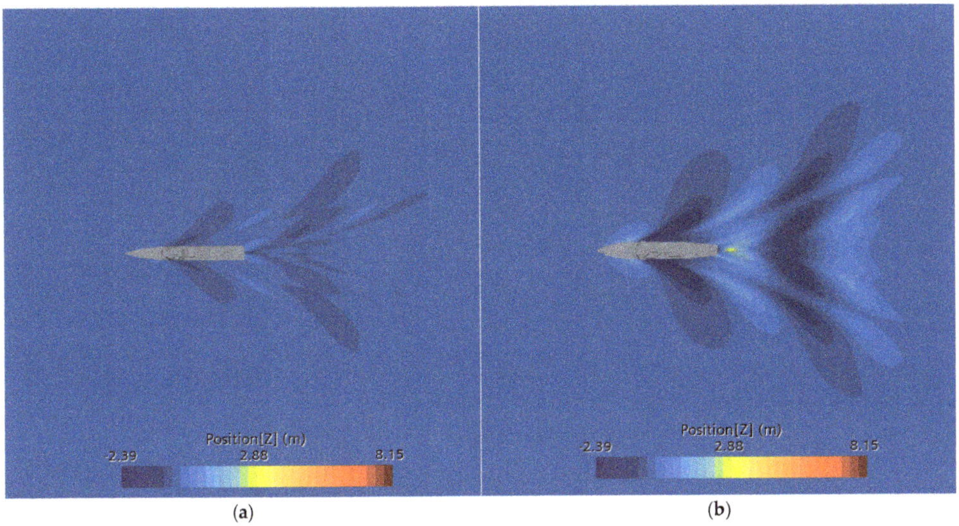

Figure 11. Wave patterns and free surface deformation generated by the two ships advancing at maximum speed (t = 50 s): (**a**) mainstream vessel model of China; (**b**) mainstream vessel model of the U.S.

The wave patterns generated after both ship models had advanced for 100 s are shown in Figure 12. The sailing distance at this point had reached ten times the overall length of both ship models. The motion of both vessels had stabilized, with the symmetry of motion being particularly pronounced. Compared to the wave patterns at 50 s, a smaller wave-making region was produced by the motion of the mainstream vessel model of China relative to that generated by the mainstream vessel model of the U.S. The Kelvin wake patterns of both were observed to originate from the bow and gradually develop towards the stern, but more intense disturbances in the wave-making were generated at the stern by the mainstream vessel model of the U.S., resulting in a noticeable drop in the Kelvin waves aft of the stern. Simultaneously, the range over which the free surface was disturbed by the motion of the mainstream vessel of the U.S. was far greater than that of the mainstream vessel of China. However, at this stage, the free-surface elevation difference of the mainstream vessel of China was slightly greater than that of the mainstream vessel

https://doi.org/10.3390/jmse14020212

of the U.S. A reduction in this elevation difference between the two vessels was noted compared to the situation in Figure 11.

Similar to Section 4.1, a steady-state representation of the free surface was gradually reached by both ship models after prolonged motion. The free surface at 180 s and 360 s for the two vessels sailing at maximum speed is shown in Figure 13. The free-surface area disturbed by the motion of both vessels showed almost no change compared to that in Figure 12. In high-speed ship motion, vessels are more likely to form a stable wave-making influence range.

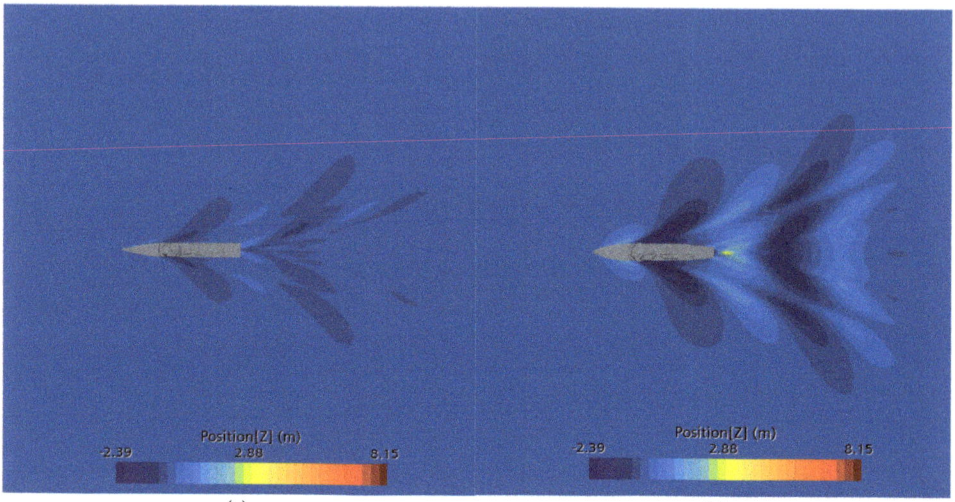

(**a**) (**b**)

Figure 12. Wave patterns and free surface deformation generated by the two ships advancing at maximum speed (t = 100 s): (**a**) mainstream vessel model of China; (**b**) mainstream vessel model of the U.S.

4.4. Analysis of Free Surface and Resistance at Maximum Cruising Speed

Figure 13 outlines the free-surface disturbance range of the two ship models after 360 s of motion at maximum speed. The data indicate that the maximum transverse disturbance distance of the mainstream vessel of China corresponds to 16.2 times its molded breadth, whereas that of the mainstream vessel of the U.S. corresponds to 20.4 times its molded breadth. Different from the economic speed condition, at maximum speed, the multiple of the molded breadth for the transverse free-surface disturbance distance generated by the mainstream vessel of the U.S. exceeds that of the mainstream vessel of China by more than four times its breadth.

The hull resistance variation experienced by the two ship models at maximum cruising speed is clearly shown in the visualization plot presented in Figure 14. From the initiation of motion until 360 s, the total resistance experienced by the hull of the mainstream vessel of the U.S. is approximately 2050 kN, while that of the mainstream vessel of China is approximately 1750 kN, resulting in a difference of about 300 kN between the total hull resistances of the two vessels. This differs from the resistance variation observed when the two vessels were moving at economic speed. In this case, the hull resistance of the mainstream vessel of the U.S. is greater than that of the mainstream vessel of China.

In accordance with Section 4.2, the slight oscillatory fluctuations in the resistance values presented here are considered numerical artifacts rather than physical phenomena. Furthermore, the amplitude of these oscillations is substantially smaller than the steady-

 https://doi.org/10.3390/jmse14020212

state mean resistance and therefore does not compromise the comparative analysis of the steady-state resistance performance between the two vessels.

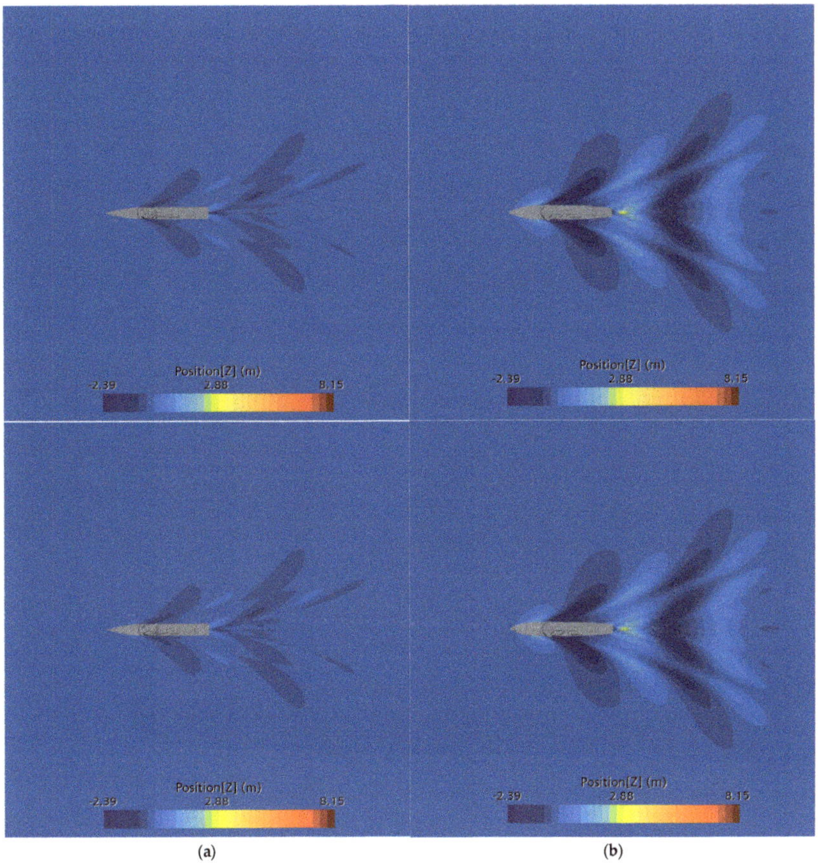

Figure 13. Wave patterns and free surface deformation generated by the two ships advancing at maximum speed (t = 180 s and t = 360 s): (**a**) mainstream vessel model of China; (**b**) mainstream vessel model of the U.S.

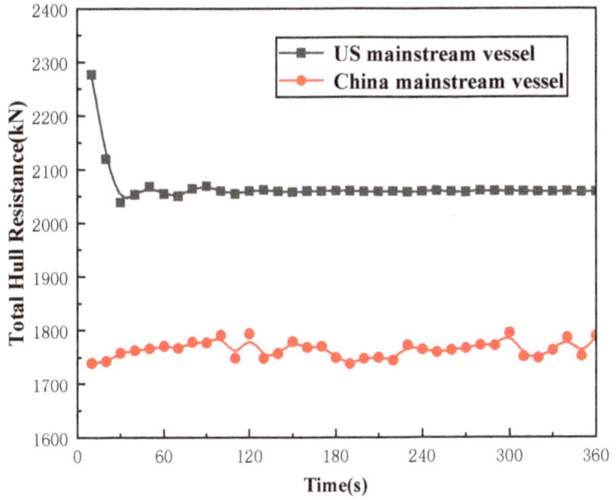

Figure 14. Variation in simulated total hull resistance with time for both vessels at maximum speed.

https://doi.org/10.3390/jmse14020212

5. Analysis of Ship Model Towing Test and Simulation Data

5.1. Experimental Design

This study conducted towing tests solely on small-scale models of Chinese ship types to verify the reliability of the numerical method. Due to limitations in experimental resources and the availability of models, experimental validation of the U.S. ship types at the same scale has not yet been conducted. Therefore, the experimental validation presented in this paper provides valuable support for the reliability of the numerical model; however, its representativeness is limited by the use of a single ship type and by scale effects associated with model testing. To validate the reliability of the numerical method, supplementary towing tank tests were designed and conducted using a 1:141 scale model (1.1 m in length) of the mainstream vessel of China. The test speed (1.5 m/s) was scaled to correspond to the full-scale ship speed through dimensional analysis. The three-dimensional form of this ship model is shown in Figure 15. The model features the internal structure of a wooden transverse-frame system. The internal frames were fabricated using laser cutting and subsequently assembled manually with precision. The superstructure was constructed employing 3D printing technology, and the hull bottom was coated with body filler to ensure the model's waterproof performance. To distinguish the waterline, the hull sections above and below it were painted in different colors. Figure 16 shows the electronic dynamometer (DMF-500N) used in the experiment to measure the resistance experienced by the hull during towing. Figure 17 presents the cable and stepper motor used for towing within the experimental platform.

Figure 15. Physical model of the mainstream vessel of China.

Figure 16. Electronic dynamometer (DMF-500N). Manufacturer: Wenzhou Aibai Measurement and Control Technology Co., Ltd. Location: Wenzhou, Zhejiang, China.

https://doi.org/10.3390/jmse14020212

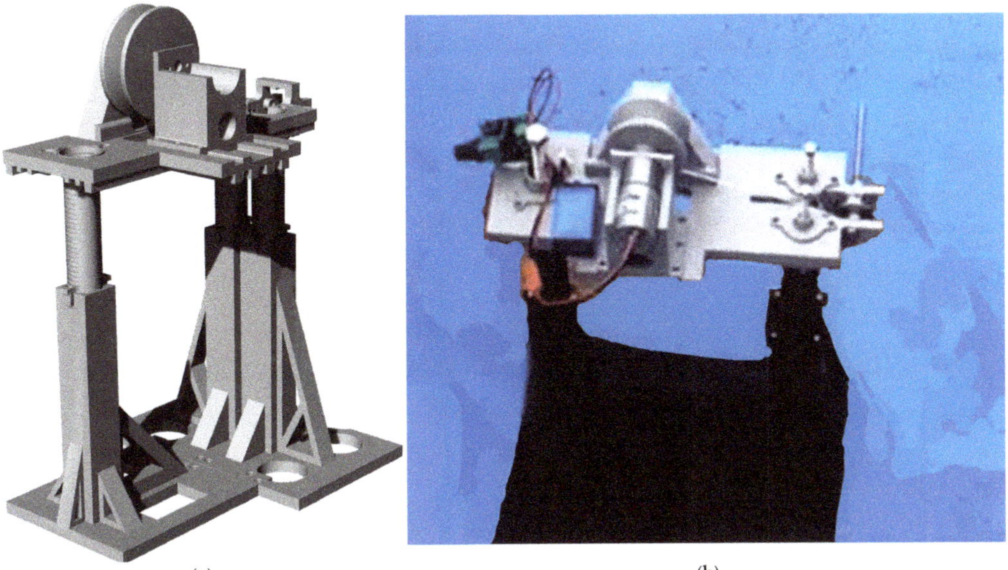

(a) (b)

Figure 17. Electronic dynamometer: (**a**) Three-dimensional design model. (**b**) Fabricated prototype.

Figure 18 presents a two-dimensional schematic of the ship model towing experimental platform. The experimental platform was designed as shown in the figure below. The stepper motor is positioned outside the water tank and is responsible for towing the ship model [25]. Two fixed pulleys are employed to transmit the force from the motor to the electronic dynamometer and simultaneously drive the hull motion. The disturbance of the free surface by the ship model's motion in water is recorded in real-time by a panoramic camera, which also clearly captures the numerical readings displayed on the electronic dynamometer.

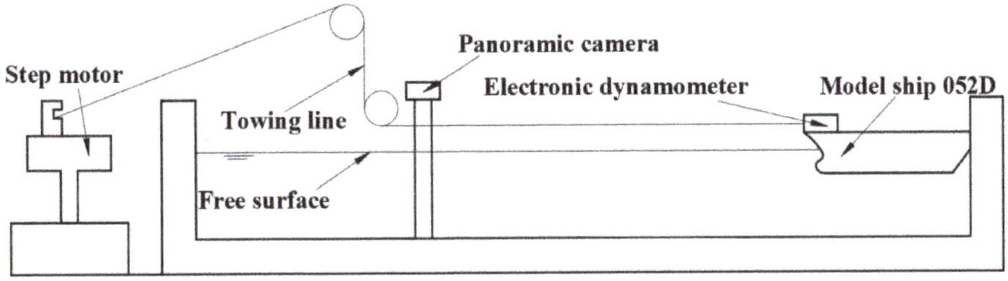

Figure 18. Schematic diagram of the ship model towing experimental platform.

The experimental configuration of the ship model towing platform is shown in Figure 19, with key geometric parameters defined as follows: The measured horizontal distance between the ship model and the panoramic camera is $L = 4.5$ m. The actual draft of the ship model is 0.5 m. All equipment within the experimental platform was arranged in the water tank in advance. The electronic dynamometer was fixed at the foremost point on the centerline of the model's upper deck at the bow. One end of the towing cable was attached to the load measurement point of the dynamometer, while the other end was connected to the motor after passing over the fixed pulleys. Before commencing the tests, it was ensured that the cable was taut. After the water surface had calmed and the model was confirmed to be in an even-keel condition, stable-speed towing

https://doi.org/10.3390/jmse14020212

was initiated using the stepper motor. The panoramic camera was responsible for recording the waves generated by the model's motion and the changes in the readings displayed on the electronic dynamometer.

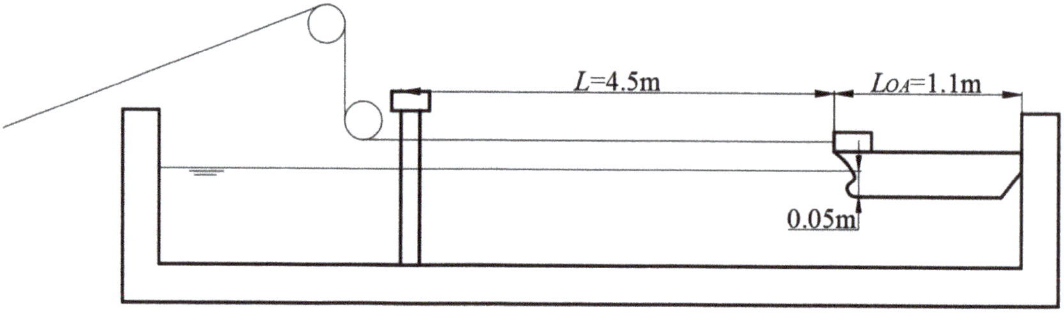

Figure 19. Other parameters.

To verify the reliability and accuracy of the simulation data against the real experiment, the Shear–Stress Transport (SST) k-ω turbulence model and the Volume of Fluid (VOF) method were again selected for simulating the three-dimensional model of the 1.1 m long ship model used in the experiment. Ship design parameters, including speed and draft, were kept identical to those adopted in the physical model test, ensuring consistency in external conditions between the experiment and the simulation.

5.2. Comparison of Experimental and Simulation Results

The wave systems and free-surface deformation generated by the simulated and the physical ship models after 3 s of motion are presented in Figure 20. Both the physical model test and the simulation indicate that the waves generated by the motion are primarily concentrated on the sides and the stern. Furthermore, the amplitude of free-surface variation is shown to decrease gradually from the midship section towards the stern.

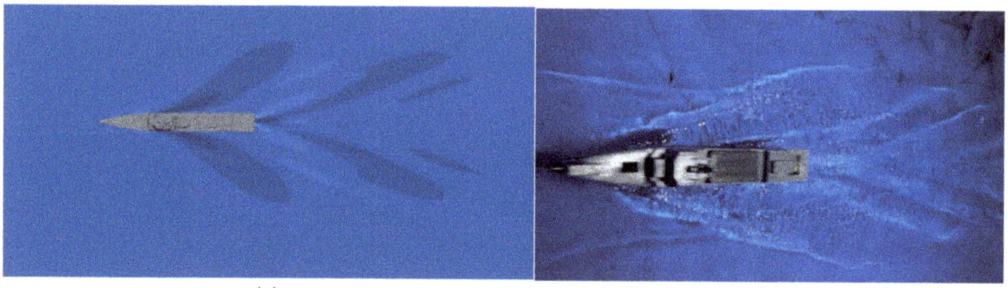

(**a**) (**b**)

Figure 20. Wave patterns and free surface deformation generated by the simulated and physical ship models: (**a**) CFD simulation; (**b**) Experiment.

Based on a comparison between the readings from the electronic dynamometer and the hull resistance values obtained from the simulation, the comparative plot shown in Figure 21 was compiled. It can be observed that after 1 s, both the resistance values derived from the simulation and those measured from the physical model tended to stabilize. With an error margin of ±0.2 N for the electronic dynamometer, the simulated resistance values were found to fall within the experimental error range. This sufficiently demonstrates that the simulation results can objectively and realistically represent both the wave generation and the resistance experienced by the actual vessel. In summary, while a full-field quantitative wave profile comparison could not be performed, the excellent

 https://doi.org/10.3390/jmse14020212

quantitative agreement on the total hull resistance serves as a robust validation of the CFD model, establishing confidence in its use for the subsequent full-scale comparative analysis.

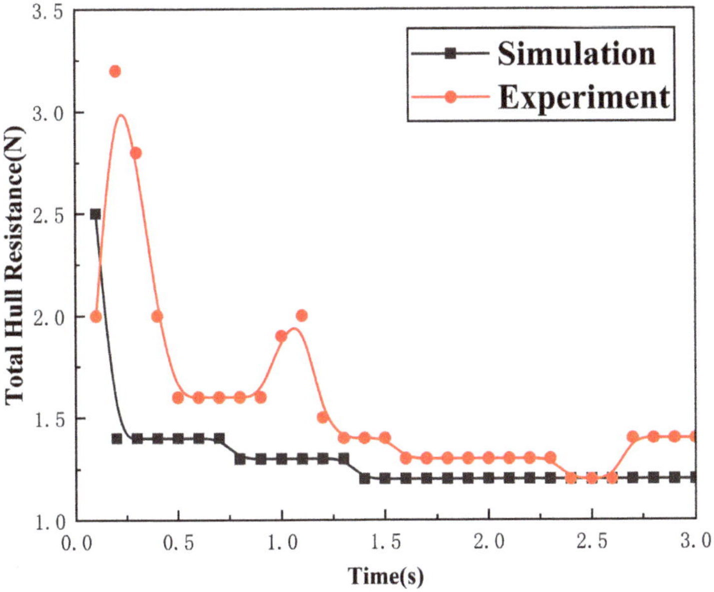

Figure 21. Variation in hull resistance with time for the simulated and physical ship model.

6. Conclusions

This study conducted a systematic comparison of the hydrodynamic performance between two representative destroyers from China and the United States, employing validated computational fluid dynamics methods complemented by scaled model tests. Through full-scale CFD simulations and experimental verification, the wave-making characteristics and resistance performance of the two ship types were analyzed under calm water conditions at both economic speed (18 knots) and maximum speed (30 knots). The main conclusions are summarized as follows:

Performance trade-off at economic speed: At 18 knots, the mainstream vessel of China exhibits superior wave stealth performance, with a maximum transverse wave-affected width of 10.5 times the beam (184.8 m), compared to 11.6 times the beam (221.6 m) for the mainstream vessel of the U.S. However, this is achieved at the cost of higher total resistance, with the mainstream vessel of China experiencing approximately 850 kN, about 240 kN greater than the 610 kN of the mainstream vessel of the U.S.

Performance reversal at maximum speed: At 30 knots, the ranking of resistance performance is reversed. The total resistance of the mainstream vessel of China is about 1750 kN, which is lower than the 2050 kN of the mainstream of the U.S. by approximately 300 kN. Meanwhile, the mainstream vessel of China maintains a notably smaller wave-affected width (16.2 times the beam) compared to the mainstream vessel of the U.S. (20.4 times the beam), indicating better wave stealth characteristics even at high speed.

Reliability of methodology and validation: The towing test of a 1:141 scale model was conducted, and the measured resistance data show excellent agreement with the corresponding CFD results, confirming the reliability and accuracy of the numerical model in predicting ship resistance and wave-making characteristics.

In summary, this research quantitatively elucidates the differences in design philosophy from a hydrodynamic perspective: the mainstream vessel of the U.S. prioritizes low

https://doi.org/10.3390/jmse14020212

resistance at economic speed, whereas the mainstream vessel of China emphasizes wave stealth across the speed range while demonstrating better overall resistance performance at high speed. These findings provide an important basis for the multi-objective optimization of future surface combatants in terms of stealth, speed, and energy efficiency.

Despite its contributions, this study has several limitations. First, the 3D geometric models of both vessels were reconstructed from public sources, which may lack certain localized details, leading to inherent uncertainties in geometric fidelity and input parameters. Second, due to resource constraints, towing tank experiments were only performed on a small-scale model of the vessel of China. While these tests provide crucial validation for the numerical methods, their representativeness is constrained by the use of a single hull type and potential scale effects. Future work should incorporate systematic uncertainty quantification to enhance the extrapolation reliability of the full-scale results.

Author Contributions: Conceptualization, Y.T. and D.Z.; methodology, D.Z.; software, H.S.; validation, Z.Z. and S.G.; investigation, Y.Z.; resources, Y.Z. and K.S.; writing—original draft preparation, Y.T.; writing—review and editing, Y.T. and D.Z.; visualization, Z.C.; supervision, D.Z.; project administration, D.Z.; funding acquisition, D.Z. All authors have read and agreed to the published version of the manuscript.

Funding: This research was funded by the National Natural Science Foundation of China, grant number 62272109, and Zhanjiang Marine Youth Talent Project-Comparative Study and Optimization of Horizontal Lifting of Subsea Pipeline, grant number 2021E5011.

Data Availability Statement: The original contributions presented in this study are included in the article. Further inquiries can be directed to the corresponding author.

Conflicts of Interest: The authors declare that they have no known competing financial interests or personal relationships that could have appeared to influence the work reported in this paper.

References

1. Zhao, G.; Zhang, Y.; Xie, Y.; Liang, Z.; Lin, Q.; Jiang, H. Submarine Maneuvering Performance Analysis: A Study of the Effect of Different Rudder Angle Settings on the Turning Dynamics of SUBOFF Submarine. *Eng. Solut. Mech. Mar. Struct. Infrastruct.* **2024**, *1*, 6. [CrossRef]
2. Ravenna, R. CFD analysis of the effect of heterogeneous hull roughness on ship resistance. *Ocean Eng.* **2022**, *258*, 111733. [CrossRef]
3. Zhang, Y.; Zhang, D.; Jiang, H. A Review of Offshore Wind and Wave Installations in Some Areas with an Eye towards Generating Economic Benefits and Offering Commercial Inspiration. *Sustainability* **2023**, *15*, 8429. [CrossRef]
4. Zhang, Y.; Zhang, D.; Jiang, H. Review of Challenges and Opportunities in Turbulence Modeling: A Comparative Analysis of Data-Driven Machine Learning Approaches. *J. Mar. Sci. Eng.* **2023**, *11*, 1440. [CrossRef]
5. Lutfi, M.A.; Prabowo, A.R.; Muttaqie, T.; Baek, S.J.; Do, Q.T.; Cho, J.H. Performance of the Designed Patrol Boat Hull: Resistance Prediction Using Savitsky Method. In Proceedings of the 9th International Conference and Exhibition on Sustainable Energy and Advanced Materials (ICE-SEAM 2023), Putrajaya, Malaysia, 1 January 2024.
6. Putranto, T.; Aryawan, W.D.; Kurniawati, H.A.; Setyawan, D.; Pribadi, S.R.W. Resistance and stability analysis for catamaran fishing vessel with solar cell in calm water. In Proceedings of the 2nd International Joint Conference on Advanced Engineering and Technology (IJCAET 2017) and International Symposium on Advanced Mechanical and Power Engineering (ISAMPE 2017), Bali, Indonesia, 24–26 August 2017.
7. Sulovsky, I.; Hauteclocque, G.; Greco, M.; Prpić-Oršić, J. Comparative Study of Potential Flow and CFD in the Assessment of Seakeeping and Added Resistance of Ships. *J. Mar. Sci. Eng.* **2023**, *11*, 641. [CrossRef]
8. Hakim, M.L.; Nugroho, B.; Chin, R.C.; Putranto, T.; Suastika, I.K.; Utama, I.K.A.P. Drag penalty causing from the roughness of recently cleaned and painted ship hull using RANS CFD. *CFD Lett.* **2020**, *12*, 78–88. [CrossRef]
9. Oyuela, S.; Ojeda, H.; Arribas, F.; Otero, A.; Sosa, R. Investigating Fishing Vessel Hydrodynamics by Using EFD and CFD Tools, with Focus on Total Ship Resistance and Its Components. *J. Mar. Sci. Eng.* **2024**, *12*, 622. [CrossRef]
10. Cheng, X.; Huang, X.; Xu, D.; Zhao, Z.; Liu, H.; Kong, M.; Ji, R. Ship Optimization Based on Fully-Parametric Models for Hull, Propeller and Rudder. *J. Mar. Sci. Eng.* **2024**, *12*, 1635. [CrossRef]

https://doi.org/10.3390/jmse14020212

11. Chen, Z.; Jiao, J.; Xu, W.; Jiang, C.; Chen, S. Numerical simulation of ship hydroelastic responses in short-crested irregular waves. *Mar. Struct.* **2025**, *103*, 103858. [CrossRef]
12. Song, S.; Demirel, Y.; De, M.; Sant, T.; Villa, D.; Tezdogan, T.; Incecik, A. Investigating the Effect of Heterogeneous Hull Roughness on Ship Resistance Using CFD. *J. Mar. Sci. Eng.* **2021**, *9*, 202. [CrossRef]
13. Chiroșcă, A.; Rusu, L. Comparison between Model Test and Three CFD Studies for a Benchmark Container Ship. *J. Mar. Sci. Eng.* **2021**, *9*, 62. [CrossRef]
14. Shi, K.; Zheng, Y.; Tian, Y.; Zhang, D. Design of Ship Model Resistance Experiment Platform. *Ocean Eng. Equip. Technol.* **2025**, *12*, 1–10. (In Chinese)
15. Shou, H.; Tian, Y.; Zhang, S.; Zhang, D. Exhibition of Modern Model Making Technology Based on the Model Making of 093 Nuclear Submarine. *Ocean Eng. Equip. Technol.* **2025**, *12*, 11–21. (In Chinese)
16. Zhang, S.; Xie, Y.; Guo, S.; Liang, Z.; Zhang, Y.; Zhang, D. Design and Construction of Aircraft Carrier Teaching Experimental Platform Based on the Fujian Ship. *Ocean Eng. Equip. Technol.* **2025**, *12*, 9–18. (In Chinese)
17. Zhang, S.; Guo, S.; Li, J.; Li, Y.; Xie, K.; Zhang, D. Design and Fabrication of a Compact Brushed Electromagnetic Catapult Based on Halbach Array. *Ocean Eng. Equip. Technol.* **2025**, *12*, 22–29. (In Chinese)
18. Zhao, G.; Zhang, S.; Xie, K.; Li, Y.; Li, J.; Li, C.; Zhang, D. Introduction to Ship Model Making: Taking the Adora Magic City Cruise as an Example. *Ocean Eng. Equip. Technol.* **2025**, *12*, 1–8. (In Chinese)
19. Xie, Y.; Zhang, S.; Guo, S.; Liang, Z.; Zhang, Y.; Zhang, D. Introduction to Ship Model Making: Taking Type 903A Composite Supply Ship as an Example. *Ocean Eng. Equip. Technol.* **2025**, *12*, 19–27. (In Chinese)
20. Li, P.; Tian, Y.; Chen, Z.; Zhang, D. Design and Implementation of a Crab-Like Underwater Robot System. *J. Unmanned Undersea Syst.* **2025**, *33*, 826–835. (In Chinese)
21. Zhang, D.; Zhao, B.; Sun, J. Hydrodynamic Performance and Wake Study of an UUV Sailing Near the Free Surface. *Front. Mar. Sci.* **2024**, *11*, 1292874. [PubMed]
22. Jiang, H.; Xie, Y.; Zeng, S.; Guo, S.; Chen, Z.; Liang, Z.; Zhang, D. Research on Lightweight Design Performance of Offshore Structures Based on 3D Printing Technology. *J. Mar. Sci. Eng.* **2025**, *13*, 2007. [CrossRef]
23. Li, J. Damage Effectiveness Evaluation of Blasting Torpedo to Large Surface Warship. Master's Thesis, North University of China, Taiyuan, China, 2020.
24. ITTC. Uncertainty Analysis in CFD Verification and Validation Methodology and Procedures. ITTC Quality Manual Recommended Procedure 7.5-03-01-01, Revision 01. Available online: https://ittc.info/media/8153/75-03-01-01.pdf (accessed on 1 January 2026).
25. Zhang, D.; Bai, Y.; Zhang, Y.; Yan, J.; Zhang, Y.; Wu, Z. Advances in Dynamic Positioning of Offshore Platform-Riser Rigid-Flexible Multi-Body Systems. *Chin. J. Appl. Mech.* **2025**, *42*, 243–263. (In Chinese)

 https://doi.org/10.3390/jmse14020212